Block Copolymers

OVERVIEW AND CRITICAL SURVEY

Block Copolymers

OVERVIEW AND CRITICAL SURVEY

Allen Noshay

Chemicals and Plastics
Union Carbide Corporation
Bound Brook, New Jersey

James E. McGrath

Department of Chemistry
Virginia Polytechnic Institute
and State University
Blacksburg, Virginia

ACADEMIC PRESS New York San Francisco London 1977

A Subsidiary of Harcourt Brace Jovanovich, Publishers

ACADEMIC PRESS, INC.
111 Fifth Avenue, New York, New York 10003

United Kingdom Edition published by
ACADEMIC PRESS, INC. (LONDON) LTD.
24/28 Oval Road, London NW1

Library of Congress Cataloging in Publication Data

Noshay, A
    Block copolymers.

    Includes bibliographical references.
    Includes index.
    1.   Block copolymers.   I.   McGrath, James E., joint
author.   II.   Title.
QD382.B5N67   547$'$.84   75-44758
ISBN 0–12–521750–1

PRINTED IN THE UNITED STATES OF AMERICA

To our wives, Florence and Marlene, and to our children. We are indebted to them for their patience and support, and for their endurance of the inconveniences associated with our absence from home on many an evening.

Contents

PART TWO

CRITICAL SURVEY

Preface

An enormous amount of interest has been generated in block copolymers in recent years. Our search of the literature, based primarily on "Chemical Abstracts" from 1960 through mid-1976, uncovered over two thousand references on block copolymers and related systems. Such a prolific outpouring can lead to considerable confusion, especially for those new to the area. This situation has generated a need for clarification of both the potentials and limitations of block copolymers. We felt that this need could best be fulfilled by writing a critical review of block copolymer technology—embracing all structural types and all facets of the field—that enunciates general principles and at the same time describes the individual systems in detail.

A bilevel strategy has been employed to achieve these dual goals and to serve two types of audiences. The book is comprised of two parts: (1) a concise, critical overview of block copolymer technology and (2) a comprehensive critical survey on the synthesis, characterization, properties, and applications of the specific block copolymer structures reported in the literature. The overview (Part One) presents a summarized and unified picture of the block copolymer field as a whole. It also attempts to place this field in perspective within the general framework of polymer science and to indicate some fruitful areas for future effort. Accordingly, this portion of the book should be useful both to those readers who want to become generally conversant with the technology and also, as a starting point, for those who need to delve more deeply into the subject. The latter audience is served by the critical survey portion of the book (Part Two), which discusses in greater detail the individual block copolymer structures. The copolymers are organized according to both segmental architecture and chemical composition. The organizational format of the book is described further in Chapter 1.

We felt that a book addressing the above-described goals should be written by a minimum number of authors in the interest of cohesiveness, and should be devoted exclusively to block copolymers, aside from discussion necessary to establish perspective. Although other books have appeared with the words "block copolymer" in their titles, they have been directed to different objectives, since they are compilations of symposium papers or collections of individually authored chapters, or they deal with nonblock as well as block copolymers.

We hope that our active involvement in a variety of block copolymer synthesis and property investigations over the past decade has helped us to properly digest and evaluate the literature data. However, the reader should recognize that this experience may also have resulted in the appearance of a few opinionated statements.

We would like to acknowledge the encouragement for writing this book given by the Union Carbide management, particularly Drs. L. Shechter, T. T. Szabo, N. L. Zutty, E. F. Cox, and L. M. Baker. We would also like to thank several of our staff associates at Union Carbide, especially Drs. M. Matzner and L. M. Robeson, for many useful discussions. In addition, a number of colleagues, including Drs. G. L. Wilkes, L. J. Fetters, and P. C. Juliano, provided helpful comments and suggestions on portions of the manuscript.

Our sincere thanks are also extended to the many secretaries who transformed our often illegible penmanship into neat type: Mary Gallo, Barbara Cochilla, Ruth Horvath, Peggy Cox, Pat Kralovich, Joan Stryker, Rosemary DeSiato, Louise DeCastro, Kathy DeFeo, Kathy Moschak, and Pam Featherson.

We are grateful to the staff of Academic Press who helped bring this book to fruition.

Allen Noshay
James E. McGrath

1

Introduction and Format

Simply stated, the central theme of this book is that block copolymers, as one form of macromolecular architecture, have both potential and limitations. In some respects they offer unique possibilities not attainable with other polymeric materials; in others they offer no advantages. It is hoped that some degree of success will be achieved in elucidating this theme by providing a perspective picture of block copolymers.

This book has an organizational format that differs significantly from those employed in previously published volumes relating to block copolymers (1–10). It is divided into two parts. The first part, comprised of Chapters 2, 3, and 4, is an overview of block copolymer technology. Chapters 5, 6, and 7 constitute the second part, which is a critical survey and encyclopedic compilation of the many block copolymer systems reported in the literature.

Part One is intended to provide a qualitative understanding of what block copolymers are, how they are made, and what they can and cannot be expected to do. In addition, it serves as a guide to the second part of the book. In Chapter 2, block copolymers are defined and compared with other types of polymer "hybrids," i.e., polymer blends, random copolymers, and graft copolymers. Chapter 3 describes the various segmental architectures that are possible with block copolymers. Chapter 4 summarizes and compares (a) the various synthesis techniques applicable to block copolymers, (b) the characterization methods capable of elucidating block copolymer structures, and (c) the physical properties displayed by these systems. In addition, some of the applications of commercially available block copolymers are discussed in this chapter. Finally, Chapter 4 also identifies some future challenges for block copolymer technology in an attempt to stimulate further thinking along these lines and to focus attention on some of the important problems that still need to be solved.

Part Two discusses in considerable detail the synthesis, characterization, and properties of the various block copolymers described in the literature. The three chapters of Part Two are arranged according to block copolymer architecture, i.e., Chapters 5, 6, and 7 are devoted to A-B, A-B-A, and $(A-B)_n$ structures, respectively. Within each of the chapters, the copolymers are further subdivided according to chemical composition.

Specific block copolymers can be located by consulting the Table of Contents under the appropriate structural heading. The following, somewhat arbitrary, "structural hierarchy" was employed in organizing those chapter sections that deal with heteroatom-containing segments: polysiloxane > polycarbonate > polyurethane > polyamide > polyester > polyether > polyvinyl. A block copolymer in question will most often be found in that section pertaining to the segment most highly placed in the "hierarchy"; e.g., polysiloxane–polycarbonate systems are described in the polysiloxane section rather than the polycarbonate section. Since it was not always possible to adhere strictly to this general rule, the surest approach would be to consult the sections pertaining to both segments.

The literature references cited in this work include both publications and patents. Emphasis was placed on the publications, the more fundamental papers often being singled out for extensive discussion in order to illustrate important points. Most of the references are cited in the Part Two chapters (5, 6, and 7). For points of reference in Part One (Chapters 2, 3, and 4), the reader is usually directed to the appropriate sections of Part Two. In addition to the citation of references within the text, summary reference tables are provided in many of the sections of Chapters 5, 6, and 7. The letter prefix of reference citations reflects the chapter section, e.g., reference C15 is the fifteenth reference cited in Section C of a given chapter. Combined sectional reference lists are presented at the end of each chapter. In addition Supplemental References are provided at the end of Part Two. Subject and Author Indexes are located at the end of the book.

In Part Two an attempt has been made to provide (a) a complete compilation of all of the specific block copolymer structures reported, (b) a discussion of the major conclusions drawn by the workers (in many cases accompanied by critical comments), and (c) a convenient collection of the references pertaining to the various structures to permit "digging deeper." However, detailed comparisons of characterization and property data (e.g., molecular weight, intrinsic viscosity, melting point) are not made unless required to illustrate a point. This was made necessary by the wide diversity that exists in the breadth and depth of the reported data. Characterization and property data are, in many cases, not available or cannot be compared from one reference to another in a meaningful way, since they had been determined by different techniques. Furthermore, in many instances, properties are attributed to an assumed but not adequately proved structure. Therefore, routine recitation of detailed data would be cumbersome and of questionable value. The original literature references provide a source of such detailed information for those seeking it.

IUPAC has not yet established a uniform system of nomenclature for block copolymers. Lacking this, it was decided to use an informal system of nomenclature in the interest of easier and more comprehensible reading. In many cases, the prefix "poly-" is omitted, resulting, for example, in the identification of a polystyrene–polybutadiene block copolymer as "styrene–butadiene" or simply "SB." Liberties such as these are only taken, however, when it is obvious from context what the abbreviated forms mean.

Another decision made in the interest of conciseness was to minimize discussion of the polymerization chemistry and polymer characterization techniques that are generally applicable to all macromolecular systems. Thus, it has been assumed that the reader has a general knowledge of polymer science. This has, we hope, allowed us to focus more intently on those aspects that are unique to block copolymers: (a) the chemistry especially suitable for the synthesis of block structures, (b) the analytical tools that are particularly useful for characterizing them, and (c) the special physical properties that can be achieved.

REFERENCES

1. Burlant, W. J., and Hoffman, A. S., "Block and Graft Copolymers." Van Nostrand-Reinhold, Princeton, New Jersey, 1960.
2. Ceresa, R. J., "Block and Graft Copolymers." Butterworth, London, 1962.
3. Ceresa, R. J., *Encycl. Polym. Sci. Technol.* **2**, 485 (1964); *C.A.* **65**, 4037e (1966).
4. Moacanin, J., Holden, G., and Tschoegl, N. W., eds., "Block Copolymers." Wiley (Interscience), New York, 1969.

5. Aggarwal, S. L., ed., "Block Copolymers." Plenum, New York, 1970.
6. Molau, G. E., ed., "Colloidal and Morphological Behavior of Block and Graft Copolymers." Plenum, New York, 1971.
7. Burke, J. J., and V. Weiss, eds., "Block and Graft Copolymers." Syracuse Univ. Press, Syracuse, New York, 1973.
8. Allport, D. C., and Janes, W. H., eds., "Block Copolymers." Appl. Sci. Publ. Ltd., Barking, England, 1973; Halsted, New York, 1973.
9. Ceresa, R. J., ed., "Block and Graft Copolymerization." Wiley, New York, 1973.
10. Sperling, L. H., ed., "Recent Advances in Polymer Blends, Grafts and Blocks." Plenum, New York, 1974.

PART ONE

OVERVIEW

2

Polymer Hybrids

A. INTRODUCTION

The variety of macromolecules commercialized over the past four to five decades is truly phenomenal. The large number of products available to materials scientists for a given end use application is bewildering. Even design engineering experts have a difficult time in selecting the best material based on cost and performance. By far, the polymers most commonly used commercially are homopolymers. However, to a continually increasing degree, sophisticated applications are being developed that demand combinations of properties not attainable with simple homopolymers. As a result of this need, several techniques for producing polymer "hybrids" have arisen.

There are essentially two general approaches for forming polymer hybrids. The first and simplest approach is that of physically blending two polymers. The second general approach is "chemical" blending, i.e., the formation of copolymers. The latter can be further subdivided into random/alternating copolymers, graft copolymers, and block copolymers. Another, though less general, hybrid type is the interpenetrating network (A1), which contains one or more thermosetting resins. Many of these polymer hybrids, in addition to combining the

properties of their components, display unique, previously unantici-
pated behavior.

One purpose of this chapter is to briefly discuss properties that can
be achieved with physical blends, random/alternating copolymers,
and graft copolymers, giving illustrative examples of each. A second
purpose is to compare these polymer hybrids with block copolymers
as a general class of materials. The overall intent is to thereby place
block copolymers in meaningful perspective vis-à-vis the other
polymer hybrids. This will serve as an introduction to the remaining
chapters of the book, where the various types of block copolymers are
discussed.

B. PHYSICAL BLENDS

From a preparative point of view, the most direct and versatile
method for producing polymer hybrids is the physical blending of two
or more polymers. In most cases, this is also the most economical
approach. Many commercially available products are blends of either
two rigid polymers, two elastomeric polymers, or combinations of the
two types. Since blends of crystalline polymers are considerably more
complex and less well understood than amorphous blends, they are
beyond the scope of this brief section. Accordingly, the discussion will
be limited to amorphous blends.

The properties and, therefore, the utility of physical blends are
strongly dependent upon the degree of compatibility of the components. The great majority of amorphous physical blends are highly
incompatible (B1–B10). A very small number of amorphous polymer–
polymer pairs are thermodynamically compatible, i.e., truly soluble in
each other. Blends that display an intermediate degree of compatibility also exist. These are usually based on random copolymer rather
than homopolymer components in order to achieve more closely
matched compositions. These systems have been termed "mechanically" compatible (B11, B12). The salient features of each of these
blend categories are described below.

1. Incompatible Blends

As already mentioned, high molecular weight polymer blends are
nearly always grossly incompatible. Polystyrene–polybutadiene,
polybutadiene–styrene/butadiene rubber (SBR), and polysulfone—
poly(dimethylsiloxane) systems exemplify this type of blend. Incom-

patibility is observed in dilute solution as well as in the solid state and in the melt. This is a direct consequence of the well-known relationship for free energy (ΔG) given by the equation (B13):

$$\Delta G = \Delta H - T\,\Delta S$$

Polymers have very small entropies of mixing (ΔS) due to their high molecular weight. Therefore, even a slightly positive enthalpy (ΔH) due to endothermic mixing is sufficient to produce a positive free energy, thus resulting in incompatibility (i.e., polymer–polymer insolubility).

The incompatibility of the blend components provides a driving force for each to aggregate in separate phases. These two-phase morphological systems are coarse dispersions in which the particles are usually (a) large, (b) inhomogeneous, and (c) characterized by poor interphase adhesion. This behavior has important ramifications for the optical and other physical properties of the blends (B11). Since the phases are usually larger than the wavelength of light, excessive light scattering occurs at the phase boundaries, resulting in opacity. An exception to this behavior occurs if the refractive indices of the blend components are tailored to be nearly identical, in which case the two-phase blends can be transparent (B14). The poor interphase adhesion characteristic of highly incompatible blends usually results in very poor mechanical properties, e.g., tensile strength, elongation, and impact strength. This behavior is presumably related to a high degree of stress concentration in the vicinity of the interface. The thermal properties of these blends reflect the transitional characteristics of both components (B11).

2. Compatible Blends

It is very rare to find a pair of polymers that are truly compatible, i.e., completely soluble in each other. However, a few systems of this type do exist. The most familiar example is that of polystyrene and poly(2,6-dimethyl-1,4-phenylene oxide) (PPO) (B15, B16). Other examples include blends of amorphous polycaprolactone and poly(vinyl chloride) (B17) and blends of butadiene–acrylonitrile copolymers with poly(vinyl chloride) (B18). The reason for the compatibility of polystyrene with PPO has not been completely elucidated. However, it is believed (B5, B16) that the two materials have essentially identical solubility parameters. There may also be some weak intermolecular interactions that contribute to compatibility. The two poly(vinyl chloride) blend examples referred to above clearly owe their compatibility to relatively strong specific interactions and/or "acid–base" interactions between the polar substituents (B19). These interactions produce exothermic heats of mixing and, hence, negative free energies, thereby resulting in compatible systems.

These compatible blends are characterized by single-phase morphology. As a result, they are transparent and exhibit physical properties intermediate to those of the components. The glass transition temperature (B20), rheology (B21), permeability (B22, B23), and mechanical properties displayed by these blends are predictable by the same techniques that are well known for random copolymers (B5).

3. Mechanically Compatible Blends

The incompatible and compatible blends described above represent the two extreme cases. However, some blends, which are most often based on random copolymer components, display intermediate behavior that can only be explained on the basis of partial miscibility or compatibility. A well-documented example of this type of mechanically compatible (B24) system is the blend of butadiene–acrylonitrile copolymer and styrene–acrylonitrile copolymer. This alloy constitutes one form of acrylonitrile-butadiene-styrene (ABS) resin (B10). Another example is the blend of polystyrene with SBR, which was one of the early but inefficient forms of impact-modified polystyrene. The borderline compatibility displayed by blends of this type is believed to be due to the similar but not identical solubility parameters of the components and also to possible weak specific interactions.

These mechanically compatible blends are essentially identical to the incompatible blends in their thermal behavior; i.e., they display two major glass transitions. However, their morphology is finer and they are more translucent. In addition, they display a higher degree of interphase adhesion, which is reflected in improved mechanical properties (B12).

C. RANDOM/ALTERNATING COPOLYMERS

The most common types of chemical hybrids are random copolymers and alternating copolymers. Random copolymers are characterized by a statistical placement of the comonomer repeat units along the backbone of the chain (see Fig. 2-1). They are the most versatile, economical, and easily synthesized type of copolymer. A wide variety of free radical and ionic addition and ring-opening polymerization

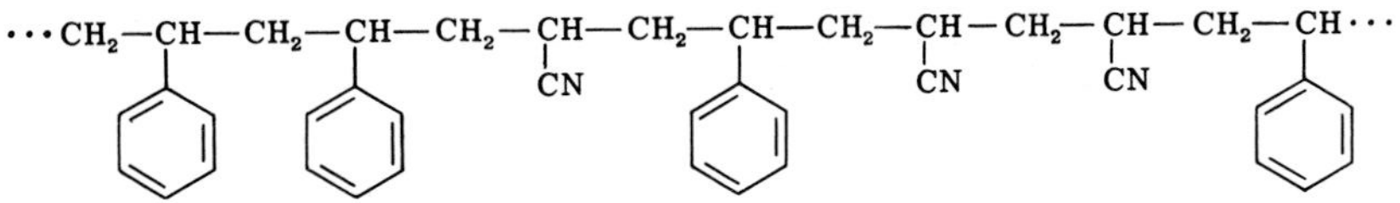

Random styrene-acrylonitrile copolymer

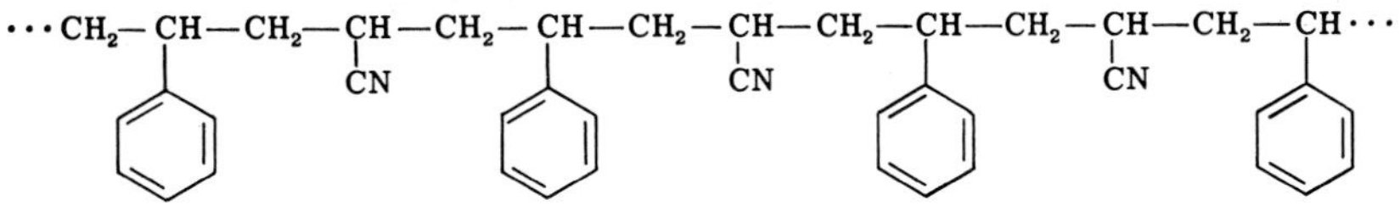

Alternating styrene-acrylonitrile copolymer

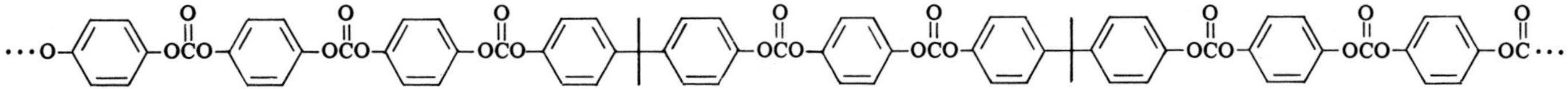

Random hydroquinone-bisphenol-A copolycarbonate

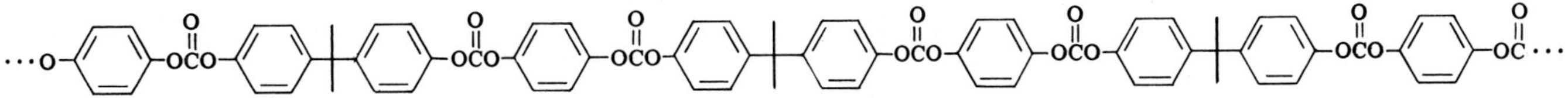

Alternating hydroquinone-bisphenol-A copolycarbonate

Fig. 2-1. Random and alternating copolymers.

techniques, as well as many step-growth reactions, are suitable. A voluminous literature concerning copolymerization reactivity ratios and other features of statistical copolymerization exists (C1). Two typical examples of random copolymers are styrene–acrylonitrile copolymers formed by free radical initiation and copolycarbonates formed by the phosgenation of mixtures of bisphenol-A and hydroquinone (see Fig. 2-1).

Alternating copolymers are characterized by the alternate, rather than statistical, placement of the comonomer repeat units along the chain (C2). This type of copolymer is relatively rare due to the requirements for (a) pairs of monomers with highly specific copolymerization reactivity ratios and/or (b) special reaction conditions. An example of the former is styrene–maleic anhydride copolymer, which is highly alternating in nature. Alternating styrene–acrylonitrile copolymers, of the type shown in Fig. 2-1, can be synthesized by free radical initiation of equimolar mixtures of the monomers in the presence of nitrile-complexing agents such as aluminum alkyls (C3). Hydroquinone–bisphenol-A copolycarbonates with an alternating structure (see Fig. 2-1) can be prepared by first forming the dichloroformate of one of the monomers and subsequently condensing this intermediate with the second bisphenol.

The major incentive for synthesizing random and/or alternating copolymers is to achieve homogeneous systems that display properties representing a weighted average of the two repeat units. This objective is, presumably, most ideally reached in the case of the alternating copolymers, since these can, in fact, be considered to be "homopolymers" in which the repeat unit consists of the comonomer pair. However, the more easily prepared random copolymers closely approximate this situation and permit the attainment of averaged physical and chemical properties such as glass transition temperatures, mechanical properties, permeability, and chemical resistance (C4).

In contrast to the incompatible physical blends, the uniform random copolymers display single-phase morphology. The sequential runs that may be present in these compositions are usually too short to induce microphase separation. The single-phase morphology results in optical transparency in amorphous systems. Furthermore, monomers that can produce crystalline homopolymers can be copolymerized to produce materials with controllably reduced crystallinity and, hence, greater transparency. A well-known example of this is the copolymerization of ethylene and propylene to produce copolymers ranging from partially crystalline plastics to amorphous elastomers.

D. GRAFT COPOLYMERS

Graft copolymers combine some of the features of physical blends and random copolymers. These copolymers may be looked upon as chemically linked pairs of homopolymers. As a result they are similar to block copolymers in many ways. As will be discussed below, the presence of the chemical linkage has important ramifications for the physical properties of these systems.

All graft copolymers are comprised of two general structural features: a backbone of polymer "A" to which a number of "B" sequences are grafted, such as shown in Scheme 2-1. There is considerable

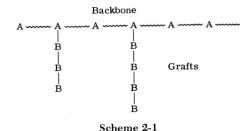

Scheme 2-1

latitude as to the chemical nature of the backbone and the grafted segments and the techniques for joining them. This class of polymer hybrid provides the basis for a number of commercially important polymeric materials, such as impact polystyrene, ABS, and methacrylate-butadiene-styrene (MBS).

Graft copolymer technology has been discussed in a number of books and reviews (D1–D3). It is our intent not to treat this subject in detail here, but rather to present the highlights of the current state-of-the-art in order to present a basis for comparison with block copolymers.

1. Synthesis and Characterization

Graft copolymers are generally prepared by the free radical, anionic, or cationic addition (or ring opening) polymerization of a monomer in the presence of a preformed reactive polymer. Appropriate types of backbone polymers useful with these methods are shown in Table 2-1. Step-growth graft polymerization reactions, due to their di- or multifunctional nature, would produce cross-linked systems and therefore will not be considered here. Free radical techniques have received the

TABLE 2-1

Typical Reactive Backbones for Grafting

Mechanism	Reactive site	Backbone structure
Radical	Allylic H	∼∼∼∼∼∼ CHCH=CHCH ∼∼∼∼∼∼ 　　　　 |　　　 | 　　　　 H　　　 H
Radical	Hydroperoxide	CH_3 　　　　　　　　　　 | ∼∼∼∼∼∼∼∼∼∼∼ CH_2C ∼∼∼∼∼∼ 　　　　　　　　　　 | 　　　　　　　　　　 O 　　　　　　　　　　 | 　　　　　　　　　　 OH
Radical	Redox	∼∼∼∼∼∼∼∼∼ CH_2CH ∼∼∼∼∼∼∼∼ + Ce^{4+} 　　　　　　　　　　 | 　　　　　　　　　　 OH
Cationic	PVC allylic and tertiary Cl	∼∼∼CHCH=CH∼∼∼∼∼∼∼ CH_2C 　　　 |　　　　　　　　　 | 　　　 Cl　　　　　　　　 Cl
Anionic	Metallated PBD	⊖ ∼∼∼∼∼∼∼ CH_2CH≕CH≕CH ∼∼∼∼∼∼ 　　　　　　　　 $M^{⊕}$
Anionic	Ester group	CH_3 　　　　　　　　　　 | ∼∼∼∼∼∼∼∼∼∼∼ CH_2C ∼∼∼∼∼∼ 　　　　　　　　　　 | 　　　　　　　　　　 C=O 　　　　　　　　　　 | 　　　　　　　　　　 O 　　　　　　　　　　 | 　　　　　　　　　　 CH_3

greatest amount of attention in the literature and are the most commonly used commercially.

Free radical techniques can be subdivided into two general categories. The first involves the polymerization of an olefinic monomer in the presence of a preformed polymer bearing labile hydrogen, e.g., grafting of styrene onto polybutadiene (D4). Initiation is achieved with peroxides, irradiation, or thermal methods (D2). Abstraction of the labile hydrogen by the peroxide initiator or the growing chain during the polymerization produces radicals on the backbone. The link between the graft and the backbone can be formed via monomer initiation by the backbone radical or by recombination reactions. The second category features initiation of the monomer by hydroperoxide or functional groups already on the preformed backbone. Examples of this technique are the initiation of styrene polymerization by hydroperoxides of polypropylene and the ceric ion redox initiated grafting of methyl methacrylate onto cellulosics or polyvinyl alcohol (D2). Commercial graft copolymer products produced by these techniques are characterized by broad compositional heterogeneity and are highly contaminated with homopolymers.

Better control of structure and less homopolymer contamination can be achieved by ionic rather than radical grafting mechanisms. The aluminum alkyl-initiated cationic grafting of isobutylene onto the allylic and/or tertiary chloride sites present in poly(vinyl chloride) is one example of this technique (D5) (see Table 2-1). However, the as yet unresolved problem of chain transfer to monomer limits the efficiency of this process. Much better efficiency is possible via the living anionic mechanism. This is possible due to the much lower propensity for spontaneous termination in anionic systems.

Anionic grafting techniques can be subdivided into two general categories involving grafting via (a) backbone initiation and (b) backbone coupling. The first type is emplified by initiation of styrene by metallated polybutadiene (see Table 2-1). This backbone can be prepared by reaction of polybutadiene with chelated organolithium compounds (D6–D18). Obviously, this approach is only useful with monomers capable of anionic polymerization, such as styrene or dienes. In order to minimize homopolymer formation, it is necessary to ensure that all of the organolithium reagent is consumed in the metallating step. Another example of grafting via backbone initiation is provided by the anionic initiation of caprolactam polymerization by macromolecular ester sites (see Table 2-1) such as those present in styrene–methyl methacrylate copolymers (D19). The ester group reacts with the caprolactam anion to generate an acylated lactam, thereby forming a macromolecular initiating species. The approach of grafting via backbone coupling is exemplified by the interaction of a living polystyrene anion with macromolecules bearing ester side groups, e.g., poly(methyl methacrylate) (D20). This interaction results in the displacement of a methoxyl group and formation of a ketone graft linkage.

Graft copolymers may be the most difficult type of noncross-linked polymer to accurately characterize. In addition to the problem of considerable homopolymer contamination, the question of the *number* of grafts per molecule and the *spacing* and average *length* of the grafts greatly complicate the picture. Superimposed on this, of course, is the additional factor of the polydispersity of the grafted segments as well as that of the backbone. To date, these problems remain largely unsolved. As a result, efforts to design graft copolymer structures for specific applications have been severely hampered. Accordingly, significant future advances in polymer characterization techniques will certainly have a large impact on graft copolymer technology.

2. Properties

The major justification for utilizing graft copolymers is that this type of hybrid is a single chemical species that displays the properties characteristic of each of the components, rather than an averaging of their properties. While single-phase morphology is possible in graft copolymers, two-phase morphology is much more commonly observed (D21). The morphology is greatly dependent upon the volume fraction of the graft and backbone species. The component present in the larger concentration will normally form the continuous phase and thereby greatly influence the physical properties of the copolymer. In compositions containing nearly equal concentrations of both components, phase continuity can be dramatically altered by varying the specimen fabrication conditions, e.g., type of casting solvent. This effect was first noted with two-phase copolymers of methyl methacrylate grafted onto natural rubber (D22). This behavior can be used to advantage in controlling properties such as modulus and permeability.

The two-phase morphology of graft copolymers is also reflected in their thermal transitional behavior. Like incompatible physical blends, these graft copolymers display two distinct glass transition temperatures. However, because of the presence of the intersegment linkage, they display a finer morphology. As a result, amorphous systems show good optical clarity if they are substantially free of homopolymer contamination.

One interesting and important characteristic of two-phase graft copolymers is their facility for blending well with their respective homopolymers. This feature has utility regardless of whether the components are rigid or elastomeric. Perhaps the best-known example of this phenomenon is the use of styrene–butadiene graft copolymer in impact modified polystyrene. Another way to utilize this property is to employ graft copolymers as interfacial emulsifying or "compatibilizing" agents in polymer blends (D2).

E. BLOCK COPOLYMERS

Subsequent chapters of this book deal at some length with the synthesis, characterization, and properties of the various types of block copolymers. Before getting into that discussion, however, it may be instructive to briefly consider the general characteristics of block copolymers to see how they relate to the other, previously described classes of polymer hybrids. The purpose of this comparison is to highlight the similarities and differences between block copolymers on one hand, and physical blends, random copolymers, and graft copolymers on the other. The overall conclusion drawn from this exercise is that block copolymer synthesis is more demanding than preparation of the other physical or chemical hybrids, but that the additional effort is often justified by the unique and novel properties attainable with block copolymers.

1. Synthesis and Characterization

Block copolymers are macromolecules comprised of chemically dissimilar, terminally connected segments. Their sequential arrangement can vary from A-B structures, containing two segments only, to A-B-A block copolymers with three segments, to multiblock $\{A\text{-}B\}_n$ systems possessing many segments. These are shown in Scheme 2-2.

It is obvious that to prepare the above well-defined structures it is necessary to resort to sophisticated synthetic techniques. The re-

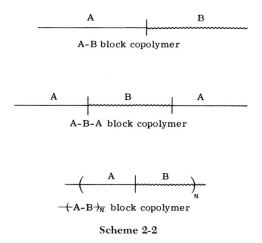

Scheme 2-2

quirements include (a) accurate knowledge and control of initiating and propagating species, (b) low impurity levels, (c) the use of low-concentration solution polymerization methods, and/or (d) the need for separately producing reactive polymeric intermediates of known functionality. These requirements can result in an economic penalty with respect to production of the other polymeric hybrids. Step-growth, anionic, cationic, and, to some extent, free radical methods have been used with varying degrees of success to prepare block copolymers (see Chapters 4–7). A number of useful reviews on synthesis and characterization have been published (E1–E14).

The best block copolymer syntheses are based on sequential anionic addition or ring opening polymerization techniques or on step-growth polymerization. These methods allow the control of block integrity and sequential architecture that is so important for achieving the ultimate attainable properties. Although block copolymers are more difficult to prepare than random copolymers, they are not hampered by the comonomer reactivity ratio restrictions of the latter. Furthermore, combinations of techniques can be used for synthesizing the various block copolymer segments.

A basic difference between block and graft copolymers is the number of intersegment linkage sites—one or two versus many. It is this difference that makes free radical techniques more applicable to graft copolymerization than to block systems. Because of the many potential grafting sites present in backbone polymers, the probability is high for forming at least some grafted segments along with significant quantities of homopolymer. However, in block systems, prepared by free radical techniques, the probability of forming copolymer structures is lower due to the nonselective nature of the polymerization process. Ionic synthetic methods are suitable for graft as well as block copolymers, but step-growth techniques do not give thermoplastic graft copolymers, since these multifunctional reactions would produce cross-linked systems. In contrast, the applicability of step-growth techniques to block copolymers permits the synthesis of a wide spectrum of high performance compositions.

The degree to which structural control is achieved can only be ascertained through the use of effective characterization tools. The characterization techniques traditionally used to analyze homopolymers can also be employed in elucidating hybrids. Simple physical blends, in principle, are the easiest hybrids to characterize. Separation of the blend components via solution fractionation or extraction techniques should lead to the easily characterized homopolymer components. Both random and block copolymers are more difficult to characterize than homopolymers or polymer blends. Since most block copolymers are, in reality, contaminated to at least a minor degree with homopolymer, they are frequently more difficult to define than uniform statistical copolymers. In addition, the question of the determination of block copolymer architecture (e.g., A-B versus A-B-A) is indeed a difficult feature to elucidate (see Chapter 4).

Graft copolymers clearly present additional obstacles to accurate structural characterization. These macromolecules are even more complex than block copolymers for several reasons. While the number of segments in a block copolymer can be deduced with some certainty from the synthetic technique employed, this is rarely possible with graft copolymers due to the multifunctional nature of the backbone and to the questionable efficiency with which these functionalities participate in the grafting reaction. The length of the graft segments and their polydispersity are also more imponderable for the same reasons. An important further complication is the unanswered question of the *spacing* of graft junction points along the backbone. In block copolymers, this parameter is more accessible. In A-B and in A-B-A block copolymers there are, by definition, one and two junction points, respectively. In $\{A\text{-}B\}_n$ block copolymers, the distance between intersegment linkages can be deduced from a knowledge of the block molecular weight.

2. Properties

Many basic characteristics of block copolymers and graft copolymers are similar. This is due to an important feature that they have in common, namely, the presence of intersegment chemical linkages. From a simplistic point of view, block and graft copolymers resemble incompatible physical blends in some respects and homogeneous random copolymers in others. Such systems usually exhibit two-phase morphology, but this occurs on a *micro*-scale rather than the *macro*-scale dimension of incompatible physical blends. This is due to the influence of the intersegment linkage, which restricts the extent to which the phases can separate. The small domain size and excellent interphase adhesion resulting from this microphase morphology can produce a high degree of transparency and a good balance of mechanical properties. This behavior is reminiscent of homogeneous copolymers and is not typical of polymer blends. Many other properties, such as permeability and chemical resistance, are dependent upon which segment exists in the continuous phase (see Chapter 4).

The thermal properties of block and graft copolymers resemble those of physical blends. They display multiple thermal transitions, such as glass transitions and/or crystalline melting points, characteristic of each of the components. In contrast, homogeneous random copolymers display a single, compositionally dependent glass transition temperature. Furthermore, while crystallinity is possible in block or graft copolymers, due to long sequences, it is diminished or eliminated in the random systems due to a disruption of chain regularity.

The presence of long segments in block and graft copolymers presents the possibility of using these materials as emulsifiers or surfactants. The incompatibility of homopolymer blends, both in solution and in the solid state, is reduced by the addition of small quantities of the respective block or graft copolymers. In addition, blending a block or graft copolymer with one of the respective homopolymers provides a means for achieving a fine dispersion of the "foreign" segment in the homopolymer matrix. The latter feature is made use of in important commercial products such as impact-modified thermoplastics.

The above discussion points out the similarities of two-phase block and graft copolymers and the features that distinguish them from physical blends and random copolymers. It is important to recognize that a high degree of structural control and integrity is necessary in order to achieve the ultimate properties inherent in such two-phase systems. It is in this respect that block copolymers offer a clear advantage over graft copolymers. Because of the greater reliability and predictability of block copolymer synthetic techniques, it is possible to achieve desired structures more precisely. This results in much better control of important parameters such as sequence architecture, segment length and spacing, polydispersity, and contamination by homopolymer or undesired copolymer architectures. These factors lead to a higher degree of morphological perfection in block copolymers than in graft copolymers, which, in turn, is reflected in superior physical properties.

A unique development resulting from block copolymer technology is the concept of thermoplastic elastomeric behavior. Block copolymer systems of this type are characterized by rubbery behavior in the absence of chemical cross-linking. This feature permits the fabrication of these materials by means of conventional thermoplastic processing techniques. The key to this unique behavior is the ability to achieve a network structure by physical rather than chemical means. This, in turn, results from finely controlled morphology in A-B-A or $(A-B)_n$ systems containing both flexible and rigid segments (see Chapters 3 and 4). In principle, this type of behavior is also attainable with graft copolymer structures. However, in actual practice, only limited success has been achieved to date (D6, E15–E17). This is undoubtedly due to the difficulties in obtaining a high degree of architectural and structural control in graft copolymers.

REFERENCES

A1. Sperling, L. H., *Polym. Prepr., Am. Chem. Soc., Div. Polym. Chem.* **14**(2), 958 (1973).

B1. Krause, S., *J. Macromol. Sci., Rev. Macromol. Chem.* **7**(2), 251 (1972); *C.A.* **77**, 34988h (1972).

B2. Bohn, L., *Kolloid Z. & Z. Polym.* **213**(1–2), 55 (1966); *C.A.* **66**, 29094g (1967).

B3. Brodsky, P. H., *Diss. Abstr. Int. B* **30**(3), 1096 (1969); *C.A.* **72**, 112170y (1970).

B4. Dobry, A., and Boyer-Kawenoki, F., *J. Polym. Sci.* **2**, 90 (1947).

B5. Gesner, B. D., *Encyl. Polym. Sci. Technol.* **10**, 694 (1969); *C.A.* **72**, 44375s (1970).

B6. Fettes, E. M., and Macray, W. N., *Appl. Polym. Symp.* **7**, 3 (1968); *C.A.* **70**, 48062u (1969).

B7. Skiest, I., and Miron, J., *Am. Paint J.* **53**(54), 31 (1969); *C.A.* **71**, 61899n (1969).

B8. Yoshimura, N., and Fujimoto, K., *Nippon Gomu Kyokaishi* **41**(3), 161 (1968); *C.A.* **69**, 78264h (1968).

B9. Vrij, A., *J. Polym. Sci., Part A-2* **6**(11), 1919 (1968); *C.A.* **69**, 97277f (1968).

B10. Rosen, S. L., *Polym. Eng. Sci.* **7**, 115 (1967).

B11. Matzner, M., Noshay, A., Schober, D. L., and McGrath, J. E., *Ind. Chim. Belge* **38**, 1104 (1973).

B12. McGrath, J. E., Robeson, L. M., and Matzner, M., *Polym. Prepr., Am. Chem. Soc., Div. Polym. Chem.* **14**(2), 1032 (1973); *in* "Recent Advances in Blends, Blocks and Grafts" (L. Sperling, ed.), p. 195. Plenum, New York, 1974.

B13. Scott, R. L., *J. Chem. Phys.* **17**(3), 279 (1949).

B14. Conaghan, J., and Rosen, S. L., *Polym. Eng. Sci.* **12**, 134 (1972).

B15. Cizek, E. P., U.S. Patent 3,383,435 (General Electric Co.) (1968); *C.A.* **69**, 28292v (1968).

B16. Bair, H. E., *Polym. Eng. Sci.* **10**(4), 247 (1970); *C.A.* **73**, 46077r (1970).

B17. Robeson, L. M., *J. Appl. Polym. Sci.* **17**(12), 3609 (1973); *C.A.* **80**, 96488r (1974).

B18. Zakrzewski, G. A., *Polymer* **14**, 347 (1973).

B19. Matzner, M., and McGrath, J. E., unpublished work.

B20. Bair, H. E., *Anal. Calorimetry, Proc. Symp., 2nd, 1970* p. 51 (1970); *C.A.* **75**, 21553w (1971).

B21. Prest, W. M., Jr., and Porter, R. S., *J. Polym. Sci., Part A-2* **10**, 1639 (1972).

B22. Jacques, C. H. M., Hopfenberg, H. B., and Stannett, V., *Polym. Eng. Sci.* **13**(2), 81 (1973).

B23. Matzner, M., Schober, D. L., Johnson, R. N., Robeson, L. M., and McGrath, J. E., *Am. Chem. Soc., Div. Org. Coat. Plast. Chem., Pap.* **34**(1), 469 (1974).

B24. Matzner, M., Noshay, A., Schober, D. L., and McGrath, J. E., *Ind. Chim. Belge* **38**, 1104 (1973).

C1. Ham, G. E., *Encycl. Polym. Sci. Technol.* **4**, 165 (1966); *C.A.* **65**, 17054a (1966).

C2. Harwood, H. J., chairman, "Symposium on Alternating Copolymers," Vol. 14(1). Am. Chem. Soc., Div. Polym. Chem., Polym. Prepr., New York, 1973.

C3. Johnston, N. W., *Polym. Prepr., Am. Chem. Soc., Div. Polym. Chem.* **14**(1), 46 (1973).

C4. Tobolsky, A. V., "Structure and Properties of Polymers." Wiley, New York, 1960.

D1. Stannett, V., *J. Macromol. Sci., Chem.* **4**(5), 1177 (1970); *C.A.* **73**, 25882j (1970).

D2. Battaerd, H., and Tregear, G. W., "Graft Copolymers." Wiley (Interscience), New York, 1967.

D3. Ceresa, R. J., "Block and Graft Copolymers." Butterworth, London, 1962.

D4. Fischer, J. P., *Angew. Chem., Int. Ed. Engl.* **12**(5), 428 (1973); *C.A.* **82**, 58220f (1975).

D5. Kennedy, J. P., Charles, J. J., and Davidson, D. L., *in* "Recent Advances in Polymer Blends, Grafts and Blocks" (L. H. Sperling, ed.), p. 157. Plenum, New York, 1974.

D6. Falk, J. C., Schlott, R. J., and Hoeg, D. F., *J. Macromol. Sci., Chem.* **7**(8), 1647 (1973).

D7. Langer, A. W., Jr., chairman, "Symposium on N-chelated Metal Compounds," Vol. 13 (2). Am. Chem. Soc., Div. Polym. Chem., Polym. Prepr., New York, 1972.

D8. Heller, J., *Polym. Eng. Sci.* **11**(1), 6 (1971); *C.A.* **74**, 42623p (1971).

D9. Brooks, B. W., and Riches, K. M., British Patent 1,097,997 (Shell Internationale Research Maatschappij N.V.) (1968).

D10. Misumi, T., Minekawa, S., and Minoura, Y. Japanese Patent 13,010 (Asaki Chemical Industry Co.) (1967); *C.A.* **67**, 109472q (1967).

D11. Hirota, K., Kuwata, K., and Matsuura, J., *Nippon Kagaku Zasshi* **83**, 503 (1962); *C.A.***58**, 5790b (1963).

D12. Gervasi, J. A., Gosnell, A. B., Woods, D. K., and Stannett, V., *J. Polym. Sci., Part A-1* **6**(4), 859 (1968); *C.A.* **68**, 96459j (1968).

D13. Yam pol'skaya, M. A., Plate, N. A., and Kargin, V. A., *Vysokomol. Soedin., Ser. A* **10**, 152 (1968); *C.A.* **68**, 69401c (1968).

D14. British Patent 1,121,195 (Sun Oil Co.) (1968); *C.A.* **69**, 97441e (1968).

D15. Naylor, F. E., U.S. Patent 3,492,369 (Phillips Petroleum Co.) (1970); *C.A.* **72**, 67777p (1970).

D16. Pope, G. A., and Peterson, W. J., S. African Patent 68/02,474 (Dunlop Co. Ltd.) (1969); *C.A.* **71**, 31203p (1969).

D17. Heller, J., and Miller, D. B., *J. Polym. Sci., Part B* **7**(2), 141 (1969); *C.A.* **71**, 4296q (1969).

D18. Dondos, A., and Rempp, P., *C. R. Hebd. Seances Acad. Sci.* **256**, 4443 (1963); **254**, 1426 (1962); *C.A.***59**, 6531g (1963).

D19. Matzner, M., Schober, D. L., and McGrath, J. E., *Polym. Prepr., Am. Chem. Soc., Div. Polym. Chem.* **13**(2), 754 (1972).

D20. Rempp, P., *Pure Appl. Chem.* **16**(2–3), 403 (1968); *C.A.* **69**, 77743b (1968).

D21. Stannett, V., *J. Macromol. Sci., Chem.* **4**(5), 1177 (1970).

D22. Merrett, F. M., *J. Polym. Sci.* **24**, 467 (1957).

E1. Morton, M., and Fetters, L. J., *Macromol. Rev.* **2**, 71 (1967); *C.A.* **68**, 50059a (1967).

E2. Fetters, L. J., *J. Polym., Sci., Part C* **26**, 1 (1969); *C.A.* 107018j (1969).

E3. Gobran, R. H., *in* "Chemical Reactions of Polymers" (E. M. Fettes, ed.), p. 295 Wiley (Interscience), New York, 1964.

E4. Szwarc, M., *Makromol. Chem.* **35**, 132 (1960).

E5. Immergut, E. H., and Mark, H., *Makromol. Chem.* **18/19**, 322 (1956).

E6. Mark, H. F., *Tex. Res. J.* **24**, 294 (1953).

E7. Mark, H., *Angew. Chem.* **67**, 53 (1955).

E8. Hoffman, A. S., and Bacskai, R., *in* "Copolymerization" (G. Ham, ed.), p. 335. Wiley (Interscience), New York, 1964.

E9. Smets, G., *Fortschr. Hochpolym.-Forsch.* **2**, 173 (1960).

E10. Molau, G. E., *N.A.S.—N.R.C., Publ.* **1573** (1968).

E11. Szwarc, M., "Carbanions, Living Polymers and Electron Transfer Processes." Wiley (Interscience), New York, 1968.

E12. Zelinski, R., and Childers, C. W., *Rubber Chem. Technol.* **41**(1), 161 (1968); *C.A.* **68**, 79257z (1968).

E13. Benoit, H., *J. Polym. Sci., Part C* **4**, 1589 (1964); *C.A.* **61**, 16157a (1964).

E14. Benoit, H., *Ber. Bunsenges. Phys. Chem.* **70**(3), 286 (1966); *C.A.* **64**, 16000a (1966).

E15. Falk, J. C., and Schlott, R. J., *J. Macromol. Sci., Chem.* **7**(8), 1663 (1973).

E16. Wells, S. C., *J. Elastoplast.* **5**, 102 (1973).

E17. Falk, J. C., Hoeg, D. F., Schlott, R. J., and Pendleton, J. F., *J. Macromol. Sci., Chem.* **7**(8), 1669 (1973).

3

Block Copolymer Architecture

In Chapter 2, block copolymers as a class of materials were distinguished from random copolymers and graft copolymers. Within the general category of block copolymers, there are several architectural variations that describe the sequential arrangement of the component segments. The importance of sequential architecture in block copolymers can not be overemphasized. It is a prime consideration in defining the synthetic technique to be used in preparing a specific block copolymer structure. Furthermore, this factor plays a dominant role in determining the inherent properties attainable with a given pair of segments.

The three basic architectural forms are shown schematically in Fig. 3-1. The simplest arrangement is the diblock structure, commonly referred to as an A-B block copolymer, which is composed of one segment of "A" repeat units and one segment of "B" repeat units. The second form is the triblock, or A-B-A, block copolymer structure, consisting of a single segment of B repeat units located between two segments of A repeat units. The third basic type is the $(A-B)_n$ multiblock copolymer, which contains many alternating A and B blocks. Another, but less common, variation is the radial block copolymer. This structure takes the form of a star-shaped macromolecule in which

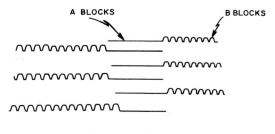

A-B ARCHITECTURE

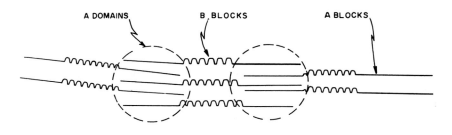

A-B-A ARCHITECTURE

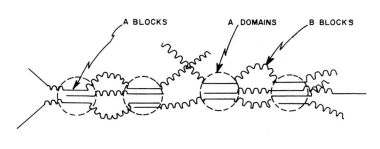

$(A-B)_n$ ARCHITECTURE

Fig. 3-1. Schematic representation of various block copolymer architectures.

three or more diblock sequences radiate from a central hub. Two such arrangements are shown schematically in Fig. 3-2.

Due to the important ramifications outlined above, the detailed discussion of individual block copolymers in this book is subdivided, in Chapters 5, 6, and 7, into architectural categories, i.e., A-B, A-B-A, and

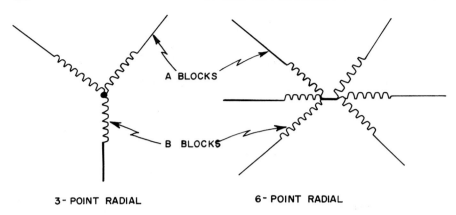

A BLOCKS

B BLOCKS

3 - POINT RADIAL   6 - POINT RADIAL

Fig. 3-2.   Radial block copolymer subarchitectures.

$\{A\text{-}B\}_n$. Detailed information and references to the original literature can be found in these chapters. The purpose of this chapter is to comment on the ways in which the synthesis and properties of block copolymers are influenced by architectural considerations.

A. SYNTHESIS

The techniques most commonly employed to prepare block copolymers are living addition polymerization schemes and step-growth condensation techniques of various types. A-B and A-B-A architectures are primarily synthesized by anionic living polymerization techniques. By contrast, $\{A\text{-}B\}_n$ structures are most often prepared via step-growth methods. It is not possible to synthesize A-B and A-B-A structures by step-growth polymerization techniques due to the statistical nature of these stoichiometrically controlled reactions. On the other hand, it is inconvenient to produce well-defined $\{A\text{-}B\}_n$ structures via living addition techniques because of the high probability of premature chain termination. This is caused by the adventitious impurities encountered during repeated sequential monomer addition cycles.

The anionic living polymerization approach allows the preparation of blocks that have predictable molecular weights and narrow molecular weight distributions. Also, long block lengths are possible, since the block molecular weight is governed only by the ratio of monomer to initiator. While these features are not easily achieved with step-growth processes, the latter offer the advantage of a wider selection of chemical structures including many "high performance" materials.

It can be concluded from the above remarks that the available synthetic techniques are often restricted to a considerable extent by the block copolymer architecture that is desired. However, by judicious choice of the available techniques and, indeed, by combinations thereof, it is possible to prepare a wide range of chemical structures encompassing all three basic architectural forms.

B. PROPERTIES

Architecture exerts a major influence on some of the properties of block copolymers, while other properties are essentially independent of sequential arrangement. For all practical purposes, the architecture-independent properties include those derived from the chemical nature of the segments, e.g., thermal transition behavior, chemical resistance, stability, and electrical and transport properties. The properties that are dramatically effected by architecture are elastomeric behavior, melt rheology, and toughness in rigid materials.

The unique elastomeric behavior of A-B-A and $\{A-B\}_n$ block copolymers is responsible for the development of an entire new technology—thermoplastic elastomers. Block copolymers of this type are characterized by a combination of features that were previously considered to be mutually exclusive, namely, thermoplasticity together with rubberlike behavior. Several commercial products of this type are already available in the marketplace. The key requirements for achieving thermoplastic elastomeric behavior is the ability to develop a two-phase physical network. Such a system is composed of a minor fraction of a hard block (i.e., T_g or T_m above room temperature) and a major fraction of a soft block (i.e., T_g below room temperature.) In these systems, the hard blocks associate to form small morphological domains that serve as physical cross-linking and reinforcement sites. These sites are thermally reversible, i.e., melt processibility is possible at temperatures above the hard block T_g or T_m. Only architectural forms that contain two or more hard blocks per macromolecule are capable of producing this effect (see Fig. 3-1). Therefore, all block copolymer thermoplastic elastomers are based on A-B-A or $\{A-B\}_n$ sequential arrangements. Diblock copolymers are incapable of producing network structures, since only one end of the soft block is chemically linked to a domain of hard segments. The phenomenal elastomeric behavior of the thermoplastic elastomers is illustrated by the

stress–strain curve shown in Fig. 3-3. In the case of the A-B-A and
{A-B}$_n$ systems, high strength and good recovery properties approaching those displayed by *chemically* cross-linked elastomers are attainable. On the other hand, A-B block copolymer elastomers resemble
weak uncured rubbers.

A-B-A and {A-B}$_n$ architectures exert an adverse influence on melt
rheology. This is also due to the ability of these structures to form
physical network systems. To varying degrees, depending on specific
chemical structure and block length, these networks can persist even
in the block copolymer melt. This results in extraordinarily high melt
viscosities and elasticities. In contrast, A-B architectures, which do not
produce network structures, display better melt processibility. Radial
structures (see Fig. 3-2) display much lower viscosities (in the melt or
in solution) than their linear A-B-A counterparts of similar composition
and molecular weight. This is presumably related to the branched
nature of the radial systems.

One technique for preparing toughened rigid systems is to synthesize a block copolymer containing a major fraction of a hard block and
a minor fraction of a soft block. This is most successful for systems
containing two or more hard segments, possibly due to the presence of
physical networks. Therefore, A-B-A, radial, and {A-B}$_n$ structures are
more suitable for this purpose than diblock systems.

In addition to the above-described well-established effects of architecture, it can be speculated that a number of additional features
may be influenced by it. These features, though not yet fully

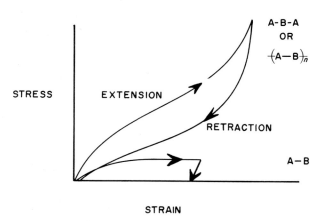

Fig. 3-3.   Effect of block copolymer architecture on stress–strain properties.

documented in the literature, include elastomeric behavior, morphology, crystallization, and compatibility.

Thermoplastic elastomers based on $(A-B)_n$ rather than A-B-A structures might be expected to display enhanced recovery properties due to the presence of a greater number of physical junction sites per macromolecular chain. For the same reason, it could also be argued that network disruption due to degradation or to the presence of physical network imperfections should be less extensive for $(A-B)_n$ than for A-B-A thermoplastic elastomers. Finally, it is interesting to consider the possibility that $(A-B)_n$ thermoplastic elastomers are more likely than comparable A-B-A structures to have block molecular weights shorter than their characteristic entanglement molecular weights (M_e). This feature could have important ramifications for the melt rheology of these systems.

It has been observed that A-B systems develop somewhat coarser morphology than A-B-A structures. Extrapolation of this trend suggests that $(A-B)_n$ copolymers should produce the "finest" morphology and therefore result in materials that have the greatest optical clarity. Furthermore, crystallization kinetics in crystallizable block copolymers would also be expected to be a function of architecture.

Essentially all morphological studies to date have been conducted on block copolymers containing only two dissimilar segments—A and B. Highly novel morphological structures would be expected to result from block terpolymers containing three mutually incompatible segments, e.g., A-B-C or $(A-B-C)_n$. Very little has appeared in the literature to date on this subject.

The use of block copolymers to compatibilize homopolymers via an emulsifying effect is a subject of great practical importance that is receiving an increasing amount of attention. It is not known to what extent block copolymer architecture is important for this function. However, it might be argued, from an entropic point of view, that simple architectures, e.g., A-B, should be optimum for this purpose.

4

Block Copolymer Synopsis

In this chapter, a synopsis of block copolymers as a class of materials is presented. For those interested only in a general knowledge of the field, it is a concise qualitative review of the entire field of block copolymers. Such a broad summary has not appeared in the literature heretofore. For those who require detailed information on block copolymers, it serves as a guide to the subsequent chapters in which each of the main types of block copolymers is discussed individually.

The chapter is divided into five sections: synthesis, structural characterization, physical properties, applications for commercially available block copolymers, and challenges for the future. The basic intent is to answer the following questions about block copolymers: (1) How are they made? (2) How are their structures identified? (3) What are their properties? (4) What are they used for? (5) What are the future opportunities? Since this chapter serves as a summary and guide, the references quoted in it are principally to the other sections of the book.

A. SYNTHESIS

1. General Background

Many approaches for synthesizing polymers with "blocklike" character have been reported. However, only a few actually are capable of producing legitimate block copolymer structures. Others do not result in predictable or well-controlled architecture.

The latter is exemplified by the random interaction of two high molecular weight homopolymers by means of mastication or interchange reactions and also by copolymers produced by free radical initiation. Mastication is the mechanochemical cleavage of polymers followed by recombination of the resulting fragments. This general technique has been described in the literature (A1, A2). The interchange route involves catalyzed reactions between two polymers that contain in their backbones functional linkages capable of undergoing exchange, i.e., polyesters, polyamides, polysiloxanes, (see Chapter 7, Sections D, F–I). Both of these approaches result in products that can contain both of the homopolymers and highly randomized copolymers as well as some ill-defined block copolymers of varying architecture. Some free radical initiated polymerization processes result in somewhat better block integrity than those obtained by mastication or interchange (see Chapter 5, Section B; Chapter 6, Sections B; C, 2; and C, 3). However, these techniques are still plagued by homopolymer contamination. For example, when the polymerization of a vinyl monomer is initiated by a hydroperoxide-terminated macromolecule, the resulting block copolymer is invariably contaminated by substantial quantities of both homopolymers. These arise from (a) the presence of nonfunctional species in the polymeric initiator, (b) vinyl polymerization initiated by nonpolymeric radicals, and (c) chain transfer and termination reactions. Recombination reactions can also produce block copolymer species other than those intended.

In contrast to the limited success achieved by the above approaches, there are techniques that do give rise to predictable structures with a high degree of block integrity. All of these techniques are forms of either living addition polymerization or step-growth (condensation) polymerization. Since they are clearly the superior synthetic methods, they have been emphasized in this chapter and also in the detailed discussion given in Chapters 5–7. The success of these techniques stems primarily from three desirable features, common to both approaches. First, the location and concentration of active sites are known. Second, homopolymer contamination is minimal. This results from the absence of terminating side reactions in living systems and from stoichiometry control in the step-growth systems. Finally, segment length and placement are controlled. This is accomplished by sequential monomer addition techniques in the living polymerizations and by the judicious selection of oligomer end groups and oligomer molecular weight in the step-growth case.

Each of these two preferred general methods offers its own relative advantages. Living polymerization techniques can be used to achieve all three types of block copolymer architecture [A-B, A-B-A, {A-B}$_n$]. On the other hand, step-growth processes are primarily limited to {A-B}$_n$ structures as a consequence of stoichiometric and statistical considerations. Long block lengths and narrow molecular weight distributions are more readily achieved in living systems than in step-growth block copolymers. This is due to the low concentrations of end groups present in high molecular weight step-growth oligomers and to the Gaussian molecular weight distributions inherent in step-growth polymers. On the other hand, the step-growth processes offer the advantages of a much wider selection of polymeric types. In addition, these processes are much less sensitive to reactive impurities than the living polymerizations.

The discussion that follows summarizes the three preferred methods for synthesizing block copolymers. In the first approach, sequential addition living polymerization processes are used to form both of the segments and also the intersegment linkage. The second method involves the interaction of two preformed functionally terminated oligomers to form the intersegment linkage. The third route involves the polymerization of a second block onto the end group of a preformed first segment. The second and third techniques allow the chemical combination of many different types of segmental structures. Numerous highly efficient chemical reactions can be utilized. They allow step-growth, addition, and ring opening blocks to be combined into a single macromolecule. Indeed, these techniques may be the only ones possible for preparing some copolymers, since random copolymerization may not be feasible due to the lack of a common polymerization mechanism.

2. Living Polymerization Sequential Addition Processes

Living polymerization processes entail only initiation and propagation steps and are essentially devoid of terminating side reactions. This feature allows the synthesis of predetermined and well-controlled structures, which is a valuable asset in the preparation of block copolymers. Structural control is often critical in order to develop the unique properties inherent in block copolymers. There is a wide spectrum of systems that can, in theory, be synthesized by this method. Some of these systems approach the above ideal situation quite closely. Others deviate from this ideal to varying degrees. This section discusses the various systems in decreasing order of achievable structural integrity. Both olefin and ring opening polymerizations can be carried out via living processes. These can proceed, at least in principle, via anionic, cationic, and coordination mechanisms. The anionic route is inherently more free of terminating reactions due to the greater stability of the anionic growing ends. Therefore, this technique is the preferred method for synthesizing block copolymers. However, progress is being made with cationic techniques, especially with polyethers and with polyisobutylene systems.

a. Well-Defined Structures

The best example of a well-defined block copolymer synthesized by anionic living techniques is the alkyllithium-initiated polymerization of styrene and butadiene (see Chapter 5, Section A; Chapter 6, Section A). This technique, which is used mainly with alkenyl aromatic and diene monomers, is carried out by the polymerization of styrene followed by the sequential addition and polymerization of butadiene as shown in reactions (4-1) through (4-4).

Initiation

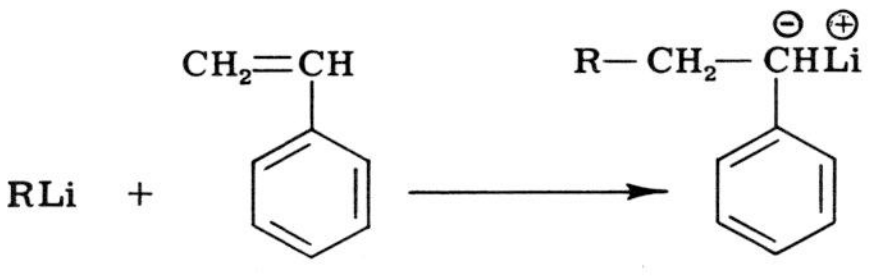

$$RLi \ + \ CH_2{=}CH \longrightarrow R{-}CH_2{-}\overset{\ominus}{\underset{}{CH}}\overset{\oplus}{Li} \tag{4-1}$$

First Propagation

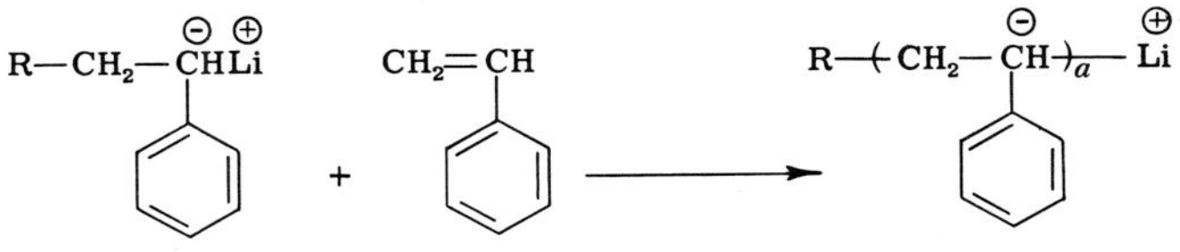

$$R{-}CH_2{-}\overset{\ominus}{CH}\overset{\oplus}{Li} \ + \ CH_2{=}CH \longrightarrow R{+}CH_2{-}\overset{\ominus}{CH}{)}_a{-}\overset{\oplus}{Li} \tag{4-2}$$

Cross-Initiation

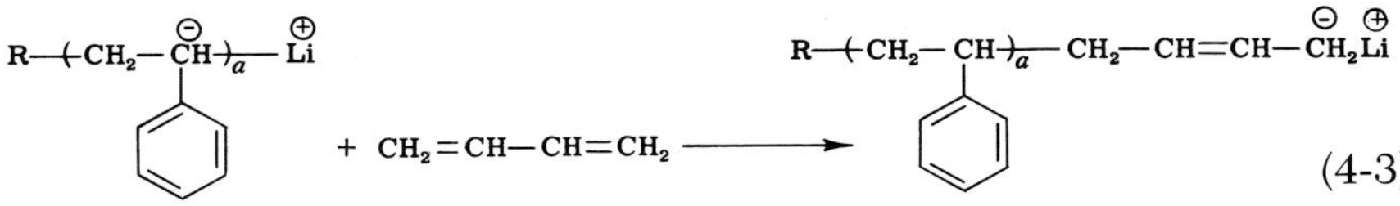

$$R{+}CH_2{-}\overset{\ominus}{CH}{)}_a{-}\overset{\oplus}{Li} \ + \ CH_2{=}CH{-}CH{=}CH_2 \longrightarrow R{+}CH_2{-}CH{)}_a{-}CH_2{-}CH{=}CH{-}\overset{\ominus}{CH_2}\overset{\oplus}{Li} \tag{4-3}$$

Second Propagation

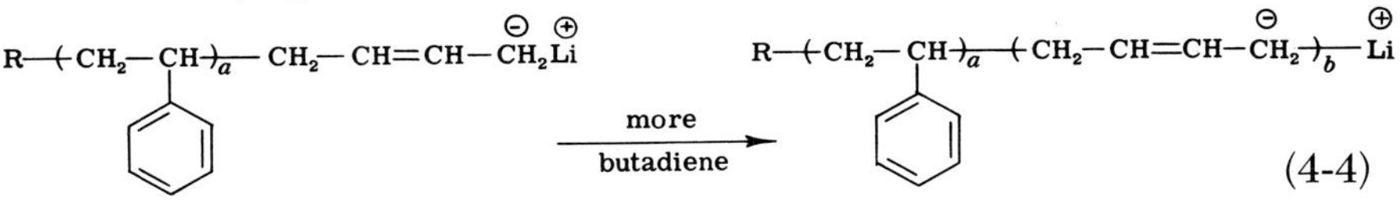

$$R{+}CH_2{-}CH{)}_a{-}CH_2{-}CH{=}CH{-}\overset{\ominus}{CH_2}\overset{\oplus}{Li} \xrightarrow[\text{butadiene}]{\text{more}} R{+}CH_2{-}CH{)}_a{+}CH_2{-}CH{=}CH{-}CH_2{)}_b{-}\overset{\oplus}{Li} \tag{4-4}$$

The reverse monomer addition sequence can also be used. The reaction proceeds in homogeneous solution in hydrocarbon solvents. It is essential that all active hydrogen impurities (water, alcohols, etc.) be scrupulously excluded in order to obtain a termination-free system. This apparent obstacle can be overcome, as is evident from the commercial availability of block copolymers made via this process. Furthermore, these materials sell in the same price range as conventional copolymers prepared by free radical processes.

The anionic active chain ends exist primarily in the form of ion pairs when hydrocarbons are used as the polymerization solvent. On the other hand, polar solvents, such as ethers, result in the appearance of free ion species and hence greatly increased reaction rates. Another important role of the solvent in lithium-initiated systems is its effect on the microstructure of the polybutadiene segment. Hydrocarbon media encourage the formation of 1,4 microstructure, which is necessary to achieve good elastomeric properties. The use of polar solvents and/or alkali metal initiators other than lithium result in high 1,2 configurations, which display much higher glass transition temperatures and hence less attractive elastomeric properties.

Since there is no termination reaction, block length, which is dependent only upon the monomer/initiator ratio, is easily controlled. Another important outgrowth of the absence of termination is the ability to achieve a very high degree of monodispersity. If initiation is rapid compared to propagation, $\overline{M}_w/\overline{M}_n$ (weight average molecular weight/number average molecular weight) ratios approaching 1.0 can be achieved. The combination of these two features in a block copolymer synthesis results in the high degree of structural uniformity necessary to achieve properties approaching theoretical limits.

Sequential monomer addition, as is shown in the above equations, is the most obvious and direct route to block copolymers. An interesting synthetic variation is possible with styrene and dienes that leads to "tapered block" structures. This is possible because of their unique anionic copolymerization behavior in hydrocarbon media with lithium initiation. Surprisingly, mixtures of butadiene and styrene monomers polymerize in such a way that very little styrene is incorporated until essentially all of the butadiene is consumed. The result of this phenomenon is the formation of a block copolymer in which a segment of essentially pure polybutadiene is linked to a segment of pure polystyrene through a "tapered" junction. This junction is comprised of a gradient composition ranging from all butadiene to all styrene.

All architectural types of block copolymers [e.g., A-B, A-B-A and $\{A-B\}_n]$ can be synthesized by living polymerization techniques. This can be achieved by using monofunctional initiators and by choosing the desired number of sequential monomer addition cycles. A-B-A and $(A\text{-}B)_n$ structures can also be prepared through the use of difunctional initiators as shown in Scheme 4-1. This route has the advantage of

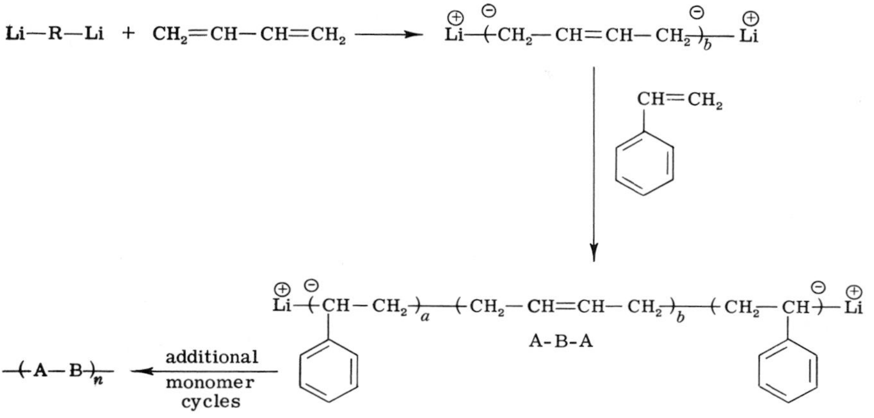

$$Li\text{—}R\text{—}Li \ + \ CH_2\text{=}CH\text{—}CH\text{=}CH_2 \ \longrightarrow \ \overset{\oplus}{Li}\text{—}(\overset{\ominus}{CH_2}\text{—}CH\text{=}CH\text{—}CH_2)_{\overline{b}}\text{—}\overset{\oplus}{Li}$$

$$CH\text{=}CH_2$$

$$\overset{\oplus}{Li}\text{—}(\overset{\ominus}{CH}\text{—}CH_2)_{\overline{a}}\text{—}(CH_2\text{—}CH\text{=}CH\text{—}CH_2)_{\overline{b}}\text{—}(CH_2\text{—}\overset{\ominus}{CH})\text{—}\overset{\oplus}{Li}$$

A-B-A

$$\text{—}(A\text{—}B)_{\overline{n}} \quad \xleftarrow[\text{monomer}]{\text{additional}} \quad$$
cycles

Scheme 4-1

requiring fewer sequential monomer additions than are necessary with monofunctional initiators. Furthermore, when used in conjunction with the tapered block technique, A-B-A block copolymers can be synthesized directly from mixtures of the monomers. A further feature of the living polymer technique is the capability for synthesizing very long segments and also very high molecular weight block copolymers. This is a desirable feature, since long block lengths are often necessary to achieve good properties.

b. Moderately Defined Structures

The discussion above pertains to the nearly ideal systems based on anionically initiated hydrocarbon monomers. Anionic living polymerization techniques can also be applied to other monomer types. For various reasons, these systems generally produce block copolymers with a less ideal, moderately defined structure. This is due to a greater susceptibility to terminating side reactions. This group includes both olefinic and heterocyclic monomers such as the acrylics, vinyl pyridines, siloxanes, lactones, episulfides, and epoxides. A further limitation with these systems is the inability of their relatively weakly basic anions to initiate the polymerization of olefinic monomers such as styrene. This limits the number of architectural types that can be synthesized by these techniques.

Acrylic monomers such as methyl methacrylate, methyl acrylate, and acrylonitrile undergo anionic polymerization, but the pendant ester and nitrile groups are sources of complicating side reactions. In unsubstituted polyacrylates, the labile tertiary hydrogen presents further problems. With methyl methacrylate, both the initiator and the propagating chain end can attack the ester moiety as shown in reactions (4-5) and (4-6). Obviously, these reactions will lead to uncertain-

$$RLi \; + \; CH_2{=}\overset{\displaystyle CH_3}{\underset{\displaystyle \underset{\displaystyle OCH_3}{|}}{\overset{|}{C}}} \qquad \longrightarrow \qquad CH_2{=}\overset{\displaystyle H_3C}{\overset{|}{C}}{-}\overset{\displaystyle O}{\overset{\|}{C}}R \; + \; Li\overset{\oplus\ominus}{OCH_3} \qquad\qquad (4\text{-}5)$$

$$PMMA \text{\Large\textasciitilde\textasciitilde\textasciitilde} CH_2{-}\overset{\displaystyle CH_3}{\underset{\displaystyle \underset{\displaystyle OCH_3}{|}}{\overset{\ominus}{\underset{|}{C}}}}\overset{\oplus}{Li} \; + \; CH_2{=}\overset{\displaystyle CH_3}{\underset{\displaystyle \underset{\displaystyle OCH_3}{|}}{\overset{|}{C}}} \qquad \longrightarrow \qquad PMMA \text{\Large\textasciitilde\textasciitilde\textasciitilde}\overset{\displaystyle O}{\overset{\|}{C}}{-}\overset{\displaystyle CH_3}{\overset{|}{C}}{=}CH_2 \; + \; Li\overset{\oplus\ominus}{OCH_3} \qquad (4\text{-}6)$$

ties in molecular weight prediction and to premature termination. Both of these features prevent the preparation of well-defined block copolymer structures. The initiator side reaction is the more serious of the two. This problem can be alleviated by the use of less basic initiators, such as alkoxides, or of sterically hindered carbanions. Furthermore, low polymerization temperatures (e.g., −78°C) favor the double bond attack relative to the ester side reaction.

Most of the acrylic block copolymers reported do not have a high degree of block integrity. However, some investigators have claimed significant advances in structural control by employing the precautions outlined above. These block copolymers are exemplified by those containing segments of methyl methacrylate together with segments of either alkyl methacrylates or styrene. The latter type of copolymer imposes the additional restrictions that (a) the styrene must be polymerized first, and (b) the resulting polystyrene must be capped with 1,1-diphenylethylene before the methyl methacrylate is polymerized, as shown in Scheme 4-2. The resulting sterically hindered, capped polymeric carbanion can polymerize methyl methacrylate, but unlike the uncapped polystyrene anion, it does not readily attack the ester group. The synthesis must proceed in the order shown, since poly(methyl methacrylate) anion is not sufficiently basic to initiate styrene polymerization. A-B-A structures can also be prepared if

polystyrene anion $CH_2-\overset{\ominus}{C}\overset{\oplus}{H}Li$ + $CH_2=C$

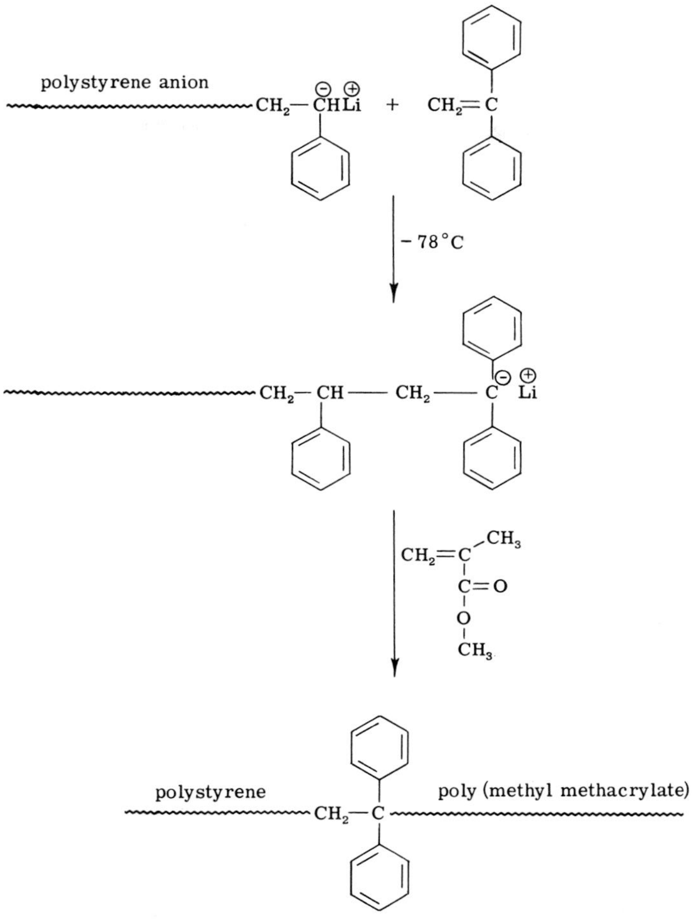

$-78°C$

$CH_2-CH---CH_2---\overset{\ominus}{C}\overset{\oplus}{Li}$

$CH_2=C\overset{CH_3}{\underset{\underset{\underset{CH_3}{O}}{C=O}}{}}$

polystyrene CH_2-C poly (methyl methacrylate)

Scheme 4-2

difunctional initiators are used (see Chapter 5, Section B; Chapter 6, Section B).

Cyclic monomers have also been utilized to prepare block copolymers with moderately well-defined structures. Anionically prepared block copolymers containing polysiloxanes, polylactones, or polyalkylene sulfides as at least one of the segments have been synthesized with predictable molecular weights and narrow molecular weight distributions. Block copolymers with polyepoxide segments have been prepared with controlled molecular weights but not in a highly monodisperse form. All of these systems suffer from an inability to produce segments in any desired sequence, since heteroatom anions cannot initiate the polymerization of hydrocarbon monomers. The result of this limitation is that the vinyl polymer segment must be synthesized before the heteroatom segment.

Some heteroatom polymers crystallize during their preparation, resulting in insolubilization. This feature leads to limitations in block integrity and molecular weight. Also, in crystalline–noncrystalline block copolymers, the insolubilization effect usually necessitates the synthesis of the soluble, noncrystalline segment prior to crystalline segment formation, unless special solvents can be used.

A large number of anionically prepared block copolymers containing heteroatom segments have been reported and are discussed in subsequent chapters of this book (see Chapter 5, Section D; Chapter 6, Section C). Anionically synthesized siloxane-containing block copolymers are illustrative of this general class and are summarized briefly below.

Siloxane polymers are normally polydisperse. However, the cyclic trimer of dimethylsiloxane can be polymerized via living anionic processes to form poly(dimethylsiloxanes) of predictable molecular weight and narrow molecular weight distribution. It is critical to use lithium-based initiators (alkyls or silanolates) and promoting solvents such as tetrahydrofuran. Under these conditions, the cyclic trimer undergoes rapid ring opening polymerization with minimal redistribution of siloxane bonds. This is not the case when the less reactive but more commonly used cyclic tetramer is employed or when other alkali metal-based initiators are used. In these latter cases, redistribution reactions are prevalent, leading to loss of monodispersity.

By use of the preferred conditions described above, block copolymers with well-defined siloxane segments have been prepared. These include diphenylsiloxane–dimethylsiloxane and alkenyl aromatic–dimethylsiloxane block copolymers. The latter is exemplified by styrene–dimethylsiloxane as shown in Scheme 4-3. A-B-A block copolymers in which the B block is polystyrene can also be made by using difunctional initiators. However, the inability of the siloxanolate anion to initiate styrene polymerization precludes the synthesis of styrene–siloxane–styrene A-B-A block copolymers and of {A-B}$_n$ structures by these sequential addition procedures (see Chapter 5, Section D,6; Chapter 6, Section C,6; Chapter 7, Section I).

c. Poorly Defined Structures

The polymerization of olefins such as ethylene and propylene initiated by Ziegler–Natta transition metal catalysts is considered to pro-

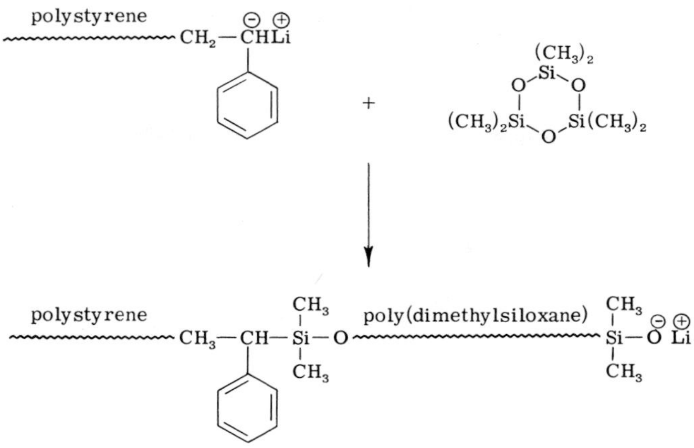
polystyrene
CH2—CHLi
(CH3)2
Si
O
O
(CH3)2Si
Si(CH3)2
O
+
polystyrene
CH3
poly(dimethylsiloxane)
CH3
CH3—CH—Si—O
Si—O Li
CH3
CH3

Scheme 4-3

ceed via an anionic-coordinated polymerization mechanism. Some catalytic systems have been reported to produce living polymers, and block copolymers prepared with these catalysts have been claimed. However, the chain lifetime achieved in these systems is short, at best, due to the occurrence of chain transfer reactions. This results in a high degree of uncertainty as to the block sequence structure of the products obtained. Furthermore, the heterogeneous nature of most Ziegler–Natta catalyst systems, together with the chain transfer phenomenon, results in highly polydisperse systems and poorly-controlled block molecular weights. Some investigators have grappled with these problems and have made products with properties suggestive of block copolymers. However, these structures have not been well characterized and are almost certainly highly contaminated with homopolymers. Despite these synthetic difficulties, quite useful olefin-containing block copolymer products are in commercial use. Further work directed toward a better understanding of these catalytic systems should be rewarding (see Chapter 5, Section C).

3. Interaction of Functionally Terminated Oligomers

The interaction of functionally terminated oligomers is a versatile method for producing block copolymers. Many functional groups can be utilized, at least in theory. By this approach, only the intersegment linkage is formed during the block copolymerization reaction. Generally, difunctional species are used leading to $(A\text{-}B)_n$ block copolymer

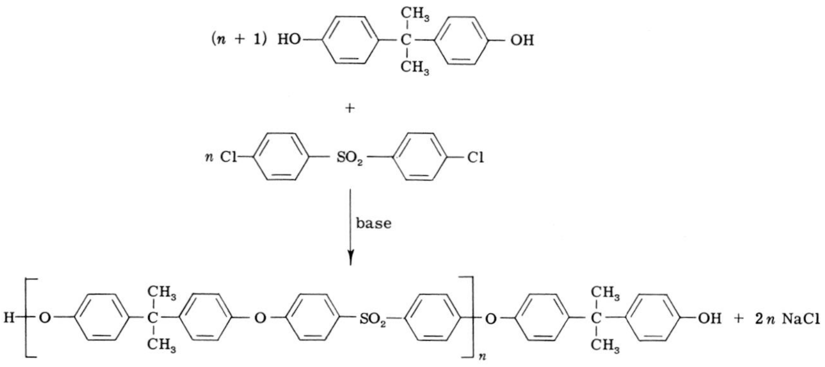

Scheme 4-4

structures only. However, in principle, monofunctional oligomers could also be used to generate A-B or A-B-A structures. The preformed oligomers can be prepared either by step-growth reactions or by appropriate addition or ring opening polymerizations. In the step-growth case, the end groups are a natural consequence of the polymerization chemistry. These polymers bear the end group of the monomer used in excess. An example for hydroxyl-terminated polysulfone is shown in

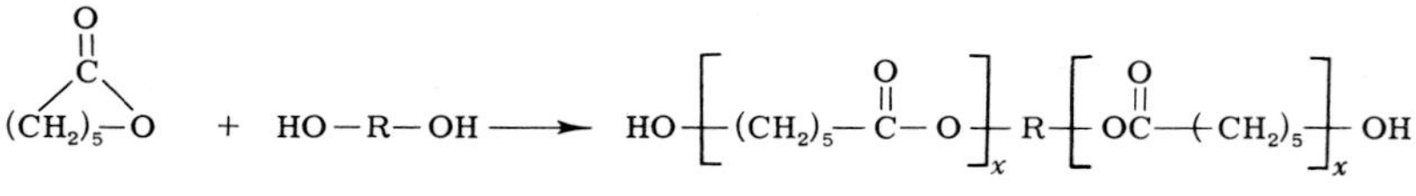

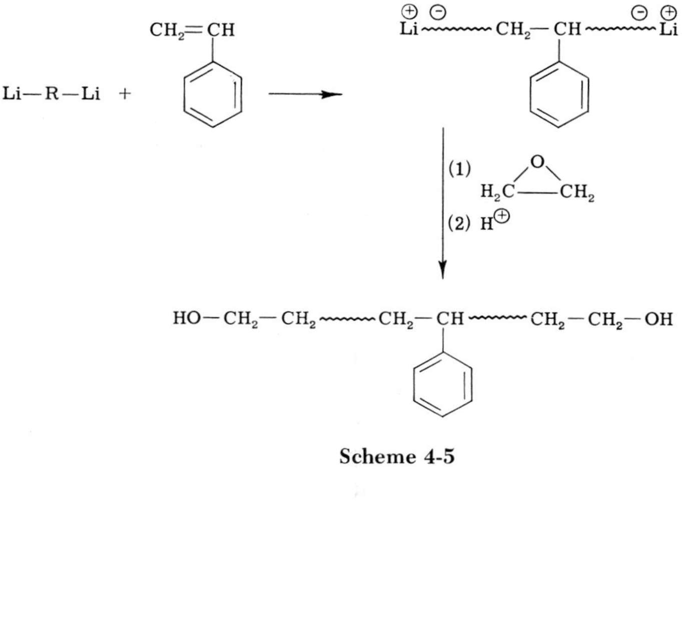

Scheme 4-5

Scheme 4-4. In the case of ring opening or addition polymerization, the end group can be predetermined via initiator choice or by end capping such as shown in Scheme 4-5 for polycaprolactone and for polystyrene. The reactions in Scheme 4-5 are merely examples of the kinds of oligomeric structures that can be employed by this technique. Other possibilities include polyamides, polycarbonates, polyure-thanes, polysiloxanes, poly(alkylene ethers) and poly(aryl esters). Useful end groups other than hydroxyl include amines, isocyanates, acid halides, silyl halides, and even carbanions. The only major requirement is that they interact in a highly efficient manner.

The above described functionally terminated oligomers can be in-corporated into $(A-B)_n$ block copolymers with perfectly alternating segments or with statistically placed segments. Perfectly alternating sequence distributions are obtained when oligomers bearing *mutually reactive* end groups are used. By definition, such oligomers can only react with each other and not with themselves as outlined in Scheme 4-6.

This general technique is typified by the synthesis of polysulfone–poly(dimethylsiloxane) block copolymers. Dimethylamino-terminated

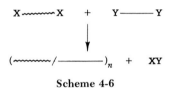

Scheme 4-6

siloxanes are reacted with hydroxyl-terminated polysulfones as shown in Scheme 4-7.

The average molecular weights of the blocks are identical to those of the corresponding oligomers. Although these segments occur in a per-fectly alternating sequence, the polydispersity of the block copolymer is, of course, a function of both the oligomer-forming and block copolymer-forming reactions. In contrast, block copolymers with less control of segment sequence are obtained when two oligomers with the same functional end group are coupled via reaction with a third component. If both oligomers are similar in reactivity toward the cou-pling agent, they are incorporated into the block copolymer molecule in a statistical or random fashion. This is shown in Scheme 4-8. A typical example of this is the coupling of hydroxyl-terminated poly(ethylene oxide) and bisphenol-A polycarbonate oligomers by

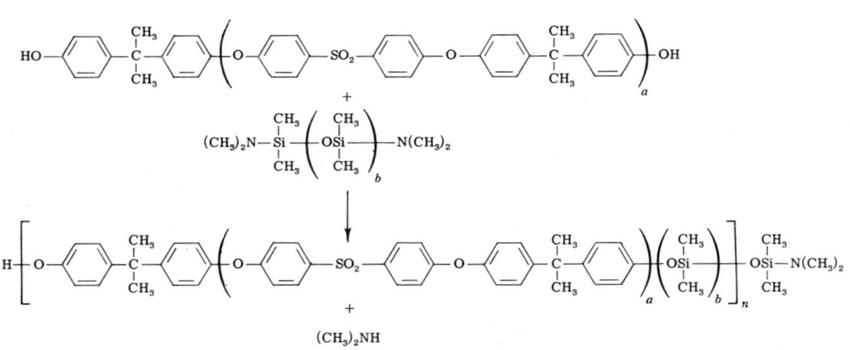

Scheme 4-7

phosgene. A result of the above reaction scheme is that the average *block* molecular weights are higher than those of the corresponding oligomers. This is due to the ability of a given oligomer to couple with itself as well as with the second oligomer. If the two oligomers are greatly different in their reactivity toward the coupling agent, it is possible to produce structures that contain long runs of similar segments. The chemical composition of the perfectly alternating systems is directly proportional to the oligomer molecular weights. By contrast, the statistically coupled systems have the advantage that compositions can be easily controlled since they are independent of block molecular weight. In the alternating systems, equimolar quantities of the functional oligomers are required in order to fulfill stoichiometric requirements. In the statistically coupled systems, any combination of the two oligomers can be stoichiometrically satisfied by the use of the appropriate quantity of coupling agent.

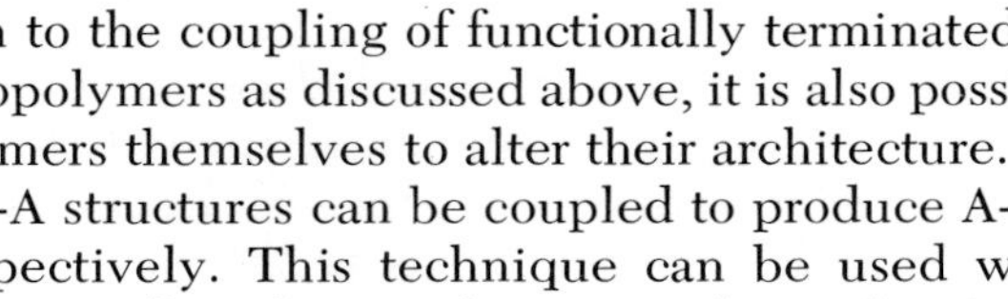

Scheme 4-8

In addition to the coupling of functionally terminated oligomers to form block copolymers as discussed above, it is also possible to couple block copolymers themselves to alter their architecture. For example, A-B and A-B-A structures can be coupled to produce A-B-A or $\{A-B\}_n$ systems, respectively. This technique can be used with relatively stable "end groups" such as carbanions, siloxanolates, and thiolates, and many coupling agents such as phosgene, alkylene dihalides, and dihalosilanes. An illustrative example is the linear coupling of polystyrene–polybutadiene as shown in Scheme 4-9. The so-called

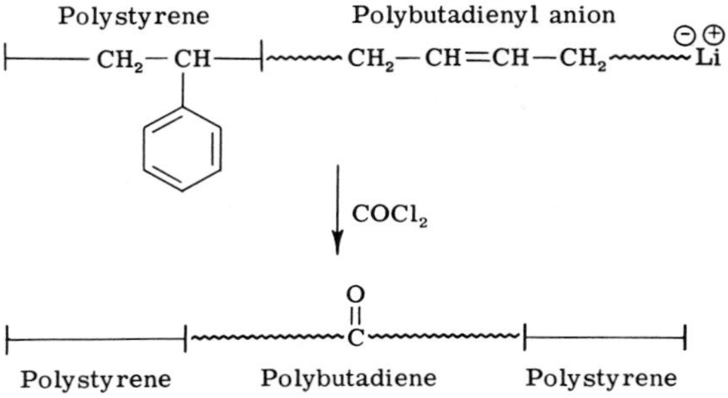

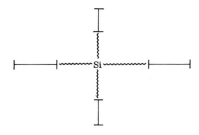

Scheme 4-9

radial block copolymers are prepared by a similar technique using a polyfunctional coupling agent such as silicon tetrachloride yielding a star-shaped structure as shown in Scheme 4-10 (see Chapter 6, Section A; Chapter 7).

Scheme 4-10

4. Polymer Growth from Oligomer End Groups

This technique is essentially a combination of the two techniques described above in that the block copolymer forming reaction employs both monomers and preformed functionally terminated oligomers. In general, this route is more economical than the oligomer–oligomer approach, since it allows the possibility of bulk polymerization and does not require the isolation of the second oligomer. Bulk polymerization is not normally possible with the oligomer–oligomer process due to the well-known phenomenon of polymer–polymer incompatibility. Furthermore, this technique has an advantage in the solution polymerization of block copolymers containing both amorphous and crystalline segments. This is difficult to achieve by the oligomer–oligomer process due to the insolubility of the crystalline segment. This can be accomplished easily by the growth of the crystallizing polymer initiated by the end groups of a preformed soluble oligomer.

Block copolymers can be prepared by this technique from various combinations of starting materials:

(a) Addition/ring opening oligomers plus addition/ring opening monomers
(b) Step-growth oligomers plus addition/ring opening monomers
(c) Addition/ring opening oligomers plus step-growth monomers
(d) Step-growth oligomers plus step-growth monomers

The linkage between the blocks is formed simultaneously with the growth of the second block. The nature of this linkage depends upon the end group of the preformed oligomer, which in turn is a function of the initiator, capping agent, or comonomer stoichiometry employed.

All three block copolymer architectural types can be synthesized by this oligomer–monomer approach. Architecture is dependent upon monomer type and oligomer functionality. The reaction of addition or ring opening monomers with mono- or difunctional oligomers produces A-B and A-B-A structures, respectively (Scheme 4-11). The situ-

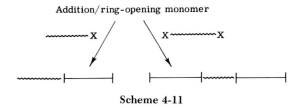

Addition/ring-opening monomer

Scheme 4-11

ation is more complicated with step-growth monomers, due to the statistical nature of their polymerization reactions. Under conditions of perfect stoichiometry, they can interact with monofunctional and difunctional oligomers to produce principally A-B-A and $(A-B)_n$ structures, respectively (Scheme 4-12). Even under conditions of perfect stoichiometry, statistical sequence imperfections can result, which are

Step-growth monomers

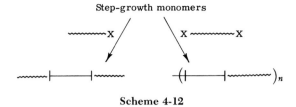

Scheme 4-12

especially disruptive with monofunctional oligomers. Furthermore, if stoichiometry deviates from this ideal situation, the resulting products will be even more heterogeneous, due to the presence of unintended, contaminating block structures (or even coupled homopolymer).

The oligomer–monomer approach can be used to prepare many block copolymers containing both addition–ring opening blocks and step-growth blocks. For the purpose of illustration, four typical syntheses are shown on pages 46 and 47.

Although this approach can be used to synthesize a large number of block copolymers, it does suffer from the drawback that the molecular characteristics of the second block cannot be determined separately from the block copolymer. Furthermore, in $\{A\text{-}B\}_n$ block copolymer structures of the type illustrated by Reactions (c) and (d) on pages 46 and 47, the performed oligomers enter the block copolymer in a statistical rather than perfectly alternating fashion. This is explained by the fact that the oligomer end-group functionality is identical to that of one of the step-growth monomers. Therefore, as in the statistical oligomer–oligomer route discussed earlier, the molecular weight of the segment in the block copolymer is higher than that of the corresponding oligomer. It is worthwhile to note that Reactions (c) and (d) are techniques that are believed to be used in the commercial production of two types of high-performance thermoplastic elastomers.

All of the synthetic procedures discussed in this chapter pertain to soluble thermoplastic block copolymers. However, some recent reports in the literature indicate that cross-linked thermosetting block copolymers can also be prepared by technique (c) above. For example, the cross-linking of epoxy resins in the presence of functionally terminated oligomers gives rise to structures that are composed of linear segments and three-dimensional cross-linked "segments." This further demonstrates the broad latitude of structures possible. Obviously, the thermosetting polyurethane materials, which are industrially important in rigid and elastomeric foam and solid applications, also fall in this same category (see Chapters 6 and 7).

(a) *Addition Oligomer–Ring Opening Monomer*

Monohydroxyl-terminated
polystyrene

Polystyrene   Poly (ε-caprolactone)
A–B block copolymer

(b) *Step-Growth Oligomer–Ring Opening Monomer*

Chlorine-terminated polysulfone

Nylon 6   Polysulfone   Nylon 6
A–B–A block copolymer

(c) *Ring Opening Oligomer–Step-Growth Monomers*

$$HO(CH_2-CH_2-CH_2-CH_2-O)_b H + CH_3O-C(=O)-\phi-C(=O)-OCH_3 + HO(CH_2)_4-OH$$

Hydroxyl-terminated poly(tetramethylene glycol)

Dimethyl terephthalate

1,4-Butanediol

Poly (tetramethylene glycol)

Poly (butylene terephthalate)

$-(A-B)_n-$ block copolymer

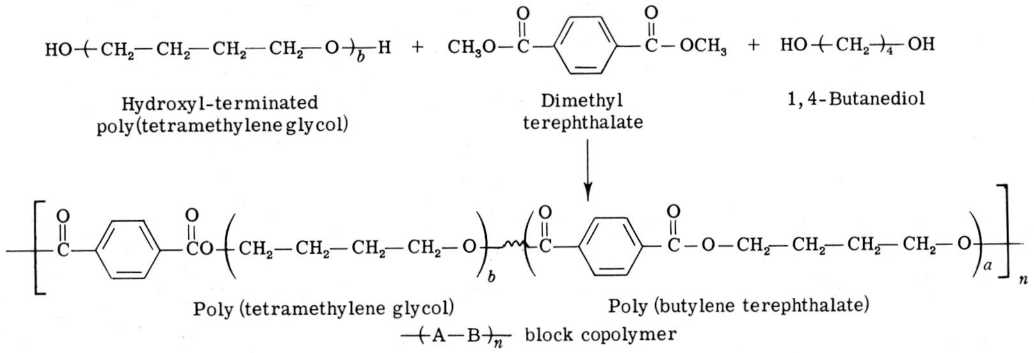

(d) *Step-Growth Oligomer–Step-Growth Monomers*

$$H[OCH_2-CH_2-CH_2-CH_2-OC(=O)-CH_2-CH_2-CH_2-CH_2-C(=O)]_b OCH_2-CH_2-CH_2-CH_2-OH$$

Hydroxyl-terminated poly(butylene adipate)

$$OCN-\phi-CH_2-\phi-NCO$$

MDI

$$+ \; HOCH_2-CH_2-CH_2-CH_2-OH$$

1,4-Butanediol

Poly (butylene adipate)-butanediol-MDI polyurethane $-(A-B)_n-$ block copolymer

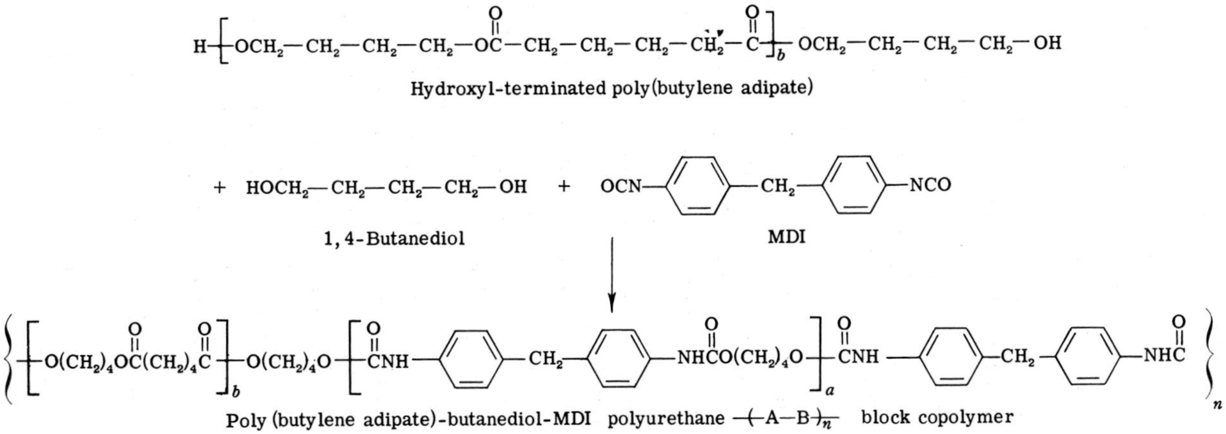

B. STRUCTURAL CHARACTERIZATION

The preparation of block copolymers has been the goal of many synthetic chemists. The degree to which this goal is actually achieved can only be ascertained by the successful utilization of appropriately discriminating characterization methods. The unequivocal determination of block copolymer structure is a formidable task. Characterization of block copolymers is not as highly developed as that of homopolymers. Accordingly, the great majority of block copolymers reported in the literature have not been well characterized. Only a few of them, most notably the styrene–diene block copolymers prepared by controllable anionic living polymer processes, have been subjected to exhaustive structural studies. Most other block copolymers have been assigned structures deduced almost entirely from the synthetic procedure employed.

Many of the analytical tools that are useful with homopolymers can also be used to elucidate block copolymer structure, e.g., determination of average composition by elemental analysis or spectroscopic techniques. However, no single individual method is capable of adequately describing the nature of these complex macromolecules. A combination of several methods, including knowledge of the mechanism of the polymerization, must be used in concert. Even when this approach is taken, many questions can still remain unanswered. All of the complexities inherent in the characterization of homopolymers are also encountered with block copolymers. Superimposed on these, however, are uncertainties associated specifically with block copolymers, such as the length, sequential placement, polydispersity, and compositional heterogeneity of the component segments. For example, it is necessary to know whether a given reaction product is, in fact, a block copolymer or whether it is a random copolymer, a blend of homopolymers, or indeed a complex mixture of two or more of these. It is also important to determine the sequential architecture of the block copolymer [i.e., A-B or A-B-A or $\{A-B\}_n$] and the molecular weights and dispersities of the segments as well as those of the total copolymer. Supermolecular structure and its ramifications are particularly important in block copolymers. The role of supermolecular structures in the morphology of semicrystalline homopolymers is well known. This feature is even more important in determining the morphological characteristics of the more complex macromolecules present in block copolymers.

In the following discussion, an attempt is made to outline the various characterization techniques that have been or could be employed to elucidate the above structural features of block copolymer systems. It is not our purpose to describe these analytical techniques in detail. This has already been done in several books and reviews (B1–B3) pertaining to homopolymers. Instead, it is intended to point out in a concise manner those techniques that can best yield useful information regarding the segmental nature of block copolymers. Pertinent references to the original block copolymer literature are referred to throughout the subsequent chapters of this book and are therefore not repeated in this section.

Chronologically speaking, the first question that must be answered is whether a reaction product is indeed a block copolymer as intended, or if the polymerization procedure fell short of its mark and resulted instead in a mixture of two homopolymers. If it can be shown that a copolymer has indeed been formed, it is then necessary to ascertain whether it is a random copolymer or a block copolymer. Assuming that a block copolymer structure can be demonstrated, the next step is to determine its architecture (i.e., the number of segments in the copolymer) and also whether or not the major component is contaminated with homopolymers or with block copolymers of a different architecture. Having reached this plateau, it is then desirable to ascertain the molecular weight and molecular weight distribution of the component segments. Finally, the supermolecular nature of the product must be determined. The discussion below outlines those techniques that are most useful in each of these steps. In addition, the techniques are listed in Table 4-1 for easy reference.

1. Block Copolymer or Homopolymer Blend?

The property most commonly relied upon to answer this question is solubility behavior. Homopolymer blends and copolymers respond differently to extraction with selective solvents and to solvent–nonsolvent fractionation techniques such as fractional precipitation, column fractionation, and turbidometric titrations. These methods are simple and reasonably effective, particularly with systems containing segments that differ significantly in their chemical structure and/or physical state, e.g., polar–nonpolar, crystalline–amorphous. However, even with these systems, techniques based on solubility characteristics are not foolproof. Homopolymer mixtures are relatively easy to separate. Sequential extraction with solvents *selective* for each of the components results in complete dissolution of the blend, leaving no insoluble residue. On the other hand, extraction data obtained on

TABLE 4-1

Techniques for Block Copolymer Characterization

Block copolymer versus homopolymer blend	*Molecular structure*

1. Solubility characteristics
    a. Solid phase extraction
    b. Solution fractionation
2. Film clarity
3. Solution compatibility
4. Molecular weight distribution
    a. Density gradient ultracentrifugation
    b. Gel permeation chromatography
5. Rheological characteristics
	1. Osmometry
	2. Solution light scattering
	3. Ultracentrifugation
	4. Gel permeation chromatography
	5. Solution viscometry
	6. Oligomer analysis
	7. Selective degradation

Block copolymer versus random copolymer	*Architecture and purity* 1. Elastic recovery 2. Rheological characteristics 3. Gel permeation chromatography 4. Density gradient ultracentrifugation
1. Proton magnetic resonance 2. Infrared spectroscopy 3. Dynamic mechanical behavior 4. Differential scanning calorimetry 5. Electron microscopy 6. Small-angle X-ray scattering 7. Mechanical properties 8. Rheological characteristics 9. Crystallinity characteristics 10. Solution light scattering 11. Thermomechanical analysis	*Supermolecular structure* 1. Dynamic mechanical behavior 2. Differential scanning calorimetry 3. Rheological characteristics 4. Electron microscopy and scanning electron microscopy 5. Wide-angle X-ray scattering 6. Small-angle X-ray scattering 7. Birefringence 8. Small-angle light scattering

block copolymers are more difficult to interpret. A similar experiment performed on an *ideal,* pure, monodisperse block copolymer containing approximately equivalent amounts of two chemically dissimilar segments of equal length might result in essentially no dissolution, i.e., 100% insoluble residue. This is due to the prevailing influence of the insoluble segment on the solubility of the second segment to which it is chemically linked. However, most *real* block copolymers are not ideal, but rather have a distribution of composition and of segment molecular weight. As a result, that fraction of a block copolymer that is comprised predominantly of segment A will be completely soluble in segment A-selective solvents. Therefore, the block copolymer residue remaining after such an extraction may not accurately reflect the true composition of the entire block copolymer actually synthesized. The extent of compositional fractionation that occurs in this way may be estimated by determining the segment B content of the extraction solution.

In solvents that are *not* highly selective for *either* segment, the intersegmental linkage *enhances* the solubility of the block copolymer. The range of solvents that would be expected to dissolve two homopolymers is broadened in block copolymers due to the partial compatibilizing effect of this chemical bond. The uncertainties discussed above are further complicated by the time-dependent diffusion phenomenon associated with the separation of two highly entangled macromolecules. Thus, the segments of a block copolymer can exert either a negative or positive influence on copolymer solubility, depending upon the above-described considerations.

Film clarity is another method frequently used to distinguish between block copolymers and homopolymer blends. The gross incompatibility characteristic of most homopolymer blends results in opaque films due to a high degree of light scattering at the interface between the two phases. An exception occurs in the relatively rare case in which the two homopolymers have similar refractive indices. On the other hand, block copolymers produce transparent films, since they usually exist in microphase-separated morphological states in which the domains are too small to scatter visible light. These optical characteristics permit rapid distinction between block copolymers and blends in amorphous systems. Obviously, this technique is of more limited value when one or both of the segments are highly crystalline. However, crystalline systems could be investigated in the melt. Under these conditions the copolymers would be clear and the blends opaque. The film clarity test, though simple to perform, is only qualitative and may not detect minor amounts of homopolymer contamination.

The clarity of high-concentration polymer solutions (e.g., >10 wt.% polymer) can distinguish between block copolymer structures and incompatible polymer blends. Since block copolymers constitute a single chemical species, they produce clear single-phase solutions. On the other hand, incompatible homopolymer blends result in cloudy solutions that eventually separate into two liquid layers due to well-known thermodynamic immiscibility phenomena.

Another distinguishing tool is based on the presence of single-mode or bimodal molecular weight distributions. A block copolymer representing a single chemical species should display essentially single-mode behavior. By contrast, homopolymer blends are more likely to exhibit bimodal characteristics. One of the better ways to investigate this phenomenon is density gradient ultracentrifugation. Assuming the two components possess different densities, a bimodal distribution will be detected in the blend, while the block copolymer will have a single intermediate density and hence display a single mode distribution. Gel permeation chromatography (GPC), a more commonly used technique for measuring molecular weight distribution, can also be used for this purpose. However, since this technique is based on separation according to molecular size, it can only be used to detect bimodal distributions in blends in which the components are widely different in molecular size.

Finally, certain rheological measurements can often reveal the presence or absence of block copolymer structures. For example, the melt viscosity of block copolymers is usually considerably higher than that of a blend of homopolymers of similar molecular weight. This is especially true for A-B-A and $\{A\text{-}B\}_n$ structures due to the persistence of some degree of physical network structure even in the melt. This effect is magnified in block copolymers consisting of segments that are long and that vary widely in solubility parameter and, hence, intersegment compatibility.

2. Block Copolymer or Random Copolymer?

Most of the techniques described in the preceding paragraphs were addressed to distinguishing copolymers from homopolymer blends. The next task is to ascertain whether the copolymer is random or blocklike. Although several methods have been used to answer this question, those pertaining to sequence distribution and supermolecular structure are most discriminating. Sequence distribution can be investigated by techniques such as proton or carbon-13 magnetic resonance or infrared spectroscopy. Supermolecular structures can be detected with various thermomechanical and morphological techniques, such as dynamic mechanical behavior, differential scanning calorimetry (DSC), microscopy, and X-ray.

Random copolymers contain short runs of the same repeat unit, e.g., diads, triads, and tetrads, as well as runs of an alternating nature. It is frequently possible to quantitatively measure the distribution of these sequence types via infrared or PMR procedures. Block copolymers represent an extreme case wherein the run sequences are very long and the alternating sequences are vanishingly small, occurring only at the segment junctions. As a result, this technique can be a very powerful tool in discerning between random and block copolymers.

Block copolymers generally display behavior ascribable to supermolecular structures, while random copolymers do not. This is a consequence of the two-phase morphology present in most block copolymers. One of the most dramatic methods for demonstrating the presence of two phases is the determination of the thermomechanical behavior by conducting dynamic mechanical or DSC studies. Two-phase systems display two separate and distinct glass transition temperatures (T_g) characteristic of each of the component segments. By contrast, random copolymers display one T_g which is normally a compositionally weighted average of the T_g values of the corresponding homopolymers. An exception to this rule is a block copolymer in which the segments are mutually compatible (e.g., styrene–α-methylstyrene). Such a material displays single-phase transition behavior even though it contains long segments.

Two-phase block copolymers can also be distinguished from single-phase materials by morphological techniques such as transmission electron microscopy (EM), scanning electron microscopy (SEM), and small-angle X-ray scattering (SAXS). The homogeneous single-phase copolymer is relatively featureless by these methods. However, electron micrographs of two-phase structures can provide direct visual evidence of the coexistence of two-phase domains. The SAXS technique is complimentary to microscopy, since it provides information on the internal morphology rather than the surface characteristics.

Other techniques can yield supplemental information useful in deciding whether a copolymer is blocklike or random. For example, those block copolymer architectures that form physical networks are discernable from random copolymers because of differences in certain mechanical and rheological properties such as elastic recovery and melt viscosity. Measurement of block copolymer crystallinity by techniques such as X-ray or DSC can be useful. If component A and/or B are crystallizable, the presence of crystallinity in a copolymer suggests the existence of block rather than random sequences. Solubility behavior may also be useful as a tool to distinguish between block and random copolymers in certain specific systems. However, this technique is much less useful here than for the case of copolymers and homopolymer blends. Measurement of solution light scattering can add information regarding the block-like nature of a copolymer. Since block copolymers behave more like homopolymers than random copolymers do, the weight average molecular weight $(\overline{M}_w)$ should be independent of the solvent medium for a well-defined block copolymer, whereas the greater compositional heterogeneity along the chain in a random copolymer causes $\overline{M}_w$ to be solvent-dependent.

3. Molecular Structure

Once it has been established that a given system is indeed a block copolymer, the next step is to determine its macromolecular structural characteristics, such as molecular weight and molecular weight distribution. These properties can be investigated with many of the same techniques that are established for homopolymers, e.g., membrane and vapor pressure osmometry, light scattering, ultracentrifugation, gel permeation chromatography, and viscometry. The data are generally interpreted in the same manner as for homopolymers. However, with some block copolymers there are interpretational uncertainties with techniques such as light scattering. Interpretation of these data to yield weight average molecular weight is straightforward for certain well-defined block copolymers. However, other, less well-defined, more highly randomized block copolymers may be subject to the deviations characteristic of random copolymers (e.g., compositional heterogeneity effects). Furthermore, the determination of molecular chain dimensions from light-scattering behavior and other solution techniques is not possible at the present state of theoretical development. This is due to the inability to determine true theta conditions simultaneously for both segments of the block copolymer.

In addition to characterizing the total block copolymer, it is important to also determine the molecular structure of the segments themselves. This is possible with some block copolymers prepared via well-controlled synthetic techniques, such as living polymerization and the interaction of mutually reactive oligomers. In living polymerizations, samples may be withdrawn after each sequential monomer addition cycle for subsequent characterization. In the mutually reactive oligomer case, the starting components can be carefully characterized before use. Because of the *mutual* reactivity of the oligomers, their molecular structure by definition will not change upon incorporation into the block copolymer. In addition to the usual homopolymer characterization techniques described above, the reactive oligomers may be characterized by chemical or physical methods such as end group analysis or nuclear magnetic resonance spectroscopy (NMR). Another way to elucidate the molecular structure of the segments is by selective degradation of the block copolymer. This is possible with block copolymers containing one segment that can be degraded completely and exclusively, leaving the other segment intact for subsequent analysis. Obviously, this approach is limited to a small number of block copolymer systems.

4. Architecture and Purity

After carrying out the previously described characterization measurements, a major question still remains as to the architecture of the block copolymer product, e.g., A-B, A-B-A, $\{A-B\}_n$. Furthermore, it is important to ascertain the purity of its architectural form, i.e., the degree to which it is contaminated with homopolymers or block copolymers of a lower architectural order. The *intended* segment sequence arrangement can be predicted from a knowledge of the polymerization procedure employed. The degree of success that is actually achieved is reflected by the type and amount of polymeric impurities in the final product. For example, the synthesis of an A-B block copolymer may result in contamination by one or both homopolymers. Similarly, an $\{A-B\}_n$ block copolymer could contain homopolymer, A-B, and/or A-B-A "impurities."

Block copolymer architecture is very difficult to ascertain without some knowledge of the polymerization technique. It may be possible, in the specific case of elastomeric compositions, to distinguish an A-B architecture from A-B-A and $\{A-B\}_n$ structures. Elastomeric A-B structures will display poor elastic recovery properties, while the latter two architectural forms will display the good recovery properties characteristic of cross-linked elastomers, due to the presence of a physical network. Rheological behavior can also distinguish diblock from triblock and multiblock two-phase amorphous structures. The two latter types generally display higher melt viscosities due to partial retention of the physical network even in the melt. The latter behavior requires long block lengths and a significant differential in segment solubility parameter in order to produce a high degree of phase separation. A further limitation of these tools is the fact that A-B-A can not be distinguished from $\{A-B\}_n$ structures.

It is obvious, therefore, that the number of tools available to define architecture is extremely limited. The situation is improved somewhat if the polymerization technique is known. This knowledge allows architectural interpretation of solution fractionation data that would be rather meaningless without it. It also allows polymeric impurities in the main product to be identified. Techniques such as gel permeation chromatography (GPC) and density gradient ultracentrifugation, which are normally used to measure molecular weight distribution, can reveal the presence of more than one macromolecular population if they differ sufficiently in either molecular size or density. Therefore, this allows a determination of the number of impurities present, their

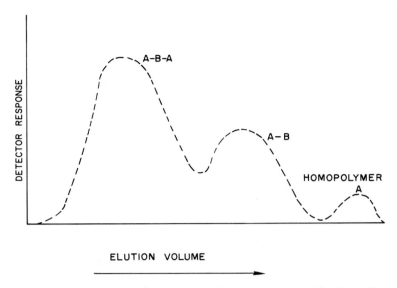

Fig. 4-1.   Gel permeation chromatogram of an impure A-B-A block copolymer.

molecular weights, and their relative concentrations. This data, along with a knowledge of the polymerization mechanism, can assist in the assignment of architectural structures to the main product and also to the contaminants. This is illustrated by the GPC analysis of an impure A-B-A block copolymer synthesized by sequential addition polymerization (see Fig. 4-1). The chromatogram of this product reveals two minor peaks at low and intermediate molecular weight levels in addition to a major high molecular weight peak. This is strong evidence that the principal product has an A-B-A architecture and that it is contaminated with two polymeric impurities—a low molecular weight homopolymer and an intermediate molecular weight A-B block copolymer. Analogous approaches could be used in $\{A\text{-}B\}_n$ systems.

5. Supermolecular Structure

The preceding discussion pertained to analytical techniques for distinguishing block copolymers from homopolymer blends and/or random copolymers and for establishing their molecular structure, architecture, and purity. In contrast to the molecular characteristics, block copolymers can also display behavior attributable to supermolecular structures. These structures result from the aggregation of segmented polymers to form complex morphological systems. These systems are comprised of two normally incompatible phases forced to coexist with each other, which produces microheterogeneous structures of colloidal dimensions. The size of the primary domains is much smaller than those observed with incompatible polymer blends (10^2–10^3 versus $>10^4$ Å). This is due to the presence of the intersegment linkage, which exerts a restraining influence on the entropy of the system. Many of the novel and useful properties of block copolymers are due to these features.

Not all block copolymers display supermolecular structures. The extent to which microphase separation (which is due to segmental incompatibility) occurs to form supermolecular structures depends on three critical features of the segments: (a) compositional dissimilarity, (b) molecular weight, and (c) crystallizability. Single-phase morphology results when the segments are compatible due to similar composition and/or short length. On the other hand, two-phase systems are produced when the segments are mutually incompatible due to appreciable differences in chemical composition. Solubility parameter (δ) allows a semiquantitative assessment of the compositional characteristics of the blocks. The differential solubility parameter (Δ), defined as δ_A-δ_B, is therefore an important parameter governing the phase separation of two amorphous high molecular weight segments, i.e., the higher the Δ, the greater the probability of phase separation.

Block molecular weight exerts a large influence on the critical minimum Δ value necessary to obtain a two-phase system. Thus, a pair of low molecular weight segments will display two-phase behavior only if the Δ value is relatively large. On the other hand, two high molecular weight segments can produce two-phase systems even when Δ values are small. The above criteria pertain primarily to wholly amorphous block copolymers. Crystallizability in one or both segments provides another dimension. Crystallization is, by definition, a strong driving force for phase separation even when the blocks are similar in chemical nature and low in block molecular weight. Indeed, this behavior is analogous to that displayed by semicrystalline homopolymers, which may be considered to be comprised of crystalline "segments" and amorphous "segments" that are compositionally identical.

The type of morphology obtained with amorphous two-phase block copolymers depends both on the segment ratio (i.e., the volume fraction of segments A and B) and on the method of specimen fabrication (see Fig. 4-2). The major component will normally exist as the continuous phase, with the minor component present as discrete domains. The latter assume spherical shapes at very low volume fraction levels,

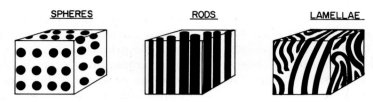

Fig. 4-2. Schematic representation of morphological forms in block copolymers.

e.g., <20%, and rod-like forms at somewhat higher levels. When the two phases are present in nearly equivalent volume fractions they can exist in co-continuous phases with lamellar structures. Obviously these morphological variations can exert a major influence on physical properties that are strongly dependent on the nature of the continuous phase. The influence of specimen fabrication on morphology is most dramatic in block copolymers containing nearly equal volumes of both segments, since these are most subject to phase continuity reversal. Selection of casting solvents is a particularly effective method for controlling morphology (see Table 4-2). The segment least soluble (i.e., most contracted) in the casting solvent utilized will tend to be the first to "precipitate" during solvent evaporation. These segments form discrete domains that become dispersed in continuous matrix of the second "expanded" segment. The reverse morphology can be produced simply by choosing a casting solvent that is preferential for the first segment.

The morphological nature of the interface in supermolecular structures is an intriguing feature of block copolymers. In principle, the interface can be either very sharp or rather diffuse, depending on the nature of the phase boundary. Presumably, relatively incompatible

TABLE 4-2

Effect of Casting Solvent on Block Copolymer Mechanical Properties[a]

Casting solvent	1% secant modulus (psi)	Tensile strength (psi)	Elongation (%)
Tetrahydrofuran	35,500	3300	410
Benzene	22,400	3000	520
Toluene	16,800	2300	310
Xylene	9,500	1900	350

[a] Polysulfone–poly(dimethylsiloxane) block copolymer containing 5000 $\overline{M}_n$ segments.

segments of high molecular weight would be expected to display sharp boundaries and insignificant interfacial volumes. More compatible segment pairs should produce a more diffuse interface. The nature of the interfacial region, or interphase, should resemble that of a single-phase block copolymer. Theoretical treatments have been derived that predict both of the above types of behavior.

There are several techniques that can be used to investigate supermolecular structure and the effect of parameters such as block molecular weight on domain size. Some of these techniques determine the effect of morphology on thermal and rheological properties, while others can yield information based on microscopical or light-scattering behavior. Thermal analysis, such as modulus–temperature relationships, differential scanning calorimetry, and rheological measurements, can detect the presence of supermolecular structures but can not readily distinguish between morphological types. As was discussed earlier, modulus–temperature and differential scanning calorimetry data can, for example, reveal the presence of two-phase morphology by showing the existence of multiple glass transition and/or melting behavior. Abnormally high melt viscosity observations are indicative of supermolecular structures characteristic of a two-phase physically cross-linked network.

Transmission and scanning electron microscopy may reveal the shape and size of domains on the surface of block copolymer specimens. Examination of specimens cut at different angles (e.g., perpendicular or parallel to the plane) can yield additional information on the shape of the domains. While some block copolymers can be investigated directly by these techniques due to large differences in the electron density of the segments, other block copolymers must be exposed to selective staining procedures.

Wide-angle X-ray scattering (WAXS) is, of course, useful for studying morphology if one or both of the segments are crystalline, but is of limited utility in amorphous block copolymers. On the other hand, small-angle X-ray scattering (SAXS) is a powerful tool for investigating the morphology of amorphous block copolymers. In contrast to microscopic methods, the SAXS technique affords information on subsurface morphological characteristics. In addition to characterizing domain size, SAXS is the most useful approach for investigating interdomain spacing and the nature of the interfacial regions.

Birefringence measurements, such as the stress optical coefficient, can be useful in determining which of the segments is the most continuous for a given composition and fabrication procedure. Form birefringence, which can be large for block copolymers with segments that vary significantly in refractive index, can reflect the anisotropic shape of phase boundaries. Small-angle light scattering (SALS) also has the potential for identifying the sizes and shapes of morphological structures. This is particularly true for semicrystalline systems.

C. PHYSICAL PROPERTIES

In previous sections, it was demonstrated that block copolymers, compared to other polymeric materials, are relatively difficult to synthesize and to characterize. Hence, these segmented materials are usually, but not always, more expensive. Therefore, a logical question would be "What unique properties do block copolymers offer that would justify their use?" This section is addressed to answering this question. Clearly, a block copolymer should be chosen only if a homopolymer, physical blend or random copolymer could *not* be used to achieve the desired properties.

As was pointed out in the previous sections, block copolymers can exist in either a single-phase or a multiphase morphology. Both of these can offer unique property advantages. Single phase block copolymers are desirable when the properties of a random copolymer are sought from pairs of monomers that can not be combined via traditional random copolymerization techniques. More commonly, however, block copolymer structures are chosen because of the unique properties possible with *two*-phase morphological systems. This feature is especially beneficial in elastomeric systems.

The following discussion is intended to point out the property advantages and limitations of block copolymers. Throughout this discussion, it is important to bear in mind that changes in the supermolecular structure of two-phase systems brought about by variation in test specimen preparation technique (e.g., solution casting, thermal history) can exert a large influence on these physical properties.

1. Thermal Properties

The modulus temperature behavior of block copolymers is basically different from that of random copolymers, as can be seen from the curves in Fig. 4-3. A random copolymer derived from monomers A and B displays behavior intermediate to that of homopolymers A and B. A single glass transition temperature (T_g) is observed that falls between those of the corresponding homopolymers. The position of the transition is related to the weight fraction of the A and B components.

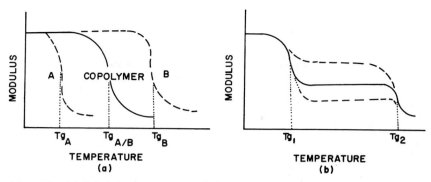

Fig. 4-3. (a) Modulus–temperature behavior of a typical amorphous random copolymer. (b) Modulus–temperature behavior of a two-phase block copolymer.

Single-phase block copolymers in which the segments are highly compatible with each other display similar behavior. On the other hand, two-phase block copolymers generally exhibit modulus–temperature characteristics somewhat analogous to those of physical blends. The identity of both segments is retained, as is evidenced by the presence of two distinct glass transition temperatures. This type of behavior is displayed by thermosetting block copolymers (such as the epoxy ester systems described in Chapter 7, Section J), as well as by linear systems. A constant modulus plateau exists between the two T_g values. The flatness of the plateau is dependent upon the degree of phase separation; the more complete the phase separation, the more temperature insensitive the modulus becomes. Some highly phase-separated block copolymer systems actually approach this ideal behavior quite closely, (e.g., polysulfone –poly(dimethylsiloxane), see Chapter 7, Section I). In contrast to the random copolymer behavior described above, the position of the two T_g values in a two-phase system is not changed significantly by compositional variations. Rather, it is the *position* of the *modulus plateau* that is dependent upon composition. Stated simply, compositional changes cause *vertical modulus* shifts in two-phase block copolymers, but *horizontal temperature* shifts in random copolymers and in single-phase block copolymers. Obviously, this pertains only to two-phase block copolymers in which the block lengths are beyond the level at which $\overline{M}_n$ affects T_g.

This behavior has important practical implications. In single-phase rigid systems, the T_g, and hence the heat distortion temperature of a segment, can be increased via block copolymerization with a second compatible high T_g segment. On the other hand, the constant modulus behavior of two-phase block copolymers can be used to great advantage in elastomeric systems. This feature permits block copolymer elastomers to emulate the thermal characteristics of chemically cross-linked elastomers, as will be discussed later.

The modulus–temperature behavior of the block copolymers discussed above is independent of their architecture [e.g., A-B, A-B-A, (A-B)$_n$], except in so far as architecture affects the efficiency of phase separation.

Thermal transition behavior can, of course, be determined by techniques other than modulus–temperature measurements. Perhaps the most sensitive technique for determining minor as well as major transitions in multiphase systems is the measurement of dynamic mechanical loss characteristics. For block copolymers containing crystalline segments, (e.g., polysulfone–nylon 6; see Chapter 6, Section C,5) calorimetric methods such as differential scanning calorimetry (DSC) provide information on the crystalline melting point (T_m), heat of fusion, and degree of crystallinity of the materials, as well as on major amorphous transitions.

The thermal and thermal-oxidative stability of block copolymers offers no advantages over comparable homopolymers or random copolymers. The reason for this, of course, is that, in general, a macromolecule is only as stable as its most sensitive bond. In fact, if "unzipping" is an important mode in the degradation of one of the components, then block copolymers would be inferior to a corresponding random copolymer, since an entire segment would "unzip" before a stable linkage was reached.

2. Processability

Processability, as used herein, is a term describing the capacity of a material to be transformed into useful shapes via solution casting or melt fabrication techniques. Block copolymers present no unusual problems in solution fabrication. However, two-phase amorphous block copolymers are generally more difficult to melt fabricate than homopolymers, random copolymers, or single-phase block copolymers of similar molecular weight. This is an outgrowth of the unusual rheological characteristics of these materials, which results from the partial retention of their two-phase morphology in the melt. The high melt viscosity and elastic character of these materials often necessitates the use of high processing temperatures and pressures. The high temperature required can often approach, or indeed exceed, the thermal stability limits of the copolymer. Furthermore, shear rate sensitivity can be a limitation in these systems. For example, it is sometimes observed that block copolymers that can be easily compression molded display severe "melt fracture" characteristics during extrusion fabrication (e.g., polysulfone–poly(dimethylsiloxane) see Chapter 7, Section I). This melt fracture behavior is a reflection of the highly elastic and viscous character of these materials. This results in the onset of melt fracture at abnormally low shear rates.

Block copolymer architecture exerts a major influence on melt processability. A-B diblock copolymers process much more readily than A-B-A or {A-B}$_n$ copolymers (e.g., styrene–butadiene; see Chapter 6, Section A). This is due to the network structures formed by the latter two architectures, which persist in the molten state. If the number of network junctions is influential in determining melt processability characteristics, then it might be predicted that {A-B}$_n$ architectures should be more difficult to process than A-B-A systems. However this hypothesis has not yet been experimentally verified.

In amorphous block copolymers, phase separation is dependent upon the differential solubility parameter (Δ) and the molecular weights of the two segments. This behavior can be assessed semiquantitatively by consideration of the product of these three terms. Melt processability is also affected by these two features. Processing ease increases as either the Δ value or the block lengths become smaller. This behavior is convincingly illustrated by organosiloxane block copolymers containing various types and lengths of organic blocks (see Chapter 7, Section I.) While processability can be improved through manipulation of both Δ and block length, control of the Δ parameter may be preferable, since the alternative approach of shortening block length is invariably accompanied by a reduction in the T_g of the segment.

The above comments on block copolymer processability were directed to completely amorphous systems. Block copolymers that contain crystalline segments display phase separation at relatively short crystalline block lengths (e.g., cyclobutylene carbonate–caprolactone; see Chapter 7, Section F). This is due to the facts that (a) crystallization, by definition, requires phase separation, and (b) crystallization can occur at shorter block lengths than would be necessary to obtain two phases in wholly amorphous systems.

The presence of short crystalline segments in A-B-A or {A-B}$_n$ block copolymers has important ramifications for processability. Above the T_m of the crystalline block, the now wholly amorphous system can approach the melt behavior of a single-phase block copolymer, which displays negligible network structure in the melt. The end result of

these features is that block copolymers containing crystalline segments can display a valuable combination of good melt processability
together with physical properties characteristic of a high degree of
phase separation.

3. Mechanical Properties

From a mechanical property point of view, block copolymers may
be conveniently divided, on the basis of room temperature modulus,
into two classes—rigid and elastomeric. Rigid materials may be comprised either of two "hard" segments or of one "hard" segment together with a minor fraction of a soft segment. A "hard" segment is
defined as one that has a T_g and/or T_m above room temperature, while a
"soft" segment has a T_g (and possibly a T_m) below room temperature.
Elastomeric block copolymers normally contain a soft segment together with a minor proportion of a hard segment. Block copolymers
containing two soft segments are also possible but do not offer significant mechanical property advantages.

Rigid block copolymers composed of two hard segments can offer
advantages in mechanical properties that are related to heat distortion
temperature, such as creep or stress–relaxation resistance, (e.g.,
aliphatic–aromatic polyamide block copolymers, Chapter 7, Section G).
Furthermore, the inherent ductility of the segments of a hard–hard
block copolymer can be retained due to fine dispersion of the phases
and to a high degree of adhesion between the phases. In contrast,
physical blends of two ductile homopolymers are often brittle due to
the absence of these features.

The toughness of an inherently brittle rigid polymer can be greatly
improved via block copolymerization with a minor fraction of a soft
segment. This improvement is due to the two-phase nature of the
resulting system and to the low temperature transition imparted to it
by the soft block. Examples of such systems are the high polystyrene
content linear or radial A-B-A styrene–butadiene–styrene block
copolymers and the epoxy–polycaprolactone thermosetting block
copolymers (see Chapter 6, Section A; Chapter 7, Section J).

Architecture is extremely important in determining the mechanical
properties of elastomeric block copolymers. Systems comprised of an
A-B architecture do not display any dramatic improvements over random copolymer elastomers in mechanical properties. Both of these
must be chemically cross-linked or vulcanized to develop good ultimate properties. However, dramatic, unique properties are displayed
by elastomeric block copolymers having an A-B-A (linear or radial) or
{A-B}$_n$ architecture. This behavioral difference is well illustrated by

styrene–butadiene (A-B) and styrene–butadiene–styrene (A-B-A) block copolymers (see Chapter 5, Section A; Chapter 6, Section A).

These A-B-A and $\{A\text{-}B\}_n$ compositions, which have been termed thermoplastic elastomers, combine the mechanical properties of a cross-linked rubber with the processing behavior characteristic of linear thermoplastic polymers. The very unusual combination of these two features has, more than any other factor, provided the driving force for the intensive investigation of block copolymers that has occurred over the last ten years. This concept has been successfully applied to block copolymer types other than styrene–diene systems, including polyurethanes, silicones, polyesters, and polyethers. These systems are discussed in detail in Chapters 6 and 7.

Thermoplastic elastomers are two-phase block copolymers comprised of a major proportion of a soft segment and a minor proportion of a hard segment. Each of the segments serve a specific purpose. The soft block affords the flexible elastomeric nature and the hard block provides both physical cross-link sites and filler reinforcement characteristics. This behavior is possible because of the novel two-phase morphology displayed by these systems. Due to microphase separation, the hard blocks associate with each other to produce small (100–300 Å) dispersed domains that are chemically attached to the rubber matrix. These domains provide the strong interchain association necessary for the development of *physical* cross-links that serve the same function as that provided by chemical cross-links in vulcanized elastomers. This is shown schematically in Fig. 4-4.

Unlike the chemically vulcanized elastomers, the domains in the network of the thermoplastic elastomers soften or melt at temperatures above their T_g or T_m transitions and thus allow fabrication via melt processing techniques. Another function of the glassy or crystalline hard domains is to provide high strength via reinforcement of the rubbery matrix. This is possible due (a) to the discrete nature of the hard domains, (b) to their ideally small size and uniformity, and (c) to the perfect interphase adhesion ensured by the chemical intersegment linkage.

As was mentioned earlier for two-phase block copolymers in general, the properties of thermoplastic elastomers are dependent upon the molecular weight and the volume fraction of hard and soft segments present. Block length must be great enough to develop the two-phase system yet not so excessive as to obviate thermoplasticity. Variation of the hard block/soft block ratio affects modulus, recovery characteristics, and ultimate properties. The volume fraction of the hard block must be sufficiently high (e.g., $\geq 20\%$) to provide an adequate level of physical cross-linking if good recovery properties

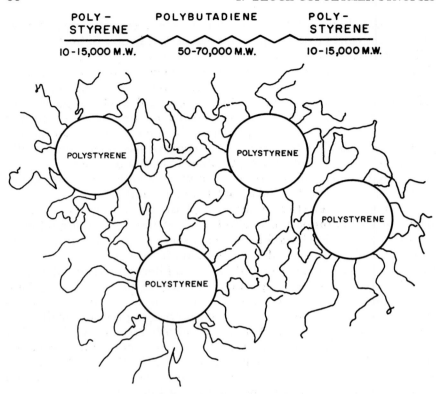

POLY −
STYRENE
10 − 15,000 M.W.

POLYBUTADIENE
50 − 70,000 M.W.

POLY −
STYRENE
10 − 15,000 M.W.

Fig. 4-4.   Morphological model of the physical network.

and high tensile strengths are to be obtained. On the other hand, excessively high (∼30%) hard block concentrations can cause the hard domains to change from a discrete spherical shape to a co-continuous lamellar form, which in turn causes deterioration in recovery properties. Since the mechanical properties of thermoplastic elastomers are so highly dependent on the degree of perfection of the network structure, any architectural impurity that disrupts this network should be minimized. A-B block copolymers have been shown to adversely affect the network structure of A-B-A thermoplastic elastomers. (See Chapter 6, Section A.)

4. Optical Properties

Block copolymers, both rigid and elastomeric, offer significant advantages over homopolymer blends in optical clarity. Homopolymer blends that are normally incompatible are opaque due to large particle size and to the refractive index differences of their constituent macrophases. Due to the entropy restrictions imposed by even one intersegment linkage, block copolymers can only undergo microphase separation to form small domain structures. These domains are much smaller (<1000 Å) than the wavelength of light and thus the copolymers appear transparent even though their segments may differ significantly in refractive index. Examples of this behavior include the organosiloxanes and the styrene dienes (see Chapters 5–7). The domain size increases predictably with molecular weight, but opacity does not occur except at very high block molecular weights. Interestingly, irridescence has been observed in some block copolymer systems containing long segments.

Transparency is surprisingly observed in some crystalline block copolymers as well as in wholly amorphous materials. Presumably, the crystallite size in block copolymers containing crystallizable segments is smaller than that found in corresponding homopolymers. This behavior may be due to either the dispersed nature of the crystalline segment or to a lower degree of crystallinity within these domains. For example, highly crystalline polymers such as polyesters (e.g., polybutylene terephthalate) or polyurethanes derived from butanediol–MDI are opaque, while block copolymers containing these segments may be transparent (see Chapter 7, Sections C and H).

5. Chemical Resistance

In general, the chemical and stress crack resistance of block copolymers is no better than that of their constituents. However, the ability to combine chemically resistant segments with segments of poor chemical resistance via block copolymerization provides a means by which a significant degree of chemical resistance can be achieved without sacrifice of ductility. The alternative approach of blending homopolymers usually results in brittle behavior due to gross incompatibility and poor interphase adhesion. Obviously, crystalline and highly hydrogen-bonded segments are most useful for imparting chemical resistance to a block copolymer. This approach can be applied to both rigid and elastomeric block copolymers. Examples of these types are polysulfone–nylon 6 rigid block copolymers and poly(butylene terephthalate)–poly(tetramethylene glycol) elastomeric block copolymers. (See Chapter 6, Section C; Chapter 7, Section C). Clearly, the volume fraction of crystalline hard segment is a large factor in determining the degree of chemical resistance. Optimum results are achieved when the volume fraction is sufficiently high to ensure some degree of co-continuity.

The hydrolytic stability normally attributed to certain chemical linkages can be greater in block copolymer structures than is displayed in homopolymers or small molecules. An example of this phenomenon is provided by two-phase organosiloxane block copolymers in which the segments are linked by a $\equiv$Si—O—C$\equiv$ bond (see Chapter 7, Section I). The enhanced stability of this bond in the block copolymer structure is due to several factors, including (a) steric hindrance provided by the segments, (b) the low concentration of this bond along the polymer backbone, and (c) the hydrophobic nature of the siloxane block.

6. Transport Properties

The permeability properties of block copolymers are highly dependent on their superstructure. In the extreme case of a single-phase block copolymer, a linear relationship exists between the logarithm of permeability and segmental volume fraction, as expected. On the other hand, two-phase systems diverge from this linear relationship due to a strong dependence on phase continuity. The compositional dependence usually takes the form of an S-shaped curve.

Two-phase block copolymers offer the advantage of allowing tough thin membranes to be prepared without the need for chemical cross-linking or the use of added fillers. This is due to the previously described filler effects and physical cross-links provided by the hard block.

The block copolymer approach to permeable membranes presents the opportunity for combining segments that compliment each other, thereby resulting in attractive combinations of properties. Examples of these are (a) high T_g–low T_g segments, (b) high permeability–low permeability segments, (c) hydrophilic–hydrophobic segments, and (d) anionic–cationic segments, e.g., amphoteric block copolymers. Silicone copolymers are good examples of this behavior (Chapter 7, Section I).

7. Blending Properties

Two-phase block copolymers have the unique ability to show partial compatibility with their corresponding homopolymers. This phenomenon permits homopolymer–block copolymer blends to be prepared that display a high degree of "mechanical" compatibility as a result of good interphase adhesion and fine dispersion. Complete compatibility (i.e., mutual solubility) is not possible in such blends due to the dissimilar nature of the second segment of the block copolymer. This feature can be put to good practical use, such as impact modification of homopolymers, by blending with elastomeric block copolymers. For example, the blending of low concentrations of polysulfone–poly(dimethylsiloxane) block copolymers with polysulfone homopolymer greatly improves the notched impact strength of the latter (see Chapter 7, Section I). Another use of this feature is in improving the chemical resistance of a homopolymer by blending it with a block copolymer containing segments of this homopolymer together with other more chemically resistant segments. An example is a blend of polysulfone with polysulfone–nylon 6 block copolymer (see Chapter 6, Section C). Yet another use of this feature is improvement in the processability of elastomers such as polybutadiene via the addition of the styrene–butadiene block copolymers (see Chapter 5, Section A).

Two-phase block copolymers also have a limited ability to partially compatibilize pairs of the corresponding homopolymers via an "emulsification" or micelle-like effect. This is achieved either in solution or in the solid state. Best results are observed when the molecular weights of the blocks are higher than those of the homopolymers (see Chapter 5, Section A).

Two-phase block copolymers also display surface active properties. Compositions containing water-soluble and oil-soluble segments such as ethylene oxide–propylene oxide are useful as nonionic detergents (See Chapter 5, Section D). Organosiloxane block copolymers are uniquely useful as foam surfactants (Chapter 7, Section I). Similarly, the ability of hydrophilic–hydrophobic polyurethanes to selectively absorb oily lipids (e.g., cholesterol) may form the basis of biologically important block copolymers of the future (C1).

D. APPLICATIONS FOR COMMERCIALLY AVAILABLE BLOCK COPOLYMERS

Block copolymer investigations, in addition to proving fruitful from a scientific point of view, have spawned an entirely new class of commercial materials. Several block copolymer products are currently available in the marketplace, most having been introduced during the last decade. They can be divided into three groups—elastomers, toughened thermoplastic resins, and surfactants. The rapid acceptance of these products is exemplified by the prediction (D1) of a one billion pound per year market by the mid-1980s for thermoplastic elastomers. These and the other currently available block copolymer products have a variety of end uses that are reviewed in this section. The discussion pertains only to those products that are available in commercial quantities. Some other block copolymer products, which are in the developmental or semicommercial stages, are mentioned in Section E.

1. Elastomers

Perhaps the most unique property of block copolymers is the two-phase morphology that is responsible for the phenomenon of thermoplastic elastomeric behavior. However, it must be recognized that not all block copolymers are thermoplastic elastomers, and conversely, not all thermoplastic elastomers are block copolymers. The requirements discussed in previous sections, such as architecture, block length, and composition, must be fulfilled in order for a block copolymer to produce thermoplastic elastomeric properties. Other thermoplastic elastomers are commercially available that do not have a block copolymer structure (e.g., two-phase graft copolymers and/or physical blends). These latter systems can approach the modulus–temperature behavior of block copolymer structures. However, they do not display optimum recovery and ultimate properties due to the absence of a well-defined physical network. The discussion below deals exclusively with thermoplastic elastomers having block copolymer structures.

The two-phase block copolymer elastomers that have been commercialized are of three structural types: (a) styrene diene A-B-A or radial block copolymers and their hydrogenated derivatives, i.e., the Kratons of Shell Chemical Co. and some grades of the Solprenes of Phillips Petroleum Co. (see Chapter 5, Section A; Chapter 6, Section A); (b) ester–ether $(A-B)_n$ block copolymers, i.e., the Hytrels of DuPont Co. (see Chapter 7, Section C); and (c) urethane–ester $(A-B)_n$ block copolymers, i.e., the Pellethanes of Upjohn Co., the Estanes of B.F. Goodrich Co., and the Texins of Mobay Co. (see Chapter 7, Section H). Spandex elastomeric fibers, e.g., Lycra of DuPont Co., represent another type of urea-urethane block copolymer. The pertinent properties typical of these thermoplastic elastomers are qualitatively compared in Table 4-3 and are further amplified below.

The first styrene–diene elastomeric block copolymer to be offered commercially was a Solprene styrene–butadiene A-B product. This is, of course, not a thermoplastic elastomer, since it has no physical network structure. It is used in vulcanized products such as footwear, in

TABLE 4-3

Property Advantages of the Various Block Copolymer Thermoplastic Elastomers[a]

Property	Styrene–diene[b]	Hydrogenated[b] styrene–diene	Ester–ether[c]	Urethane–ester[d]
Tensile			+	+
Recovery	+	+		
Upper use temperature			+	+
Lower use temperature	+	+		
Aging stability		+		
Acid–base resistance	+	+		
Oil resistance			+	+
Electrical	+	+		
Abrasion resistance				+
Melt processability			+	
Cost	+			

[a] A designation of + indicates a performance strong point.
[b] See Chapter 6, Section A.
[c] See Chapter 6, Section C.
[d] See Chapter 7, Section H.

which it offers improved hardness compared to conventional SBR random copolymers. Furthermore, it is used as an additive to improve the hardness and processability of other conventional elastomers.

In contrast to the above A-B system, all of the other commercial elastomeric block copolymers have a physical network structure due to an A-B-A or $\{A\text{-}B\}_n$ block copolymer architecture. The thermoplastic elastomeric behavior of these materials make them uniquely useful in a variety of application areas such as automotive, mechanical goods, electrical and electronic, sealants, caulks and adhesives, and footwear. In contrast to conventional, chemically cross-linked, thermoset rubbers, these materials can be economically fabricated into end-use articles by processes similar to those used for thermoplastics, e.g., injection or blow molding, extrusion, vacuum forming, and solution casting. Since no vulcanization step is necessary, reprocessing is possible. In addition to competing with conventional rubbers due to their more economical processability, they also compete on another front with flexible but nonelastomeric thermoplastics, e.g., low modulus ethylene copolymers, and plasticized polyvinyl chloride. In the latter area, the relative attractiveness of the thermoplastic elastomers is derived from superior recovery properties, a more constant modulus–temperature profile, broader use temperature range, and in many cases better mechanical properties. In many of the more demanding applications, these properties justify the generally higher cost of the thermoplastic elastomers.

Many of the properties considered in making a decision of this type on a cost–properties basis are compared in Table 4-3. The tensile strengths and elongations are good for all four block copolymer systems. However, the polyester and polyurethane products crystallize on deformation and thus show particularly high strengths. On the other hand, the glassy polystyrene segments are less prone to undergo permanent set and hence display better recovery. The ester–ether and urethane–ester systems exhibit higher upper use temperatures by virtue of their high crystalline melting points. The styrene–diene systems, by contrast, have better low temperature properties due to the lower T_g of their soft blocks.

The relatively poor aging stability of styrene–diene systems is significantly improved in their hydrogenated counterparts as a result of the elimination of the unsaturation in the diene segment. The all-hydrocarbon nature of the styrene–diene systems result in superior resistance to dilute acids and bases, but the more polar ester–ether and urethane–ester copolymers display better oil resistance. The all-hydrocarbon materials also have better electrical insulation properties.

The polyurethanes are clearly outstanding in abrasion and mar resistance. The ester–ether copolymers have an advantage over the styrene–diene systems in melt processing behavior. This is due to the short length of their crystalline hard blocks and to the small differential in solubility parameter of the hard and soft segments in the melt. The urethane–ester block copolymers also have these features, but their upper processing temperature is somewhat limited by the thermal stability of the aromatic urethane bond. The styrene–dienes are soluble in a wide variety of solvents and thus are particularly easy to fabricate via solution casting techniques.

The above-described properties of the various types of thermoplastic elastomers provide leads as to which of them are most suitable for various end-use applications. Automotive markets are subdivided into three categories—interior, exterior, and under the hood. The plasticized vinyls used currently in interior flexible applications could be replaced by any of the four types of thermoplastic elastomers based on their mechanical and elastomeric properties. Of course, flame retardant formulations would probably be needed. Exterior automotive applications are more demanding, since the parts must often be weather resistant and paintable and must fulfill structural functions, e.g., shock absorbing bumpers. The paintability requirement necessitates the use of relatively polar block copolymers that show good coating adhesion, as well as resistance to the high bake temperatures often employed. Under-the-hood applications, e.g., hoses and tubing, require oil and heat resistance in addition to good elastomeric mechanical properties.

The mechanical goods market include articles such as flexible couplings, O-rings, seals, gaskets, and extruded hydraulic and industrial hoses. The block copolymer properties most important in these applications are dimensional stability, recovery, compression set, utility at high and low temperatures, and, in some cases, oil, chemical, and abrasion resistance. Electrical and electronic applications include wire and cable insulation and transformer encapsulation. Clearly, low dielectric constant and dissipation factor levels and, in some cases, weatherability are of prime importance here.

Sealants, caulks, and adhesives represent an important end-use area for block copolymer elastomers. Important advantages for these materials over conventional adhesives are the ability to apply them either via solution or melt techniques and their ability to subsequently develop high strength and recovery characteristics without the need for a curing step. The footwear industry is a rapidly growing market for flexible thermoplastic resins in general and especially for thermoplastic elastomers. The combination of melt processability, good elastomeric properties, high dynamic coefficient of friction (e.g., the Kratons), and excellent abrasion resistance (e.g., the polyurethanes) make block copolymers elastomers especially attractive for footwear applications. The conventional thermosetting polyurethane block copolymers find even greater use than their more expensive thermoplastic counterparts in footwear. Although these materials have not usually been considered block copolymers, they are indeed complex segmented structures derived from a polyol, a diisocyanate (e.g., TDI), and a diamine extender–cross-linking agent. Depending on the formulation utilized, a large variety of products can be produced, ranging from foams to solid products and from rigid to elastomeric compositions. These complex structures represent both the oldest and the largest commercial application for block copolymer structures.

The well-known Spandex elastomeric fibers (see Chapter 7, Section H) are also based on polyurethanes. These essentially linear elastomeric block copolymers have found wide use in many apparel applications. Certain polyurethane structures containing carbamate hard blocks are thermoplastic elastomers and can be converted into fibers via melt spinning techniques. More often, however, higher melting polyurethanes with urea hard segments are used for fiber applications because of their superior recovery characteristics. These materials must be converted into fibers via solution spinning techniques because of their thermal instability at the high processing temperatures that would be required for melt spinning.

2. Toughened Thermoplastic Resins

The most common route to the impact modification of a rigid but brittle polymer is by physical blending with or solution grafting onto a rubber. Another approach is the production of block copolymers containing a high volume fraction of a hard block and a minor concentration of a soft block. One example of such a commercially available system is an amorphous radial styrene–butadiene block copolymer containing ~75 wt. % polystyrene, e.g., Phillips K resins, (see Chapter 6, Section A). This material is nearly as tough as conventional rubber-modified polystyrene but has the additional advantage of optical clarity due the small domain size of the polybutadiene phases. This combination of properties makes these resins quite attractive in transparent packaging applications in spite of their somewhat higher selling price.

Crystalline rigid thermoplastics can also be toughened by block copolymer techniques. Block copolymers containing segments of polypropylene and segments of either polyethylene or ethylene–propylene random copolymer exemplify this type of material (see Chapter 5, Section C). Even at very low ethylene contents (e.g., <5 wt. %), these copolymers display much better toughness and low-temperature properties than propylene homopolymers or poly-ethylene–polypropylene blends. Several materials of this type are commercially available, an early example of which is the polyallomers of Eastman. These materials are used in place of polypropylene homopolymer in applications that require improved toughness and that can tolerate the somewhat lower modulus levels of the block copolymers.

3. Surfactants

Two types of block copolymer surfactants are in commercial use. One type contains hydrophobic segments and hydrophilic segments. An example of such a material is the class of polypropylene oxide–polyethylene oxide A-B/A-B-A block copolymers, e.g., the Pluronics of Wyandotte (see Chapter 6, Section C). Some silicone–ethylene oxide block copolymers (see Chapter 7, Section I) are also illustrative of this class. The properties of these surfactants make them very useful in applications requiring the emulsification of aqueous and nonaqueous components and/or the wetting of substrate surfaces. These nonionic surfactants are especially useful in applications that can not tolerate the more commonly used nonpolymeric anionic or cationic surfactants.

Another type of block copolymer surfactant is represented by the silicone–alkylene oxide copolymers which are used as polyurethane foam stabilizers. In this application, the alkylene oxide segment is soluble in the urethane matrix, while the highly incompatible siloxane segment resides at the gas–urethane interface. This results in greater control of bubble nucleation and cellular growth, thus producing more uniform foam structure.

E. CHALLENGES FOR THE FUTURE

In the preceding sections of this chapter, an attempt was made to assess the state of the art with regard to the synthesis, characterization, properties, and applications of block copolymers. It is evident from these discussions that several limitations and/or shortcomings remain. These provide a considerable incentive for intensive research, development, and marketing activities directed to solving these problems. In short, only a start has been made toward realization of the full potential inherent in block copolymer materials. In this section, an effort is made to identify some of these remaining problem areas as worthy challenges for future effort.

1. Synthesis

The development of more economical synthetic processes would certainly enhance the rate of growth of block copolymer markets. Many current synthetic routes suffer economically, since they are carried out via relatively dilute solution polymerization processes. This situation could be improved by the development of bulk (e.g., no solvent) copolymerization techniques. Bulk polymerizations which can be carried out in a continuous manner (e.g., in an extruder) could be especially desirable. Another type of bulk polymerization is the casting or molding of reactive intermediates to form block copolymer structures directly during the fabrication of an article. This approach has already been used with linear crystalline systems, such as those based on the anionic polymerization of caprolactam (see Chapter 6,

Section C). It has also been used with cross-linked systems such as the thermosetting segmented polyurethanes and the thermosetting epoxy–polyol block copolymers (see Chapter 7, Section J). The application of techniques such as these to other block copolymer structures could be very rewarding.

Most synthetic techniques currently used to prepare block copolymers are based on anionic sequential addition polymerization or on a few step-growth polymerizations that utilize highly efficient well-known classic organic reactions. The development of appropriate *cationic* sequential techniques and of other classic or nonclassic efficient step-growth reactions would offer the opportunity to prepare hitherto unknown block copolymer structures.

The properties of existing block copolymers could be improved significantly by refinement of presently used synthetic techniques. For example, the block integrity and purity of styrene–diene block copolymers prepared by anionic routes are very good due to the absence of complicating side reactions. Such is not the case, however, when this synthetic technique is applied to polar monomers, e.g., the acrylics and vinyls (see Chapter 5, Section B; Chapter 6, Section B). In addition, the heterogenous coordinated anionic techniques used to synthesize α-olefin block copolymers (e.g., ethylene–propylene) also lead to poor block integrity and purity due to uncertainties in chain transfer and active site lifetime characteristics (see Chapter 5, Section C). Elucidation of these mechanistic features should produce a high return in terms of physical property improvement.

2. Characterization

As was pointed out in the characterization section, there is often considerable uncertainty as to the architecture, block integrity, and purity of block copolymers as characterized by currently available analytical tools. There is a great need for new and/or improved methods for quantitatively determining these aspects of block copolymer structure. Important questions still remain as to whether an unknown product is a random copolymer, a block copolymer, a physical blend, or some mixture thereof. This is especially true when a knowledge of the method of preparation is unavailable.

3. Properties

The thermal, mechanical, and stability properties displayed by the commercially available block copolymers cover a fairly wide range.

However, further improvement is possible by the judicious choice of the component segments. For example, in the case of block copolymer elastomers, greater latitude in the useful temperature range and better thermal stability have been achieved in developmental materials through the use of high T_g hard blocks (e.g., polysulfone, polycarbonates) together with low T_g silicone soft blocks (Chapter 7, Section I). Furthermore, the incompatibility of these segments results in a desirably flat modulus–temperature profile. It is logical to expect that segments of other types will similarly result in property improvements. For example, all-acrylic block copolymers (see Chapter 5, Section B; Chapter 6, Section B), if they could be synthesized efficiently, should display superlative weatherability. In addition, excellent chemical and environmental resistance would be expected from copolymers containing crystalline segments (e.g., polyamides, polyimides) (see Chapter 7, Section G). Other property improvements can similarly be envisioned by the appropriate choice of segmental structure.

One problem common to all of the thermoplastic elastomers is their inferior recovery, compression set, and creep properties relative to those of well-vulcanized chemically cross-linked rubbers. A worthy goal is to define those parameters that are critical to this behavior in an effort to emulate the behavior of the thermosetting rubbers.

Good melt processability is more difficult to achieve with two-phase block copolymers than with homopolymers. The development of techniques to improve block copolymer melt processability without undue sacrifice of the properties dependent on two-phase morphology would represent a significant contribution to the field. The recognition of the importance of the solubility parameter differential to predict block copolymer processability is a good start in this direction.

4. Applications

Many of the new block copolymer structures suggested in this section should be useful in the same end-use applications enumerated earlier for the commercially available block copolymers. However, new applications are conceivable that would take maximum advantage of the inherent capabilities of block copolymers. For example, in the surface coatings area, block copolymer structures should offer the opportunity to achieve an uncommonly good combination of substrate adhesion through one of the segments and overall coating properties characteristic of the second segment. In a similar manner, block copolymer structures should offer enhanced performance in adhesive applications. Enhanced adhesion between two dissimilar substrates could be achieved by the use of block copolymers containing segments selectively adhesive to the respective substrates.

The advantages that can be achieved with two-phase block copolymers are not as well recognized in thermosetting systems as they are in linear structures. The epoxy–polyol system is an example of the possible advantages that can be realized through the proper control of morphology in such cross-linked products (see Chapter 7, Section J). It should be possible to apply this principle to other thermosetting systems as well.

The utilization of block copolymers in semipermeable membrane applications, such as gas or liquid separations, desalination, ultrafiltration, and pollution control, represents an important new technology. The advantages offered by block copolymers in these applications include (a) high strength in thin films, (b) tailored permeability, diffusivity, and solubility toward a particular penetrant, and (c) good dimensional stability imparted by high-temperature-resistant hard blocks. For example, organosiloxane block copolymers are in developmental stages for blood oxygenator applications (see Chapter 7, Section I).

Block copolymers are also highly suited to other biomedical applications. These include lipid control by hydrophobic–hydrophilic block copolymers, artificial organs, and reconstructive prosthesis and cosmesis (C1). For example, the mechanical properties and appearance of human skin can be simulated by pigmented low-modulus–high-strength block copolymer thermoplastic elastomers. Needless to say, the potential for other applications of this type is very high and of great social significance.

Block copolymers can be utilized in alloying applications in two ways. In the first, the block copolymer is blended with a homopolymer corresponding to one of the segments. In such blends, it is the other segment of the block copolymer that is expected to impart some special property, while the common segment provides mechanical compatibility. For example, the impact strength or chemical resistance of rigid polymers can be improved by the incorporation of elastomeric block copolymers (e.g., polysulfone plus polysulfone–silicone block copolymer) (see Chapter 7, Section I) or semicrystalline block copolymers (e.g., polysulfone plus polysulfone–nylon 6) (Chapter 7, Section C). The second way in which block copolymers can be used in alloys is as a compatibilizing agent. The mechanical compatibility, and therefore the mechanical properties, of blends of two homopolymers can be enhanced by the incorporation of small quantities of the corresponding block copolymer. This is due to the emulsifying ability of the block copolymer at the homopolymer–homopolymer interface. This important application is still relatively new, and a more fundamental understanding is required of the critical parameters necessary for optimum performance.

REFERENCES

A1. Ceresa, R. J., "Block and Graft Copolymers." Butterworths, London, 1962.

A2. Allport, D. C., and Janes, W. H., ed., "Block Copolymers." Appl. Sci. Publ. Ltd., Barking, England, 1973; Halstead, New York, 1973.

B1. Morawetz, H., "Macromolecules in Solution," High Polym. XXI. Wiley (Interscience), New York, 1965. (See also 2nd. edition, 1975.)

B2. Ke, B., ed., "Newer Methods of Polymer Characterization." Wiley (Interscience), New York, 1964.

B3. Ezrin, M., ed., "Recent Advances in the Measurement of Molecular Weight," *Adv. Chem. Ser.* **125**, (1973).

C1. H. E. Marsh, Jr., G. C. Hsu, C. J. Wallace, and D. H. Blankenhorn, Applications for biomedical polymers. *In* "Polymer Science and Technology" (H. P. Gregor, ed.) vol. VII, p. 33. Plenum, New York, 1975.

D1. Kossoff, R. M., *Mod. Plast.* 50 (1974).

PART TWO

CRITICAL SURVEY

Chapters 5, 6, and 7 are devoted to a relatively detailed discussion of specific block copolymer structures, all of which are subdivided into the respective three main architectural forms: A-B, A-B-A, and $(A-B)_n$ block copolymers. The few instances in which this general rule is superceded are amply explained at the appropriate point.

5

A-B
Diblock Copolymers

This chapter deals primarily with A-B diblock copolymers, the simplest architectural form. These compositions have been organized into four categories: (A) polystyrene-containing systems, (B) polyacrylic or poly(vinylpyridine) systems, (C) poly(α-olefins), and (D) heteroatom-containing block copolymers.

A. POLYSTYRENE BLOCK COPOLYMERS

One of the segments of the block copolymers described in this section is polystyrene, the second segment being either polybutadiene, polyisoprene, or an alkenyl aromatic polymer.

1. Styrene–Butadiene

a. Synthesis

The synthesis of true block copolymers of styrene and butadiene was made possible by discovery of Szwarc and co-workers in 1956 of homogeneous "living" anionic polymerization (A1, A2). The unique behavior of these polymerization systems is due to the occurrence of initiation and propagation steps but no termination steps (A3–A10, A22). The molecular weight is then governed by the ratio of monomer to initiator. If the initiation reaction rate is comparable to the propagation rate, it is possible to achieve a very narrow molecular weight distribution. When the stable polymeric anion thus formed is capable of initiating a second pure monomer, well-defined block copolymers of predictable molecular weight and molecular weight distribution can be obtained. Butadiene and styrene are suitable monomers for such a sequential approach (Reaction 5-1).

$$CH_2=CH-CH=CH_2 \quad + \quad \overset{CH_2=CH}{\bigcirc} \qquad\qquad (5\text{-}1)$$

$$\xrightarrow[\text{solvent}]{\substack{\text{RLi} \\ \text{hydrocarbon}}}$$

$$+CH_2-CH=CH-CH_2\!+_x\ +CH_2-CH\!+_y$$

It is also important to note that lithium counterion in hydrocarbon media is unique (A34), since it produces a high 1,4-polybutadiene configuration that displays a very low glass transition temperature of about −96°C (A11, A12). The other alkali metals result in mixed structures (A14).

Ionic copolymerization of certain monomer mixtures can lead directly to essentially pure block polymers without having to resort to the sequential monomer addition technique (A5–A8, A11). Mixtures of butadiene and styrene polymerized by alkyllithium initiators (A14, A29–A33) in hydrocarbon media display this characteristic (A16–A18). The segments are not completely pure and have been referred to as tapered blocks (A13). Although the homopolymerization rate of styrene is much faster than that of butadiene (A19, A20), the butadiene is essentially all depleted before any of the styrene is polymerized (A13, A21). The explanation usually offered for this unusual behavior is that the polymerization kinetics very strongly favor the addition of polystyryl anion to butadiene monomer rather than to styrene monomer. Polybutadienyl anion also adds to butadiene monomer more rapidly than to styrene monomer (A19, A20). One can even visually observe in the polymerization of styrene–butadiene mixtures that the orange color of the styrene anion does not appear until the butadiene monomer is consumed. Therefore, polystyrene segments are not formed in significant quantities until all of the butadiene monomer has polymerized. Very careful NMR studies of the active polydiene chain end have demonstrated that in hydrocarbon solvents the species may be identified as a "4,1 covalent polybutadienyllithium" (A22–A26). The model proposed by Morton and co-workers (A23) is shown in Figure 5-1. They suggested that the concerted reaction between the

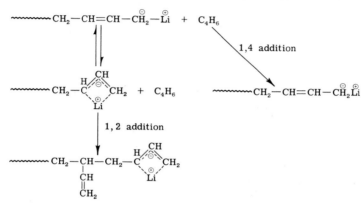

Fig. 5-1. Localized–delocalized equilibrium of chain ends (A23).

chain ends and the monomer could also account for the strong preference of the chain ends for dienes in copolymerization with styrene. It was reasoned that the mechanism utilizes the 1,4 system of the dienes and would pose difficulty for the styrene.

Numerous patents have been issued covering the block copolymerization of butadiene and styrene. It has been reported that thousands of tons of rubbery block polymers are now synthesized annually by the anionic polymerization of butadiene and styrene (A13, A15). The use of polar solvents, modified initiators, and other methods to produce random copolymers from styrene–butadiene mixtures has been extensively studied and has been discussed in several reviews (A7, A8, A12, A21, A28). The random structure is of greater interest in tire applications (A132), where low heat generation characteristics are important. In general, polar solvents or alkali metal counterions other than lithium disrupt the high 1,4-butadiene configuration and lead to 1,2 structures with relatively high glass transition temperatures. The stability of the polymeric anion and its subsequent ability to initiate the polymerization of monomer can be decreased in polar media.

Styrene–butadiene copolymerization can sometimes be achieved by purifying the monomer–solvent mixture through a desiccant such as silica gel, alumina, or molecular sieves. A review discusses the purification of solvents for solution polymerization (A27). Materials of construction are also discussed. However, the kinetic molecular weights expected on the basis of one mono-functional initiator starting one chain will not be achieved without very rigorous purification methods (A11). A convenient laboratory procedure using alkylmagnesium reagents has been described (56).

Any discussion of anionic block copolymerization would be incomplete without consideration being given to some of the problems associated with scale up of this "living" type of polymerization. The most difficult problem has been to purify the ingredients for the reaction process. This means that streams containing less than 5 ppm water, alcohols, sulfur compounds, acetylenes, and other active compounds must be provided. The technology has been evolving for 15 years. A simplified flow sheet is shown in Fig. 5-2 (A15). The solution polymerization plant can have quite a bit of flexibility. It can, for example, produce random copolymers or homopolymers as well as various types of block copolymers.

Much of the detailed technology is proprietary and is revealed only partially in the patent literature. A number of process variables, such as continuous polymerization, are discussed in the patent literature (A35–A45). It may be noted, in particular, that it has been claimed that one can make the block copolymers continuously in an extruder (A46, A47).

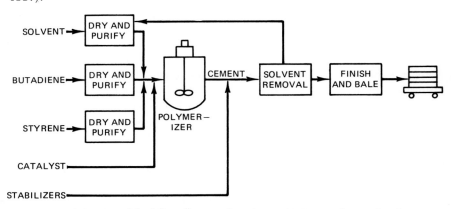

Fig. 5-2. Simplified flow diagram for solution SBR manufacture (A15).

It is well known that butadiene can be polymerized to a high (>90%) 1,2 configuration by alkali or organoalkali initiators in tetrahydrofuran (THF) solvent. Usually, this structure is to be avoided in elastomeric copolymers because of its relatively high T_g. Block copolymers of polystyrene–1,2-polybutadiene have been reported (A48, A49). The viscoelastic behavior of the copolymer has been studied over a wide temperature–frequency range (A50). Chlorination of SB blocks has been reported (A51, A52).

Syntheses of block copolymers of butadiene–styrene by techniques other than anionic polymerization have been reported in the literature and have been discussed in earlier books and reviews (A55). The most common approach (A53, A54), has been to try to incorporate on the initial polymer segment a terminal hydroperoxide or other reactive group, which can later be used to initiate polymerization of the second monomer. This is seldom a highly efficient process and results in a difficult to characterize composition that contains large amounts of homopolymers (A66).

b. Characterization

The composition of butadiene–styrene block polymers can be established by a variety of techniques, such as infrared, ultraviolet, or NMR (nuclear magnetic resonance) spectroscopy. The last method has been especially popular due to differences in chemical shift between the aromatic and diene protons (A11, A57).

An NMR method for determining the "block styrene" in copolymers by using an analog computer to resolve overlapped aromatic proton peaks has been developed (A58, A78). It was reported that sequences as small as two or three units are measurable. Techniques based on differential thermal analysis or infrared spectroscopy (A13) have also been employed to distinguish block from total polystyrene. In addition to the gross composition of the block polymer, it is also frequently necessary to learn information about the molecular weight and homopolymer content. The number average molecular weight of styrene–butadiene block copolymers can be measured by membrane osmometry methods (A11). This should agree with the predicted value if impurities have not greatly decreased the effective initiator concentration. The weight average molecular weights can be determined by light scattering, but one must recall that compositional variations in copolymer molecules can lead to an erroneous result (A59, A60, A130).

Homopolymer impurities can sometimes be observed by gel permeation chromatography (GPC) (A11, A61–A65). Molecular weights by GPC were shown to agree with osmometry (A11, A128).

Even when very rigorous precautions are taken and good agreement between predicted and measured number average molecular weights is obtained, it is possible for small amounts of homopolymer to be detected. The data in Table 5-1 and Fig. 5-3 illustrate this point.

Note that there is excellent agreement between both the predicted $(\overline{M}_k)$ and measured $(\overline{M}_n)$ molecular weights and styrene contents Nevertheless, GPC studies (A11) demonstrated the presence of small quantities (<5 wt%) of polystyrene homopolymer (Fig. 5-3).

The small peak near GPC count 26 was independently determined to be polystyrene of the expected homopolymer molecular weight.

Another technique that shows promise for separating homopolymers from block polymers is density gradient ultracentrifugation (A67). There have, of course, also been chemical separation techniques such as fractional precipitation or fractional solution reported in the earlier literature (A68–A70, A78). Other techniques to isolate homopolystyrene have involved oxidative degradation of the butadiene segment by osmium tetroxide to demonstrate that the isolated homopolystyrene was present in amounts close to the initial styrene content of the feed (A13).

c. Morphology

Many of the most interesting aspects of block polymers are related to their morphology. The colloidal and morphological behavior of block and graft copolymers is the subject of several reviews (A71–A75, A79–A81). Above a certain block length, butadiene–styrene block polymers display microphase separation, i.e., they coexist in separate distinct microdomains. Meier was the first to discuss the requirements for domain formation theoretically (A76). One of the most important conclusions from this work was that the critical block molecular

TABLE 5-1

Synthesis and Characterization of Poly(styrene–butadiene) Block Polymers[a]

Predicted molecular weight $(\overline{M}_k)$	Number average molecular weight $(\overline{M}_n)$	Styrene (wt%)		1,4-diene structure (%)
		Charged	Found	
144,000	135,000	24.6	24	92
77,400	80,000	72.1	73	92
84,500	89,000	25	24.8	92

[a] From Juliano (A11).

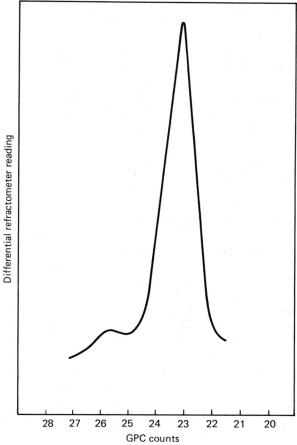

Fig. 5-3. Gel permeation chromatogram of a styrene–butadiene block copolymer (A11).

weights needed for domain formation are much greater than those required for phase separation in physical mixtures of the corresponding homopolymers. The reason for this is the loss of configurational entropy due to the constraints on the spacial placement of chains in a domain structure. A similar conclusion has recently been reached on the basis of polymer solution theory (A77). When the block molecular weights of styrene and butadiene are greater than approximately 5000–40,000 (A63, A72, A79, A82), a two phase system is obtained that has domains of colloid dimensions.

Vanzo reported some unusual phenomena one can observe with high molecular weight styrene–butadiene block polymers (A83). Concentrated solutions in ethylbenzene exhibited irridescent colors that change with the concentration of the polymer. This is a result of the incompatible chains undergoing a degree of microphase separation or two-dimensional ordering great enough to interfere with visible light. It was possible to carefully cast films from these block polymer solutions that retained their layer structure. Electron microscopy studies revealed patterns that varied with the monodispersity of the blocks. Further work showed that the width of the copolymer layer spacing increased in a predictable way with increasing molecular weight (A84, A93).

The morphology of diene-containing block or graft copolymers has been greatly elucidated in recent years by the use of selective staining agents such as osmium tetroxide (A83a). With this technique, one can study the sizes of selectively stained polydiene domains. An example of the typical morphology reported by Bradford (A84, A85) is shown in Fig. 5-4. The black areas in the micrograph represents the butadiene phase stained selectively by osmium tetroxide. The block copolymer shown in this figure had a number average molecular weight $(\overline{M}_n)$ of ~630,000 and contained 60% styrene. The film was prepared by casting from dilute toluene solution.

It is important to point out the effect of composition on morphology. Molau (A86) has concluded that butadiene–styrene block copolymers seem to be capable of forming three basic morphological units, namely, spheres, cylinders, and lamellae. The major variables are believed to be the volume ratio of the phases and the block molecular weight. Thus, as the styrene content decreases, one observes first butadiene spheres, then butadiene cylinders or rods, then lamellae of both phases, then styrene cylinders, and finally styrene spheres. This aspect was also discussed by Matsuo (A87, A88) and others.

As in the case of graft copolymers (A89), it is often possible to selectively prepare hard (high modulus) or soft (low modulus) forms of the butadiene–styrene block copolymers. Molau (A86, A90) has demonstrated two different morphologies from butadiene–styrene copolymers containing 70 wt% styrene. Selective solvents are employed in the film casting process to expand or contract either or both of the segments. If the styrene segment is contracted, for example, by using a poor polystyrene solvent, the soft form is obtained.

In addition to electron microscopy, a second technique, namely, small-angle X-ray scattering, shows promise as a tool for studying the morphology of block polymers (A131). It is especially useful in the investigation of block polymer systems that cannot be selectively stained. McIntyre and co-workers (A91) have very carefully de-

Fig. 5-4. Morphology of styrene–butadiene block copolymers. Configurations of a 60 : 40 styrene–butadiene diblock copolymer. (A) Slowly dried film; (B) original film stained before embedment (A84, A85).

veloped the small-angle X-ray technique to the point where they can distinguish the scattering due to intraparticle distances from the size of the particles. A particular detailed study of long range order and supramolecular structures has been reported by Kaempf *et al.* (A92).

d. Solution Properties

The best work available suggests that the solution properties of styrene–butadiene block copolymers are quite similar to the styrene–butadiene–styrene triblocks (A94). These will be discussed in detail in Chapter 6.

e. Mechanical and Thermal Properties

The mechanical behavior of a block copolymer is governed to a considerable degree by the glass transition temperature of each block. The butadiene segment transition is directly related to the ratio of 1,4 versus 1,2 configuration.

For a 90% 1,4 structure, the T_g will be approximately $-100°C$, whereas the <90% 1,2 polymer has a T_g near 0°C. Polystyrene's principal transition temperature is about 100°C as long as its molecular weight is 15,000–20,000 or higher. The individual transitions are clearly evident (A129). The number of blocks or the sequence arrangement of the blocks does not significantly change the nature of these curves. The effect of composition changes is to shift the modulus plateau vertically at temperatures (or frequencies) between the two transitions. This is, of course, in contrast to the case of random copolymers, where the effect of compositional changes is to shift the curve horizontally along the temperature axis. This approach has been used to differentiate block and random copolymers as well as copolymers possessing intermediate degrees of randomness (A95, A96). Both the loss modulus and the storage modulus peaks are affected by these variations. The loss modulus is broadened by the presence of block sequences. At a given polystyrene content, the modulus of a block copolymer is much higher than that of a random copolymer at temperatures above the T_g of polybutadiene and below that of polystyrene. This is due to reinforcement by the dispersed polystyrene phase.

The mechanical properties of these block copolymers have been thoroughly studied (A13). In particular it has been of interest to compare block copolymers with random copolymers prepared in solution or emulsion. Since there is no physical network in these copolymers, nearly all of the comparisons have been made with chemically crosslinked vulcanizates cured with accelerated sulfur systems (A13, A29).

Oil extension is also possible (A97). Mechanical properties of tread stock vulcanizates are shown in Table 5-2. It should also be noted that at low degrees of crosslinking unfilled styrene–butadiene copolymers show some similarity to styrene–butadiene–styrene materials due to the radical coupled polydiene segments (A29).

The properties of the block copolymer have suggested applications where increased hardness and low brittleness temperature are important (A98). When compared to the random copolymer, the block structure appears to confer excellent extrusion (A29, A107, A109) characteristics. The block copolymers also have better electrical properties, abrasion resistance (A29, A99), rheological properties, and thermal stability than the emulsion-prepared random copolymers (A100–A103). Applications in shoe soles, floor tile foam (A104), wire and cable (A105), artificial leather (A106), and adhesives (A29, A108) have been considered.

Rigid, transparent, impact-resistant block copolymers have been prepared (A133, A134). Ten to thirty weight percent polybutadiene was utilized. Transparency was better for sequentially prepared triblocks than for one- or two-step diblock materials.

An important application area for block copolymers is in blends with other polymers to produce improved modified compositions. Several publications (A110) and patents discuss impact modification of rigid brittle polymers such as polystyrene with these block copolymers (A111, A112). Further improvements are observed with peroxide cross-linked blends (A113–A119). A hydrogenated styrene–butadiene

TABLE 5-2

Properties of 75/25 Butadiene–Styrene Block and
Random Copolymers in Tread Vulcanizates[a]

Property	Block	Random
300% modulus (psi)	1600	1400
Tensile strength (psi)	2500	3400
Heat build-up (°F)	90	57
Resilience (%)	60	66
Hardness, shore A	80	62
Freeze point (°F)	−85	−55

[a] Containing fifty parts of carbon black per hundred of rubber and ten parts per hundred of rubber processing oil. After Zelinski and Childers (A13).

block displayed Izods of 10 foot-pounds per inch and heat distortions of 96°C (A120).

It is possible to speculate on the reasons for these improvements. The polystyrene block should be compatible with the homopolystyrene and thus allow the polybutadiene segment to become much more intimately dispersed than it could as homopolybutadiene. The role of the peroxide may be to partially cross-link the polybutadiene phase. This probably prevents the modifier particles from being shear degraded to an inefficient size. It is quite well established that the particle size of the elastomer phase is very important in impact polystyrene (A121). Impact modified styrene acrylonitrile (ABS) and impact-modified polypropylene (A122, A124) have also been reported to be prepared using butadiene–styrene block copolymers as the impact modifier (A123).

In contrast to the impact modification of the rigid polymers, styrene–butadiene copolymers have also been used to improve the hardness, strength, and processability of polybutadiene elastomers (A29, A125). Other blends of styrene–butadiene block copolymers with neoprene rubber possessing improved ozone resistance have also been reported (A126). The utilization of styrene–butadiene in the sizing of glass fibers has also been claimed (A127).

2. Styrene–Isoprene

a. Synthesis

The anionic copolymerization of isoprene (2-methyl-1,3-butadiene) with styrene has many features that are similar to the butadiene–styrene system (A3, A5–A9, A12, A13). Most patents concerning butadiene also claim isoprene. However, there are some significant differences. For example, isoprene can be polymerized stereospecifically to a high (up to 90%) cis-1,4 configuration via lithium or organolithium initiation in hydrocarbon solvents (A12). Butadiene, under identical conditions, yields high 1,4, but the cis content is much lower (30–40%).

Early studies on the sodium-catalyzed copolymerization of mixtures of isoprene and styrene demonstrated that the isoprene quite selectively polymerized first in hydrocarbon solvents, whereas the styrene polymerizes preferentially in polar solvents such as THF (A135). The results were interpreted in terms of the ionic character of the propagating (C^-M^+) ion pair. A nearly free anion produced an initial copolymer containing 80% styrene (A135). The ionic character of the propagating site was reduced by using lithium in place of sodium and by employing a hydrocarbon rather than a polar solvent.

The 1,4-diene configuration is greatly reduced in polar solvents. It was demonstrated by Livigni and co-workers (A136) that the block structure of isoprene–styrene could be randomized without serious alteration of the diene microstructure through the use of aromatic ether additives. Earlier it had been shown (A137) that the diene configuration was essentially retained in the presence of weakly basic ethers such as diphenyl ether or anisole, in contrast to the total loss of 1,4-diene specificity in the presence of THF.

As was pointed out earlier, the anionic polymerization of mixtures of butadiene and styrene in hydrocarbon media leads to block copolymer formation. The same effect is observed with the isoprene–styrene system. However, the polymerization of isoprene–styrene mixtures does not yield block structures as "pure" as those obtained with butadiene mixtures (A9, A136, A138). These results have been rationalized on the basis of the measured individual kinetics of the reaction steps involved (A19–A21, A139). The crossover reaction of polyisoprene anion with styrene monomer is faster than the reaction of polybutadienyl anion with styrene (A6, A7, A9, A22). In addition, although the propagation rate of isoprene is faster than that of butadiene, the crossover reaction of the polystyrene anion is faster with butadiene than with isoprene monomer. Thus, as a result of the kinetics, larger amounts of styrene are incorporated in the diene segment when isoprene rather than butadiene is the diene monomer employed. The continuous anionic copolymerization has been also studied in a tubular reactor (A140).

The formation of monodisperse "pure" blocks of styrene and isoprene requires, in addition to rigorous purification of monomers and solvents, both rapid initiation and a sequential monomer addition procedure (A9, A56, A63, A64, A160, A161). The rapid initiation step is critical if one desires a low molecular weight block (e.g., 10,000). Otherwise, residual initiator remains that can lead to homopolymer impurities. The use of branched alkyllithium initiators (A56, A67) or polar additives (A63, A64, A143) can surmount the problem of slow initiation without seriously altering the diene microstructure.

The synthesis of model A-B block copolymers of isoprene and styrene has been reported by several investigators (A63, A64, A67, A141, A142). It is entirely possible to prepare well-defined structures with a predictable molecular weight and narrow distribution of molecular size and composition. Characterization methods for the block structures are similar to those discussed for styrene–butadiene systems.

b. Solution Properties

The availability of model, well-defined A-B block copolymers of styrene and isoprene has encouraged workers to measure the solution properties of these systems (A144–A151). The intrinsic viscosity has been reported to be close to a weighted average of the corresponding homopolymers of equivalent molecular weight (A144, A151). In a study of molecular dimensions by viscosity and light-scattering methods, it was reported (A152) that the displacement length parameter $\langle r^2 \rangle^{1/2}/Z^{1/2}$ (where r is the root mean square end-to-end distance, and Z is the number of links in the chain) reaches a maximum (as compared to either homopolymer) in the middle of the compositional scale. This shows that the copolymer is more expanded. The implication is that at these compositions there is a maximum of heterocontacts between the different chain segments.

Binary cluster integrals that represent the excluded volume of a pair of segments were calculated from second virial coefficients. These data showed that the magnitude of the volume effect increases over the range of solvents studied from isobutyl methyl ketone to cyclohexane to toluene (e.g., progressively "better" solvents for the copolymer).

A study of the thermodynamic parameters derived from osmotic pressure data has also been reported (A153). The second virial coefficients were determined over a range of temperatures in toluene and cyclohexane. The "theta" temperatures, heat and entropy of solution, and solvent–solute interaction parameters were derived. Concepts such as the unperturbed configuration of a macromolecule are inapplicable to block or graft copolymers except for a particular solvent where the theta point of each block coincides (A154). One can measure the particular temperature where the second virial coefficient, A_2 is reduced to zero. However, since both blocks contribute to the polymer–solvent interaction, the effects must cancel each other. Intramolecular phase transitions have been observed during viscosity–temperature measurements (A155). These investigators proposed that this transition was more important for estimating chain dimensions than A_2. Unfortunately, other investigators were not able to observe the same phenomenon under similar conditions (A156). Sedimentation velocity studies have also been reported (A192).

A study of the molecular weights of styrene–isoprene block

copolymers as determined by gel permeation chromatography (GPC), light scattering, and osmometry showed that the GPC values deviated somewhat from those obtained by the other techniques (A157, A158). The extent of the deviation depended on composition, and comparison of GPC and osmometry values could be rationalized if heterogeneity is taken into account. A linear relationship was found for this system between the logarithm of the product of limiting viscosity number and the molecular weight, and the peak count. This approach was originally suggested by Benoit and co-workers (A159). The linearity of this plot was taken to be good evidence that the fractionation during the GPC analysis is taking place according to hydrodynamic volume. Other studies (A156) have shown diblock styrene–isoprene, triblock styrene–isoprene–styrene, and polyisoprene fit on an $\overline{M}_n$ versus elution volume plot. Small differences with the $[\eta]M$ universal calibration were attributed to the latter procedure being able to take into account small heterocontact effects (A156).

c. Morphology

Some of the most interesting and revealing studies of the morphology of block copolymers have been conducted with the styrene–isoprene system (A162, A163, A193). Most investigations have involved the use of electron microscopy, but small-angle X-ray and light-scattering techniques are also important tools.

Kawai, Inoue, and co-workers (A164–A173) prepared block polymers of styrene–isoprene by the high vacuum polymerization of styrene in tetrahydrofuran at dry ice–methanol temperatures, followed by sequential polymerization of isoprene. Although the microstructure was not discussed, this procedure must have produced a high 3,4 configuration in the polyisoprene segment. The purity of the block structure was characterized by osmometry, ultraviolet spectroscopy, and ultracentrifuge sedimentation (A169). The molecular weights of the block polymers were high (0.5–1.0×10^6). The fine structure of films cast from dilute solution was investigated by light scattering (A165, A166) and by an electron microscope using the osmium tetroxide staining method (A174). Typical micrographs are shown in Fig. 5-5 (A169). The dark portions are the selectively stained polydiene segments. Clearly, as the diene concentration increased, the texture changed from that of discrete diene particles in a polystyrene matrix to that of two more or less continuous phases or lamellar arrangements. These very high molecular weight block copolymers also displayed the irridescent colors reported earlier for the styrene–butadiene system (A83). This

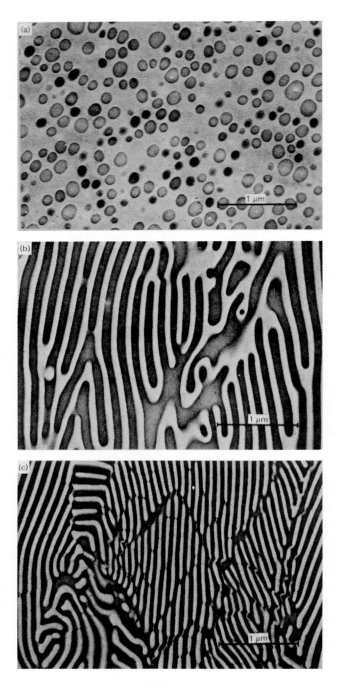

Fig. 5-5.

phenomenon was again related to the periodic nature of the alternating lamellar structure.

The morphology of styrene–isoprene block copolymers has also been investigated by Sadron and co-workers (A175, A176). Solution cast copolymer films were described as being organized and as having structures similar to cylinders, thin sheets, or spheres, depending on solvent type and solution concentration. Irridescent films were prepared. The dimensions of the structures were found to depend on the total molecular weight, casting solvent, and the composition. The structure of liquid crystalline phases from amorphous block copolymers such as styrene–isoprene has been studied by small-angle X-ray diffraction. Structures identified as lamellar, hexagonal, and cubic were reported. Effects similar to those already described were observed when solvents preferential to one block were employed.

Reiss *et al.* (A177–A183) have studied stained styrene–isoprene block copolymers by electron microscopy. They also investigated the compatibility of the block copolymer with the corresponding homopolymers. Compatibility was defined by constructing ternary diagrams based on observations of whether or not cast films were transparent. Transparency was attributed to the microheterogeneous phase (domain) sizes of less than ~0.1 μm. They found that the main parameters are (a) the molecular weights of the homopolymers, (b) the molecular weight and composition of the copolymer, and (c) the composition of the ternary mixture. It was concluded that it is possible to "solubilize" the homopolymer in the block domains if the molecular weight of the block and the concentrations of block copolymer are higher than those of the homopolymer. It is important to point out that the system still exists in two phases. The block copolymer is simply able, under the conditions cited, to behave like an emulsifier in restraining phase separation of the homopolymers into macroscopic domains.

Further studies by Inoue and co-workers verified this concept (A164, A168). They also demonstrated that, when the conditions for solubilization of the homopolymers into the block domains are met, the original block copolymer of domain structure (spheres, lamellae, etc.) may be maintained even for the binary and ternary systems. Thus, the formation of the following five types of fundamental domain structures may

---

Fig. 5-5. Electron micrographs of ultrathin sections about 350 Å thick, cut normal to the surface of films cast from about 5% toluene solutions of (a) 30:70 isoprene–styrene block copolymer, (b) 50/50 isoprene–styrene block copolymer, (c) 60:40 isoprene–styrene block copolymer (A169).

be achieved, depending on the total ratio of A sequences to B sequence in the system: (1) A spheres in B matrix; (2) A rods in B matrix; (3) alternate lamellar arrangement; (4) B rods in A matrix; (5) B spheres in A matrix.

A thermodynamic interpretation of the domain structure in solution-cast films has been reported (A168). The assumption was made that the domain structure originates from micellar structure at a critical concentration in relatively dilute solution during film casting. The formation of the three types of domain structure (spheres, rods, lamellae) and their sizes were treated in terms of the equilibria expected to govern the formation of micelles at a critical concentration. Their analysis took into account such thermodynamic and molecular parameters as the incompatibility of the A and B segments, the solvation of the segments, the casting temperature, the total chain length of the block copolymer and the weight fraction composition of the block copolymer. These investigators concluded that the block segments are preferentially oriented in the direction perpendicular to the interface between the two phases. They also suggested that this orientation–aggregation of the segments must make the bulk properties of the block copolymer much different from those of mechanical mixtures of homopolymers, even if it is possible to form semimicroheterogeneous structures in the latter case.

Reiss and co-workers (A177) have also attempted to study the relationships between shock resistance and morphology for a ternary system of polystyrene, polyisoprene, and a styrene–isoprene block copolymer. The morphology of systems that were solution blended, coagulated, and then compression molded was examined by interference microscopy. Ternary diagrams were constructed to show the effect of the molecular weight of the two homopolymers on impact strength. If 10–20 wt% of block copolymer was incorporated, it was possible to obtain good impact strength. In order to obtain acceptable hardness, it was necessary that the homopolymer molecular weights be >150,000. The block copolymer was again considered to be an emulsifier that promotes interfacial adhesion in the ternary system.

d. Mechanical Properties

In contrast to the styrene–butadiene A-B block co-polymers, there are no commercial elastomers based on styrene–isoprene A-B systems. In addition, there is surprisingly little information available in the published literature on the mechanical properties of isoprene–styrene A-B block elastomers. This is probably so because butadiene monomer is more readily available and more economical than isoprene and because there are no outstanding property advantages for the isoprene–styrene copolymer relative to the butadiene–styrene materials. Another consideration could be that the glass transition temperature is about 30°C higher for polyisoprene than for polybutadiene, thus resulting in poorer low-temperature properties. Unvulcanized isoprene–styrene block polymers high in isoprene content have been reported to have very low strengths (A72). It would be expected that satisfactory strengths could be achieved for chemically cross-linked vulcanizates containing active reinforcing fillers such as carbon black or finely divided silica. There are reports in the literature (A184, A185) that isoprene–styrene block copolymers can be successfully compounded by methods known for natural rubber.

Styrene–isoprene block copolymers high in styrene content (85%) have been reported (A186). A copolymer having an intrinsic viscosity of 0.86 had a heat distortion of 73.5°C, a Rockwell hardness of 98, and an Izod impact of 0.26 ft lbs/inch. Since high impact polystyrene resins (Izod values ~1–3 ft lbs/inch) are prepared commercially with much less than 15% rubber phase, it is apparent that this system is far from optimum. It is probable that the small domain size of the isoprene block segments is at least partially responsible for the low impact strength of these block copolymers.

Studies of the deformation and fracture mechanisms in rubber-toughened plastics based on styrene–isoprene or styrene–isoprene–styrene compositions have been reported (A187). These systems permit the investigation of well-defined domain regions, as well as the molecular arrangement of the domain. Two types of toughening mechanisms associated with macro- and micronecking phenomena were observed depending on the rubber domain size and volume fraction. These were, respectively, localized heating of the glassy phase and craze formation. The latter mechanism was enhanced by the presence of homopolymer.

Hydrogenated block copolymers of isoprene and styrene have been claimed in a patent (A188). Improved weathering characteristics were reported for these materials. A discussion of coordination catalysts useful for the selective and quantitative hydrogenation of polymeric unsaturation has been presented (A189–A191). Further references on hydrogenation of styrene–diene–styrene systems are cited in Chapter 6, Section A.

3. Styrene–Alkenylaromatics

Block copolymers of styrene with α-methylstyrene display unique and interesting thermal properties. These systems are the subject of this section.

Polymers based on α-methylstyrene have been examined in great detail. It is well known that this sterically hindered monomer does not respond well to free radical polymerization, although random copolymers are possible. Anionic polymerization can produce predictable molecular weights and quantitative conversions providing that the polymerization temperature is far below the ceiling temperature of this monomer. One of the major incentives for investigating poly(α-methylstyrene) is the fact that its glass transition temperature is some 70–80°C higher than that of polystyrene.

Baer (A194) was perhaps the first to investigate block copolymers of polystyrene with poly(α-methylstyrene). Difunctional initiators were utilized so that essentially A-B-A structures were prepared. It was noted that apparently one-phase (single T_g) copolymers were obtained. This feature was rather surprising, since "normally" one observes microphase separation in block copolymers.

Robeson and co-workers (A195, A196) studied the same system in great detail. However, in contrast to Baer, the "simpler" A-B block copolymer system was chosen for investigation. Anionic sequential polymerizations under high vacuum conditions were utilized in order to produce nearly monodisperse block copolymers that possessed a high degree of structural integrity. The synthesis reaction is illustrated in Reaction 5-2. A random copolymer was also prepared for comparative

$$(5\text{-}2)$$

Polystyrene   Poly (α-methylstyrene)

purposes. All materials were characterized with respect to molecular weight and block purity by osmometry and gel permeation chromatography. The results confirmed Baer's data and showed that one-phase block copolymers were obtained even at block molecular weights greater than 200,000 in these A-B systems. The mechanical loss and

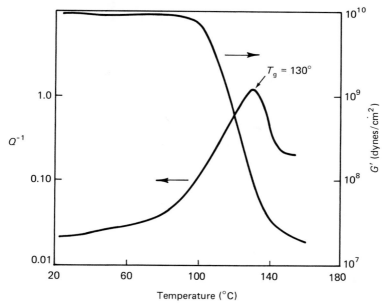

Fig. 5-6. Mechanical loss and shear modulus temperature data on polystyrene–poly(α-methylstyrene) A-B block copolymer (A195, A196).

shear modulus temperature data for a styrene–α-methylstyrene block copolymer is shown in Fig. 5-6.

The experimental results of Robeson and co-workers (A195, A196) are entirely consistent with the theoretical predictions of Krause (A199), who calculated that microphase separation in block copolymers should be more difficult to achieve than for simple polymer–polymer blends. The single-phase morphology of this system is undoubtedly due, to a significant extent, to the very small difference in the solubility parameters of polystyrene (9.1) and poly(α-methylstyrene) (8.9). Subsequently, experimental results by Dunn and Krause (A197) and by Shen and co-workers (A198) further confirmed and expanded the conclusions of Robeson *et al.*

B. POLYACRYLIC AND POLYVINYLPYRIDINE
BLOCK COPOLYMERS

1. Acrylics

Acrylic polymers such as poly(methyl methacrylate) and polyacrylonitrile are very useful as weatherable transparent thermoplastics or as textile fibers. It is well known that high molecular weight homopolymers may be prepared via either free radical or anionic mechanisms. Not surprisingly, block copolymers have been prepared by both techniques and will be the subject of this section. Their synthesis, solution properties and mechanical properties will be reviewed. This class of A-B block copolymers include those containing two acrylic segments and those with one acrylic subchain and a second block of another chemical type. References to radical and anionic block copolymer synthesis and solution behavior are tabulated for further detailed investigation by interested readers.

Techniques for synthesis of well-defined acrylic block copolymers are not as advanced as those for the styrene–diene systems (B1, B134, B135). Furthermore, the characterization of block structure and homopolymer content is not as well documented as could be desired.

a. Synthesis

i. Anionic Polymerization. The anions derived from polar monomers such as the methacrylates or acrylonitrile are not basic enough to initiate the polymerization of hydrocarbon monomers such as styrene or the dienes (B2, B3). This aspect has been studied by Franta (B4), Graham (B5), and others (B6). Thus, block copolymers of hydrocarbons (e.g., styrene) and acrylics can only be prepared by first forming the nonpolar segment and utilizing the resulting polymeric anion to initiate polymerization of the acrylic monomer.

In principle, the nonterminated anionic sequential polymerization should produce well-defined copolymers. Narrow distribution poly(methyl methacrylates) having a predictable molecular weight have been prepared at low temperatures (B7–B9). Highly purified reactants and sodium biphenyl initiation were required. Polyacrylonitrile has also been prepared with a narrow molecular weight distribution (B10). However, more side reactions are encountered in polymerization of acrylic monomers than in the case of styrene or diene polymerization. For example, the ester or nitrile group can react with the initiator or the macromolecular anions, thus resulting in inefficient initiator utilization and homopolymer formation via premature termination. See Reaction 5-3 for an example of such a reaction. It has been reported that only about 10% of the alkyllithium initiator is utilized in high polymer formation (B6). Presumably, similar side reactions can occur to some extent between growing acrylic chain anions and the monomer or even the ester groups of the polymer (see Reaction 5-4). However, the lower basicity of the acrylic chain anion, compared to that of the polystyryl anion, reduces the extent of this reaction. If

$$n\ C_4H_9-Li\ +\ \underset{\substack{\displaystyle | \\ C=O \\ | \\ O \\ | \\ CH_3}}{\overset{\substack{CH_3 \\ | \\ CH_2=C}}{}}\ \longrightarrow\ \underset{\substack{\displaystyle | \\ C=O \\ | \\ C_4H_9}}{\overset{\substack{CH_3 \\ | \\ CH_2=C}}{}}\ +\ LiOCH_3$$

or   (5-3)

Polystyrene
~~~~~~~~~~ $CH_2-\overset{\ominus}{C}H\overset{\oplus}{L}i$

$$+\ \underset{\substack{\displaystyle | \\ CO \\ | \\ O \\ | \\ CH_3}}{\overset{\substack{CH_3 \\ | \\ CH_2=C}}{}}\ \longrightarrow\ \text{Polystyrene} \sim\sim\sim CO-\underset{}{\overset{\substack{CH_3 \\ | \\ C}}{}}=CH_2\ +\ LiOCH_3$$

one conducts the polymerization at low temperatures (B3, B22), the stability of the anion is greatly increased, and in some cases, predictable molecular weights and narrow distributions are achieved. Similar behavior was qualitatively reported for acrylonitrile polymerization. At $-78°C$, colorless polymers were obtained. At higher temperatures, the solutions became colored, which suggests that anion attack on the nitrile function was taking place.

$$\sim\sim\sim\underset{\substack{\displaystyle | \\ C=O \\ | \\ OCH_3}}{\overset{\substack{CH_3 \\ | \\ \overset{\ominus}{C}\ \overset{\oplus}{Li}}}{}}\ +\ \underset{\substack{\displaystyle | \\ CO \\ | \\ OCH_3}}{\overset{\substack{CH_3 \\ | \\ CH_2=C}}{}}\ \longrightarrow\ \sim\sim\sim CH_2-\underset{\substack{\displaystyle | \\ C=O \\ | \\ OCH_3}}{\overset{\substack{H_3C \\ | \\ C}}{}}-\overset{\substack{O \\ || \\ C}}{}-\overset{\substack{CH_3 \\ | \\ C}}{}=CH_2\ +\ LiOCH_3$$

(5-4)

The alkyl group apparently plays an important role in the anionic polymerization of acrylic monomers. For example, isopropyl and n-hexyl acrylates readily yield high polymers, whereas ethyl (B11) or methyl (B12) acrylate do not. The reason for such behavior is not clear, although steric considerations or undetected impurities may be responsible. A detailed study has elucidated the chemistry involved in the anionic polymerization of methyl acrylate (B12).

Early investigators could not produce linear block copolymers of styrene and methyl methacrylate (B13). Light scattering measurements demonstrated that the molecular weight of the styrene segments was too high considering the chain length of polystyryl anion before addition of the methyl methacrylate monomer. This was attributed to reaction of the polymeric anion with the ester group to produce a branched molecule. It was later reported (B24) that this branching reaction could be avoided by addition of 1,1-diphenylethylene prior to the acrylate polymerization. Presumably, the bulky and less basic diphenylethylene anion is less able to attack the ester while still retaining sufficient reactivity to react with the carbon–carbon double bond of the acrylic monomer. Methyl methacrylate–alkyl methacrylate block copolymers were subsequently prepared (B14–B16). Styrene–methacrylic acid block copolymers have been prepared by hydrolysis of the ester (B17).

The polar monomers utilized for block copolymers via anionic polymerization have been reviewed (B2, B22). Difunctional electron transfer initiators such as sodium naphthalene have been employed most often. These result, of course, in A-B-A triblock systems, which are discussed in Chapter 6. However, some monofunctional initiators, such as cumylpotassium and n-butyllithium, have been employed to prepare A-B structures. The A-B block copolymers containing acrylic segments prepared using these initiators are shown in Table 5-3. Other reports in the patent literature (B18–B21) have claimed that alkyllithium modified by certain sulfur-containing compounds show improved block copolymerization of polar monomers relative to the unmodified system.

It must be concluded that extreme care is required in order to synthesize acrylic block copolymers of predictable chain length, block purity, and narrow molecular weight distribution. There are no acrylic block copolymers produced commercially at this time.

ii. Free Radical Polymerization. Many investigators have employed free radical technique for the preparation of block copoly-

TABLE 5-3

A-B Block Copolymers Containing Acrylic Segments Prepared by Anionic Polymerization

Monomer A	Monomer B	References
Methyl methacrylate	Acrylonitrile	B26–B28
	Alkyl methacrylates	B14–B16
	Styrene	B4, B17, B23–B25
	p-Bromostyrene	B27
	Isoprene	B29
	4-Vinylpyridine	B27
	2-Vinylpyridine	B36
	Ethylene	B30–B32
Methacrylonitrile	Styrene	B27, B33, B35
Acrylonitrile	n-Butylisocyanate	B22
1,4-Acrolein	1,2-Acrolein	B34

mers. There are numerous methods for preparing macromolecular radicals that may be utilized for the initiation of a second monomer. For
example, Ceresa (B37, B38) has reviewed this approach wherein even
mechanical stress on a polymer chain can rupture it homolytically and
thus initiate the polymerization of a second monomer. More commonly, the second monomer is initiated by chemical means. Thus,
terminal groups are introduced that are stable during the polymerization of the first monomer but that may be decomposed under the conditions used for the formation of the second block.

Dihydroperoxides may be used for the synthesis of the first block
under conditions which cause only one of the hydroperoxide groups to
decompose (see Scheme 5-1) (B39–B41, B73). Note that A-B and A-B-A

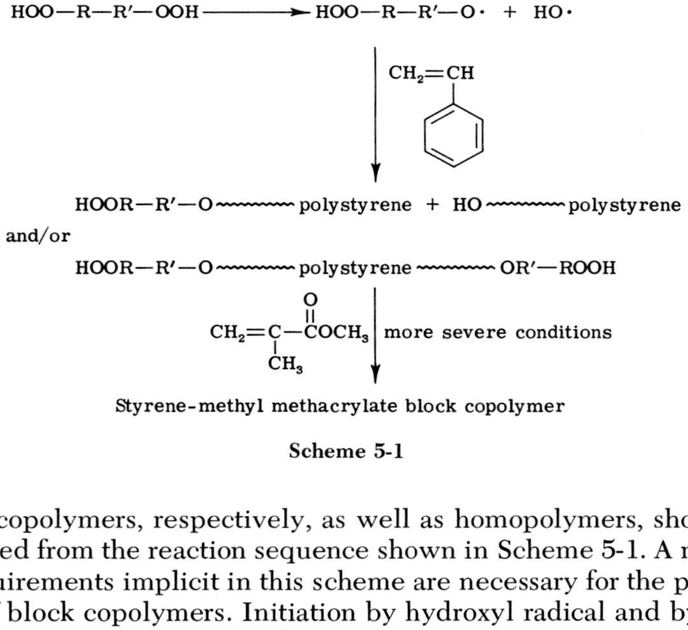

$$HOO-R-R'-OOH \longrightarrow HOO-R-R'-O\cdot \ + \ HO\cdot$$

$$CH_2{=}CH$$

$$HOOR-R'-O\,\wwww\, polystyrene \ + \ HO \,\wwww\, polystyrene$$

and/or

$$HOOR-R'-O\,\wwww\, polystyrene \,\wwww\, OR'-ROOH$$

$$CH_2{=}\underset{\underset{CH_3}{|}}{C}-\overset{\overset{O}{||}}{C}OCH_3 \quad \text{more severe conditions}$$

Styrene-methyl methacrylate block copolymer

Scheme 5-1

block copolymers, respectively, as well as homopolymers, should be
expected from the reaction sequence shown in Scheme 5-1. A number
of requirements implicit in this scheme are necessary for the preparation of block copolymers. Initiation by hydroxyl radical and by chain
transfer to monomer and/or solvent would have to be minimal to avoid
excessive homopolymer formation. Induced decomposition by, for
example, ferrous ion is said to reduce homopolymerization caused by
the competing hydroxyl radical (B73). Termination by coupling, which
is expected for styrene radicals, would produce the difunctional
polymeric hydroperoxide shown in Scheme 5-1. These same radical
side reactions can, of course, also occur during the synthesis of the
second block.

Peroxide groups have been introduced into homopolymer fragments by initiating with polymeric phthaloyl peroxide (Structure **I**). (B42–B44). Similar problems of homopolymer formation encountered with dihydroperoxide initiators were observed.

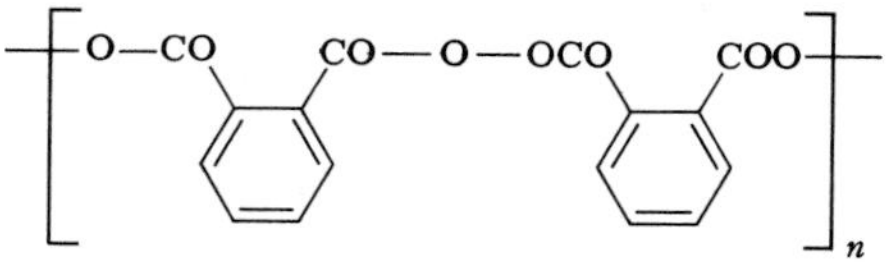

Polymeric phthaloyl peroxide

I

Chain transfer reactions have been maximized in some systems in another approach to prepare block copolymers (B45, B94). Tertiary amines, such as triethylamine (B46), display a high chain transfer rate in methyl methacrylate and acrylonitrile polymerization to produce a terminal group on the first block (Structure **II**).

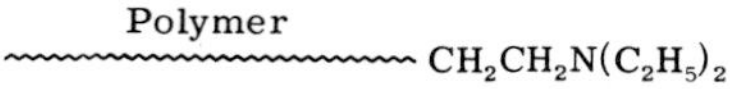

Polymer
$\sim\sim\sim\sim\sim\sim\sim\sim\sim\sim\sim\sim$ $CH_2CH_2N(C_2H_5)_2$

II

The above oligomer is then used as a macromolecular chain transfer agent for polymerization of the second monomer. In order to minimize homopolymer formation, the lifetime of the kinetic chain must be appreciable (B46). Phosphines (B47–B49) and sulfur (B50) have also been similarly utilized as transfer agents.

All of the above approaches as well as many others have been satisfactorily reviewed (B73). Unfortunately, there are no highly efficient routes to pure block copolymer by this technique. Because of their previous coverage, this discussion will be limited to a tabulation of some of the more representative and recent papers on this subject (see Table 5-4). Other sources may also be consulted (B116–B126).

b. Solution Properties

The solution properties of A-B block copolymers, graft copolymers, and random copolymers of styrene and methacrylate (S–MMA) have received considerable attention from a number of authors. A review of the solution behavior of block and graft copolymers, including the S–MMA system, has been published (B114). Characterization techniques are also described in this reference.

Subsequently (B77), the chain conformation of S–MMA and poly(2-

TABLE 5-4

Free Radical Initiated Acrylic Block Copolymerizations

Monomer A	Monomer B	References
Methyl methacrylate	Styrene	B39, B42–B44, B50–B66, B75
	Methacrylonitrile	B67
	Acrylonitrile	B45, B46, B94
Methacrylic acid	Styrene	B74
Acrylonitrile	Styrene	B47–B49, B68
	Ethylene	B69
	Acrylamide	B70
	Acrylates	B71, B72
Acrylic acid	Acrylamide	B70

vinylpyridine–methyl methacrylate) block copolymers and S–MMA random copolymers were studied in dilute solution. Since chemically dissimilar homopolymers are generally incompatible in solution as well as in bulk, it had been postulated by some (B78–B81) that block copolymers could exhibit segregated conformations. According to this theory, A and B blocks of the same molecule have distinct locations and take configurations in which the intramolecular contacts are as few as possible. Other workers (B82–B84) assumed that pseudo-Gaussian statistics are followed and that intramolecular heterocontact interactions make significant contributions to the overall expansion of the coil. Another study (B77) agrees with the view that the segments are segregated. It concludes, on the basis of viscosity behavior, that for block copolymers, the unperturbed dimensions are not affected by the heterocontacts and that the conformation must be segregated. Intermolecular heterocontacts do exist, and these affect the value of the second virial coefficient in a manner similar to that observed in binary mixtures of homopolymers. In contrast to this behavior, the chain dimensions of random copolymers are affected by unavoidable heterocontact interactions. Although this careful study (B77) seems to show that intramolecular contacts are not affecting the chain dimensions of the block copolymer, further elucidation will be necessary to completely define the microstructure of these systems. Papers on the subject of solution behavior are tabulated in Table 5-5.

c. Morphology

Mesomorphic phases have been observed with block copolymers of different alkyl methacrylates (B104). The materials were mixed with

TABLE 5-5

Solution Properties of Styrene–Methyl
Methacrylate Block Copolymers [a]

Subject	References
Colloidal dispersions	B76
Structure by dielectric polarization and NMR	B95
Turbidometric titrations	B91
Light scattering, viscometry, etc.	B77–B90, B93, B98–B100, B102, B103, B113, B115
Solution behavior (review)	B114
Thin-layer chromatography	B97
Intramolecular rearrangement	B92
Adsorption	B96, B101

[a] Random copolymers are often also discussed in these references.

solvents selective for one of the blocks and then studied by small-angle X-ray diffraction. Lamellar, inversed cylindrical, cylindrical, or spherical structures were noted depending upon the alkyl groups, copolymer–solvent ratio, and block lengths.

d. Mechanical Properties

The physical properties of methyl methacrylate–styrene and methyl methacrylate–acrylonitrile block copolymers and those of the corresponding random copolymers have been reported (B105, B106). The block copolymers were prepared by the tertiary base (B45) method discussed in Section a above.

The glass transition behavior of the methyl methacrylate–acrylonitrile (MMA–AN) random copolymer is unusual. A minimum is noted as a function of composition. Block copolymerization did not significantly change the copolymer properties. However, the block copolymers had one intermediate T_g value which is in itself rare and may indicate that the block segments are mutually compatible. X-Ray studies (B107) of the MMA–AN materials showed that the block copolymer had sharper peaks than the random copolymer and was presumably more crystalline. Styrene–methyl methacrylate block copolymers that can be selectively plasticized have been reported (B108, B109). The emulsifying effect of similar two and three block systems has also been discussed (B110–B112).

Perry has studied the properties of random and A-B block copolymers of styrene and acrylonitrile as well as homopolymer blends and block copolymer–homopolymer blends (B49). The synthesis utilized dialkyl phosphines analogously to the tertiary base method discussed earlier. The block purity was estimated to be 60–100% based on extraction and fractional precipitation data. Homopolymer blends containing only 1–2% polyacrylonitrile in polystyrene were opaque, whereas block copolymers containing 14% or less polyacrylonitrile were transparent and had a domain size of <500 Å, as determined by microscopy.

The mechanical properties of the block copolymers were, in this case, inferior to the random copolymer. The difference was attributed to stress concentrations at the phase boundary of the block copolymer. It is of interest to note that the block copolymer could be easily mixed with polystyrene whereas the two homopolymers were grossly incompatible.

2. Vinylpyridine

These A-B block copolymers are synthesized by anionic living polymer techniques. All but one of the references found in this category pertain to block copolymers containing a polystyrene segment and a poly(2-vinylpyridine) or poly(4-vinylpyridine) segment (B127–B132). The other reference discusses poly(2-vinylpyridine)–poly(4-vinylpyridine) block copolymers.

The styrene–vinylpyridine copolymers are prepared by initiating styrene polymerization with butyllithium (B127) or cumylpotassium (B128–B131) followed by addition of the vinylpyridine monomer and polymerization at −78°C (B127). The sequence for the case of a 2-vinylpyridine copolymer is shown in Reaction 5-5. The styrene

$$(5\text{-}5)$$

block must be formed first, since polyvinylpyridine is susceptible to attack by styryl carbanions. Copolymers containing 2-vinylpyridine were more soluble in tetrahydrofuran than those containing 4-vinylpyridine.

Solution-cast films of the polystyrene–polyvinylpyridine block copolymers were colorless, but after immersion in methanol, they exhibited dichroism (bright blue in reflectance) (B127). This was due to swelling of the polyvinylpyridine domains. By comparison, random copolymers disintegrated upon swelling with methanol. Electron microscopy showed the presence of two phases, with the polystyrene phase the continuous one.

The structure of the mesomorphic gels obtained with these block copolymers in the presence of octanol (a preferential solvent for the vinylpyridine blocks,) and toluene (a solvent selective for the styrene segments) were studied by small-angle X-ray diffraction (B128). Three types of structures were detected (B128–B130): (a) cylindrical domains arranged in a hexagonal network, (b) a periodic stacking of lamellae, and (c) an assembly of spheres according to a cubic network.

Quaternization of films of the block copolymers with methyl bromide and with dihalides gave interesting chemical modifications (B127). The quaternized films displayed dichroism when swelled either in water or in toluene. Films quaternized with dihalides gave products with ionically cross-linked hydrophilic domains. Membranes prepared from these materials, in reverse osmosis experiments with 3.5% NaCl solutions at 1500 psi pressure, exhibited 80% salt rejection.

Further chemical modification of quaternized block copolymers by sulfonation resulted in products with both polyanionic domains (sulfonated polystyrene blocks) and polycationic domains (quaternized polyvinylpyridine segments) in a mosaic arrangement (B127). These films were clear and slightly blue in color, and they were insoluble due to salt formation at the phase boundaries. These "charge mosaic membranes," after exposure to radiation to form covalent as well as ionic cross-links, were said to be of interest for water desalination by piezodialysis.

Poly(2-vinylpyridine)–poly(4-vinylpyridine) block copolymers were synthesized via diphenylmethylsodium initiation (B133). Either monomer can be polymerized first in this synthesis. The structure of the mesomorphic gels obtained from these copolymers in the presence of octanol, tetrahydrofuran, dioxane, and benzene were investigated by X-ray diffraction. This study showed the presence of two types of structure—stacked lamellae and a hexagonal arrangement of cylinders.

C. POLY(α-OLEFIN) BLOCK COPOLYMERS

This chapter is entitled A-B block copolymers, and, accordingly, this section should deal solely with α-olefin block copolymers containing only two segments. Such A-B block structures are synthesized by polymerizing monomer A in the presence of specific α-olefin catalysts and subsequently feeding monomer B. It should be possible to similarly prepare A-B-A or $\{A\text{-}B\}_n$ block copolymers by changing the monomer feed once more or many times to obtain any desired value for n. However, in the polymerization of α-olefins, there are complications due to chain transfer reactions. As a result, polymerizations intended to produce A-B-A or $\{A\text{-}B\}_n$ block copolymers may actually result in some A-B structures and even homopolymer fractions. There may also be continuous initiation in these heterogeneous systems.

These complications will be discussed in greater detail below. However, they are mentioned briefly here, at the outset of this section, to point out the uncertainty of assigning a block structure to a given α-olefin block copolymer product based on method of preparation. For this reason, no attempt has been made in reviewing the α-olefin block copolymers reported in the literature to separate them into A-B, A-B-A, and $\{A\text{-}B\}_n$ structures. Instead, they are discussed together in the present A-B block copolymer section as a single class of materials. It is to be understood, however, that some of the products treated in this section may actually have, wholly or in part, A-B-A or $\{A\text{-}B\}_n$ block structures as well as simple A-B sequence arrangements.

There are some reports in the literature of products with some degree of "blockiness" that results from the polymerization of mixtures of two monomers with heterogeneous Ziegler–Natta catalyst systems. Since the block structure of these materials is even more questionable than that of copolymers made by the sequential monomer polymerization technique described above, they are not discussed here except for a few examples as noted.

Most of the literature on α-olefin block copolymers deals with ethylene–propylene compositions. These will be discussed first, followed by a survey of other α-olefin block copolymers.

1. Ethylene–Propylene

a. Synthesis

i. Lifetime of Growing Chains. Ethylene–propylene block copolymers are synthesized using Ziegler–Natta-type (C1) coordination catalysts, which are based on combinations of transition metal compounds such as the chlorides of titanium, vanadium, or zirconium and organometallic compounds such as aluminum alkyls. The most commonly used catalysts are based on $TiCl_3$. The actual structures of the catalytically active sites are still not universally agreed upon. Some workers are proponents of the monometallic theory, according to which the aluminum alkyl component serves only as an alkylating agent and the alkylated titanium chloride is the active species. Others believe that the aluminum alkyl combines with the titanium chloride to form a catalytically active bimetallic complex. This subject is beyond the scope of this book; detailed discussions of it can be found in the literature (C2–C7).The block copolymer products prepared using these catalysts range widely in properties, from rigid plastics to elastomers (C8), depending on the chemical composition and the degree of structural regularity within the blocks.

In a typical block copolymer synthesis, monomer A (e.g., propylene) is fed into a reactor containing the catalyst and the reaction medium (e.g., heptane) at the polymerization temperature. In some processes, monomer A is completely polymerized. In others, the monomer remaining unreacted after a given period of time is removed from the reactor by evacuation techniques. Monomer B (e.g., ethylene) is then charged and polymerized. In some syntheses, this completes the reaction cycle and the product is isolated (e.g., by coagulation). Such a reaction sequence should result in an A-B block structure containing one polypropylene segment and one polyethylene segment. Other synthesis procedures entail *repeated* changes in monomer feed to produce A-B-A or $\{A\text{-}B\}_n$ structures containing many polyethylene and polypropylene segments. A modification of this procedure is to continuously feed monomer A to the reactor and to add monomer B periodically. This results in a product with an $\{A\text{-}[A\text{-}B]\}_n$ block structure; i.e., one segment is a homopolymer of A units and the other is a random copolymer of A and B units. Random copolymer segments of this type can also result unintentionally if the removal of monomer A (e.g., by flushing) is not done efficiently.

The successful attainment of the desired structures by the above procedures requires that the lifetime of the active, propagating sites be sufficiently long (C8–C11). Obviously, if the lifetime of the growing chains is shorter than the time required to perform all of the operations of the block copolymer synthesis cycle, the expected block copolymer structure will not be obtained. For example, if the intended product is an $\{A\text{-}B\}_n$ multiblock structure containing three polypropylene blocks and three polyethylene blocks (i.e., $n = 3$) but the lifetime of the growing chains is equivalent to the time required to polymerize only the first propylene and ethylene monomer charges, the result will be an A-B structure containing only one segment of polypropylene and one of polyethylene. Similarly, if an A-B diblock structure is desired but the lifetime of the active growing sites is shorter than the period of time allowed for reaction of each of the monomer charges, then the product will be primarily a mixture of propylene and ethylene hompolymers.

This problem can be avoided only when the system is entirely free of chain transfer reactions. Such is the case with the styrene–diene block copolymers synthesized by Szwarc's (C181) homogeneous "living" polymer process (see Chapter 5, Section A). However, unlike this behavior, α-olefin polymerization systems do undergo chain transfer reactions (5-6 through 5-9) to give new active catalytic sites (C10, C12).

$$\overset{\oplus \ominus}{M}\text{---}CH_2\text{---}CH \text{\small \char`\~\char`\~} + CH_2{=}CH\text{---}R \longrightarrow \overset{\oplus \ominus}{M}\text{---}CH_2\text{---}CH_2\text{---}R + CH_2{=}C\text{\small \char`\~\char`\~} \tag{5-6}$$

$$\overset{\oplus \ominus}{M}\text{---}CH_2\text{---}CH \text{\small \char`\~\char`\~} + CH_2{=}CH\text{---}R \longrightarrow \overset{\oplus \ominus}{M}\text{---}H + CH_2{=}C\text{---}CH_2\text{---}CH\text{\small \char`\~\char`\~} \tag{5-7}$$

$$\overset{\oplus \ominus}{M}\text{---}CH_2\text{---}CH \text{\small \char`\~\char`\~} \longrightarrow \overset{\oplus \ominus}{M}\text{---}H + CH_2{=}C\text{\small \char`\~\char`\~} \tag{5-8}$$

$$\overset{\oplus \ominus}{M}\text{---}CH_2\text{---}CH \text{\small \char`\~\char`\~} + M\text{---}R' \longrightarrow \overset{\oplus \ominus}{M}\text{---}R' + M\text{---}CH_2\text{---}CH\text{\small \char`\~\char`\~} \tag{5-9}$$

(where R subscripts appear below the CH groups)

(active)   (nonactive)   (active)   (nonactive)

The lifetime of growing chains in ethylene–propylene polymerization systems has been the subject of considerable debate (C8, C9, C70). Chain lifetime studies have been based primarily on the relationship between polymerization conversion and polymer molecular weight. For example, it has been reported in the patent literature that molecular weight increases steadily with increasing time and conversion. This has been cited as evidence for the formation of block copolymers with many segments (see Table 5-6). However, this may not be unassailable proof of block copolymer formation. Increasing molecular weight with time might be caused by changes in conditions at the catalyst surface, change in the number of active sites, or changes in the transfer–propagation rate ratio, as well as by the growth of exist-

TABLE 5-6

ηsp/C Value as a Function of Experiment Time in the Block Copolymerization of Ethylene and Propylene [a]

Number of blocks	Experimental interval (minutes)	Solid product yield (gm)	ηsp/C [b]
6	90	316	1.4
12	180	800	2.3
24	360	1613	2.9
40	600	2700	4.0

[a] Taken from Bier (C10, C14), Berger *et al.* (C13), and Caunt (C15).
[b] 0.1 gm of polymer in 100 ml decahydronaphthalene, measured at 135°C.

ing molecules (C10, C13–C15). There appears to be fair agreement, however, that those catalysts based on $TiCl_3 + (C_2H_5)_2AlCl$ produce growing chains with some finite lifetime. The length of the lifetime may depend to a large extent on the level of impurities introduced during the polymerization.

Although long chain lifetimes have been claimed at temperatures as high as 80°C, (C10), the lifetime of the growing chain has been reported in general to be shortened by increasing temperature, but to be unaffected by pressure (C16–C18).

Hydrogen also shortens the lifetime of growing chains (C16–C18). Hydrogen is used in α-olefin homopolymerization and in some block copolymerizations to prevent the formation of very high molecular weight polymers (which display poor melt processability) and at the same time to obtain high catalyst efficiency (C19). Chain termination caused by the hydrogen is followed by the growth of new chains on the hydrogenated form of the catalyst (see Reaction 5-10). However,

$$M\text{---}CH_2\text{---}CH\overset{\oplus\ \ominus}{\sim\sim\sim} \xrightarrow{H_2} M\text{---}H + CH_3\text{---}CH\sim\sim\sim$$

$$\downarrow CH_2{=}CHR \qquad (5\text{-}10)$$

$$M\text{---}CH_2\text{---}CH\sim\sim\sim$$

this technique can only be successful in block copolymer preparations (a) if the chain lifetime in the absence of hydrogen is long enough to permit the formation of the desired number of segments in the block copolymer product, and (b) if the hydrogen is fed intermittently. For example, the repeated sequential feeding of ethylene, followed by propylene, followed by hydrogen should result in the continuous formation of A-B block copolymers. However, continuously feeding hydrogen along with the monomer (as is reported in some of the literature) will not result in efficient block copolymer formation due to premature termination.

Although it may be possible to achieve long chain lifetimes with certain catalyst systems and synthesis conditions, the results of many investigators vary from one extreme to the other. Some report lifetimes of the order of a few seconds (C20), while others claim lifetimes of hours (C10) or even hundreds of hours (C8, C9) leading to polymers containing as many as forty-eight segments (C10). It is for this reason that there is considerable uncertainty about the structures of the various ethylene–propylene block copolymers reported in the literature. Furthermore, adequate structural proof by appropriate molecular characterization techniques is not provided in these references. Therefore, no attempt will be made here to categorize the block copolymers by block structure. Instead, as discussed earlier, a description will be given of the synthesis and properties of these materials only as a function of composition.

ii. Reaction Conditions. Ethylene–propylene block copolymers have been synthesized under a wide variety of conditions. These have been summarized in Table 5-7 according to the type of catalyst system used. References are provided in this table for those who wish to use them to obtain more detailed information.

In most cases, the block copolymers were synthesized using Ziegler–Natta catalysts prepared by mixing organoaluminum compounds with titanium or vanadium compounds. Polymerization behavior is greatly affected by the composition of the catalyst (ratio of components) and by the method of preparing and isolating the catalyst system. In some instances, the catalyst systems listed in Table 5-7 were modified with compounds such as diethyleneglycol dimethyl ether (C21–C26), tetraoctyltitanate (C29, C30), tetra(dimethylamino)silane (C31), ethyl silicates (C27, C28), and hexaethyl melamine (C32).

Polymerizations were carried out at relatively low temperatures (20°–150°C) and pressures (5–1000 psi), and usually resulted in linear products (C10). Aliphatic, alicyclic, or aromatic hydrocarbons were most frequently used as the reaction media. Both soluble and insoluble catalysts have been employed. Some syntheses were carried out in

TABLE 5-7

Catalysts and Synthesis Conditions for Preparation of Ethylene–Propylene Block Copolymers

Catalyst	Medium	Temperature (°C)	Pressure (psi)	Polymerization time per block (minutes)	References
$TiCl_3 + AlR_3$	Benzene, toluene, xylene	15-120	4-1000	1-95	C1, C10-C12, C36, C47-C50
$TiCl_3 + LiAlR_4$	Mineral oil	150-160	600-1800	10-60	C51, C52
$TiCl_3 + R_2AlCl$ (or $RAlCl_2$)	Hexane, heptane, propane	25-71	5-750	2-196	C21, C22, C24, C25, C27-C29, C31, C38, C42, C53-C62
$TiCl_3 + R_2AlOEt$	Heptane	0	5	15-60	C63
$TiCl_3 \cdot AlCl_3 + AlR_3$	Aliphatic hydrocarbon, xylene	63-79	60-150	60-360	C40, C41, C64
$TiCl_3 \cdot AlCl_3 + R_2AlCl$	Benzene, hexane, Pentane, propane	49-90	75-400	30-120	C10, C19, C23, C26, C30, C33-C35, C65-C68
$TiCl_3 + R_4Al_2SO_4$	Heptane	50	45	40	C32, C69
$TiCl_4 + Al$	Heptane	90-120	150-600	—	C10,
$TiCl_4 + AlR_3$	Isooctane	50	—	—	C71
$TiCl_4 + AlR_3$ (or Al) + R_2AlCl (or $RAlCl_2$)	Hexane, octane	20-120	<15-600	8-30	C10, C72-C76
$TiCl_4 + R_2AlCl$	Hydrocarbons	50	—	70	C77, C78
$TiCl_4 + LiAlR_4$	Heptane	25-30	—	2-20	C8, C10, C79, C80
$TiCl_4 + AlR_3 + LiR_4$	—	25-30	<15	—	C10, C79
$ZrCl_3 + AlR_3$	Benzene	94-121	100-1000	5	C10, C47
$VOCl_3 + R_2AlCl$ (or $RAlCl_2$)	Aliphatic hydrocarbons	25-35	15-100	2-8	C8, C81
$VOCl_3 + AlR_3 + LiR$	Aliphatic hydrocarbons	20-35	<15	3-4	C10, C79
$V(OOCR)_2 + RAlCl_2$	Heptane	0	3	—	C82

liquid propylene as the reaction medium, for at least part of the reaction cycle (C33–C35). Polymerization times per segment varied from 1 to 360 minutes.

The order of monomer addition has an effect on polymerization rate and on product properties. Charging ethylene first gives faster rates, due to the higher reactivity of ethylene (C10, C16, C36). This difference in reactivity was compensated for in some cases by using higher temperatures or longer times for the propylene polymerization cycles (C10, C36).

Most of the work reported involved one-pot, batch, solution processes. However, some exceptions to these conditions have also been reported. Some patents (C37–C41) claim that block copolymer synthesis can be carried out in a double reactor process in which propylene is polymerized in the first reactor and then transferred to a second reactor where ethylene is charged. Other processes were carried out continuously by means of fluidized bed techniques in which the initial fluidized bed was composed of preformed block copolymer (C42–C44). The specific processes claimed by various companies, as described in their patents, is discussed in a review by Bier and Lehmann (C10).

b. Characterization

Little, if any, data was found in the literature that elucidates the block structure of ethylene–propylene block copolymers. The number of segments present is assumed from the method of preparation. However, this is unreliable, as was discussed earlier, because of the questionable nature of the growing chain lifetimes. However, some work has been done that demonstrates that some of the products display the properties of block copolymers, as opposed to random copolymers or homopolymer blends. This work and the general structural characterization data reported for the block copolymers is discussed below.

Block structures were attributed to some ethylene–propylene copolymers because their mechanical properties were different from those of homopolymer blends and random copolymers of equivalent chemical composition. Products of intended block copolymer structure were shown to display much higher elongations and impact strengths than homopolymer blends of corresponding chemical composition (C10, C45, C46).

Extractability behavior (in ether and in heptane) has also been cited as evidence for block structure, as opposed to random copolymers and homopolymer mixtures. Block structures are reported to be less soluble in ether than random copolymers and more soluble in heptane than homopolymer mixtures (C10). This is presumably due to the intermediate degree of crystallinity displayed by the block copolymers, i.e., more crystalline than random copolymers and less crystalline than homopolymer blends. Cloud point determination (in ethyl benzoate solution) and turbidimetric titration are other methods used to demonstrate differences in the solubility behavior of block copolymer structures (C10, C83) as compared to random copolymers and homopolymer mixtures.

The differential thermal analysis behavior of ethylene–propylene block copolymers and homopolymer mixtures has been studied by several workers (C84–C86). Both block copolymers and mixtures display two endothermal transitions as the temperature is raised. The polyethylene melting point is observed at 126°–138°C and that of the polypropylene at 162°–163°C. However, the cooling curves of the block copolymers differ from those of mixtures. Mixtures of homopolymers showed a single exotherm on cooling (at 110°C), while block copolymers of >10% ethylene content displayed two exotherms (at 110°C and 140°C). This observation was explained by proposing that the polypropylene segments in the block copolymer recrystallize sharply near the normal melting point of polypropylene, whereas in the mixtures, the homopolypropylene undergoes supercooling.

The degree of X-ray crystallinity of polyethylene and polypropylene is considerably lower in block copolymers than in homopolymer blends (C87). The degree of crystallinity of the blocks is proportional to the concentration of those blocks in the copolymer. The dimensions of the crystalline regions are also smaller in block copolymers. This reduced crystallinity in block copolymers was attributed to the introduction of a new type of defect into the polymeric crystalline lattice. Increasing ethylene content in block copolymers over the 10–40% range did not increase the defects of the block copolymer crystallites, suggesting that the materials were actually mixtures of true copolymer and the two homopolymers (C88). A correlation between degree of crystallinity and hardness has been reported (C9, C89).

The X-ray diffraction patterns of elastomeric ethylene–propylene random copolymers and of block copolymers containing polyethylene segments and ethylene–propylene random copolymer segments were found to be dependent upon the state of the sample (C8). In the unstretched form, random (amorphous) and block (crystallizable) copolymers of the same composition exhibited essentially identical diffraction patterns. Upon stretching, the random copolymers gave no change in the X-ray pattern, but the block copolymers displayed the discrete spots or arcs characteristic of orientation or strain-induced crystallization. This effect was reversible; i.e., the arcs were not present in the X-ray patterns of stretched and subsequently relaxed specimens. This behavior is analogous to that observed in natural rubber and high *cis*-polyisoprene vulcanizates.

Infrared, near infrared, and Raman spectral techniques have been used to analyze for the ethylene and propylene contents of block copolymers (C10, C49, C90–C94). The effects of crystallinity were studied by running the spectra both at room temperature and at 180°C. This technique does not elucidate the block structure of the copolymers.

High-resolution proton magnetic resonance has been reported to give qualitative information on ethylene–propylene copolymer sequencing by using a series of reference polymers characterized by radio tracer methods (C95). Analysis of block copolymer pyrolysis products by mass spectrometry or by gas chromatography has also been used as a method to obtain data on monomer distribution in block copolymers (C10, C96).

c. Properties

i. Rigid Plastic Compositions. Polyethylene has good low temperature properties, but it is relatively low in melting point and in surface hardness. Polypropylene has a higher melting point and is harder, but it is brittle at low temperatures due to a relatively high glass transition temperature. Random copolymers of ethylene–propylene are elastomeric materials with greatly reduced crystallinity. They therefore have poorer tensile (see Fig. 5-8) and hardness properties than the corresponding homopolymers. In block copolymers, the desirable properties of each of the homopolymers are combined with minimum sacrifice.

The effect of composition on rigid block copolymer properties, such as impact strength, brittle point, heat distortion temperature, reduced viscosity, extractability, density, tensile strength, elongation, hardness, and crystallinity has been reported (C10, C13, C97–C100). The most dramatic effect is that brittleness temperature decreases and impact strength increases with increasing ethylene content. Rigid plastic ethylene–propylene block copolymers with the best properties are obtained at the 1–17% ethylene content level. Crystalline block copolymers of low (<3%) ethylene content, called "polyallomers," have been prepared using a $Al(C_2H_5)_3$–$TiCl_3$ catalyst at 70°–80°C and 450 psig pressure in mineral oil medium. These materials have several property advantages over ethylene and propylene homopolymers (C49). These properties are presented in summary form in Table 5-8.

TABLE 5-8

Comparison of Polyallomer and Homopolymer Properties

	Polyallomer in comparison to[a]	
	Polypropylene	Polyethylene
Brittleness temperature	−	+
Impact strength	+	+
Density	−	−
Vicat softening point	−	+
Stiffness	−	0
Tensile strength	−	0
Elongation	+	+
Hardness	−	+
Mold shrinkage	0	−

[a] Key: 0, the indicated property is approximately equivalent for the polyallomer and the homopolymer; −, the indicated property is lower for the polyallomer; +, the indicated property is higher for the polyallomer.

Some numerical comparisons are made in Table 5-9 between the homopolymers and block copolymers containing 0.6, 2.0, and 3.0% ethylene (C49).

It can be seen from these data that the polyallomers overcome the most serious deficiencies of polypropylene—i.e., brittleness temperature and notched and unnotched impact strength—at some sacrifice in Vicat softening point, stiffness, strength, and hardness. It is especially noteworthy that the incorporation of a low level of ethylene results in an inordinate drop in stiffness. In comparison to polyethylene, the polyallomers are superior in impact strength, Vicat softening point, hardness, melt flow, and mold release characteristics, at some sacrifice in brittleness temperature. In addition, the stress-crack resistance, transparency, and surface quality properties of the polyallomers are superior to those of polyethylene.

The effect of ethylene content within the range of 0–5% on the brittleness temperature, stiffness, strength, impact resistance, and hardness of polyallomers is illustrated in Figs. 5-7 and 5-8. Brittleness temperatures as low as −50°C and notched Izod impact strengths as high as 10 foot-pounds per inch have been attained.

The effect of molecular weight on polyallomer properties is shown in Table 5-10. As would be expected, the higher molecular weight

TABLE 5-9

Comparison of Properties of Ethylene and Propylene Homopolymers, Block Copolymers, and Homopolymer Blends [a]

	Propylene homopolymer	Block copolymers			Ethylene homopolymer	Polyethylene–Polypropylene blends		
						(1)	(2)	(3)
Ethylene (%)	0.0	0.6	2.0	3.0	100	5	10	25
Flow rate at 230°C (dg/minute)	2.4	2.8	2.9	2.5	2.4	1.5	2.5	1.9
Density (annealed) (gm/ml)	0.9100	0.9093	0.9044	0.9010	0.9724	0.9132	0.9164	0.9210
Brittleness temperature (°C)	+8	−5	−22	−35	<−78	+1	0	−3
Tensile strength at yield (psi)	4700	3870	3560	3050	3100	4400	4400	4200
Elongation (%)	360	550	500	>650	290	110	200	150
Stiffness in flexure (psi)	142,000	99,500	92,400	80,000	101,000	134,000	133,000	130,000
Hardness (Rockwell R scale)	93	87	68	60	54	91	88	83
Vicat softening point (°C)	145.0	140.3	127.2	124.5	122.8	143.1	139.4	139.1
Notched Izod impact strength (23°C) (ft-lb/in)	0.5	0.9	1.9	3.5	1.3	0.6	0.6	0.7
Unnotched Izod impact strength (23°C) (ft-lb/in)	No break	No break	No break	No break	No break	No break	No break	No break
Tensile impact strength (ft-lb/in^2)	32	56	79	90	63	35	38	44

[a] Taken from Hagemeyer and Edwards (C49).

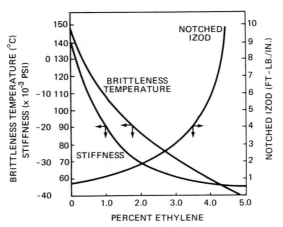

Fig. 5-7. Effect of ethylene content on brittleness temperature, stiffness, and impact strength of a propylene–ethylene polyallomer (C49).

material exhibited a lower brittleness temperature and a higher impact strength. Table 5-9 also compares the properties of mechanical blends of polypropylene–polyethylene with those of polyallomers and the homopolymers. The blends were stronger, stiffer, and higher in Vicat softening point than the polyallomers. However, the brittleness temperature and impact strength of polypropylene were improved only slightly by blending with as much as 25% polyethylene, as compared to the large improvements obtained with polyallomers at much lower ethylene levels.

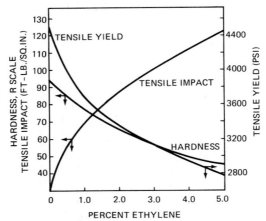

Fig. 5-8. Effect of ethylene content on hardness, tensile impact, and tensile yield of a propylene–ethylene polyallomer (C49).

TABLE 5-10

**Effect Of Molecular Weight on Properties of
Propylene–Ethylene Polyallomers** [a]

Ethylene (%)	2.0	2.2
Flow rate at 230°C (dg/minute)	2.9	0.6
Inherent viscosity in tetralin	1.80	2.38
Density (annealed) (gm/ml)	0.9044	0.9030
Brittleness temperature (°C)	-22	-32
Tensile strength at yield (psi)	3560	3250
Elongation (%)	500	600
Stiffness in flexure (psi)	92,400	78,800
Hardness (Rockwell R Scale)	68	64
Vicat softening point (°C)	127.2	128.0
Notched Izod impact strength (23°C) (ft-lb/in)	1.9	10.8
Unnotched Izod impact strength (23°C) (ft-lb/in)	No break	No break
Tensile impact strength (ft-lb/in ^{2})	79	125

[a] From Hagemeyer and Edwards (C49).

Tensile stress–relaxation measurements indicated that the relaxation modulus is somewhat smaller in ethylene–propylene block copolymers than in mechanical blends of the two homopolymers, and that a minimum value is observed in compositions containing ~30 wt% of propylene (C101).

The ethylene–propylene block copolymers have been reported to be quite useful in several application areas. The polyallomers retain the "built-in hinge" effect characteristic of polypropylene (C49). Blends of polypropylene with rubbers also display good impact properties, but these materials are inferior to the polyallomers in transparency, processability, and strength (C16). The block copolymers are also reported to display good stress-crack resistance (C107).

The excellent ambient and low-temperature impact properties and clarity of ethylene–propylene block copolymers are discussed in several references (C21, C24, C25, C27, C40, C41, C49, C53, C78, C102–C104). Impact is reportedly further improved by blending the block copolymers with small amounts (~25%) of random ethylene-propylene copolymers (C105) and with ethylene and propylene homopolymers (C104, C106). Improvement in the low-temperature impact strength of propylene–ethylene block copolymers was reportedly achieved by thermally degrading high molecular weight materials (e.g., at temperatures up to 390°C) prior to molding (C108, C109).

Such degradation presumably causes a narrowing in molecular weight distribution, resulting in improved impact strength. Thermal treatment in the presence of peroxides was also claimed to improve melt processability (C110).

The block copolymers are useful in fabricating containers, blow molded bottles, films, tubing, wire coatings, paper coatings, and molded objects in general (C111). They have also been claimed to be useful as adhesives for bonding polyethylene to polypropylene (C10, C112). It has also been reported that blends of the block copolymers with polypropylene give melt spun fibers with good crimping performance (C113, C114). Blends of ethylene–propylene block copolymers with polyethylene–polypropylene mixtures were claimed to display superior impact strength (C78).

ii. Elastomeric Compositions. In addition to the highly crystalline (>40% X-ray crystallinity) ethylene–propylene block copolymers discussed above, elastomeric block copolymers of lower degree of crystallinity have been prepared by Kontos *et al.* (C8). Block copolymers of high (~50%) ethylene content have been reported in which one segment is a random ethylene–propylene copolymer and the other segment is either polyethylene or polypropylene (C8, C10, C82, C115). These products synthesized using $VOCl_3$–$ClAlR_2$ or $TiCl_4$–$LiAlR_4$ catalyst systems, had varying degrees of crystallinity, as determined by X-ray analysis. "Crystalline" compositions displayed >15% degree of crystallinity at 25°C, "semicrystalline" products were 3–15% crystalline, and "crystallizable" materials were amorphous in the relaxed state but crystallized when stretched. These were contrasted to "amorphous" random copolymers, which were not crystalline even in the stretched state. These polymers exhibited some very dramatic differences in stress–strain performance, as is shown in Figure 5-9. The amorphous polymer (A) exhibited high elongation and low, constant stress. On the other hand, crystallizable and semicrystalline polymers displayed higher tensile strengths due to strong intermolecular forces, i.e., crystallinity. By way of comparison, a 50/50 blend of polyethylene–polypropylene was incompatible and broke at low (50–100%) elongation.

The stress –strain behavior of these unfilled, uncross-linked elastomeric block copolymers containing at least two polyethylene segments was reported to be similar to that of filled natural rubber vulcanizates (C116, C117). The crystalline regions (or domains) of the block copolymers, composed of polyethylene segments, were proposed to behave like filler particles to produce this effect. This analogy

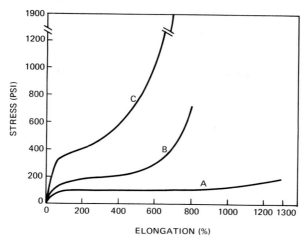

Fig. 5-9. Stress–elongation curves of uncured polymers. (A) Noncrystallizable rubber; (B) crystallizable rubber; (C) semicrystalline material (C8, C115).

has also been applied to styrene–diene A-B-A block copolymers and is discussed more fully in the section of this book dealing with those compositions (Chapter 6, Section A).

A few references report the synthesis of elastomeric ethylene–propylene block copolymers that specifically have an A-B-A structure. The B block is a random copolymer of ethylene–propylene and both A blocks are either crystalline polypropylene (C59, C63, C81, C118, C119) or crystalline polyethylene (C59). These uncross-linked products are reported to have properties characteristic of vulcanized rubbers (C63, C118) (e.g., ⟨2000 psi tensile strength and 600% elongation) (C59). They are also claimed to be useful in adhesives applications (C81).

A new family of polyolefin-based thermoplastic elastomers, designated TPR, is commercially available from Uniroyal Chemical Co. (C120–C125). Although the structure of these materials has not been announced, they may be ethylene–propylene block copolymers, graft copolymers, or blends containing these copolymers. The materials are claimed to have a low density, rubberlike compression set and elongation characteristics, a wide useful temperature range ($-40°$ to $+275°$F), good electrical properties, and good weather resistance. Paintability, after an etching operation, is also claimed for these products. Since they are thermoplastic and do not require a curing operation, they offer considerable injection molding, extrusion, calendering, and blow molding processing advantages over traditional rubbers. On the other hand, they are inferior to cured rubbers in compression set, tear resistance, and capacity for loading with oil and fillers. The properties of the currently availably four grades of TPR are represented in Table 5-11, and they are compared with ethylene–propylene random copolymer rubber, styrene–butadiene random copolymer rubber, and butyl rubber in Table 5-12. The combination of properties displayed by the TPR materials makes them potentially attractive for use in nontire applications such as automotive, appliance, closure, hose and tubing, houseware, toys, medical, and cable end uses.

While the crystallizable, rubbery block copolymers have useful properties as "plastic rubbers" without being cross-linked, they can also be reinforced and cured in the same way that random copolymers are processed. Table 5-13 shows the typical properties of vulcanized polyethylene–poly(ethylene–propylene) and polypropylene–poly-(ethylene–propylene) block copolymers.

d. Miscellaneous

Ethylene–propylene block copolymers are discussed in the following references, in addition to those specifically cited in the text of the

TABLE 5-11

Properties of Uniroyal's TPR Polymers[a]

Property	Grade 010	Grade 023	Grade 020	Grade 019
Hardness (A)	74	82	87	92
Melt flow at 450°F	2	4	6	8
Tensile strength (psi)	1100	1100	1350	2100
Elongation (%)	220	200	170	260
Modulus of elasticity in flexure (psi)	2000	6000	15,000	20,000
Rebound (%) (Bayshore)	45	40	35	30
Heat deflection (2000 gm/in², 1 hr, 250°F)	0	0	0	0
Comp Set "B"				
22 hr, 73°F	27	35	40	33
22 hr, 158°F	40	70	78	73
Hardness retention (%)				
200°F	90	85	90	95
300°F	85	75	85	90

[a] From C122.

TABLE 5-12

Comparison of Uniroyal's TPR Polymers with Other Rubbers[a]

	TPR[b]	EPDM	SBR	Butyl
Specific gravity	0.88	0.85–0.89	0.94	0.92
Hardness (Shore A)[c]	70–90	30–100	40–100	15–90
Tensile strength	1000–2000	1000–3000	1000–3500	1000–3000
Elongation (%)	200–350	100–600	100–700	100–700
300% modulus	600–1400	200–300	200–3200	200–5000
Compression set resistance	Fair	Good	Good	Good
Abrasion/tear resistance	F–G	F–G	VG	F–G
Oil/fuel resistance	P–F	VP	VP	VP
Solvent resistance	VG–E	VG–E	F	G
Hydrocarbon solvent resistance	F–P	P	P	P
Aging resistance	VG–E	G–E	F–VG	G–E
Electrical properties	G–E	VG	G	G
Adhesion	P–F	F	E	G–F
Bayshore resilience (%)	40–50	40–50	45–60	20–30
Min./Max. Service Temperature (°F)	−75/300	−60/300	−65/225	−50/350
Melt flow at 450°F (ASTM D-1238-51T)	2–8	—	—	—
Torsional Modulus (psi)	500–3000	—	—	—
Ross flex (Kc. to 100% cut growth)	0.2–35	—	—	—
Heat deflection (%) (1 hr at 250°F, 2000 gm load/in²)	0	—	—	—

[a] From C120.

[b] Includes four grades.

[c] Hardness drops off only 25–30% over a temperature range spanning −50° to +300°F.

foregoing section (C43, C50, C55–C58, C65–C67, C77, C101, C112, C126–C139).

2. Other α-Olefins

The previous section dealt specifically with ethylene–propylene block copolymers. In addition to these materials, copolymers have also been reported that contain blocks of polyethylene or polypropylene together with other types of polymer segments. These copolymers fall into four categories: those in which the "other" polymer segments are derived from (a) higher linear and branched α-olefins, (b) styrene or α-methylstyrene, (c) dienes, trienes, or alkynes, and (d) polar mono-

TABLE 5-13

Physical and Mechanical Properties of Rubbery Block Copolymers[a]

Property	Copolymer A	Copolymer B
Polymerization technique	Random E-P and E blocks	Random E-P and P blocks
Ethylene content (Infrared) (%)	50	48
Crystallinity (X-ray) (%)	2	1
Density (gm/ml)	0.864	0.860
Intrinsic viscosity (Tetralin, 135°C)	3.6	2.8
Vulcanizate Properties		
Stress at 300% elongation	1500	1400
Tensile strength (psi)	4450	3900
Elongation (%)	540	570
Permanent set (%) (1 min. after rupture)	25	20
Hardness (Shore A)	67	68

[a] Data from Kontos *et al.* (C8, C115).

mers. By and large, the synthetic procedures and general comments made in Section 1, above, also apply to these materials. Exceptions to this generalization and the properties of specific structures are discussed below.

a. Linear and Branched α-Olefins

Kontos *et al.* (C8) reported the molecular weight of polybutene-1 (initiated with $TiCl_4$–$LiAlR_4$) to increase with increasing conversion, thus indicating reasonably long active chain lifetimes. As was discussed earlier for ethylene and propylene polymerization, this feature is a prerequisite for the synthesis of block copolymers. Table 5-14 lists the block copolymers that have been reportedly prepared from butene-1 and higher linear and branched α-olefins, along with the catalyst systems used in their synthesis.

Clegg *et al.* (C140) determined the heat capacity, entropy, enthalpy, and free energy properties of a rigid butene-1–ethylene block copolymer containing 15 mole% of butene-1. This polymer had a glass temperature of −73°C, a melting transition at 127°C, and an intermediate transition at 62°C. Another butene-1–ethylene block copolymer, containing 12 wt% butene-1, was reported to have a melting point of 124°–127°C and to display good low-temperature notched

TABLE 5-14

Linear and Branched Poly-α-Olefin Block Copolymers

Block A	Block B	Catalyst system	Reference
Butene-1	Ethylene	$TiCl_3–Al(n\text{-}C_6H_{13})_3$	C143
	Ethylene	$VOCl_3$ to Al	C72
	Ethylene	$TiCl_4$ to $Al(i\text{-}C_4H_9)_3$ to $n\text{-}C_4H_9Li$	C8, C115
	Ethylene	—	C10, C140
	Ethylene–butene-1	$TiCl_4–Al(i\text{-}C_4H_9)_3–n\text{-}C_4H_9Li$	C8
	Ethylene–propylene	$TiCl_4–Al(i\text{-}C_4H_9)_3–n\text{-}C_4H_9Li$	C8
	Propylene	$TiCl_3–AlCl_3–(C_2H_5)_2AlCl$ or $TiCl_3–LiAlH_4$	C19, C102, C103, C144
Butene-1–propylene	Propylene	$TiCl_3–AlCl_3–(C_2H_5)_2AlCl$ or $TiCl_3–LiAlH_4$	C19, C102, C103, C144
Butene-1–ethylene	Propylene	$TiCl_3–AlCl_3–(C_2H_5)_2AlCl$ or $TiCl_3–LiAlH_4$	C19, C102, C103, C144
Butene-1	Propylene	$TiCl_3–C_2H_5AlCl_2[(CH_3)_2N]_3PO$	C141
Hexene-1 or octene-1	Ethylene or propylene	Transition metal–AlR_3	C102, C103, C105, C115, C145–C147
4-Methylpentene-1	Ethylene	$TiCl_4–(C_2H_5)_2AlCl$	C10
	Ethylene	$TiCl_3–Al(C_2H_5)_3–Ti(OOCC_2H_5)_4$	C149
	Hexene-1	$TiCl_3–(C_2H_5)_2AlCl$	C150
	Decene-1	—	C151
4-Methylpentene-1 or 3-methylbutene-1	4-Methylpentene-1/butene-1/ethylene	$TiCl_3–Al(C_2H_5)_3–Ti(OOCC_2H_5)_4$	C152
4-Methylpentene-1 (1,2-linked)	4-Methylpentene-1 (1,3- and 1,4-linked)	$TiCl_4–(C_2H_5)_3Al_2Cl_3–HCl +$ $(C_2H_5)_2AlCl$	C142

Izod impact strength (no break at −20°C) (C10). Low-temperature impact strength was improved further by blending with ethylene–propylene random copolymer rubbers (C105).

Kontos *et al.* (C8) prepared elastomeric block copolymers consisting of polybutene-1 sequences and ethylene–butene-1 random copolymer segments, and also some terpolymers containing polybutene-1 blocks and ethylene–propylene random copolymer blocks. X-Ray diffraction data showed that the polybutene-1 blocks crystallized when the samples were stretched. The mechanical properties of some block copolymers and terpolymers containing butene-1 have also been presented (C115). For example, one butene-1–ethylene block copolymer containing 26 wt% butene-1 was 22% crystalline (by X-ray analysis) and had a tensile strength of 1900 psi and an elongation of 600%.

Butene-1–propylene block copolymers have also been reported (see Table 5-12). One of these, containing 4% butene-1, was claimed to have a brittle temperature of −35°C, about the same as obtained with a propylene–ethylene block copolymer.

In contrast to the above polymers, all of which were prepared by the sequential monomer feed technique, Coover *et al.* (C141) claimed that they had prepared propylene–butene-1 "costereosymmetric" multi-block copolymers directly from mixtures of the two monomers. They suggested that the block structure of the products was the result of using a highly stereospecific three-component catalyst system consisting of $C_2H_5AlCl_2$, $TiCl_3$, and $[(CH_3)_2N]_3PO$. Block structures were deduced for these products based on (a) the observation of two melting points by differential thermal analysis (lower than for the corresponding homopolymers), (b) X-ray diffraction patterns showing peaks characteristic of polypropylene and of polybutene-1, and (c) the inability to detect the presence of homopolymers by fractionation techniques. It was concluded that the polymers contained stereoregular segments of at least twenty monomer units. The explanation offered for this unusual behavior was based on the argument that the probability of enchainment of one comonomer relative to another is determined mainly by the nature of the preceding triad of monomer units in the growing polymer chain. Copolymers of 3–80% butene-1 content were prepared. Increasing the butene-1 content decreased the hardness, strength, stiffness, and melting point and increased the impact strength, low-temperature properties, and clarity.

Copolymers containing blocks of linear α-olefins other than butene-1 (i.e., hexene-1 and octene-1) have also appeared in the literature. These materials, prepared by sequential monomer feed, are listed in Table 5-12.

Block copolymers containing segments of branched α-olefins such as 4-methylpentene-1 and 3-methylbutene-1 have also been synthesized by sequential monomer feed (see Table 5-14). One such product, a

4-methylpentene-1–ethylene block copolymer containing 20–30% ethylene, had a melting point of 204°–208°C (C10).

An unusual approach was used to prepare "block copolymers" from a single monomer that is able to polymerize in more than one way, e.g., 4-methylpentene polymers containing one segment formed by 1,2-polymerization and another segment produced by 1,3- and 1,4-enchainment (C142). This was reportedly achieved by first polymerizing the monomer with a $TiCl_3$ cationic catalyst [prepared from $TiCl_4$ + $(C_2H_5)_3Al_2Cl_3$ + HCl] to produce a 1,3- and 1,4-linked polymer, followed by addition of $(C_2H_5)_2AlCl$ and additional monomer to stereospecifically produce 1,2-linked segments. Improved impact strength was claimed for these products (which contained <15% of the "cationic" polymer) compared to exclusively 1,2-polymerized 4-methylpentene "homopolymers." Similar products were also claimed to have been prepared from 3-methylbutene-1, 3-methylpentene-1, and 5-methylhexene-1.

b. Styrene and α-Methylstyrene

Block copolymers have been reported that are composed of segments of polystyrene or poly-α-methylstyrene and segments of polyethylene or polypropylene (see Table 5-15). Langer (C153, C153a)

TABLE 5-15

Styrene- and α-Methylstyrene-Containing Block Copolymers

Block A	Block B	Catalyst system	References
Styrene	Ethylene	$TiCl_3$–$(C_2H_5)_3Al$	C155
	Ethylene–styrene	$TiCl_4$–$(C_2H_5)_3Al$	C73
	Ethylene	C_4H_9Li–N,N,N',N'-tetramethyl-1,2-ethanediamine	C153, C154, C153a
	Propylene	$TiCl_3$–$(C_2H_5)_2AlCl$	C129
	Propylene	$TiCl_3$–$(i\text{-}C_4H_9)_3Al$	C157
Styrene–propylene	Propylene–ethylene	$TiCl_3 \cdot AlCl_3$–$(C_2H_5)_2AlCl$–diglyme	C165
α-Methylstyrene	α-Methylstyrene–ethylene	C_4H_9Li	C160
α-Methylstyrene	Ethylene	Oligomer condensation (see text)	C158, C159
Vinylcyclohexane	Propylene	$TiCl_3$–$(C_2H_5)_2AlCl$	C129, C161–C163
Styrene	Isobutylene	C_4H_9Li + BrRBr + $TiCl_4$	C164
	Isobutylene	Cl—R—Br + $(C_2H_5)_3Al$ + $(C_2H_5)_2AlCl$	C182

disclosed the synthesis of styrene–ethylene block copolymers using a novel $C_4H_9L_i/N,N,N'N'$-tetramethyl-1,2-ethanediamine catalyst system. Styrene was first polymerized at 25°C in the presence of this organometallic–chelating Lewis base catalyst system, followed by the addition of ethylene to the resulting "living" polymer at 40°C and 2000 psi. Fractionation data reportedly indicated that the product was a block copolymer. These results are surprising, since they require the initiation of ethylene polymerization by resonance stabilized styrene carbanions to form the more reactive primary alkyl carbanion. Other workers (C154) reported that the melting point of copolymers synthesized by this technique increased from 120°C to 135°C as the polyethylene content increased from 30% to 70%. Langer also claimed that the polymerization of mixtures of styrene and ethylene with this catalyst system resulted in copolymers of a somewhat "blocky" nature. Both of the above synthetic schemes were reported to be unsuccessful in preparing copolymers of propylene rather than ethylene.

The synthesis of ethylene–styrene block copolymers using a $TiCl_3–Al(C_2H_5)_3$ catalyst has also been reported (C155). The addition of an ethylene–styrene block copolymer to a polyethylene–polystyrene blend was reported to give heat-sealable films of improved mechanical properties (C156).

Another article claims the preparation of a styrene–propylene block copolymer and its conversion to a quaternized crystalline product that is useful as a permselective membrane (C157). The block copolymer was synthesized by sequential monomer addition using a $TiCl_3–(i-C_4H_9)_3Al$ catalyst system, after which the styrene blocks were quaternized by (a) reaction with $CH_3OCH_2Cl–ZnCl_2$ and (b) treatment with N,N'-dimethylpiperazine or trimethylamine.

α-Methylstyrene-ethylene $\{A\text{-}B\}_n$ block copolymers containing short (one to seven monomer units), well-characterized segments were prepared by initiating α-methylstyrene polymerization with sodium metal and condensing the resulting disodio-terminated oligomers with ω-dibromoalkanes (C158). The solution properties (intrinsic viscosity, osmotic pressure, light scattering, and ultracentrifugation) of these materials were investigated. It was concluded from this work that the intrinsic viscosity–molecular weight relationship of regular sequence copolymers are affected not only by the average composition of the copolymer, but also by the segment length in the copolymer molecule. It was also suggested that the conformation of the segments in the copolymer are not always the same as that of the corresponding homopolymer.

Block copolymers similar to the above were also prepared by other workers (C159). These products had segments containing two to seventeen monomer units. The infrared and ultraviolet spectra of the copolymers were determined. Polyethylene segment melting points were observed dilatometrically in copolymers containing polyethylene segments of >10 repeat units.

A patent (C160) claims the preparation of block copolymers containing segments of poly-α-methylstyrene and segments of ethylene–α-methylstyrene random copolymers. These products were reportedly obtained by polymerizing mixtures of α-methylstyrene and ethylene, using n-butyllithium as the catalyst, first at 0°C, 30,000 psi for 1 hour to produce the random copolymer segment, followed by 24 hours at −50°C to −80°C to produce segments containing predominantly α-methylstyrene. The products were claimed to be tough, transparent, flexible thermoplastic resins.

Vinylcyclohexane–propylene block copolymers prepared using a $TiCl_3$–$(C_2H_5)_2AlCl$ catalyst system, were reported to have good impact strength (C129, C161) and thermal stability (C162, C163). The synthesis of isobutylene–propylene block copolymers by the same method have also been claimed (C129).

A styrene–isobutylene block copolymer was reportedly prepared by a combination of two types of catalytic processes (C164). Styrene was polymerized by butyllithium initiation, after which the living polystyrene chains were terminated with α,α-dibromo-p-xylene. The resulting bromine-terminated polystyrene was then reacted with $TiCl_4$, and isobutylene was added to form a polystyrene–polyisobutylene block copolymer. The product, containing 8.5 wt% isobutylene, was claimed to be superior to styrene homopolymer in impact strength.

Another method reported recently (C182) for synthesizing styrene–isobutylene block copolymers, is of especial interest, since the technique is based on cationic polymerization. Compounds containing both tertiary chloride and tertiary bromide groups were used as initiators, e.g., 2-bromo-6-chloro-2,6-dimethylheptane. The more reactive tertiary chloride group preferrentially initiates styrene polymerization in the presence of $(C_2H_5)_3Al$ to produce a polystyrene oligomer with a tertiary bromide end group. The subsequent addition of isobutylene and $(C_2H_5)_2AlCl$ results in the growth of a polyisobutylene segment from the terminal of the polystyrene block. Selective extraction studies indicated that the products contained appreciable quantities of A-B block structures. Of course, considerable amounts of both homopolymers were also formed. This work represents an interesting beginning to the synthesis of hydrocarbon block copolymers via cationic techniques.

c. Dienes, Trienes, and Alkynes

Butadiene–ethylene block copolymers were prepared using a butyl lithium–N,N,N'N'-tetramethyl-1,2-ethanediamine catalyst system in a manner similar to that described earlier for styrene–ethylene copolymers (C153). This patent claims that block copolymers can be produced by feeding the butadiene monomer either before or after the ethylene.

Isoprene–ethylene block copolymers were synthesized using Ziegler catalysts (C166). Modulus–temperature studies revealed transitions at $-95°C$ (due to the polyethylene segment) and $-45°C$ (due to the polyisoprene segment). The transition temperatures were independent of composition. Birefringence studies indicated that the strain–optical coefficient for the polyethylene segment was time-dependent due to orientation of the crystalline phase.

A patent reference claims the synthesis of butadiene–ethylene block copolymers by first polymerizing butadiene via butyllithium initiation, followed by the addition of a calcined mixture of molybdenum oxide–alumina together with ethylene (C167). Isoprene–ethylene and styrene–ethylene block copolymers were also claimed to have been prepared by this technique.

Isoprene–propylene block copolymers (C16), tetramethyl-butadiene–propylene block copolymers (C16), and piperylene–ethylene block copolymers (C168) are referred to in the literature as additives to be blended with polypropylene or ethylene–propylene random copolymers to obtain improved properties. Low-temperature impact resistance is claimed for a block copolymer prepared from propylene, ethylene, and 1,5,9-cyclododecatriene (C169).

The synthesis of copolymers containing blocks of poly-α-olefins and blocks of polyalkynes (e.g., polyethylene–polyacetylene) have been claimed using $TiCl_3$–$(C_2H_5)_3Al$ and related catalyst systems (C16, C170). The resulting products, which presumably have segments containing conjugated double bonds, are said to be intensely colored, high melting, and vulcanizable with sulfur.

d. Polar Monomers

Several copolymers with poly(α-olefin) blocks and segments of polar polymers have been prepared, as is shown in Table 5-16. These in-

TABLE 5-16
Copolymers Containing Polar Blocks

Block A	Block B	Catalyst system	References
4-vinylpyridine or 2-methyl-4-vinyl-pyridine	Propylene	$TiCl_3 \cdot AlCl_3$–$(C_2H_5)_2AlCl$	C175–C177
Methyl methacrylate–propylene oxide	Propylene	$TiCl_3 \cdot AlCl_3$–$(C_2H_5)_2AlCl$–$CH_3O(CH_2CH_2O)_2CH_3$	C178
Dimethylaminoethyl methacrylate	Propylene	$TiCl_3$–$(C_2H_5)_2AlCl$; peroxides	C171, C179
Methyl methacrylate	Ethylene	$TiCl_3$–$Al(C_2H_5)_3$–$Zn(C_2H_5)_2$; ROOH	C172, C173
Acrylonitrile	Propylene	$TiCl_4 \cdot AlCl_3$–$(C_2H_5)_2AlCl$; $(NH_3 + O_2)$	C174, C180
	Ethylene	C_4H_9Li–N,N,N',N'-tetramethyl-1,2-ethanediamine	C153
Allyl chloride	Isoprene	$TiCl_4$–$Al(i$-$C_4H_9)_3$	C148

clude copolymers containing vinylpyridines, acrylic esters, and acrylonitrile. The main purpose for making these copolymers was to improve polyolefin dyeability.

Poisoning of the organometallic catalysts by the polar monomers is a problem in the preparation of this type of block copolymer. Some workers have tried to circumvent this problem by using a combination of Ziegler-type coordination polymerization and free radical polymerization. For example, propylene–acrylic block copolymers were reportedly prepared by polymerizing propylene in the presence of a $TiCl_3$–$(C_2H_5)_2AlCl$ catalyst system, followed by the addition of dimethylaminoethyl methacrylate and benzoyl peroxide (C171). The function of the benzoyl peroxide presumably is to oxidize the metal–polypropylene bond to the metal salt of the polymeric hydroperoxide, which can then initiate the radical polymerization of the acrylate monomer.

A similar approach was taken by Agouri et al. (C172, C173). These workers recognized the limitations imposed by the short chain lifetimes in α-olefin polymerizations and attempted to overcome it. Their approach was to polymerize ethylene with a $TiCl_3$–$Al(C_2H_5)_3$ catalyst system in the presence of diethylzinc, which acts as a chain transfer agent. According to the authors, this procedure results in a larger number of metal-terminated chains, which can subsequently be activated with hydroperoxides to initiate the radical polymerization of methylmethacrylate. The reaction presumably proceeds as shown in Scheme 5-2.

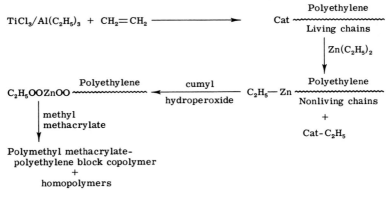

Scheme 5-2

It is obvious that considerable quantities of both homopolymers will result from such a sequence. However, the authors claim that higher yields of block copolymer are obtained with this process than when diethylzinc is not used (e.g., 25% versus 1% "fixed" polymethyl methacrylate).

A two-step process has also been reported to give polypropylene– polyacrylonitrile block copolymers (C174). Propylene was polymerized with a $TiCl_4 \cdot AlCl_3–(C_2H_5)_2AlCl$ catalyst system. The resulting product was treated with ammonia and subsequently with oxygen. Acrylonitrile was then added and polymerized free radically, presumably via initiation by activated end groups.

D. HETEROATOM BLOCK COPOLYMERS

The block copolymers discussed in previous sections contain only carbon atoms in their backbone structures. The block copolymers in this section are characterized by the presence of at least one atom other than carbon in the backbone repeat unit of one or both of the segments of the copolymer. These heteroatoms include oxygen, sulfur, nitrogen, and silicon. They are found in ether, ester, sulfide, amide, N-acylalkylene imine, and siloxane structures. The preparation and properties of these block copolymers are discussed below.

1. Ether–Ether

Several A-B block copolymers have been prepared in which both segments are polyethers. These products were synthesized from epoxide, higher alkylene oxide, aldehyde, or phenol monomers by various polymerization schemes. Table 5-17 lists the compositions that have been reported along with some of the synthesis conditions used.

a. Olefin Oxide–Olefin Oxide

One of the first references to block copolymers in the patent literature is the work of Jackson and Lunsted (D2, D3) on poly(propylene oxide)–poly(ethylene oxide) copolymers. Since poly(propylene oxide) is water-insoluble and poly(ethylene oxide) is water-soluble, these block copolymers have very useful nonionic surfactant properties. They are normally synthesized by the sodium alkoxide initiated polymerization of sequentially added propylene oxide and ethylene oxide in bulk at about 140°C (see Reaction 5-11). Other initiators (e.g., mercaptans, amines) can also be used. These low molecular weight

TABLE 5-17

Ether–Ether A-B Block Copolymers

Block A	Block B	Initiator system	Conditions	References
Propylene oxide	Ethylene oxide	RONa	Bulk, 140°C	D1– D3
Butylene oxide	Ethylene oxide	RONa	Bulk, 140°C	
Styrene oxide	Ethylene oxide	RONa	Bulk, 140°C	
Propylene oxide	Ethylene oxide	$(C_4H_9)_2Zn$	Solution, 110°C	D4, D5
Acetaldehyde	Propylene oxide	$(C_2H_5)_3Al–H_2O$	25°C and −78°C	D7
Formaldehyde	Propylene oxide	$(C_2H_5)_3Al/CH_3\overset{O}{\overset{\|}{C}}CH_2\overset{O}{\overset{\|}{C}}CH_3$	25°C	D8
	Ethylene oxide	$BF_3 \cdot O(C_4H_9)_2$	60°C	D14
Tetrahydrofuran	3,3-Bis(chloromethyl)oxacyclobutane	BF_3	+18, −50	D9, D15
2,6-Diphenylphenol	2,6-Dimethylphenol	$(CH_3)_2N(CH_2)_4N(CH_3)_2$ + CuBr + O_2	25°C	D11– D13, D16
Trioxane	1,3-Dioxolane	BF_3 or $(C_2H_5)_3Al$	—	D17

(~3000) liquid block copolymers are used as nonionic detergents and dispersing agents.

$$RO^{\ominus} Na^{\oplus} + H_2C \overset{O}{\underset{CH_3}{\diagdown}} CH \longrightarrow RO \left(CH_2 - \underset{CH_3}{\underset{|}{CHO}} \right)_a Na^{\oplus}$$

$$\Big\downarrow \begin{array}{l} (1) \ H_2C \overset{O}{-} CH_2 \\ (2) \ H^{\oplus} \end{array} \qquad (5\text{-}11)$$

$$RO \left(CH_2 - \underset{CH_3}{\underset{|}{CHO}} \right)_a \left(CH_2 - CH_2 O \right)_b H$$

The order of monomer sequence has an effect on surface active properties. Polymers of the type

$$RO \left(CH_2 CH_2 O \right)_a \left(CH_2 - \underset{CH_3}{\underset{|}{CHO}} \right)_b H$$

reportedly produce lower cloud points and foam heights than

$$RO \left(CH_2 - \underset{CH_3}{\underset{|}{CHO}} \right)_a \left(CH_2 - CH_2 O \right)_b H$$

structures of comparable composition. On the other hand, the detergency of the two types of material are about the same (D1, D3). Block molecular weight and total copolymer molecular weight also have an effect on surfactant properties. Olefine oxides other than propylene oxide (e.g., butylene oxide, styrene oxide) and mixtures of these have also been used to form the hydrophobic block. The effect of these and other structural variations on surface active properties are discussed in a review article (D1).

In addition to the above-described base-catalyzed bulk synthesis, propylene oxide–ethylene oxide block copolymers have also been prepared by dibutylzinc-catalyzed solution polymerization. Propylene oxide was polymerized first in toluene solution at 110°C, for 26 hours, after which ethylene oxide was introduced and polymerized for 21 hours. The products were solid, water-insoluble materials of reasonably high molecular weight (0.5 reduced viscosity in benzene) (D4).

Skoulios et al. (D5, D6) found that high-concentration solutions of propylene oxide–ethylene oxide block copolymers in solvents preferential to one of the blocks gave birefringent gels. These gels exhibited the anisotropic texture of mesomorphic phases. X-Ray investigation of the gels indicated the existence of domains about 100–500 Å in size. The configuration of the macromolecular chains are discussed in these papers.

b. Aldehydes

A polyacetaldehyde–poly(propylene oxide) block copolymer was reported to have been prepared by means of a triethylaluminum–water cocatalyst system (D7). Propylene oxide was polymerized first for 1.5 minutes at room temperature, after which the temperature was reduced to −78°C to inhibit further polymerization of unreacted propylene oxide. Acetaldehyde was then added and selectively polymerized at −78°C for a period of 45 hours (Reaction 5-12). The products, which were obtained in low yields ($\leq 25\%$), contained about 90% polyacetaldehyde and had intrinsic viscosities of about 3 (in methyl ethyl ketone at 25°C). They were claimed to have block structures based on solubility and hydrolysis behavior. A polyformaldehyde–poly(propylene oxide) block copolymer was also claimed to have been synthesized using a $(C_2H_5)_3Al$–acetylacetone cocatalyst system at a reaction temperature of 25°C (D8). This product was a white powder containing 80% polyformaldehyde and had a reduced viscosity of 0.7.

$$
\begin{array}{c}
H_2C\!\!-\!\!CH \xrightarrow{\ AlR_3\ } R\!\!\left(\!CHCH_2O\!\right)_a\!\!-\!\!AlR_2 \\
\ \ \ \ \ \ \ \ CH_3 \qquad\qquad\quad\ \ CH_3
\end{array}
$$

$$\downarrow CH_3CHO \qquad\qquad (5\text{-}12)$$

$$
R\!\!\left(\!CHCH_2O\!\right)_a\!\!\left(\!CHO\!\right)_b\!\!-\!\!AlR_2 \\
\ \ \ \ \ \ CH_3 \qquad\ \ \ CH_3
$$

c. Higher Alkylene Oxides

Saegusa et al. (D9) reported the preparation of a block copolymer from tetrahydrofuran and 3,3-bis(chloromethyl)oxacyclobutane by what was called a cationic living polymerization. Tetrahydrofuran was polymerized first, using a $BF_3 \cdot$ epichlorohydrin catalyst, followed by addition of the oxacyclobutane monomer. Under the conditions used (−50°C), chain transfer between the propagating oxacyclobutane growing chain and the poly(tetrahydrofuran) segments was claimed to occur only to a very small extent. Solubility, infrared spectral, and differential scanning calorimetry data were offered as proof of the block structure of the product. The infrared spectrum showed bands indicating crystallinity and the differential scanning calorimetry data showed endothermic peaks at 29°C and 169°C, corresponding to the melting points of the tetrahydrofuran and oxacyclobutane homopolymers.

d. Phenylene Oxides

2,6-Disubstituted phenols are polymerized by oxidative coupling (D10) in the presence of oxygen and a copper–amine catalyst. A 2,6-diphenylphenylene ether–2,6-dimethylphenylene ether block co-polymer was prepared by polymerizing 2,6-diphenylphenol in the presence of tetramethylbutanediamine–cuprous bromide catalyst and oxygen in benzene solution at 25°C. After 3 hours, 2,6-dimethylphenol was added and the reaction was continued for another hour (see Scheme 5-3) (D11–D13). This procedure resulted in a block

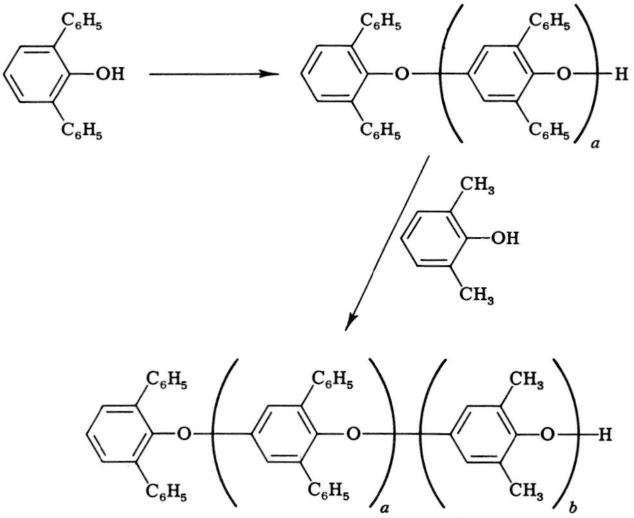

Scheme 5-3

copolymer structure, as indicated by NMR data. However, reversing the procedure—i.e., polymerizing the dimethylphenol rapidly first, followed by the relatively slower polymerization of the diphenyl-phenol—resulted in a random copolymer. This was due to a re-distribution side reaction that occurred to a significant extent during the second slow polymerization step. The block copolymer, which contained 45 mole% of dimethylphenylene ether, was amorphous with a T_g at 227°C (the homopolymer T_g values are 225°C for the dimethyl polymer and 230°C for the diphenyl derivative). The diphenyl-phenylene ether block crystallized at 290°C, but decomposed at 450°C before reaching the 480°C melting point characteristic of the diphenyl homopolymer.

2. Ether–Vinyl

This section is comprised of block copolymers containing one segment of a polyether and one segment of a vinyl polymer such as polystyrene. The structure and properties of ethylene oxide–styrene block copolymers have received the most attention in this category. Aldehyde–vinyl block copolymers and miscellaneous epoxide–vinyl block copolymers were also investigated, but less extensively. The various systems are discussed below and listed in Table 5-18.

a. Olefin Oxide–Styrene

Poly(ethylene oxide)–polystyrene block copolymers were synthesized by a two-step process (D18, D19). In the first step, styrene is polymerized by, for example, cumylpotassium initiation in tetrahydrofuran solution at 0°–80°C for about an hour. Then ethylene oxide is added to the living polystyrene at −80°C to convert the carbanions to alkoxide end groups. The relatively slow ethylene oxide polymerization is then continued for 1–2 days at 25°–75°C in order to reach high conversion (Reaction 5-13). The rate of polymerization of the ethylene oxide is proportional to the size of the counterion (i.e., Cs > K > Na > Li). Block copolymer yields, after solvent extraction to remove homopolymers, were ≥80%. Block molecular weights ranged from 10,000 to 50,000.

The properties of ethylene oxide–styrene block copolymers are very interesting because they contain a crystalline block (polyethylene oxide) and an amorphous block (polystyrene). Differential scanning calorimetry indicated a crystalline melting point (T_m) of 53°–56°C due to the poly(ethylene oxide)segment, and a glass transition temperature (T_g) of 96°–98°C for the polystyrene blocks (D19). Thermal data also indicated that the degree of crystallinity of these block copolymers was proportional to the ethylene oxide content (D19). Franta (D20) carried out light-scattering, birefringence, and dielectric constant measurements on low-concentration solutions of poly(ethylene oxide)–polystyrene block copolymers in solvents preferential for the

TABLE 5-18

Ether–Vinyl A-B Block Copolymers

Block A	Block B	Initiation system	References
Ethylene oxide	Styrene	$C_9H_{11}K$	D18, D19, D36, D37, D112
Formaldehyde	Styrene	C_4H_9Li	D42, D48–D50
	Methyl methacrylate	9-Fluorenyl-Li	D42, D48, D49
	Butyl Methacrylate	9-Fluorenyl-Li	D42, D48, D49
	Acrylonitrile	C_4H_9Li	D42, D48, D49
	Isoprene	C_4H_9Li	D42, D48, D49
	N,N-Di-n-butyl-acrylamide	C_4H_9Li	D42
Formaldehyde–phenyl isocyanate random copolymer	Styrene	C_4H_9Li	D42
Acetaldehyde	Styrene	C_4H_9Li	D51
	α-Methylstyrene	C_4H_9Li	D51
	Methyl methacrylate	C_4H_9Li	D51
	Isoprene	C_4H_9Li	D51
	Butadiene	C_4H_9Li	D51
n-Butyraldehyde	Styrene	C_4H_9Li	D51
	2-Methyl-5-vinyl-pyridine	C_4H_9Li	D51
Isobutyraldehyde	Styrene	C_4H_9Li	D51
	2-Methyl-5-vinyl-pyridine	C_4H_9Li	D51
Ethylene oxide	Ethylene	Telomers	D43
Acetaldehyde	Ethylene	$TiCl_4$–$(C_2H_5)_3Al$	D44
Epichlorohydrin	Styrene	$BF_3 \cdot O(C_2H_5)_2$	D45
1,3-Dioxolane	1,3-Dioxolane–styrene	$(C_2H_5)_3OBF_4$	D46
1,3-Dioxolane–3,3-bis(chloro-methyl)oxetane	1,3-Dioxolane–styrene	$(C_2H_5)_3OBF_4$	D46, D47
Tetrahydrofuran	Styrene	Cationic–anionic oligomer coupling	D52

$$CH_2{=}CH\text{(phenyl)} \;+\; \text{(phenyl)}C(CH_3)_2{}^{\ominus}\,K^{\oplus} \;\longrightarrow\; \text{(phenyl)}C(CH_3)_2{\sim\sim}CH_2{-}CH^{\ominus}\text{(phenyl)}\,K^{\oplus}$$

$$\big\downarrow\; H_2C\overset{O}{\diagdown\!\!\diagup}CH_2$$

$$\text{(phenyl)}C(CH_3)_2{\sim\sim}CH_2{-}CH\text{(phenyl)}{-}CH_2{-}CH_2O^{\ominus}\,K^{\oplus} \tag{5-13}$$

$$\big\downarrow\; H_2C\overset{O}{\diagdown\!\!\diagup}CH_2$$

$$\text{(phenyl)}C(CH_3)_2{-}\big({-}CH_2{-}CH\text{(phenyl)}{-}\big)_a\big({-}CH_2{-}CH_2O{-}\big)_b{-}H$$

polystyrene segments. These solutions were found to contain anisotropic aggregates the size of which are temperature dependent and that
bear a permanent dielectric moment (D20). The aggregates were
proposed to consist of poly(ethylene oxide) microcrystallites surrounded and protected by solvated polystyrene blocks.

The nucleation and growth of single crystals of poly(ethylene
oxide)–polystyrene block copolymers from dilute solution using
polystyrene-preferred solvents was investigated microscopically
(D21–D23), dilatometrically (D21), and by low-angle X-ray diffraction
(D22). The conclusion was drawn that the copolymers crystallized as
square platelets consisting of poly(ethylene oxide) folded-chain lamellae sandwiched between surface layers of amorphous polystyrene.

Concentrated solutions of poly(ethylene oxide)–polystyrene block
copolymers in solvents preferential to one of the segments have been
investigated microscopically and by X-ray diffraction by Skoulios et al.
(D24–D27). Under these conditions, mesomorphic phases are formed,
which are cylindrical, lamellar or spherical, depending upon the con-

centration. Sadron *et al.* (D29–D31), interestingly, were able to preserve these microstructures by using polymerizable monomers, such as acrylic acid or methyl methacrylate, as "solvents." Polymerization of the "solvents" (initiation with mercury-vapor lamps) resulted in solid systems in which the mesomorphic block copolymer gels were imbedded, and thus preserved.

Microscopic and X-ray studies showed the bulk morphology of solution-cast films of poly(ethylene oxide)–polystyrene block copolymers to be dependent upon copolymer composition and casting solvent (D32, D33). The morphology is dictated by the block component present in the highest concentration, i.e., the one that forms the "continuous phase." The effect of the casting solvent is due to the fact that the least soluble segment will "precipitate" from solution before the more soluble segment. Solvents that are not preferential, i.e., a good solvent for both segments, will give an intermediate morphology. The dielectric constant of block copolymer films cast from solvent systems that are good solvents for both blocks was much higher than those observed for films cast from solvent systems preferential to the polystyrene segment (D34).

These variables, and solvent evaporation rate as well, also have an effect on the surface morphology of cast films, as determined by scanning electron microscopy (D35). The films displayed birefringence, the extent of which was composition dependent. Birefringence was evident even at temperatures as high as 250°C.

Gervais and Gallot (D36, D37) determined the phase diagrams of poly(ethylene oxide)–polystyrene block copolymers in the presence of solvents preferential for either of the segments by investigating these compositions via low-angle X-ray scattering and differential scanning calorimetry.

A dilatometric study of the phase transition behavior of poly(ethylene oxide)–polystyrene block copolymers of varying composition indicated that the blocks separate into incompatible domains, at least one dimension of which is comparable to the molecular dimensions of the blocks (D38, D39).

The viscoelastic properties of these copolymers are interesting (D40). The storage shear modulus of an A-B block copolymer was two orders of magnitude higher than that of a physical blend of the homopolymers of comparable composition and molecular weight. This was interpreted as meaning that the block copolymers exist in aggregated form even in the melt. The melt viscosity and melt elasticity of the block copolymers were higher than those of either homopolymer of similar molecular weight or blends of the homopolymers (D41).

b. Aldehyde–Vinyl

Copolymers containing blocks of polyformaldehyde or polyacetaldehyde together with vinyl polymer segments have been prepared by an anionic process similar to that used to synthesize poly(ethylene oxide)–polystyrene block copolymers.

Polyformaldehyde–polystyrene block copolymers have been synthesized by polymerizing styrene via butyllithium initiation in benzene solution for 1 hour at 30°C, followed by the addition and polymerization of formaldehyde at 10°C(D42). These block copolymers were capped with acetic anhydride to obtain stable products (Reaction 5-14). Solvent extraction data indicated the product to have a block copolymer structure. Good properties were observed when the products (which were high in polyformaldehyde content) had reasonably high molecular weight (>1.0 inherent viscosity). Similar procedures were used to synthesize other block copolymers of formaldehyde. These compositions are listed in Table 5-18.

$$(5\text{-}14)$$

Block copolymers of acetaldehyde and butyraldehyde and various vinyl monomers are also reported to have been prepared by a similar procedure. These copolymers are also listed in Table 5-18.

c. Miscellaneous

Poly(ethylene oxide)–polyethylene block copolymers were reported to have been synthesized from hydroxyl-terminated polyethylene telomers made by peroxide initiation of ethylene in the presence of oxygen-containing telogens. These were converted to the sodium alkoxide form and used to initiate the polymerization of ethylene oxide (D43). The products, which were mixtures of block copolymers and homopolymers, were claimed to be useful in aqueous emulsion coating applications.

A polyacetaldehyde–polyethylene block copolymer (D44) was prepared by sequential polymerization of ethylene and acetaldehyde via TiCl$_4$–(C$_2$H$_5$)$_3$Al catalysis at $-78°C$. The product was concluded to have a block structure based on the formation of an oily polyolefin upon acid hydrolysis (which presumably cleaved the polyaldehyde). The thermal stability of the product was said to be superior to that of a homopolymer blend.

The cationic polymerization of a mixture of epichlorohydrin and styrene monomers in the presence of BF$_3$·O(C$_2$H$_5$)$_2$ was claimed to result in block copolymeric products (D45). This claim was based on the observation that epichlorohydrin has a higher monomer reactivity than styrene. However, the structure of these materials was not determined by product characterization.

The cationic polymerization of 1,3-dioxolane with (C$_2$H$_5$)$_3$OBF$_4$ was reported to give a "living" polymer system (D46). Addition of styrene was claimed to result in a product containing one poly(dioxolane)block and another segment of a styrene–1,3-dioxolane random copolymer (D47). Similarly, random polymerization of a mixture of 1,3-dioxolane and styrene followed by the addition of 3,3-bis(chloromethyl)oxetane was reported to result in a terpolymer containing a segment of 1,3-dioxolane–styrene random copolymer and a segment of 1,3-dioxolane–3,3-bis(chloromethyl)oxetane random copolymer (D46, D47).

3. Lactones

This section discusses A-B block copolymers containing one block of a polylactone and a block of another composition, such as a polyether, polyamide, or vinyl polymer. The various compositions are shown in Table 5-19.

a. Lactone–Ether

A patent reference (D53) reports the preparation of a poly(ε-caprolactone)–poly(propylene oxide) block copolymer. Propylene oxide was polymerized in toluene solution in the presence of dibutylzinc for 65 hours at 108°C, after which ε-caprolactone was added and the reaction continued for 79 hours at the same temperature. The product was a white solid of low molecular weight (0.3 reduced viscosity.) Similar copolymers were also made from methyl-ε-caprolactone and ethylene oxide or butylene oxide.

Another patent (D54) claims block polymers from β-propiolactone and tetrahydrofuran via SbCl$_5$ initiation. Tetrahydrofuran was polymerized first, in bulk at 0°C, after which the lactone was added

TABLE 5-19

Polylactone-Containing Block Copolymers

Block A	Block B	Initiator system	Conditions	References
ε-Caprolactone	Propylene oxide	$(C_4H_9)_2Zn$	108°C	D53
Methyl-ε-caprolactone	Ethylene oxide	$(C_4H_9)_2Zn$	108°C	D53
	Butylene oxide	$(C_4H_9)_2Zn$	108°C	D53
β-Propiolactone	Tetrahydrofuran	$SbCl_5$	0°C	D54
Pivalolactone	ε-Caprolactam	Na + N-Acetyl-caprolactam	—	D55
	Ethylene oxide	$(C_4H_9)_4NOH$	—	D63
Caprolactone	Butadiene	C_4H_9Li	70°C	D56
	Isoprene	—	—	D58
	Methyl methacrylate	—	—	D58
	Styrene	C_4H_9Li	25°C or 70°C	D59, D60
Pivalolactone	Styrene	C_4H_9Li	25°C	D57
	Styrene	NaOOC-terminated polystyrene	—	D64
β-Propiolactone	Styrene	Amine-terminated polystyrene telomer	130°C; 50°C	D61
Pivalolactone	Acrylonitrile	NaCN	<20°C	D62
	DL-α-Methyl-α-n-propyl-β-Propiolactone	$C_6H_5COO^-N^+(C_6H_{13})_4$	35°C	D65

and polymerization continued for 16 hours. The product was moldable and was claimed to be useful in film, fiber, adhesive, and coating applications.

b. Lactone–Amide

A pivalolactone–caprolactam block copolymer was prepared (D55) by initiating the bulk polymerization of caprolactam with sodium catalyst and N-acetylcaprolactam initiator at 240°C followed by the addition of pivalolactone. The product, which had an intrinsic viscosity of 0.5, was claimed to improve the dyeability of nylon.

c. Lactone–Vinyl

Several block copolymers have been reported (D56–D60) that contain a segment of polycaprolactone or polypivalolactone and a segment of polybutadiene, polyisoprene, poly(methyl methacrylate), or polystyrene. These copolymers, which are listed in Table 5-18, were prepared by initiating polymerization of the lactone with the "living" anionically polymerized vinyl polymer, with or without the intermediate step of capping the vinyl polymer with ethylene oxide to form an alkoxide end group before initiation of the lactone polymerization.

A β-propiolactone–styrene block copolymer (D61) was claimed to have been prepared by first initiating styrene polymerization with azobis(isobutyronitrile) in the presence of CCl_4. The resulting chlorine-terminated polystyrene was converted to an amine-terminated telomer by reaction with liquid ammonia, and this was used to initiate polymerization of the lactone. Due to the method of preparation, the block structure of this product is questionable.

Polyacrylonitrile–polypivalolactone block copolymers have been reportedly prepared by initiating acrylonitrile polymerization with NaCN in dimethylformamide solution at <20°C, followed by the addition of the lactone monomer (D62). The products are claimed to be useful in fiber applications.

4. Sulfides

In this class of A-B block copolymers, one or both of the blocks are prepared from episulfides or trimethylene sulfides (thicyclobutanes). The second segment of these block copolymers is another poly(alkylene sulfide), a polyether, or a vinyl polymer. The compositions that have been reported are listed in Table 5-20. Coordination, anionic, and cationic catalysts have been used.

a. Olefin Sulfide–Olefin Sulfide

A diethylzinc–water coordination cocatalyst system was used to synthesize ethylene sulfide–propylene sulfide block copolymers (D66, D67). The products were elastomeric, with melting points of >145°C. These materials were extrudable at 185°C and displayed good hardness, hysteresis, and permanent set properties. A boron trifluoride–etherate catalyst has also been used to prepare ethylene sulfide–propylene sulfide block copolymers cationically (D68). A sodium carbazyl anionic initiator was used to prepare an isobutylene sulfide–propylene sulfide block copolymer (D69).

b. Olefin Sulfide–Alkylene Oxide

Triethyloxonium tetrafluoroborate was used to make block copolymers of tetrahydrofuran and either propylene sulfide, trimethylene sulfide, or 3,3-dimethyltrimethylene sulfide (D75). The tetrahydrofuran was polymerized first, after which the sulfide

TABLE 5-20

Polyalkylene Sulfide-Containing A-B Block Copolymers

Block A	Block B	Initiator system	Conditions	References
Ethylene sulfide	Propylene sulfide	$(C_2H_5)_2Zn–H_2O$	27°C	D66, D67
Isobutylene sulfide	Propylene sulfide	Na carbazyl	35°C	D69

Block A	Block B	Initiator system	Conditions	References
Ethylene sulfide	Propylene sulfide	$BF_3 \cdot O(C_2H_5)_2$	20°C	D68
Propylene sulfide	Tetrahydrofuran	$(C_2H_5)_3OBF_4$	−30°C and +20°C	D75
Trimethylene sulfide	Tetrahydrofuran	$(C_2H_5)_3OBF_4$	−30°C and +20°C	D75
3,3-Dimethyltrimethylene sulfide	Tetrahydrofuran	$(C_2H_5)_3OBF_4$	−30°C and +20°C	D75
Ethylene sulfide	Ethylene oxide	K carbazyl	—	D70
1,1-Dimethylethylene sulfide	Ethylene oxide	K carbazyl	—	D70
Ethylene sulfide or propylene sulfide	Styrene	C_4H_9Li	−78°C and room temp.	D71, D73
Ethylene sulfide or propylene sulfide	α-Methylstyrene	C_4H_9Li	−78°C and room temp.	D71
Ethylene sulfide or propylene sulfide	Methyl methacrylate	C_4H_9Li	−78°C and room temp.	D71
Propylene sulfide	Isoprene	C_4H_9Li	−78°C and room temp.	D72, D72a, D113
	2-Vinylpyridine	Diphenylmethyl Na	20°C	D74

monomer was added. The crossover reaction was stated to be essentially nonreversible, but the products were said to be mixtures of block copolymer and homopolymers.

Ethylene oxide–ethylene sulfide block copolymers and ethylene oxide–1,1-dimethylethylene sulfide block copolymers were synthesized in good yield via initiation with potassium carbazyl (D70). The ethylene oxide segment was formed first. No block copolymer was obtained with 1-methylethylene sulfide due to chain transfer.

c. Olefin Sulfide–Vinyl

Block copolymers containing one segment of ethylene sulfide or propylene sulfide and another of styrene, α-methylstyrene, methyl methacrylate, or isoprene were synthesized by butyllithium initiation (D71, D72, D113). The vinyl monomers were formed first at low temperatures, followed by room temperature polymerization of the sulfide monomer. Amorphous, atactic homopoly(propylene sulfide) has a T_g at −40°C and homopoly(ethylene sulfide) has a crystalline melting point at 203°C. Differential thermal analysis of the block copolymers showed transitions near these temperatures, indicating microincompatibility. These observations, together with extraction results, were given as proof of the block structure of these products. The ethylene sulfide copolymers were insoluble materials. The propylene sulfide compositions were soluble and displayed the poor stability (even at room temperature), which is characteristic of propylene sulfide homopolymer.

5. Amides and Imines

The amide block copolymers of this category contain either one or two polyamide segments, synthesized by the anionic polymerization of amino acid N-carboxyanhydrides, isocyanates, or lactams. The N-acylalkylene imine block copolymers are prepared from cyclic imino ether monomers. Note that a review of a polyamide chemistry is available (D76).

a. Amide–Amide

Polyamide block copolymers have been reportedly prepared by many investigators through the sequential addition of amino acid N-carboxyanhydride monomers (D77–D84). Sequential polypeptides were also prepared via treatment of a dipeptide pentachlorophenyl ester hydrobromide with tertiary base (D85, D86). Extensive, elegant investigation of the derived polypeptides have been reported as summarized in Table 5-21. However, relatively little *molecular* characterization of block copolymer structural integrity has been reported. This important aspect of these systems does not appear to have received adequate attention. Relatively little characterization work to differentiate homopolymer blends from true segmented structures can be cited (D82). NMR and optical spectroscopy studies of benzyl

TABLE 5-21

Typical Polypeptide Block Copolymer Studies

Subject	Reference
Review of N-carboxyanhydride polymerizations	D77, D89, D90
Synthesis in heterogeneous systems	D78
Conformational studies by Raman and infrared spectroscopy	D79–D81, D83
Synthesis	D82, D85, D86, D91, D92, D97
Commercial production	D86
Thermogravimetric analysis	D88
NMR and optical spectroscopy	D87

L-aspartate–benzyl L-glutamate systems have shown that when the glutamate blocks were synthesized first, unreacted glutamate could cause at least a portion of the aspartate blocks to switch from left-handed to right-handed helices (D87). Thermogravimetric analysis of several polypeptide materials (D88) showed that the degradation rate of random copolymers was quite smooth and lay between the corresponding curves for the two homopolymers. Degradation of the "block" copolymers was quite similar to the homopolymers.

Enantiomorphic block copolymers of γ-methyl glutamate have been produced commercially (D84). It was noted that the conformational change of the stretched block copolymer films was low, compared with that of the corresponding homopolymers.

Due to the lack of characterization data, other than solubility and infrared spectra (D91, D97), considerable emphasis has been placed on the polymerization mechanism in inferring product structures. The generally accepted mechanism for primary amine ("weak base") initiated systems is shown in Reaction 5-15. The N-carboxyanhydride

$$
C_4H_9NH_2 + \underset{HN-CO}{\overset{\overset{\displaystyle R}{|}}{\underset{|}{HC-CO}}}\!\!\!\!\!\!\!\!\!\!\diagdown O \longrightarrow C_4H_9NH\!\!-\!\!(COCHNH)_a\!\!-\!\!H + CO_2
$$

$$
C_4H_9NH\!\!-\!\!(COCHNH)_a\!\!-\!\!(COCHNH)_b\!\!-\!\!H \quad + \quad CO_2 \tag{5-15}
$$

(NCA) adds to chains containing terminal amine units. Addition of a second monomer produces sequential polypeptides. Polymerizations initiated by strong bases, e.g., tertiary amines and alkali metal alkoxides, were thought (D89) to involve addition of anions derived from NCA molecules to chain ends containing N-acyl-NCA moieties. Recent mechanistic studies by Seeney and Harwood (D93) have provided strong evidence that these mechanisms may be incorrect and that the propagating species in *both* strong base- and weak base-initiated polymerizations is actually a carbamate ion. Thus, model compounds of the type:

$$(C_2H_5)_2NCOCHRNHCOO^{\ominus} (C_2H_5)_2NH_2^{\oplus}$$

were clearly identified (D93). Polymer spectra were consistent with this finding. Further proof included radiotracer studies of end groups and the role of carbon dioxide or aprotic counter ions in stabilizing the carbamate species.

Further detailed studies of the mechanism of sequential polypeptide formation are needed. Furthermore, detailed molecular characterization data similar to those available for styrene–diene systems are necessary before meaningful structure–property relationships can be established for these important biopolymers.

b. Amide–Vinyl

Block copolymers of isocyanates and vinyl monomers have been synthesized via anionic polymerization (D98). This was accomplished by initiating the polymerization of the vinyl monomer (e.g., styrene, isoprene, or methyl methacrylate) with n-butyllithium in toluene–tetrahydrofuran solution at −50°C, followed by the addition of the isocyanate monomer (e.g., n-butyl isocyanate, toluene diisocyanate, or 1-methyl-2,4-diisocyanatocyclohexane), as shown in Reaction 5-16.

$$\text{(5-16)}$$

The low polymerization temperature is necessary to minimize the formation of the cyclic trimer of the isocyanate (i.e., isocyanurates). In addition to the use of monoisocyanates, as illustrated in Reaction 5-16, diisocyanates in which one —NCO group is sterically hindered were also used in the preparation of block copolymers (Reaction 5-17). Further reactions are possible through the pendant —NCO groups.

$$(5\text{-}17)$$

These products were concluded to have block structures based on infrared spectra and solubility behavior. The molecular weight distribution (and presumably block purity) as determined by gel permeation chromatography, was narrow for those copolymers containing n-butyl isocyanate segments, but somewhat broader for those containing diisocyanate segments (possibly due to some branching and/or cross-linking in the latter systems).

Polystyrene–nylon 6 block copolymers have been prepared via suitable termination of living polystyrene followed by the initiated anionic polymerization of ε-caprolactam onto the polystyrene end groups (D94). Thus, butyllithium-initiated styrene was end-capped with an excess of bisphenol-A dichloroformate. Addition of sodium ε-caprolactam catalyst and ε-caprolactam monomer to the purified chloroformate-terminated polystyrene generated a block structure as evidenced by selective solvent treatments. Poly(ethylene oxide)–nylon 6 copolymers have also been prepared via the chloroformate end capping–ε-caprolactam technique (D95).

c. N-Acylalkylene imines

A block copolymer of poly(N-phenethylaziridine) and poly(tetrahydrofuran) was reported to be formed by initiating tetrahydrofuran polymerization with triethyloxonium tetrafluoroborate followed by the addition of the aziridine monomer (D75). However, no proof of block structure was presented, and there were considerable quantities of homopolymers formed.

A block copolymer was reportedly prepared that contains two different acylalkylene imine segments (D96). The corresponding cyclic imino ethers were used as the monomers for this synthesis, and polymerization was initiated with methyl p-toluene sulfonate, as shown in Scheme 5-4. The properties of the two segments are quite different. Poly(N-lauroylethyleneimine) has a crystalline melting point at 150°C and is water insoluble, while poly(N-acetyltrimethyleneimine) is amorphous, with a T_g at 30°C, and is water soluble. Block copolymers were reported to have melting points ranging from 79°C to 120°C with increasing block length. The critical surface tension and static electricity characteristics were found to depend on block length and copolymer composition.

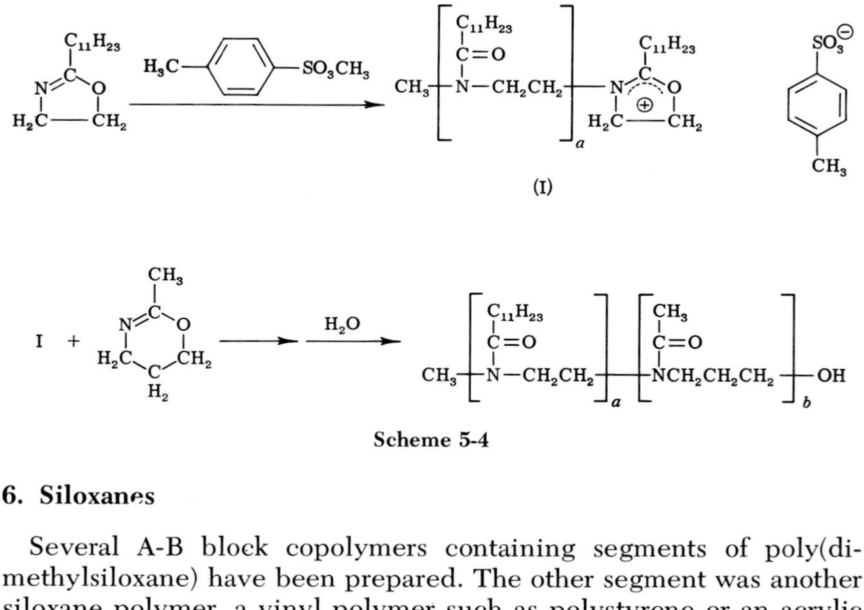

(I)

Scheme 5-4

6. Siloxanes

Several A-B block copolymers containing segments of poly(dimethylsiloxane) have been prepared. The other segment was another siloxane polymer, a vinyl polymer such as polystyrene or an acrylic polymer, or a polyether block. More will be said about polyether–poly(dimethylsiloxane) block copolymers in Chapter 7. The A-B block copolymers reported are listed in Table 5-22.

a. Siloxane–Siloxane

Poly(dimethylsiloxane)–poly(diphenylsiloxane) block copolymers were prepared by the initiation of octamethylcyclotetrasiloxane

TABLE 5-22

Polysiloxane-Containing A-B Block Copolymers

Block A	Block B	Initiator system	Conditions	References
Dimethylsiloxane	Diphenylsiloxane	$(CH_3)_3SiOK$	−40 to	D99
	Styrene	C_4H_9Li	+100°C	D100, D101 D103, D110, D111
	Styrene	AIBN + condensation	80°C	D107, D108
	Methyl methacrylate	C_4H_9Li	—	D100
	Methyl methacrylate	$Li^+(C_6H_5)_2C^-\!\!—O^-Li^+$	—	D105, D106
	Acrylonitrile	C_4H_9Li	—	D100
	Ethylene oxide– propylene oxide	Condensation	150°C	D109
3,3,3-Trifluoro- propylmethyl- siloxane	Styrene	C_4H_9Li	—	D104

polymerization with $(CH_3)_3SiO^-K^+$ in tetrahydrofuran solution to form the poly(dimethylsiloxane) block, followed by the addition of hexaphenylcyclotrisiloxane to form the poly(diphenylsiloxane) segment (D99). Low concentrations of tetravinyltetramethylcyclotetrasiloxane in the monomer feed of the first step led to the incorporation of vinyl groups that allowed the products to be cross-linked to form elastomers with good oil resistance.

b. Siloxane–Vinyl

Block copolymers of styrene and dimethylsiloxane have reportedly been prepared by initiating the polymerization of the cyclic tetramer of dimethylsiloxane with "living" polystyrene (Reaction 5-18) (D100). Similar products were also prepared using methyl methacrylate or acrylonitrile rather than styrene. The polystyryl anion is capable of initiating siloxane polymerization. The reverse procedure, initiation of styrene polymerization with silanolate-terminated polysiloxanes, was not successful due to the relatively low base strength of the silanolate anion.

$$\text{Styrene} \xrightarrow{\text{C}_4\text{H}_9\text{Li}} \underset{\overset{|}{a}}{\left(\text{C}_4\text{H}_9 \left[\text{CH}_2\text{CH} \right] \overset{\ominus}{} \text{Li}^{\oplus} \right)} \tag{5-18}$$

$$\downarrow [(\text{CH}_3)_2\text{SiO}]_4$$

$$\text{C}_4\text{H}_9 \left[\text{CH}_2\text{CH} \right]_a \left[(\text{CH}_3)_2\text{SiO} \right]_b \overset{\ominus}{} \text{Li}^{\oplus}$$

Well-defined block copolymer products are not obtained when cyclic tetramer is used as the siloxane monomer. This is due to the slow propagation rate of the tetramer, which allows a side reaction, cleavage of the linear siloxane chains, to occur to a significant extent. This causes redistribution of the siloxanes and results in the formation of siloxane homopolymer and the reformation of cyclic tetramer and other cyclic species.

This situation is changed dramatically when the cyclic trimer is used as the dimethylsiloxane monomer. The propagation rate of the trimer is much faster than that of the tetramer, perhaps because of a greater degree of ring strain in the former. Well-defined polystyrene–poly(dimethylsiloxane) block copolymers have been synthesized by initiating cyclic trimer polymerization with polystyryl anion and lithium counterion (D101). Especially good results were obtained when a promoter such as tetrahydrofuran was used. Since the rate of propagation using cyclic trimer is greater than the rate of depolymerization or interchange, products were obtained that were contaminated only slightly by homopolymers (<7%) and cyclic tetramer (<1%). The copolymers prepared ranged in molecular weight from 7000 to 300,000, and narrow molecular weight distributions were obtained.

Solutions of these block copolymers displayed a blue iridescence above 5% concentration and were thixotropic above 15% concentration. Both features are due to aggregation of the blocks into common domains in solution.

Casting solvent had an effect on film morphology, especially at intermediate compositions. A polystyrene-preferred solvent such as bromobenzene gave hard brittle films due to polystyrene phase continuity. On the other hand, soft, rubbery films were obtained from the same polymer using cyclohexane, a siloxane-preferred solvent, which produced films with a continuous polysiloxane phase. The resulting thin films also displayed iridescence. Electron microscopic investigation of the films indicated domains of 60–300 Å in size in the form of spheres, rods, and lamellae, depending on the conditions used to cast the film. Special staining techniques, such as those used in styrene–diene block copolymer electron microscopic investigations, were not necessary with these polymers due to greater difference in the degree of electron absorption and scattering of the polystyrene and polysiloxane domains (D111).

Other election micrograph studies by Saam *et al.* (D102, D111) indicated that cyclohexane (a siloxane-preferred solvent) gave cell-like structures of polystyrene surrounded by the darker regions of siloxane, while monomeric styrene (a polystyrene-preferred solvent) gave the opposite arrangement. Toluene, which solvates both blocks, gave a "spaghettilike" structure interpreted as a rod–sphere hybrid structure. The spaghettilike morphology was observed over a broad range of composition (18–68% polystyrene content) in copolymers of constant molecular weight. On the other hand, the morphology changed as a function of block copolymer molecular weight at constant composition. Figure 5-10 shows the election micrographs obtained for copolymer films (cast from toluene) of $70 \pm 2\%$ polystyrene content in which the copolymer $\overline{M}_n$ is 48,000, 29,000, and 7000. The last appears to be close to the lower limit of copolymer molecular weight necessary for phase separation and displayed ill-defined spheres. The intermediate $\overline{M}_n$ copolymer displayed a lamellar morphology, while the higher $\overline{M}_n$ composition exhibited the spaghettilike structure. This behavior differs from that observed in styrene–diene block copolymers. It was proposed that this difference may be due to a high interfacial contact energy (γ) in the styrene–siloxane block copolymers due to the solubility or compatibility of the two blocks. This would influence the molecular weight dependence of the end-to-end distance of the chains within the micelles and consequently alter entropy requirements in placing a copolymer chain within a particular type of micellar structure. The γ value for a molten polystyrene–poly(dimethylsiloxane) interface was estimated to be 12 dynes/cm compared to 1 dyne/cm for polystyrene–polybutadiene. Micelle dimensions in the styrene–siloxane copolymers of 7000–48,000 $\overline{M}_n$ were found to be in the range of 100–300 Å.

These polymers were claimed to have surfactant properties. In addition, termination of the silanolate anion with a vinyl silane, as shown in

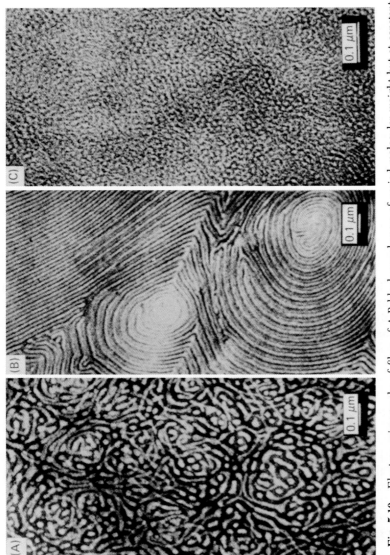

Fig. 5-10. Electron micrographs of films of A-B block copolymers of varied molecular weight but at constant composition. (A) $\overline{M}_n$ 47,600, 68 wt% polystyrene; (B) $\overline{M}_n$ 28,600, 70.9 wt% polystyrene; (C) $\overline{M}_n$ 7100, 71.6 wt% polystyrene. Soft segment = poly(dimethylsiloxane) (D102).

Reaction 5-19, gave a polymer which could be cured with peroxides to produce crosslinked elastomers.

$$R \text{--} (CH_2 \text{--} CH)_a \text{--} [(CH_3)_2SiO]_b^{\ominus} \ Li^{\oplus}$$

$$+ \quad Cl\text{--}\underset{\underset{CH_3}{|}}{\overset{\overset{CH_3}{|}}{Si}}\text{--}CH{=}CH_2$$

$$\downarrow \qquad\qquad (5\text{-}19)$$

$$R \text{--} (CH_2 \text{--} CH)_a \text{--} [(CH_3)_2SiO]_b \text{--} \underset{\underset{CH_3}{|}}{\overset{\overset{CH_3}{|}}{Si}}\text{--}CH{=}CH_2$$

Styrene–dimethylsiloxane block copolymers have been reported to be useful as additives for improving the lubricity of styrene homopolymer. The incorporation of 0.05–0.5% of a block copolymer containing styrene and siloxane blocks of 75 and 77 degree of polymerization, respectively (42 wt% siloxane; overall $\overline{M}_n = 13,000$) into 900 $\overline{M}_n$ styrene homopolymer reduced the surface tension of the polystyrene (D103). The surface tension reduction was time-dependent, due to diffusion of the siloxane segments to the surface.

Block copolymers of styrene and 3,3,3-trifluoropropylmethylsiloxane have been prepared by initiation of styrene polymerization with butyl-lithium followed by addition of 2,4,6-tris(3,3,3-trifluoropropyl)-2,4,6-trimethylcyclotrisiloxane (D104). These materials were claimed to be useful as additives to improve the physical properties of plastics and rubbers.

A novel approach was used by Juliano (D105, D106) to prepare methyl methacrylate–dimethylsiloxane block copolymers via a homogeneous anionic process. This technique involved the use of an "ambident" initiator prepared from benzophenone and lithium, which possesses two anionic sites of unequal reactivity, one of which is selective for siloxane polymerization and the other for acrylic polymerization. Hexamethylcyclotrisiloxane was polymerized first at 0°C, after which hexamethylphosphoramide was added and methyl methacrylate was polymerized at −40°C, as shown in Scheme 5-5. The products displayed a high degree of optical clarity due to a domain size (siloxane domains in a continuous acrylic phase) of <400 Å.

The above discussion pertains to the anionic synthesis of siloxane–

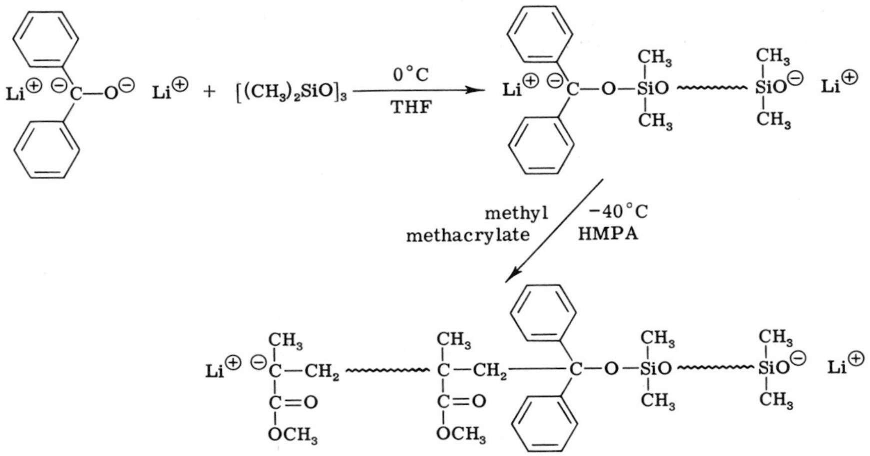

Scheme 5-5

vinyl block copolymers. Their preparation has also been claimed by free radical routes (D107, D108). Initiation of styrene polymerization with azobis(isobutyronitrile) in the presence of dimethyldichlorosilane produces a chlorosilyl-terminated polystyrene by means of chain transfer with the silane. Hydrolysis of this in the presence of additional silane reportedly gave block copolymer, as shown in Scheme 5-6. Ob-

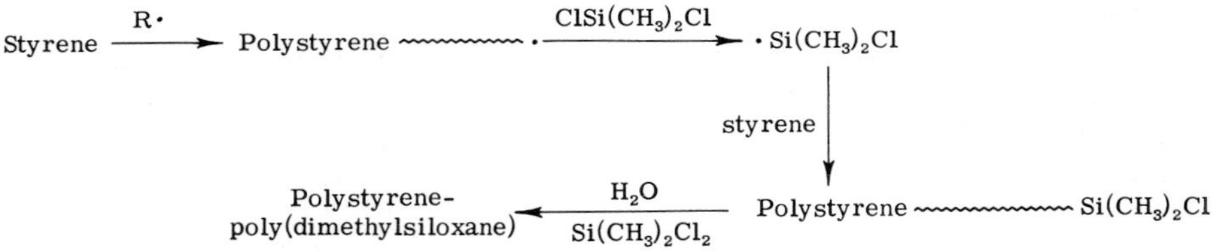

Scheme 5-6

viously, however, this approach results in a large amount of homopolymer formation and is not as satisfactory as the anionic approach discussed above.

c. Siloxane–Ether

A poly(dimethylsiloxane)–polyether block copolymer was prepared by the condensation of a monohydroxyl-terminated ethylene oxide–propylene oxide random copolymer and a monoacetoxy-terminated dimethylsiloxane polymer (Reaction 5-20) (D109). The molecular

$$R(OCH_2CH_2)_x \left(OCH_2\overset{CH_3}{\underset{|}{CH}} \right)_y OH \quad + \quad CH_3COO \left(\overset{CH_3}{\underset{\underset{CH_3}{|}}{\overset{|}{SiO}}} \right)_b Si(CH_3)_3$$

$$\downarrow 150°C$$

$$(5\text{-}20)$$

$$\left[R(OCH_2CH_2)_x \left(OCH_2\overset{CH_3}{\underset{|}{CH}} \right)_y \right]_a O \left(\overset{CH_3}{\underset{\underset{CH_3}{|}}{\overset{|}{SiO}}} \right)_b Si(CH_3)_3 \quad + \quad CH_3COOH$$

weight of the polyether block was ~1700 and that of the siloxane ~800. These products are useful as water–hydrocarbon emulsifiers and as stabilizers for urethane foams. Polysiloxane-polyether block copolymers are discussed in greater detail in Chapter 7.

REFERENCES

A1. Szwarc, M., Levy, M., and Milkovich, R., *J. Am. Chem. Soc.* **78**, 2656 (1956).

A2. Szwarc, M., *Nature (London)* **178**, 1168 (1956).

A3. Szwarc, M., "Carbanions, Living Polymers, and Electron Transfer Processes." Wiley, New York, 1968.

A4. Fetters, L. J., *J. Res. Natl. Bur. Stand., Sect. A* **70**, 421 (1966).

A5. Fetters, L. J., *J. Polym. Sci., Part C* **26**, 1 (1969).

A6. Morton, M., *Phys. Chem., Ser. One* **8**, 1 (1972); *C.A.* **78**, 30204x (1973).

A7. Hsieh, H. L., and Glaze, W. H., *Rubber Chem. Technol.* **43**(1), 22 (1970); *C.A.* **72**, 121944f (1970).

A8. Hsieh, H. L., *in* "Block and Graft Copolymers" (J. J. Burke and V. Weiss, eds.), p. 51. Syracuse Univ. Press, Syracuse, New York, 1973.

A9. Morton, M., and Fetters, L. J., *Macromol. Rev.* **2**, 71 (1967).

A10. Bywater, S., *Adv. Polym. Sci.* **4**, 66 (1965).

A11. Juliano, P. C., Ph.D. Thesis, University of Akron, Akron, Ohio (1968); *C.A.* **70**, 88372b (1969).

A12. Forman, L. E., *in* "Polymer Chemistry of Synthetic Elastomers" (J. P. Kennedy and E. G. M. Tornqvist, eds.), Part II, p. 491. Wiley (Interscience), New York, 1969.

A13. Zelinski, R., and Childers, C. W., *Rubber Chem. Technol.* **41**(1), 161 (1968); *C.A.* **68**, 79257z (1968).

A14. Hoffmann, M., Pampus, G., and Marwede, G., *Kautsch. Gummi, Kunstst.* **22**(12), 691 (1969); *C.A.* **72**, 67303f (1970).

A15. Crouch, W. W., and Short, J. N., *Inst. Rubber Ind.* May 1967, *Symp. Newer Elastom.*, p. 43 (1967).

A16. Kuntz, I., *J. Polym. Sci.* **54**, 569 (1961).

A17. Korotkov, A. A., and Chesnokova, N. N., *Vysokomol. Soedin.* **2**, 365 (1960).

A18. Korotkov, A. A., *Int. Symp. Macromol. Chem.*, 1957 Paper 66; *Angew. Chem.* **70**, 85 (1958).

A19. Johnson, A. F., and Worsfold, D. J., *Makromol. Chem.* **85**, 273 (1965).

A20. Morton, M., and Ells, F. R., *J. Polym. Sci.* **61**, 25 (1962).

A21. Bouton, T. C., and Futamura, S., *Detroit Am. Chem. Soc. Rubber Div. Meet.*, Paper No. 8 *1973* (1973).

A22. Fetters, L. J., "Kinetic and Mechanistic Aspects of Polymerizations Involving Organolithium Species." N.Y. Acad. Sci., New York, 1974; *Rubber Chem. Technol.* (in press).

A23. Morton, M., Sanderson, R. D., and Sakata, R., *Polym. Lett.* **9**, 61 (1971).

A24. Morton, M., Sanderson, R. D., and Sakata, R., *Macromolecules* **6**(2), 181 (1973); *C.A.* **78**, 160673k (1973).

A25. Morton, M., Falvo, L. A., and Fetters, L. J., *Macromolecules* **6**, 190 (1973).

A26. Morton, M., and Falvo, L. A., *Macromolecules* **6**, 190 (1973).

A27. Cinadr, B. F., and Schooley, A. T., *Chem. Eng.* (*N.Y.*), **77**(2), 125 (1970); *C.A.* **72**, 79501g (1970).

A28. Snavely, K. E., Wilson, T. R., and Railsback, H. E., *Rubber World* 45 (1973).

A29. Kraus, G., and Railsback, H. E., *Polym. Prep., Am. Chem. Soc., Div. Polym. Chem.* **14**(2), 1051 (1973).

A30. French Demande 2,008,521 (Farbenfabriken Bayer A.-G.) (1970) *C.A.* **73**, 46416a (1970).

A31. Yamaguchi, K., *Nippon Setchaku Kyokai Shi* **6**(3), 228 (1970); *C.A.* **73**, 99765v (1970).

A32. Yamaguchi, K., Yuji, K., Ibaragi, T., Toyomoto, K., and Sakamoto, K., German Offen. 1,963,038 (Asahi Chemical Industry Co., Ltd.) (1970); *C.A.* **73**, 78321g (1970).

A33. De Jong, K. J., *Ned. Rubberind.* **31**(10), 1 (1970); *C.A.* **73**, 46363f (1970).

A34. Glaze, W. H., *J. Organomet. Chem.* **68**, 1 (1974).

A35. Hinton, R. A., U.S. Patent 3,452,119 (Phillips Petroleum Co.) (1969); *C.A.* **71**, 71704j (1969).

A36. Dollinger, R. E., and Huxtable, R. L., U.S. Patent 3,356,763 (Phillips Petroleum Co.) (1967); *C.A.* **68**, 22360y (1968).

A37. Dollinger, R. E., U.S. Patent 3,297,793 (Phillips Petroleum Co.) (1967).

A38. Cooper, R. N., Jr., U.S. Patent 3,030,346 (Phillips Petroleum Co.) (1962); *C.A.* **57**, 2434b (1962).

A39. Crouch, W. W., and Hammer, R. S., *Mater. Plast.* (*Bucharest*) **10**(1), 589 (1973); *C.A.* **80**, 84412y (1974).

A40. Zelinski, R. P., Belgian Patent 661,095 (Phillips Petroleum Co.) (1965); *C.A.* **65**, 4080f (1966).

A41. British Patent 895,980 (Phillips Petroleum Co.) (1962); *C.A.* **58**, 14140a (1963).

A42. Hsieh, H. L., U.S. Patent 3,402,159 (Phillips Petroleum Co.) (1968); *C.A.* **69**, 97570w (1968).

A43. Keckler, N. F., S. African Patent 6705,819 (Firestone Tire and Rubber Co.) (1968); *C.A.* **70**, 58765q (1969).

A44. British Patent 884,974 (Esso Research and Engineering Co.) (1961); *C.A.* **58**, 11482g (1963).

A45. Miles, J. M., U.S. Patent 3,502,746 (Phillips Petroleum Co.) (1970); *C.A.* **72**; 112512m (1970).

A46. British Patent 1,218,147 (Farbenfabriken Bayer A.-G.) (1971); *C.A.* **74**, 88383p (1971).

A47. Sutter, H., and Beck, M., German Offen. 1,939,420 (Farbenfabriken Bayer A.-G.) (1971); *C.A.* **74**, 126440q (1971).

A48. McGrath, J. E., Ph.D. Thesis, University of Akron, Akron, Ohio (1967); *C.A.* **70**, 88289e (1969).

A49. Marsh, J. F., British Patent 1,109,616 (Esso Research and Engineering Co.) (1968); *C.A.* **69**, 3324u (1968).

A50. Morton, M., Tschoegl, N. W., and Froelich, D., "Heterogenous Block Copolymers" Technical Report TR-67-408, Part II. Air Force Mater. Lab., Wright Patterson AFB, Ohio.

A51. Ibarra, L., and Royo, J., *Rev. Plast. Mod.* **25**(203), 711 (1973); *C.A.* **79**, 43417q (1973).

A52. Ibarra, L., and Royo, J., *Rev. Plast. Mod.* **25**(200), 243 (1973); *C.A.* **78**, 148773k (1973).

A53. British Patent 862,507 (Polymer Corp., Ltd.) (1961); *C.A.* **55**, 20517e (1961).

A54. Orr, R. J., U.S. Patent 3,137,681 (Polymer Corporation Limited) (1964).

A55. Kolesnikov, G. S., and Yaralov, L. K., *Russ. Chem. Rev. (Engl. Transl.)* **34**(3), 195 (1965).

A56. Fetters, L. J., *in* "Block and Graft Copolymerization" (R. J. Ceresa, ed.), Vol. 1, p. 99. Wiley, New York, 1973.

A57. Mochel, V. D., *Rubber Chem. Technol.* **40**(4), 1200 (1967); *C.A.* **67**, 109445h (1967).

A58. Mochel, V. D., and Claxton, W. E., *J. Polym. Sci., Part A-1* **9**, 345 (1971).

A59. Bushuk, W., and Benoit, H., *Can. J. Chem.* **36**, 1616 (1958).

A60. Molau, G. E., *N.A.S.—N.R.C., Pub.* **1573**, 245 (1968).

A61. Pavelich, W. A., and Livigni, R. A., *J. Polym. Sci., Part C* **21**, 215 (1968); *C.A.* **68**, 96154f (1968).

A62. Gomoll, M., and Siede, H., *Plaste Kautseh.* **16**(6), 412 (1969); *C.A.* **71**, 82338u (1969).

A63. Morton, M., *Polym. Prepr., Am. Chem. Soc., Div. Polym. Chem.* **10**(2), 512 (1969).

A64. McGrath, J. E., Ph.D. Thesis, University of Akron, Akron, Ohio (1967).

A65. Morton, M., McGrath, J. E., and Juliano, P. C., *J. Polym. Sci., Part C* **26**, 99 (1969).

A66. Orr, R. J., and Williams, H. L., *J. Amer. Chem. Soc.* **79**, 3137 (1957).

A67. Prudhomme, J., and Bywater, S., *Polym. Prepr., Am. Chem. Soc., Div. Polym. Chem.* **10**(2), 518 (1969).

A68. Ceresa, R. J., "Block and Graft Copolymers." Butterworth, London, 1962.

A69. Burlant, W. J., and Hoffman, A. S., "Block and Graft Copolymers," Van Nostrand-Reinhold, Princeton, New Jersey, 1960.

A70. Ceresa, R. J., *in* "Techniques of Polymer Characterization" (P. W. Allen, ed.), Chapter 8. Butterworth, London, 1959.

A71. Matsuo, M., Sagae, S., and Asai, H., *Polymer* **10**, 79 (1969).

A72. Morton, M., *Encycl. Polym. Sci. Technol.* **15**, 508(1971); *C.A.* **76**, 100848v (1972).

A73. Matzner, M., Robeson, L. M., Noshay, A., and McGrath, J. E., *Encycl. Polym. Sci. Technol.* (1976) (in press).

A74. Molau, G. E., *Polym. Prepr., Am. Chem. Soc., Div. Polym. Chem.* **10**(2), 700 (1969); "Colloidal and Morphological Behavior of Block and Graft Copolymers." Plenum, New York, 1972.

A75. Allport, D. C., *Rep. Prog. Appl. Chem.* **54**, 479 (1969); *C.A.* **76**, 142376d (1971).

A76. Meier, D. J., *J. Polym. Sci., Part C* **26**, 81 (1969).

A77. Krause, S., *in* "Block and Graft Copolymers" (J. J. Burke and V. Weiss, eds.), p. 143. Syracuse Univ Press, Syracuse, New York, 1973.

A78. Irako, K., Anzai, S., and Onishi, A., *Bull. Chem. Soc. Jpn.* **41**(2), 501 (1968); *C.A.* **68**, 87980v (1968).

A78a. Mochel, V. D., *Macromolecules* **2**(5), 537 (1969); *C.A.* **71**, 113875y (1969).

A79. Folkes, M. J., and Keller, A., *in* "Physics of Glassy Polymers" (R. N. Hayward ed.), p. 235. Appl. Sci. Publ. Ltd., London, 1973.

A80. Folkes, M. J., and Keller, A., *in* "Block and Graft Copolymers" (J. J. Burke and V. Weiss, eds.), p. 87. Syracuse Univ. Press, Syracuse, New York, 1973.

A81. LaFlair, R. T., *IUPAC, 23rd, 1972* Vol. 8, p. 195 (1971).

A82. Holden, G., Bishop, E. T., and Legge, N. R., *J. Polym. Sci., Part C* **26**, (1969).

A83. Vanzo, E., *J. Polym. Sci., Part A-1* **4**, 1727 (1966).

A83a. Kato, K., *Polym. Eng. Sci.* **7**, 38 (1967).

A84. Bradford, E. B., and McKeever, L. D., *Prog. Polym. Sci.* **3**, 109 (1971).

A85. Bradford, E. B., *in* "Colloidal and Morphological Behavior of Block and Graft Copolymer" (G. E. Molau, ed.), p. 21. Plenum, New York, 1971.

A86. Molau, G. E., *in* "Block Polymers" (S. L. Aggarwal, ed.), p. 79. Plenum, New York, 1970.

A87. Matsuo, M., *Jpn. Plast.* **2**, 6 (1968).

A88. Matsuo, M., and Sagaye, S., *in* "Colloidal and Morphological Behavior of Block and Graft Copolymers" (G. E. Molau, ed.), p. 1. Plenum, New York, 1971.

A89. Merrett, F. M., *Trans. Faraday Soc.* **50**, 759 (1954).

A90. Molau, G. E., and Wittbrodt, W. M., *Macromolecules* **1**(3), 260 (1968; *C.A.* **69**, 27992m (1968).

A91. McIntyre, D., and Campos-Lopez, E., *in* "Block Polymers" (S. L. Aggarwal, ed.), p. 19. Plenum, New York, 1970.

A92. Kaempf, G., Kroemer, H., and Hoffmann, M., *J. Macromol. Sci., Phys.* **6**(1), 167 (1972); *C.A.* **76**, 4706b (1972).

A93. Bradford, E. B., and Vanzo, E., *J. Polym. Sci., Part A-1* **6**, 1661 (1968); *C.A.* **69**, 3264z (1968).

A94. Utracki, L. A., Simha, R., and Fetters, L. J., *J. Polym. Sci., Part A-2* **6**(12), 2051 (1968); *C.A.* **70**, 47974f (1969).

A95. Kraus, G., Childers, C. W., and Gruver, J. T., *J. Appl. Polym. Sci.* **11**, 1581 (1967).

A96. Angelo, R. J., Ikeda, R. M., and Wallach, M. L., *Polymer* **6**, 141 (1965).

A97. Wilson, P. I., and Hodgkinson, G. T., British Patent 1,180,567 (Dunlop Co. Ltd.) (1970); *C.A.* **72**, 101654h (1970).

A98. Railsback, H. E., Biard, C. C., Haws, J. R., and Wheat, R. C., *Rubber Age* **94**, 583 (1964).

A99. Railsback, H. E., and Haws, J. R., *Rubber Plast. Age* **48**(10), 1063 (1967); *C.A.* **68**, 13876u (1968).

A100. Railsback, H. E., and Kraus, G., *Kautsch. Gummi, Kunstst.* **22**(9), 497 (1969); *C.A.* **71**, 113912h (1969).

A101. Vandendael, J., *Rev. Belge Matieres Plast.* **10**(3), 189 (1969); *C.A.* **71**, 39987t (1969).

A102. Simpson, B. D., *Indian Rubber Bull.* **231**, 12 (1968); *C.A.* **69**, 20169q (1968).

A103. Haws, J. R., *Rubber Plast. Age* **46**, 1144 (1965); *C.A.* **64**, 897h (1966).

A104. Netherlands Patent Appl. 6,516,490 (Phillips Petroleum Co.) (1966); *C.A.* **65**, 17181e (1966).

A105. Svetlik, J. F., Ross, E. F., and Dearmont, D. D., *U.S.C.F.S.T.I., AD Rep.* **AD-646535** (1966).

A106. Bouquin, J. C., *Rev. Gen. Caoutch. Plast.* **47**(2), 157 (1970); *C.A.* **72**, 122587d (1970).

A107. Black, D. M., *Aust. Plast. Rubber* **24**(11), 31 (1973); *C.A.* **80**, 146787k (1974).

A108. Middlebrook, T. C., U.S. Patent 3,513,056 (Phillips Petroleum Co.) (1970); *C.A.* **73**, 36323d (1970).

A109. Netherlands Patent Appl. 6,405,713 (Shell Internationale Research Maatschappij N.V.) (1964); *C.A.* **62**, 11998h (1965).

A110. Molau, G. E., Wittbrodt, W. M., and Meyer, V. E., *J. Appl. Polym. Sci.* **13**, 2735 (1969).

A111. French Patent 1,548,192 (Asahi Chemical Industry Co., Ltd.) (1968); *C.A.* **71**, 22756u (1969).

A112. Netherlands Appl. 6,612,517 (Phillips Petroleum Co.) (1967); *C.A.* **67**, 54730q (1967).

A113. Childers, C. W., U.S. Patent 3,429,951 (Phillips Petroleum Co.) (1969).

A114. Gruver, J. T., French Patent 1,510,979 (Phillips Petroleum Co.) (1968); *C.A.* **70**, 69026a (1969).

A115. Childers, C. W., and Gruver, J. T., U.S. Patent 3,499,949 (Phillips Petroleum Co.) (1970); *C.A.* **72**, 112214r (1970).

A116. Woodhead, D. A., S. African Patent 69 05,330 (British Petroleum Co. Ltd.) (1970); *C.A.* **73**, 46415z (1970).

A117. Childers, C. W., and Kraus, G., U.S. Patent 3,491,166 (Phillips Petroleum Co.) (1970); *C.A.* **72**, 79885k (1970).

A118. Hall, W. S., and Ross, E. F., U.S. Patent 3,440,304 (Phillips Petroleum Co.) (1969); *C.A.* **71**, 14015m (1969).

A119. Childers, C. W., Kraus, G., Gruver, J. T., and Clark, E., *in* "Colloid and Morphological Behavior of Block Copolymers" (G. E. Molau, ed.), p. 193. Plenum, New York, 1971.

A120. Netherlands Patent Appl. 6,613,019 (Borg Warner) (1967); *C.A.* **67**, 100655u (1967).

A121. Hayward, R. N., ed., "The Physics of Glassy Polymers." Wiley, New York, 1973.

A122. Pritchard, J. E., Belgian Patent 662,778 (Phillips Petroleum Co.) (1965); *C.A.* **65**, ·2451c (1966).

A123. Stafford, O. L., Wing, D. V., and Stolsmark, D. E., French Patent 1,511,330 (Dow Chemical Co.) (1968); *C.A.* **70**, 68854g (1969).

A124. Akutin, M. S., Salina, Z. I., and Andrianov, B. V., *Tr. Mosk. Khim.—Tekhnol. Inst.* **61**, 229 (1969); *C.A.* **73**, 110569g (1970).

A125. French Patent 1,567,877 (Dunlop Co. Ltd.) (1969); *C.A.* **71**, 125760v (1969).

A126. Ross, E. F., and Svetlik, J. E., U.S. Patent 3,417,044 (Phillips Petroleum Co.) (1968); *C.A.* **70**, 48441s (1969).

A127. Vanderbilt, B. M., and Fasnacht, J. J., U.S. Patent 3,484,223 (Esso Research and Engineering Co. and PPG Industries, Inc.) (1969); *C.A.* **72**, 56354x (1970).

A128. Chang, F. S. C., *Polym. Prepr., Am. Chem. Soc., Div. Polym. Chem.* **12**(2), 835 (1971); *C.A.* **79**, 42933t (1973).

A129. Anderson, J. N., Weissert, F. C., and Hunter, C. J., *Rubber Chem. Technol.* **42**(3), 918 (1969); *C.A.* **71**, 92107g (1969).

A130. Sachse, H., and Seide, H., *Plaste Kautsch.* **16**(6), 416 (1969); *C.A.* **71**, 82339v (1969).

A131. Douy, A., Gervais, M., and Gallot, B., *C.R. Hebd. Seances Acad. Sci., Ser C* **270**(20); 1646 (1970); *C.A.* **73**, 35878b (1970).

A132. Willis, J. M., and Barbin, W. W., *Rubber Age (N.Y.)* **100**(7), 53 (1968); *C.A.* **69**, 68118z (1968).

A133. Trukenbrod, K., and Weber, G., *Chim. & Ind., Genie Chim.* **104**(10); 1271 (1971); *C.A.* **75**, 99133d (1971).

A134. Trukenbrod, K., Weber, H., and Dasch, J., German Offen. 2,026,308 (Chemische Werke Huels A-G) (1971); *C.A.* **76**, 114136y (1972).

A135. Kelley, D. J., and Tobolsky, A. V., *J. Am. Chem. Soc.* **81**, 1597 (1959).

A136. Livigni, R. A., Marker, L., Shkapenko, G., and Aggarwal, S. L., *Pap., Div. Rubber Chem.*, presented May 2, 1967.

A137. Tobolsky, A. V., and Rogers, C. E., *J. Polym. Sci.* **38**, 205 (1959).

A138. Crammond, D. N., and Urwin, J. R., *Aust. J. Chem.* **21**(7), 1835 (1968); *C.A.* **69**, 59914d (1968).

A139. Worsfold, D. J. *Polym. Sci., Part A-1* **5**(11), 2783 (1967); *C.A.* **68**, 40131d (1968).

A140. Rowell, G. F., III, Ph.D. Thesis, University of Pennsylvania, Philadelphia (1971); *C.A.* **77**, 153581e (1972).

A141. Cramond, D. N., Lawry, P. S., and Urwin, J. R., *Eur. Polym.* **2**(2), 107 (1966); *C.A.* **65**, 5535e (1966).

A142. Franta, E., and Rempp, P., *C. R. Hebd. Seances Acad. Sci.* **254**, 674 (1962).

A143. Fetters, L. J., *J. Res. Nat. Bur. Stand., Sect. A* **70**(5), 421 (1966).

A144. Prud'homme, J., Roovers, J. E. L., and Bywater, S., *Eur. Polym. J.* **8**(7), 901 (1972); *C.A.* **77**, 102355b (1972).

A145. Urwin, J. R., and Girolamo, M., *Makromol. Chem.* **160**, 183 (1972); *C.A.* **78**, 5181e (1973).

A146. Girolamo, M., and Urwin, J. R., *Eur. Polym. J.* **8**(10), 1159 (1972); *C.A.* **77**, 165219f (1972).

A147. Urwin, J. R., and Girolamo, M., *Eur. Polym. J.* **7**(7), 785 (1971); *C.A.* **75**, 152281d (1971).

A148. Girolamo, M., and Urwin, J. R., *Eur. Polym. J.* **8**(2), 299 (1972); *C.A.* **76**, 141476r (1972).

A149. Urwin, J. R., and Girolamo, M., *Makromol. Chem.* **150**, 179 (1971); *C.A.* **76**, 100280d (1972).

A150. Urwin, J. R., and Girolamo, M., *Makromol. Chem.* **160**, 183 (1972).

A151. Prud'homme, J., Roovers, J. E. L., and Bywater, S., *Eur. Polym. J.* **8**, 901 (1972).

A152. Cramond, D. N., and Urwin, J. R., *Eur. Polym. J.* **5**, 35 (1969).

A153. Cramond, D. N., and Urwin, J. R., *Eur. Polym. J.* **5**, 45 (1969).

A154. Urwin, J. R., *Aust. J. Chem.* **22**, 1649 (1969).

A155. Urwin, J. R., and Girolamo, M., *Makromol. Chem.* **150**, 179 (1971).

A156. Ho-Duc, N., and Prud'homme, J., *Macromolecules* **6**, 472 (1973).

A157. Urwin, J. R., and Cramond, D. N., *Aust. J. Chem.* **22**, 543 (1969).

A158. Urwin, J. R., and Girolamo, M., *Aust. J. Chem.* **25**(9), 1869 (1972); *C.A.* **77**, 127687z (1972).

A159. Benoit, H., Grubisic, Z., and Rempp, P., *J. Polym. Sci., Part B* **5**, 753 (1967).

A160. Henderson, J. F., and Szwarc, M., *Macromol. Rev.* **3**, 317 (1968); *C.A.* **69**, 97187b (1968).

A161. Szwarc, M., *Polym. Eng. Sci.* **13**(1), 1 (1973); *C.A.* **78**, 58810r (1973).

A162. Douy, A., Mayer, R., Rossi, J., and Gallot, B., *Mol. Cryst. Liq. Cryst.* **7**, 103 (1969); *C.A.* **71**, 92010v (1969).

A163. Girolamo, M., and Urwin, J. R., *Eur. Polym. J.* **7**(3), 225 (1971); *C.A.* **75**, 21663g (1971).

A164. Inoue, T., Soen, T., Hashimoto, T., and Kawai, H., *Polym. Prepr., Am. Chem. Soc., Div. Polym. Chem.* **10**(2), 538 (1969).

A165. Inoue, T., Moritani, M., Hashimoto, T., and Kawai, H., *Polym. Prepr., Am. Chem. Soc., Div. Polym. Chem.* **11**(2), 414 (1970).

A166. Moritani, M., Inoue, T., Motegi, M., and Kawai, H., *Macromolecules* **3**(4), 433 (1970); *C.A.* **73**, 88495f (1970).

A167. Kawai, H., and Inoue, T., *Jpn. Plast.* **4**(3), 12 (1970); *C.A.* **73**, 120923g (1970).

A168. Inoue, T., Soen, T., Hashimoto, T., and Kawai, H., *J. Polym. Sci., Part A-2* **7**(8), 1283 (1969); *C.A.* **71**, 113371f (1969).

A169. Inoue, T., Soen, T., Kawai, H., Fukatsu, M., and Kurata, M., *J. Polym. Sci., Part B* **6**, 75 (1968).

A170. Kawai, H., Soen, T., Inoue, T., Ono, T., and Uchida, T., *"Prog. Polym. Sci."*, *Jpn.* **4**, 145 (1970).

A171. Kawai, H., Soen, T., Inoue, T., Ono, T., and Uchida, T., *Mem. Fac. Eng., Kyoto Univ.* **33**(4), 383 (1971); *C.A.* **77**, 75846d (1972).

A172. Inoue, T., Soen, T., Hashimoto, T., and Kawai, H., *Macromolecules* **3**(1), 87 (1970).

A173. Kawai, H., and Inoue, T., *Jpn. Plast.* **3**(3), 16 (1969); *C.A.* **72**, 3950g (1970).

A174. Kato, K., *Polym. Eng. Sci.* **7**, 38 (1967).

A175. Gallot, B., Mayer, R., and Sadron, C., *C. R. Habd. Seances Head. Sci., Ser. C* **263**(1), 42 (1966).

A176. Sadron, C., *Chim. & Ind., Genie Chim.* **96**(3), 507 (1966); *C.A.* **66**, 18954m (1967).

A177. Kohler, J., Riess, G., and Banderet, A., *Eur. Polym. J.* **4**(1), 187 (1968); *C.A.* **68**, 96624j (1968).

A178. Riess, G., Periard, J., and Banderet, A., *Polym. Prepr., Am. Chem. Soc., Div. Polym. Chem.* **11**(2), 541 (1970).

A179. Kohler, J., Banderet, A., Riess, G., and Job, C., *Rev. Gen. Caoutch. Plast.* **46**(11), 1317 (1969); *C.A.* **72**, 44756s (1970).

A180. Kohler, J., Riess, G., and Banderet, A., *Eur. Polym. J.* **4**(1), 173 (1968); *C.A.* **68**, 87815v (1968).

A181. Periard, J., Banderet, A., and Riess, G., *J. Polym. Sci., Part B* **8**(2), 109 (1970); *C.A.* **72**, 133531g (1970).

A182. Riess, G., *Kinet. Mech. Polyreactions, IUPAC Int. Symp. Macromol. Chem., Plenay Main Lect., 1969* p. 607 (1971).

A183. Periard, J., Riess, G., and Neyer-Gomez, M. J., *Eur. Polym. J.* **9**(8), 687 (1973); *C.A.* **80**, 37548n (1974).

A184. British Patent 888,624 (Phillips Petroleum Co.) (1962); *C.A.* **57**, 1017h (1962).

A185. Netherlands Patent Appl. 66 16,691 (Shell Internationale Research Maatschappij N.V.) (1968); *C.A.* **69**, 78287t (1968).

A186. Porter, L. M., U.S. Patent 3,149,182 (Shell Oil Co.) (1964); *C.A.* **61**, 14808g (1964).

A187. Inoue, T., Ishihara, H., Kawai, H., Ito, Y., and Kato, K., *Mech. Behav. Mater., Proc. Int. Conf., 1st, 1971* Vol. 3, p. 419 (1972); *C.A.* **77**, 115223g (1972).

A188. Netherlands Patent Appl. 6,612,959 (Borg-Warner Corp.) (1967); *C.A.* **67**, 54958v (1967).

A189. Wald, M. M., and Quam, M. G., U.S. Patent 3,700,633 (Shell Oil Co.) (1972); *C.A.* **78**, 17353a (1973).

A190. Falk, J. C., *Makromol. Chem.* **160**, 291 (1972); *C.A.* **78**, 4727a (1973).

A191. DeVault, A. N., U.S. Patent 3,696,088 (Phillips Petroleum Co.) (1972); *C.A.* **78**, 44503f (1973).

A192. Ho-Duc, N., Daoust, H., Chabot, M.-A., and Prud'homme, J., *Polym. Prepr., Am. Chem. Soc., Div. Polym. Chem.* **15**(2), 164 (1974).

A193. Terrisse, J., and Le-Meur, F., French Demande 2,138,465 (Agence Nationale de Valorisation de la Recherche) (1973); *C.A.* **79**, 79494b (1973).

A194. Baer, M., *J. Polym. Sci., Part A* **2**(1), 417 (1964).

A195. Robeson, L. M., Matzner, M., Fetters, L. J., and McGrath, J. E., *Polym. Prepr., Am. Chem. Soc., Div. Polym. Chem.* **14**(2), 1069 (1973).

A196. Robeson, L. M., Matzner, M., Fetters, L. J., and McGrath, J. E., *in* "Recent Advances in Polymer Blends, Grafts and Blocks" (L. H. Sperling, ed.), p. 281. Plenum, New York, 1974.

A197. Dunn, D. J., and Krause, S., *J. Polym. Sci., Polym. Lett. Ed.* **12**, 591 (1974).

A198. Shen, M., and Hansen, D., *Macromolecules* **8**(6), 903 (1975).

A199. Krause, S., *Macromolecules* **3**, 84 (1970).

B1. Tsuruta, T., *Prog. Polym. Sci., Jpn.* **3**, 1 (1972).

B2. Morton, M., and Fetters, L. J., *Macromol. Rev.* **2**, 71 (1967).

B3. Szwarc, M., "Carbanions, Living Polymers and Electron Transfer Processes." Wiley (Interscience), New York, 1968.

B4. Franta, E., and Rempp, P., *C. R. Hebd. Seances Acad. Sci.* **254**, 674 (1962).

B5. Graham, R. K., Dunkelberger, D. L., and Goode, W. E., *J. Am. Chem. Soc.* **82**, 400 (1960).

B6. Bywater, S., *Adv. Polym. Sci.* **4**, 66 (1965).

B7. Roig, A., Figueruels, J. E., and Liano, E., *J. Polym. Sci., B* **3**, 171 (1965).

B8. Guzman, G. M., and Bello, A., *Makromol. Chem.* **107**, 46 (1967).

B9. Mita, I., Watabe, Y., Akatsu, T., and Kambe, H., *Polym. J.* **4**(3), 271 (1973).

B10. Chiang, R., Rhodes, J. H., and Evens, R. A., *J. Polym. Sci., Part A-1* **4**, 3089 (1966).

B11. Graham, R. K., Panchak, J. R., and Kampf, M. J., *J. Polym. Sci.* **44**, 411 (1960).

B12. Busfield, W. K., and Methven, J. M., *Polymer* **14**, 137 (1973).

B13. Bushuk, W., and Benoit, H., *Can. J. Chem.* **36**, 1616 (1958).

B14. Ailhaud, H., Gallot, Y., and Skoulios, A., *Makromol. Chem.* **140**, 179 (1970); *C.A.* **74**, 54226z (1971).

B15. Ailhaud, H., Gallot, Y., and Skoulios, A., *IUPAC, 23rd, 1971* Vol. 1, p. 533 (1971).

B16. Ailhaud, H., Gallot, Y., and Skoulios, A., *Kolloid Z. & Z. Polym.* **248**(1-2), 889 (1971); *C.A.* **76**, 127554h (1972).

B17. Brown, C. W., and White, I. F., *J. Appl. Polym. Sci.* **16**(10), 2671 (1972); *C.A.* **78**, 4568z (1973).

B18. Niemann, T. F., U.S. Patent 3,609,101 (B. F. Goodrich Co.) (1971); *C.A.* **76**, 25836j (1972).

B19. Niemann, T. F., U.S. Patent 3,609,100 (B. F. Goodrich Co.) (1971); *C.A.* **76**, 25837k (1972).

B20. Niemann, T. F., U.S. Patent 3,699,191 (B. F. Goodrich Co.) (1972); *C.A.* **78**, 59020v (1973).

B21. Niemann, T. F., U.S. Patent 3,700,756 (B. F. Goodrich Co.) (1972); *C.A.* **78**, 30541e (1973).

B22. Fetters, L. J., *J. Polym. Sci., Part C* **26**, 1 (1969).

B23. Finaz, G., Gallot, Y., Parrod, J., and Rempp, P., *J. Polym. Sci.* **58**, 1363 (1962).

B24. Freyss, D., Rempp, P., and Benoit, H., *J. Polym. Sci., B* **2**(2), 217 (1964).

B25. Freyss, D., Leng, M., and Rempp, P., *Bull. Soc. Chim. Fr.* **2**, 221 (1964); *C.A.* **61**, 16001a (1964).

B26. Yamaguchi, T., and Goto, T., German Offen. 1,934,593 (Idemitsu, Kosan Co., Ltd.) (1970); *C.A.* **72**, 79651f (1970).

B27. Rempp, P., *Polym. Prepr., Am. Chem. Soc., Div. Polym. Chem.* **7**(1), 141 (1966); *C.A.* **66**, 76297k (1967).

B28. Rempp, P., *Pure Appl. Chem.* **16**, 403 (1968); *C.A.* **69**, 77743b (1968).

B29. Schepers, H. A. J., and Roest, B. C., German Offen. 2,231,993 (Stamicarbon N.V.) (1973); *C.A.* **78**, 98283c (1973).

B30. Savelief, M., Lotz, B., and Kovacs, A. J., *Inf. Chim.* **118**, 199 (1973); *C.A.* **79**, 5800b (1973).

B31. Agouri, E., Favie, C., LaPutte, R., Philardeau, Y., and Rideau, J., *Inf. Chim.* **118**, 185 (1973); *C.A.* **79**, 79314t (1973).

B32. Agouri, E., LaPutte, R. P., and Rideau, J., German Offen. 2,339,508 (Aquitaine Organico) (1974); *C.A.* **81**, 78522p (1974).

B33. Donat, F. J., U.S. Patent 3,621,077 (B. F. Goodrich Co.) (1971).

B34. Calvayrac, H., Thivollet, P., and Gole, J., *J. Polym. Sci., Polym. Chem. Ed.* **11**(7), 1631 (1973); *C.A.* **79**, 79387u (1973).

B35. Creasy, W. S., Barnabeo, A. E., and Robeson, L. M., *J. Polym. Sci.* **13**, 1979 (1975).

B36. Kamachi, M., Kurihara, M., and Stille, J. K., *Macromolecules* **5**(2), 161 (1972); *C.A.* **77**, 35333c (1972).

B37. Ceresa, R. J., *Polymer* **1**, 72 (1960).

B38. Ceresa, R. J., *Encycl. Polym. Sci. Technol.* **2**, 485 (1964); *C.A.* **65**, 4037e (1966).

B39. Prisyazhnyuk, A. I., and Ivanchev, S. S., *Vysokomol. Soedin., Ser. A* **12**(2), 450 (1970); *C.A.* **72**, 122016s (1970).

B40. Smets, G., and Hart, R., *Adv. Polym. Sci.* **2**, 173 (1960).

B41. Smets, G., *Ber. Bunsenges. Phys. Chem.* **70**(3), 248 (1966); *C.A.* **64**, 15992g (1966).

B42. Smets, G., and Woodward, A. E., *J. Polym. Sci.* **14**, 126 (1954).

B43. Sugimura, T., Suda, I., and Minoura, Y., *Kogyo Kagaku Zasshi* **69**(4), 718 (1966); *C.A.* **65**, 15508a (1966).

B44. Woodward, A. E., and Smets, G., *J. Polym. Sci.* **47**, 51 (1955).

B45. Bamford, C. H., and White, E. F. T., *Trans. Faraday Soc.* **52**, 716 (1956).

B46. Bamford, C. H., Jenkins, A. D., and White, E. F. T., *J. Polym. Sci.* **34**, 271 (1959).

B47. Huff, T., and Perry, E., U.S. Patent 3,262,995 (Monsanto Co.) (1966); *C.A.* **65**, 13844a (1966).

B48. Laszkiewicz, B., *Wiad. Chem.* **19**(9), 629 (1965); *C.A.* **64**, 828d (1966).

B49. Perry, E., *J. Appl. Polym. Sci.* **8**(6), 2605 (1964); *C.A.* **62**, 6584d (1965).

B50. Otsu, T., *J. Polym. Sci.* **26**, 239 (1957).

B51. Borsig, E., Lazar, M., Capla, M., and Florian, S., *Angew. Makromol. Chem.* **9**, 89 (1969); *C.A.* **72**, 55933y (1970).

B52. Sheppard, C. S., Bafford, R. A., and MacLeay, R. E., German Offen. 1,905,915 (Wallace and Tiernan, Inc.) (1969); *C.A.* **72**, 13201c (1970).

B53. Urwin, J. R., *J. Polym. Sci.* **37**, 580 (1958).

B54. Pokrikyan, V. G., Sergeev, V. A., and Korshak, V. V., *Izv. Akad. Nauk SSSR, Otd. Khim. Nauk* p. 1106 (1963); *C.A.* **59**, 7650g (1963).

B55. Smets, G., De Winter, W., and Delzenne, G., *J. Polym. Sci.* **55**, 767 (1961).

B56. Dunn, A. S., Stead, B. D., and Melville, H. W., *Trans. Faraday Soc.* **50**, 279 (1954); *C.A.* **48**, 14290 (1954).

B57. Kolesnikov, G. S., Yaralov, L. K., and Tsyurupa, M. P., *Vysokomol. Soedin., Ser. A* **9**(7), 1570 (1967); *C.A.* **67**, 73910q (1967).

B58. Tsvetkov, M. S., and Markovskaya, R. F., *Izv. Vyssh. Uchebn. Zaved., Khim. Khim. Tekhnol.* **11**(8), 936 (1968); *C.A.* **70**, 29399t (1969).

B59. Bamford, C. H., and Jenkins, A. D., *Nature (London)* **176**, 78 (1955).
B60. Twigg, G. H., and Benton, J. L. B., British Patent 857,145 (Distillers Co., Ltd.) (1960).
B61. Otsu, T., Japanese Patent 5985 (Nippon Catalytic Chemical Industries, Ltd.) (1960); *C.A.* **55**, 6048f (1961).
B62. Brepoels, J., and Smets, G., *J. Polym. Sci.* **56**, 359 (1962).
B63. Molyneux, P., *Makromol. Chem.* **38**, 31, (1960).
B64. Bamford, C. H., and Jenkins, A. D., *Trans. Faraday Soc.* **56**, 907 (1960).
B65. Allen, P. E. M., Downer, J. M., Hastings, G. W., Melville, H. W., Molyneux, P., and Urwin, J. R., *Nature (London)* **177**, 910 (1956); *C.A.* **50**, 12528e (1956).
B66. Burnett, G. M., Meares, P., and Paton, C., *Trans. Faraday Soc.* **58**, 723 (1962).
B67. Nozaki, K., U.S. Patent 3,189,663 (Shell Oil Co.) (1965); *C.A.* **63**, 5848f (1965).
B68. Minoura, Y., and Ogata, Y., *J. Polym. Sci., Part A-1* **7**(9), 2547 (1969); *C.A.* **71**, 124974f (1969).
B69. Guillet, J. E., U.S. Patent 3,472,918 (Eastman Kodak Co.) (1969); *C.A.* **72**, 4027y (1970).
B70. Miller, M. L., *Can. J. Chem.* **36**, 309 (1958).
B71. Kobayashi, S., Ide, F., and Nakatsuka, K., *Kobunshi Kagaku* **24**(261), 17 (1967); *C.A.* **67**, 33028b (1967).
B72. Nakatsuka, K., Ide, F., and Kobayashi, S., Japanese Patent 70/06,941 (Mitsubishi Rayon Co.) (1970); *C.A.* **73**, 36436t (1970).
B73. Kolesnikov, G. S., and Yaralov, L. K., *Russ. Chem. Rev. (Engl. Transl.)* **34**(3), 195 (1965).
B74. Brown, C. W., and Taylor, G. A., *J. Appl. Polym. Sci.* **13**, 629 (1969); *C.A.* **71**, 61814f (1969).
B75. Belgian Patent 538,241 (Gevaert Photo-Producten N.V.) (1955); *C.A.* **52**, 4245b (1958).
B76. Newman, S., *J. Appl. Polym. Sci.* **6**, 515 (1962).
B77. Dondos, A., Rempp, P., and Benoit, H., *Makromol. Chem.* **130**, 233 (1969); *C.A.* **72**, 44241v (1970).
B78. Burnett, G. M., Meares, P., and Paton, C., *Trans. Faraday Soc.* **58**, 737 (1962).
B79. Dondos, A., Froelich, D., Rempp, P., and Benoit, H., *J. Chim. Phys.* **64**(6), 1012 (1967); *C.A.* **67**, 100490m (1967).
B80. Dondos, A., Rempp, P., and Benoit, H., *Eur. Polym. J.* **3**(4), 657 (1967); *C.A.* **68**, 50182 (1968).
B81. Gallot, Y., Franta, E., Rempp, P., and Benoit, H. *J. Polym. Sci., Part C* **4**, 473 (1963).
B82. Krause, S., *J. Phys. Chem.* **65**, 1618 (1961).
B83. Krause, S., *J. Phys. Chem.* **68**(7), 1948 (1964).
B84. Utracki, L. A., and Simha, R., *Macromolecules* **1**(6), 505 (1968); *C.A.* **70**, 38136g (1969).
B85. Kotaka, T., Ohnuma, H., and Inagaki, H., *Bull. Inst. Chem. Res., Kyoto Univ.* **46**(2), 107 (1966); *C.A.* **70**, 12010m (1969).
B86. Urwin, J. R., and Stearne, J. M., *Makromol. Chem.* **78**, 204 (1964).
B87. Gallot, Y., Leng, M., Benoit, H., and Rempp, P., *J. Chim. Phys.* **59**, 1093 (1962); *C.A.* **58**, 6936b (1962).
B88. Utracki, L., *Polym. Prepr., Am. Chem. Soc., Div. Polym. Chem.* **11**(2), 572 (1970).
B89. Tanaka, T., Kotaka, T., and Inagaki, H., *Macromolecules* **7**(3), 311 (1974); *C.A.* **81**, 92142w (1974).
B90. Kotaka, T., and Ohnuma, H., *in* "Colloidal and Morphological Behavior of Block

and Graft Copolymers" (G. E. Molau, ed.), p. 259. Plenum, New York, 1971; *C.A.* **81**, 78419k (1974).

B91. Melville, H. W., and Stead, B. D., *J. Polym. Sci.* **16**, 505 (1955).

B92. Brendle, M., Koeppel, R., and Banderet, A., *J. Polym. Sci., Part C* **22**(1), 349 (1968); *C.A.* **69**, 52550y (1968).

B93. Utiyama, H., Takenaka, K., Mizumori, M., and Fukuda, M., *Macromolecules* **7**, 28 (1974).

B94. Bamford, C. H., and White, E. F. T., *Trans. Faraday Soc.* **54**, 268 (1958).

B95. Mikhailov, G. P., Burshtein, L. L., Kol'tsov, A. I., Malinovskaya, V. P., Platonov, M. P., and Shibaev, L. A., *Vysokomol. Soedin., Ser. B* **10**(11), 828 (1968); *C.A.* **70**, 68836c (1969).

B96. Botham, R. A., Shank, C. P., and Thies, C., *Polym. Prepr., Am. Chem. Soc., Div. Polym. Chem.* **12**(2), 655 (1971).

B97. Kamiyama, F., Matsuda, H., and Inagaki, H., *Makromol. Chem.* **125**, 286 (1969); *C.A.* **71**, 81805g (1969).

B98. Duplessix, R., Picot, C., and Benoit, H., *J. Polym. Sci., Part B* **9**(5), 321 (1971); *C.A.* **75**, 36849w (1971).

B99. Kotaka, T., Tanaka, T., Ohnuma, H., Murakami, Y., and Inagaki, H., *Polym. J.* **1**(2), 245 (1970); *C.A.* **75**, 77441a (1971).

B100. Benoit, H., Dondos, A., and Froelich, D., *J. Phys. (Paris), Colloq.* **5**, 287 (1971); *C.A.* **77**, 49005j (1972).

B101. Hopkins, A., and Howard, G. J., *J. Polym. Sci., Part A-2* **9**(5), 841 (1971); *C.A.* **75**, 36854u (1971).

B102. Ohnuma, H., Kotaka, T., and Inagaki, H., *Polym. J.* **1**(6), 716 (1970); *C.A.* **75**, 21250v (1971).

B103. Dondos, A., *Makromol. Chem.* **147**, 123 (1971); *C.A.* **75**, 141291f (1971).

B104. Ailhaud, H., Gallot, Y., and Skoulios, A., *Makromol. Chem.* **151**, 1 (1972); *C.A.* **76**, 113782u (1972).

B105. Beevers, R. B., *Trans. Faraday Soc.* **58**, 1465 (1962); *C.A.* **58**, 1542h (1963).

B106. Beevers, R. B., and White, E. F. T., *Trans. Faraday Soc.* **56**, 1529 (1960).

B107. Beevers, R. B., White, E. F. T., and Brown, L., *Trans. Faraday Soc.* **56**, 1535 (1960).

B108. Reiss, G., Periard, J., and Banderet, A., French Patent 1,576,598 (Centre National de la Recherche Scientifique) (1969); *C.A.* **72**, 91173j (1970).

B109. Reiss, G., *Kinet. Mech. Polyreactions, IUPAC Int. Symp. Macromol. Chem., Plenary Main Lect., 1969* p. 607 (1971); *C.A.* **77**, 35369u (1972).

B110. Periard, J., and Riess, G., *Kolloid Z. & Z. Polym.* **248**(1-2), 877 (1971); *C.A.* **76**, 127905y (1972).

B111. Riess, G., *Kinet. Mech. Polyreactions, IUPAC Int. Symp. Macromol. Chem., Plenary Main Lect., 1969* p. 607 (1971).

B112. Banderet, A., Tournut, C., and Riess, G., *J. Polym. Sci., Part C* **16**, 2601 (1967).

B113. Serdyuk, I. N., and Fedorov, B. A., *J. Polym. Sci., Polym. Lett. Ed.* **11**(10), 645 (1973); *C.A.* **80**, 83751w (1974).

B114. Molau, G. E., *N.A.S.—N.R.C., Publ.* **1573**, 245 (1968).

B115. Bushuk, W., and Benoit, H., *Can. J. Chem.* **36**, 1616 (1958).

B116. Ivanchev, S. S., Tolpygina, T. A., Prisyazhnyuk, A. I., and Galibei, V. I., *Vysokomol. Soedin., Ser. B* **12**(6), 466 (1970); *C.A.* **73**, 56466f (1970).

B117. Black, P. E., and Worsfold, D. J., *J. Appl. Polym. Sci.* **14**(7), 1671 (1970); *C.A.* **73**, 77924u (1970).

B118. Funt, B. L., and Collins, E., *J. Polym. Sci.* **28**, 359 (1958).

B119. Hicks, J. A., and Melville, H. W., *J. Polym. Sci.* **12**, 461 (1954).

B120. Hicks, J. A., and Melville, H. W., *Proc. R. Soc. (London), Ser. A* **226**, 314 (1954).

B121. Yaralov, L. K., and Kolesnikov, G. S., *Vysokomol. Soedin.* **8**(5), 870 (1966); *C.A.* **65**, 3977g (1966).

B122. D'Alelio, C. F., and Hoffend, T. P., *J. Polym. Sci., Part A-1* **5**, 323 (1967).

B123. Zherebin, Yu. L., Ivanchev, S. S., and Domareva, N. M., *Vysokomol. Soedin., Ser. A* **16**(4), 893 (1974); *C.A.* **81**, 121098u (1974).

B124. Nozaki, K., U.S. Patent 3,069,380 (Shell Oil Co.) (1962); *C.A.* **58**, 5804g (1963).

B125. Shimomura, T., and Kudo, K., German Offen. 2,009,066 (Sumitomo Chemical Co., Ltd.) (1970); *C.A.* **73**, 110471u (1970).

B126. Nakatsuka, K., Ide, F., and Kobayashi, J., Japanese Patent 70/21,114 (Mitsubishi Rayon Co., Ltd.) (1970); *C.A.* **73**, 88583h (1970).

B127. Schindler, A., and Williams, J. L., *Polym. Prepr., Am. Chem. Soc., Div. Polym. Chem.* **10**(2), 832 (1969); *C.A.* **74**, 142719t (1971).

B128. Grosius, P., Gallot, Y., and Skoulios, A., *Makromol. Chem.* **127**, 94 (1969); *C.A.* **71**, 102270w (1969).

B129. Grosius, P., Gallot, Y., and Skoulios, A., *Eur. Polym. J.* **6**(2), 355 (1970); *C.A.* **72**, 122081j (1970).

B130. Grosius, P., Gallot, Y., and Skoulios, A., *Makromol. Chem.* **132**, 35 (1970); *C.A.* **72**, 11873t (1970).

B131. Fontanille, M., and Sigwalt, P., *Bull. Soc. Chim. Fr.* **1**, 4095 (1967); *C.A.* **68**, 69393b (1968).

B132. Grosius, P., Gallot, Y., and Skoulios, A., *C.R. Hebd. Seances Acad. Sci., Ser. C* **270**(16), 1381 (1970); *C.A.* **73**, 25991u (1970).

B133. Grosius, P., Gallot, Y., and Skoulios, A., *Makromol. Chem.* **136**, 191 (1970); *C.A.* **73**, 77649h (1970).

B134. Wiles, D. M., *In* "Structure and Mechanism in Vinyl Polymerization," (T. Tsuruta, and K. F. O'Driscoll, eds.) p. 223, Dekker, New York (1969).

B135. Allen, P. E. M., Patrick, C. R., "Kinetics and Mechanism of Polymerization Reactions," Halstead Press, New York (1974).

C1. Natta, G., *J. Polym. Sci.* **34**, 531 (1959).

C2. Boor, J., *Macromol. Rev.* **2**, 115 (1967).

C3. Boor, J., *Am. Chem. Soc., Div. Org. Coat. Plast. Chem., Prepr.* **30**, 158 (1970).

C4. Boor, J., *Ind. Eng. Chem., Prod. Res. Dev.* **9**(4), 437 (1970).

C5. Cossee, P., *in* "Stereochemistry of Macromolecules" (A. D. Ketley, ed.), Vol. I, p. 145. Dekker, New York, 1967.

C6. Jordan, D. O., *in* "Stereochemistry of Macromolecules" (A. D. Ketley, ed.), Vol. I, p. 1. Dekker, New York, 1967.

C7. Hoeg, D. F., *in* "Stereochemistry of Macromolecules" (A. D. Ketley, ed.), Vol. I, p. 47. Dekker, New York, 1967.

C8. Kontos, E. G., Easterbrook, E. K., and Gilbert, R. D., *J. Polym. Sci.* **61**, 69 (1962).

C9. Pasquon, I., Valvassori, A., and Sartori, G., *in* "Stereochemistry of Macromolecules" (A. D. Ketley, ed.), Vol. I, p. 177. Dekker, New York, 1967.

C10. Bier, G., and Lehmann, G. *in* "Copolymerization" (G. E. Ham, ed.), p. 149. Wiley (Interscience), New York, 1964.

C11. British Patent 1,010,740 (Mitsubishi Petrochemical Co., Ltd) (1965); *C.A.* **64**, 12833f (1966).

C12. Natta, G., Pasquon, I., and Giachetti, E., *Makromol. Chem.* **24**, 258 (1957).

C13. Berger, M. N., Boocock, G., and Haward, R. N., *Adv. Catal. Relat. Subj.* **19**, 211 (1969); *C.A.* **72**, 32222b (1970).

C14. Bier, G., *Makromol. Chem.* **70**, 44 (1964).

C15. Caunt, A. D., *J. Polym. Sci., C* **4**, 66 (1963).

C16. Amerik, V. V., Yakobson, F. I., and Krentsei, B. A., *Plast. Massy* No. 8, p. 10 (1969), *C.A.* **71**, 92090w (1969).

C17. Tanako, S., and Morikawa, H., *J. Polym. Sci., A* **3**, 9 (1965).

C18. British Patent 986,189 (1957).

C19. Hostetler, D. E., S. African Patent 68/02,418 (1968); *C.A.* **70**, 78539x (1969).

C20. Wesslau, H., *Makromol. Chem.* **26**, 118 (1958).

C21. Jezl, J. L., Khelghatian, H. M., and Hague, L. D., U.S. Patent 3,296,338 (Avisun Corp.) (1967); *C.A.* **66**, 56171m (1967).

C22. Ashida, M., Yoshioka, T., Fukuhara, K., and Iwamitsu, A., Japanese Patent 72/26,113 (Tokuyama Soda Co., Ltd.) (1972); *C.A.* **78**, 4787v (1973).

C23. Shirai, I., Muraki, M., and Kunihiro, K., Japanese Patent 71/06,629 (Chisso Corp.) (1971); *C.A.* **75**, 64621j (1971).

C24. Netherlands Patent Appl. 6,402,203 (Avisum Corp.) (1965); *C.A.* **64**, 8404a (1966).

C25. Khelghatian, H. M., Jezl, J. L., and Hague, L. D., U.S. Patent 3,442,978 (Avisun Corp.) (1969); *C.A.* **71**, 13511h (1969).

C26. Shirai, I., Kunimune, K., and Muraki, M., S. African Patent 68/06,130 (Chisso Corp.) (1969); *C.A.* **71**, 102602f (1969).

C27. Jezl, J. L., Khelghatian, H. M., and Hague, L. D., U.S. Patent 3,268,624 (Avisun Corp.) (1966); *C.A.* **65**, 15540e (1966).

C28. Netherlands Patent Appl. 6,402,202 (Avisun Corp.) (1965); *C.A.* **64**, 8403g (1966).

C29. Schrage, A., and Pless, E. J., Belgian Patent 645,950 (Rexall Drug and Chemical Co.) (1964); *C.A.* **63**, 8520d (1965).

C30. Hostetler, D. E., German Offen. 2,162,539 (Sinclair-Koppers Co.) (1972); *C.A.* **77**, 115059h (1972).

C31. Hatakeyama, H., Shin, Y., and Fukumoto, O., Japanese Patent 26,693 (Toyo Rayon Co., Ltd.) (1967); *C.A.* **68**, 87738x (1968).

C32. Okada, H., and Matsumura, K., Japanese Patent 69/06,274 (Toyo Rayon Co., Ltd.) (1969); *C.A.* **71**, 50642x (1969).

C33. British Patent 1,074,383 (Rexall Drug) (1968); *C.A.* **68**, 69565j (1968).

C34. Netherlands Patent Appl. 6,506,708 (Phillips Petroleum Co.) (1965); *C.A.* **65**, 823h (1966).

C35. British Patent 1,100,233 (Phillips Petroleum Co.) (1968).

C36. British Patent 889,230 (Sun Oil) (1960).

C37. Scoggin, J. S., U.S. Patent 3,525,781 (Phillips Petroleum Co.) (1970); *C.A.* **73**, 88420c (1970).

C38. Netherlands Patent Appl. 6,407,952 (Rexall Drug and Chemical Co.) (1965); *C.A.* **63**, 760b (1965).

C39. Scoggin, J. S., U.S. Patent 3,454,675 (Phillips Petroleum Co.) (1969); *C.A.* **71**, 71185r (1969).

C40. Jones, A. M., Planchard, J. A., Jr., and Speed, R. A., U.S. Patent 3,378,607 (Esso Research and Engineering Co.) (1968); *C.A.* **68**, 105664b (1968).

C41. Jones, A. M., Planchard, J. A., Jr., Speed, R. A., and Claybaugh, B. E., U.S. Patent 3,414,637 (Esso Research and Engineering Co.) (1968); *C.A.* **70**, 29897d (1969).

C42. DiDrusco, G., French Patent 1,386,838 (Montecatini) (1965); *C.A.* **63**, 8514k (1965).

C43. British Patent 1,045,221 (duPont) (1966); *C.A.* **66**, 2989r (1967).

C44. Hill, R. W., U.S. Patent 3,776,979 (Gulf Research and Development Co.) (1973); *C.A.* **80**, 146611y (1974).

C45. Bier, G., Gumboldt, A., and Lehmann, G., *Plast. Inst., Trans. J.* **28**, 98 (1960).

C46. Bier, G., *Angew. Chem.* **73**, 186 (1961).

C47. French Patent 1,220,947 (Phillips) (1959).

C48. Natta, G., Giachetti, E., and Pasquon, I., French Patent 1,220,573 (Montecatini) (1959).

C49. Hagemeyer, H. J., and Edwards, M. B., *J. Polym. Sci., Part C* **4**, 731 (1963); *C.A.* **60**, 5705d (1964).

C50. British Patent 915,622 (Montecatini) (1963); *C.A.* **58**, 8117h (1963).

C51. Hagemeyer, H. J., Jr., and Edwards, M. B., Belgian Patent 624,652 (Eastman Kodak) (1963); *C.A.* **59**, 2969b (1963).

C52. Hagemeyer, H. J., Jr., and Edwards, M. B., U.S. Patent 3,639,515 (Eastman Kodak Co.) (1972); *C.A.* **77**, 20660h (1972).

C53. Renando, S., U.S. Patent 3,347,955 (Phillips) (1967); *C.A.* **67**, 117534b (1967).

C54. Scoggin, J. S., U.S. Patent 3,474,158 (Phillips Petroleum Co.) (1969); *C.A.* **71**, 125247h (1969).

C55. Yoshioka, S., Kurashima, K., and Kinoshita, Y., Japanese Patent 68/15,620 (Mitsui Chemical Industry Co., Ltd.) (1968); *C.A.* **70**, 20484j (1969).

C56. Takashima, N., Sakata, R., Fukuda, J., Tashiro, K., Sennari, M., and Nagashima, K., Japanese Patent 11,623 (Mitsubishi Petrochemical Co., Ltd.) (1965); *C.A.* **63**, 16498c (1965).

C57. Crawford, J. W. C., and Oakes, W. G., German Patent 1,217,608 (Imperial Chemical Industries, Ltd.) (1966); *C.A.* **65**, 9126d (1966).

C58. French Patent 1,560,565 (Asahi Chemical Industry Co., Ltd.) (1969); *C.A.* **71**, 113607n (1969).

C59. Netherlands Patent Appl. 6,613,955 (Shell Internationale Research) (1967); *C.A.* **68**, 30878p (1968).

C60. Delbouilie, A., and Leblon, E., German Offen. 2,109,924 (Solvay) (1971); *C.A.* **76**, 60289e (1972).

C61. Asada, M., Uchida, R., and Tomura, K., Japanese Patent 71/32,414 (Sumitomo Chemical Co., Ltd.) (1971); *C.A.* **76**, 46698f (1972).

C62. British Patent 1,233,106 (Phillips Petroleum Co.) (1971); *C.A.* **75**, 64604f (1971).

C63. Netherlands Patent Appl. 67 15,397 (Shell Internationale) (1968); *C.A.* **69**, 44483p (1969).

C64. Griffin, J. R., Jones, A. M., and Speed, R. A., U.S. Patent 3,401,212 (Esso Research and Engineering Co.) (1968); *C.A.* **69**, 87602u (1968).

C65. Yoshioka, S., Shiomura, T., and Kinoshita, Y., Japanese Patent 69/20,621 (Mitsui Toatsu Chemicals Co., Ltd.) (1969); *C.A.* **71**, 125197s (1969).

C66. British Patent 1,124,644 (Rexall Drug and Chemical Co.) (1968); *C.A.* **69**, 97461m (1968).

C67. Khelghatian, M. H., Jezl, J. L., and Hague, L. D., British Patent 1,058,011 (Avisun Corp.) (1967); *C.A.* **66**, 66115k (1967).

C68. Hermann, J. P., Leblon, E., and Jones, M. L., U.S. Patent 3,732,335 (Solvay) (1973).

C69. Kamaishi, T., Iwamoto, M., Tanaka, S., and Kirimura, S., Japanese Patent 71/32,410 (Toray Industries, Inc.) (1971); *C.A.* **76**, 46689d (1972).

C70. Bier, G., Gumboldt, A., and Lehmann, G., *Plast. Inst., Trans. J.* **28**, 98 (1960).

C71. Suzuki, K., and Yoshioka, O., Japanese Patent 71/32,411 (Kanegafuchi Chemical Industry Co., Ltd.) (1971); *C.A.* **76**, 34740a (1972).

C72. Fukui, K., Yuasa, Y., Shimizu, T., and Hirooka, M., Japanese Patent 685('62), (Sumitomo Chemical) (1962); *C.A.* **58**, 11480 (1963).

C73. Kern, R. J., and McManimie, R. J., U.S. Patent 3,478,129 (Monsanto Co.) (1969); *C.A.* **72**, 32473j (1970).

C74. Suzuki, K., Iwamoto, K., and Yoshioka, H., Japanese Patent 69/20,459 (Kanegafuchi Chemical) (1969); *C.A.* **72**, 32469n (1970).

C75. Belgian Patent 591,623 (Hoechst) (1960).

C76. Belgian Patent 602,681 (Hoechst) (1961).

C77. Bond, J. A., Clark, M., and Rayner, L. S., British Patent 970,478 (Imperial Chemical Industries) (1964); *C.A.* **61**, 14806f (1964).

C78. British Patent 1,199,669 (Asahi Chemical) (1970); *C.A.* **73**, 67219d (1970).

C79. Kontos, E. G., Belgian Patent 609,239 (Naugatuck) (1962).

C80. Kontos, E. G., Belgian Patent 637,093 (United States Rubber Co.) (1963);

C81. Gobran, R., Kremer, L. V., and Ethier, D. O., German Offen. 1,908,749 (Minnesota Mining and Manufacturing Co.) (1969); *C.A.* **71**, 125462z (1969).

C82. Hassell, H. L., U.S. Patent 3,378,698 (Shell Oil) (1968); *C.A.* **68**, 115542z (1968).

C83. Melville, H. W., and Stead, B. D., *J. Polym. Sci.* **16**, 505 (1955).

C84. Ke, B., *J. Polym. Sci.* **61**(171), 47 (1962); *C.A.* **58**, 1545b (1963).

C85. Barrall, E. M., Porter, R. S., and Johnson, J. F., *Polym. Prepr., Am. Chem. Soc., Div. Polym. Chem.* **5**(2), 816 (1964); *C.A.* **64**, 12795b (1966).

C86. Clegg, G. A., Gee, D. R., and Melia, T. P., *Makromol. Chem.* **119**, 184 (1968).

C87. Martynov, M. A., Pilipovskii, V. I., and Grigor'ev, V. A., *Plast. Massy* No. 10, p. 58 (1968); *C.A.* **70**, 12061d (1969).

C88. Martynov, M. A., Pilipovskii, V. I., and Leosko, E. A., *Vysokomol. Soedin., Ser. B* **14**(8), 623 (1972); *C.A.* **77**, 165198y (1972).

C89. Bier, G., Lehmann, G., and Leugering, H. J., *Makromol. Chem.* **44–46**, 347 (1961).

C90. Lomonti, J. N., and Tirpak, G. A., *J. Polym. Sci., Part A* **2**(2), 705 (1964); *C.A.* **60**, 9358b (1964).

C91. Bly, R. M., Kiener, P. E., and Fries, B. A., *Anal. Chem.* **38**(2), 217 (1966); *C.A.* **64**, 12795a (1966).

C92. Takeuchi, T., Tsuge, S., and Sugimura, Y., *Anal. Chem.* **41**(1), 184 (1969); *C.A.* **70**, 48084c (1969).

C93. Popov, V. P., and Duvanova, A. P., *Zh. Prikl. Spektrosk.* **18**(6), 1077 (1973); *C.A.* **79**, 66951y (1973).

C94. Fraser, G. V., Hendra, P. J., Walker, J. H., Cudby, M. E. A., and Willis, H. A., *Makromol. Chem.* **173**, 205 (1973); *C.A.* **80**, 133953m (1974).

C95. Porter, R. S., *J. Polym. Sci., Part A-1* **4**(1), 189 (1966); *C.A.* **64**, 14278a (1966).

C96. Nencini, G., Giuliani, G., and Salvatori, T., *J. Polym. Sci., Part B* **3**(6), 483 (1965).

C97. Bier, V. G., Lehmann, G., and Leugering, H. J., *Makromol. Chem.* **44–46**, 354 (1961).

C98. Aijima, I., Sakurai, H., Kitaoka, A., and Katayama, Y., Japanese Patent 70/36,419 (Asahi Chemical) (1970); *C.A.* **75**, 141547u (1971).

C99. Frailey, N. E., and Welch, R. A., U.S. Patent 3,793,283 (Shell Oil) (1974).

C100. Ishizuka, K., Manako, H., Yamauchi, A., Iwatani, K., Kakogawa, G., Goko, N., Matsuura, S., and Sawano, K., Japanese Patent (Kokai) 73/25,781 (Mitsubishi Chemical) (1973); *C.A.* **79**, 92836j (1973).

C101. Horino, T., Iwami, E., Ban, K., Soen, T., and Kawai, H., *Kogyo Kagaku Zasshi* **73**(7), 1615 (1970); *C.A.* **73**, 110383s (1970).

C102. Hostetler, D. E., French Patent 1,352,024 (W. R. Grace) (1964); *C.A.* **61**, 9642a (1964).

C103. Hagemeyer, H. J., Jr., and Edwards, M. B., British Patent 989,724 (Eastman Kodak) (1965); *C.A.* **63**, 3074C (1965).

C104. Hara, I., Shiroyama, K., Kitamura, T., and Hirukawa, M., Japanese Patent 70/17,545 (Sumitomo Chemical) (1970); *C.A.* **73**, 121165y (1970).

C105. Hagemeyer, H. J., Jr., and Edwards, M. B., British Patent 1,156,030 (Eastman Kodak) (1969); *C.A.* **71**, 50821e (1969).

C106. British Patent 1,154,447 (Asahi Chemical) (1969); *C.A.* **71**, 50792w (1969).

C107. Tusch, R. L., *Polym. Eng. Sci.* **6**(3), 255 (1966); *C.A.* **65**, 10681 (1966).

C108. British Patent 1,124,644 (Rexall Drug & Chemical) (1968).

C109. Hoestetler, D. E., U.S. Patent 3,435,095 (Rexall Drug and Chemical) (1964).

C110. Nishimura, M., and Kobayashi, N., Japanese Patent (Kokai) 73/21,731 (Chisso Corp.) (1973); *C.A.* **79**, 54505t (1973).

C111. Friedlander, H. N., *Encycl. Polym. Sci. Technol.* **6**, 338 (1967); *C.A.* **69**, 10722a (1968).

C112. Cau, P., and Anelli, E., German Offen. 2,004,379 (Montecatini) (1970); *C.A.* **73**, 88867d (1970).

C113. Aijima, I., Fukuma, N., Sakurai, H., Chiaya, M., Okamoto, T., Tsuchii, Y., and Itsumi, H., Japanese Patent 69/17, 567 (Asahi Chemical) (1969); *C.A.* **72**, 13750f (1970).

C114. Sevcik, F., Suchanek, J., Ondrejmiska, K., Fejedelem, M., Kovac, J., and Prokopec, J., Czech Patent 120,845 (1966); *C.A.* **69**, 44530b (1968).

C115. Kontos, E. G., U.S. Patent 3,378,606 (Uniroyal) (1962).

C116. Puett, D., Smith, K. J., Jr., Ciferri, A., and Kontos, E. G., *J. Chem. Phys.* **40**(1), 253 (1964); *C.A.* **60**, 6944a (1964).

C117. Puett, D., Smith, K. J., Jr., and Ciferri, A., *J. Phys. Chem.* **69**, 141 (1965).

C118. Hassell, H. L., and Hayter, R. G., U.S. Patent 3,480,696 (Shell Oil) (1969).

C119. Netherlands Patent Appl. 6,613,955 (Shell Internationale) (1967); *C.A.* **68**, 30878p (1968).

C120. *Rubber World* **167**, No. 5, p. 49 (1973).

C121. *Materials Engineering*, 38 (1972).

C122. *Process Engineering News* p. 11 (1972).

C123. *Rubber and Plastics News* p. 1 (1972).

C124. Morris, H. L., *Soc. Plast. Eng., Tech. Pap.* **19**, 88 (1973); *C.A.* **79**, 20011p (1973).

C125. Morris, H. L., *J. Elastomers Plast.* **6**, 121 (1974).

C126. Pilipovskii, V. I., Grigor'ev, V. A., Shibalovskaya, S. A., and Nalivalko, E. I., *Plast. Massy* No. 5, 56 (1970); *C.A.* **73**, 35892b (1970).

C127. Okada, H., and Fukumoto, O., Japanese Patent 70/03, 934 (Toyo Rayon Co., Ltd.) (1970); *C.A.* **73**, 26193x (1970).

C128. Leibson, I., and Erchak, M., Jr., U.S. Patent 3,514,501 (Dart Industries) (1970); *C.A.* **73**, 15721s (1970).

C129. Yakobson, F. I., Amerik, V. V., Petrova, V. F., Shteinbak, V. Sh., and Ivanyukov, D. V., *Plast. Massy* No. 3, p. 11 (1970); *C.A.* **72**, 112111e (1970).

C130. British Patent 1,202,554 (Dart Industries) (1970); *C.A.* **73**, 110446q (1970).

C131. Hagemeyer, H. J., Jr., and Edwards, M. B., U.S. Patent 3,529,037 (1970); *C.A.* **73**, 131649y (1970).

C132. Tsujino, T., Umitani, A., Kuroda, N., and Mieda, T., Japanese Patent 71/34,091 (Japan Oil Co.) (1971); *C.A.* **76**, 60317n (1972).

C133. Bernard, J. C., and Pascal, P., German Offen. 2,032,382 (Naphtachimie) (1971); *C.A.* **74**, 112841j (1971).

C134. Yakobson, F. I., Ivanyukov, D. V., Amerik, V. V., Petrova, V. F., Krymov, P. V., Shteinbak, V. Sh., and Krentsel, B. A., U.S.S.R. Patent 300,481 (1971); *C.A.* **75**, 98944g (1971).

C135. Downing, S. B., and Seaborn, J. R., German Offen. 2,049,497 (Imperial Chemical Industries) (1971); *C.A.* **75**, 37021g (1971).

C136. Bond, J. A., Clark, M., and Rayner, L. S., British Patent 970,479 (Imperial Chemical Industries) (1964); *C.A.* **61**, 14806h (1964).

C137. British Patent 957,777 (Farbwerke Hoechst) (1964); *C.A.* **61**, 14803f (1964).

C138. Netherlands Patent Appl. 6,402,284 (Avisun Corp.) (1965); *C.A.* **64**, 8404b (1966).

C139. Italian Patent 594,018 (Montecatini) (1959); *C.A.* **55**, 12937h (1961).

C140. Clegg, G. A., Gee, D. R., and Melia, T. P., *Makromol. Chem.* **132**, 203 (1970); *C.A.* **72**, 122034w (1970).

C141. Coover, H. W., Jr., McConnell, R. L., Joyner, F. B., Slonaker, D. F., and Combs, R. L., *J. Polym. Sci., Part A-1* **4**, 2563 (1966).

C142. French Patent 1,580,554 (Imperial Chemical Industries) (1969); *C.A.* **73**, 15526g (1970).

C143. Suzuki, K., Japanese Patent 69/04,787 (Kanegafuchi Chemical) (1969); *C.A.* **71**, 13512j (1969).

C144. Guillet, J. E., and Combs, R. L., U.S. Patent 3,519,586 (Eastman Kodak) (1970); *C.A.* **73**, 56826y (1970).

C145. British Patent 986,189 (Farbwerke Hoechst) (1965); *C.A.* **63**, 3073c (1965).

C146. Cash, G. O., Jr., Defensive Publication U.S. Patent Office 746,313 (1969); *C.A.* **71**, 81873c (1969).

C147. Takida, H., and Maruhashi, M., Japanese Patent (Kokai) 73/64,189 (Japan Synthetic Chemical) (1973); *C.A.* **80**, 134071j (1974).

C148. British Patent 838,996 (Goodrich-Gulf Chemicals) (1960); *C.A.* **54**, 26016b (1960).

C149. Tanaka, S., Kamagata, K., Morikawa, H., and Masukawa, S., Japanese Patent 70/33,903 (Mitsubishi Petrochemical) (1970); *C.A.* **74**, 42869y (1971).

C150. Alderson, G. W., and Richards, C. T., British Patent 1,093,344 (British Petroleum) (1967); *C.A.* **68**, 30387c (1968).

C151. Netherlands Patent Appl. 6,516,232 (British Petroleum) (1966); *C.A.* **66**, 3160d (1967).

C152. British Patent 1,094,838 (Mitsubishi Petrochemical) (1967); *C.A.* **68**, 40297n (1968).

C153. Langer, A. W., Jr., U.S. Patent 3,450,795 (Esso Research and Engineering) (1969).

C153a. Langer, A. W., Jr., ed., "Polyamine-Chelated Alkali Metal Compounds," Adv. Chem. Ser., No. 130. Am. Chem. Soc: Washington, D.C., 1974.

C154. Suzuki, Y., Saito, M., and Kosaka, Y., *Kogyo Kagaku Zasshi* **74**(3), 481 (1971); *C.A.* **75**, 21077u (1971).

C155. Gehrke, K., Schmidt, B., and Ulbricht, J., *Plaste Kautsch.* **18**(5), 329 (1971); *C.A.* **75**, 21142m (1971).

C156. Agouri, E., Catte, R., and Dauba, J. L., German Offen. 2,165,367 (Société Anon. Aquitaine-Organico) (1972); *C.A.* **77**, 165701g (1972).

C157. Chabert, H., German Offen. 1,949,594 (Rhone-Poulenc) (1970); *C.A.* **72**, 133598j (1970).

C158. Tanzawa, H., Tanaka, T., and Soda, A., *J. Polym. Sci., Part A-2* **7**(5), 929 (1969); *C.A.* **71**, 71062y (1969).

C159. Yamazaki, N., Shirakawa, H., and Kambara, S., *J. Polym. Sci., Part C* **16**(3), 1685 (1967); *C.A.* **67**, 44136m (1967).

C160. Anderson, W. S., U.S. Patent 3,290,414 (Shell Oil) (1966); *C.A.* **66**, 29400d (1967).

C161. Amerik, V. V., Kleiner, V. I., Ivanyukov, D. V., and Krentsel, B. A., French Demande 2,027,194 (A. V. Topechiev Institute of Petrochemical Synthesis) (1970); *C.A.* **74**, 142662u (1971).

C162. Amerik, V. V., Invayukov, D. V., Stotskaya, L. L., and Shteinback, V. Sh., German Offen. 1,964,883 (Topchiev, A. V., Institute of Petrochemical Synthesis) (1970); *C.A.* **73**, 121218t (1970).

C163. Americk, V. V., Inayukov, D. V., Yakobson, F. I., and Krentsel, B. A., *Vysokomol. Soedin., Ser. B* **15**(7), 500 (1973); *C.A.* **80**, 37481k (1974).

C164. Minekawa, S., Yamaguchi, K., and Toyomoto, K., Japanese Patent 68/21,069 (Asahi Chemical) (1968); *C.A.* **70**, 48022f (1969).

C165. Kohotsune, S., Shirai, I., and Watanabe, K., Japanese Patent 69/19,252 (Chisso Corp.) (1969); *C.A.* **71**, 125229d (1969).

C166. Asada, T., Sugiyama, H., and Onogi, S., *Zairyo* **20**(212) 595 (1971); *C.A.* **75**, 77409w (1971).

C167. Aoyama, T., Japanese Patent 72/26,114 (Nisseki Jushi Kagaku Co.) (1972); *C.A.* **78**, 4783r (1973).

C168. Leugering, H. J., and Schaum, H., S. African Patent 69/04,328 (Farbwerke Hoechst) (1970); *C.A.* **73**, 26190u (1970).

C169. Hostetler, D. E., S. African Patent 68/02,419 (Rexall Drug and Chemical) (1968); *C.A.* **70**, 78541s (1969).

C170. British Patent 1,054,764 (1963).

C171. Chu, N. S., and Jezl, J. L., French Patent 1,531,409 (Avisun) (1968); *C.A.* **71**, 13734h (1969).

C172. Agouri, E., Parlant, C., Mornet, P., Rideau, J., and Teitgen, J. F., *Makromol. Chem.* **137**, 229 (1970); *C.A.* **73**, 88521m (1970).

C173. Agouri, E., Parlant, C., Mornet, P., Rideau, J., and Teitgen, J. F., *Makromol. Chem.* **137**, 229 (1970).

C174. Craven, W. J., S. African Patent 67/06,649 (Rexall Drug and Chemical) (1968); *C.A.* **70**, 38297k (1969).

C175. British Patent 1,110,995 (Rexall Drug and Chemical) (1968); *C.A.* **69**, 11149f (1968).

C176. Nakatsuka, K., Ide, F., and Jo, Y., Japanese Patent 68/06,535 (Mitsubishi Rayon) (1968); *C.A.* **69**, 78405e (1968).

C177. Nakatsuka, K., Ide, F., and Jo, Y., Japanese Patent 68/06,536 (Mitsubishi Rayon) (1968); *C.A.* **69**, 52907y (1968).

C178. Konotsune, S., and Matsumoto, S., Japanese Patent 69/08,679 (Chisso Corp.) (1969); *C.A.* **71**, 81881d (1969).

C179. Jezl, J. L., Chu, N. S., and Khelghatian, H. M., *Adv. Chem. Ser.* **91**, 268 (1969); *C.A.* **72**, 21990z (1970).

C180. Hostetler, D. E., French Patent 1,508,588 (Rexall Drug and Chemical) (1968); *C.A.* **70**, 20941f (1969).

C181. Szwarc, M., "Carbanions, Living Polymers, and Electron Transfer Processes." Wiley, New York, 1968.

C182. Kennedy, J. P., and Melby, E. G., *Polym. Prepr., Am. Chem. Soc., Div. Polym. Chem.* **15**(2), 180 (1974).

D1. Schmolka, I. R., *in* "Nonionic Surfactants" (M. J. Schick, ed.), Chapter 10, p. 300. Dekker, New York, 1967.

D2. Jackson, D. R., and Lundsted, L. G., U.S. Patent 2,677,700 (Wyandotte Chemicals) (1954).

D3. Jackson, D. R., and Lundsted, L. G., U.S. Patent 3,036,130 (Wyandotte Chemicals) (1962).

D4. Bailey, F. E., Jr., and France, H. G., U.S. Patent 3,029,216 (Union Carbide Corp.) (1962); *C.A.* **57**, 2440c (1962).

D5. Skoulios, A. E., Tsouladze, G., and Franta, E., *J. Polym. Sci., Part C* **4**, 507 (1964).

D6. Tsouladze, G., and Skoulios, A., *J. Chim. Phys.* **60**(5), 626 (1963); *C.A.* **59**, 7713h (1963).

D7. Fujii, H., Saegusa, T. T., and Furukawa, J., *Makromol. Chem.* **63**, 147 (1963); *C.A.* **59**, 2966b (1963).

D8. Netherlands Patent Appl. 6,408,971 (Farbwerke Hoechst) (1965); *C.A.* **63**, 3076f (1965).

D9. Saegusa, T., Matsumoto, S., and Hashimoto, Y., *Macromolecules* **3**(4), 377 (1970); *C.A.* **73**, 77644c (1970).

D10. Hay, A. S., Blanchard, H. S., Endres, G. F., and Eustance, J. W., *J. Am. Chem. Soc.* **81**, 6335 (1959).

D11. Bennett, J. G., Jr., and Cooper, G. D., *Macromolecules* **3**(1), 101 (1970); *C.A.* **72**, 101140n (1970).

D12. Cooper, G. D., Bennett, J. G., Jr., and Katchman, A., *Adv. Chem. Ser.* **99**, 431 (1971); *C.A.* **75**, 88955x (1971).

D13. Cooper, G. D., Bennett, J. G., Jr., and Factor, A., *Adv. Chem. Ser.* **128**, 230 (1973).

D14. Kray, R. J., and Stevenson, R. W., U.S. Patent 3,754,053 (Celanese Corp. of America) (1973); *C.A.* **80**, 15638q (1974).

D15. Matsumoto, S., and Mitsueda, T., Japanese Patent 73/07,719 (1973); *C.A.* **80**, 122135v (1974).

D16. Cooper, G. D., Bennett, J. G., Jr., and Factor, A., *Nuova Chim.* **48**(9), 93 (1972); *C.A.* **77**, 165126y (1972).

D17. Konstantinov, C., and Kabaivanov, V., *Polymer* **12**(6), 358 (1971); *C.A.* **75**, 77324q (1971).

D18. Finaz, G., Rempp, P., and Parrod, J., *Bull. Soc. Chim. Fr.* **1962**, 262 (1962); *C.A.* **57**, 1055f (1962).

D19. O'Malley, J. J., Crystal, R. G., and Erhardt, P. F., *Polym. Prepr., Am. Chem. Soc., Div. Polym. Chem.* **10**(2), 796 (1969).

D20. Franta, E., *J. Chim. Phys.* **63**(4), 595 (1966); *C.A.* **65**, 3983a (1966).

D21. Kovacs, A. J., Manson, J. A., and Levy, D., *Kolloid Z. & Z. Polym.* **214**(1), 1 (1966); *C.A.* **66**, 46708c (1967).

D22. Lotz, B., Kovacs, A. J., Bassett, G. A., and Keller, A., *Kolloid Z. & Z. Polym.* **209**(2), 115 (1966); *C.A.* **65**, 15530a (1966).

D23. Lotz, B., and Kovacs, A. J., *Kolloid Z. & Z. Polym.* **209**(2), 97 (1966); *C.A.* **65**, 15529h (1966).

D24. Skoulios, A., *C.R. Hebd. Seances Acad. Sci.* **255**(23), 3189 (1962); *C.A.* **59**, 2988e (1963).

D25. Skoulios, A., and Finaz, G., *J. Chim. Phys.* **59**, 473 (1962); *C.A.* **57**, 12701f (1962).

D26. Franta, E., Skoulios, A., Rempp, P., and Benoit, H., *Makromol. Chem.* **87**, 271 (1965); *C.A.* **63**, 18363h (1965).

D27. Skoulios, A., and Finaz, G., *C.R. Hebd. Seances Acad. Sci.*, **252**, 3467 (1961); *C.A.* **55**, 22995e (1961).

D28. Gervais, M., and Gallot, B., *C.R. Hebd. Seances Acad. Sci., Ser. C* **270**(9), 784 (1970); *C.A.* **72**, 12217h (1970).

D29. Finaz, G., Skoulios, A., and Sadron, C., *C.R. Hebd. Seances Acad. Sci.* **253**, 265 (1961); *C.A.* **56**, 7485g (1962).

D30. Sadron, C., *Pure Appl. Chem.* **4**, 347 (1962); *C.A.* **57**, 8717f (1962).

D31. Sadron, C., *Angew. Chem.* **75**(11), 472 (1963); *C.A.* **59**, 6523a (1963).

D32. Crystal, R. G., O'Malley, J. J., and Erhardt, P. F., *Polym. Prepr., Am. Chem. Soc., Div. Polym. Chem.* **10**(2), 804 (1969).

D33. Short, J. M., and Crystal, R. G., *Appl. Polym. Symp.* **16**, 137 (1971); *C.A.* **75**, 77606h (1971).

D34. Pochan, J. M., *Polym. Prepr., Am. Chem. Soc., Div. Polym. Chem.* **12**(1), 212 (1971); *C.A.* **78**, 16612x (1973).

D35. Crystal, R. G., *Polym. Prepr., Am. Chem. Soc., Div. Polym. Chem.* **11**(2), 668 (1970).

D36. Gervais, M., and Gallot, B., *Makromol. Chem.* **171**, 157 (1973); *C.A.* **80**, 71238w (1974).

D37. Gervais, M., and Gallot, B., *Makromol. Chem.* **174**, 193 (1973); *C.A.* **80**, 121429g (1974).

D38. Lotz, B., and Kovacs, A. J., *Polym. Prepr., Am. Chem. Soc., Div. Polym. Chem.* **10**(2), 820 (1969).

D39. Kovacs, A. J., *Chim. Ind., Genie Chim.* **97**(3), 315 (1967); *C.A.* **66**, 95717p (1967).

D40. Erhardt, P. F., O'Malley, J. J., and Crystal, R. G., *Polym. Prepr., Am. Chem. Soc., Div. Polym. Chem.* **10**(2), 812 (1969).

D41. Erhardt, P. F., O'Malley, J. J., and Crystal, R. G., *Block Polym., Proc. Symp., 1969* p. 195 (1970); *C.A.* **75**, 6487j (1971).

D42. Smith, W. E., Galiano, F. R., Rankin, D., and Mantell, G. J., *J. Appl. Polym. Sci.* **10**(11), 1659 (1966); *C.A.* **66**, 2838j (1967).

D43. Smith, J. J., and Reichle, W. T., U.S. Patent 2,921,920 (Union Carbide Corp.) (1960); *C.A.* **54**, 23445h (1960).

D44. Ota, S., Saegusa, T., and Furukawa, J., *Kogyo Kagaku Zasshi* **67**(6), 947 (1964); *C.A.* **61**, 13427h (1964).

D45. Minoura, Y., and Mitoh, M., *Makromol. Chem.* **99**, 186 (1966); *C.A.* **66**, 38257q (1967).

D46. Yamashita, Y., *Adv. Chem. Ser.* **91**, 350 (1969); *C.A.* **72**, 21986c (1970).

D47. Yamashita, Y., Okada, M., and Hirota, M., *Angew. Makromol. Chem.* **9**, 136 (1969); *C.A.* **72**, 55931w (1970).

D48. Netherlands Patent Appl. 6,411,511 (Chemical Investors) (1965); *C.A.* **63**, 10089d (1965).

D49. Kirkland, E. V., and Roberts, W. J., U.S. Patent 3,219,725 (Celanese Corp.) (1965).

D50. Mantell, G. J., Smith, W. E., Galiano, F. R., and Rankin, D., U.S. Patent 3,732,333 (Gulf Oil Corp.) (1973); *C.A.* **79**, 67066a (1973).

D51. Takida, H., and Noro, K., *Kobunshi Kagaku* **21**(234), 459 (1964); *C.A.* **62**, 10537e (1965).

D52. Takahashi, A., and Yamashita, Y., *Polym. Prepr., Am. Chem. Soc., Div. Polym. Chem.* **15**(1), 184 (1974).

D53. Bailey, F. E., Jr., and France, H. G., U.S. Patent 3,312,753 (Union Carbide Corp.) (1967); *C.A.* **67**, 3393z (1967).

D54. Busler, W. R., U.S. Patent 3,489,819 (Phillips Petroleum Co.) (1970); *C.A.* **72**, 56271t (1970).

D55. Netherlands Patent Appl. 69/12,915 (Shell Internationale) (1969); *C.A.* **72**, 79672p (1970).

D56. Hsieh, H. L., Mueller, F. X., Jr., and Busler, W. R., German Offen. 1,805,864 (Phillips Petroleum Co.) (1970); *C.A.* **73**, 78331k (1970).

D57. Kobayashi, F., Iwasaki, T., Utsumi, N., and Asahra, N., Japanese Patent 68/26,629 (Japan Rayon Co., Ltd.) (1968); *C.A.* **70**, 78742h (1969).

D58. Tabuchi, T., Nobutoki, K., and Sumitomo, H., *Kogyo Kagaku Zasshi* **71**(11), 1926 (1968); *C.A.* **70**, 58488b (1969).

D59. Mueller, F. X., Jr., and Hsieh, H. L., U.S. Patent 3,585,257 (Phillips Petroleum Co.) (1971).

D60. Nobutoki, K., and Sumitomo, H., *Bull. Chem. Soc. Jpn.* **40**, 1741 (1967).

D61. Katayama, S., and Horikawa, H., German Offen. 1,928,080 (Denki Onkyo Co.) (1970); *C.A.* **72**, 112080u (1970).

D62. King, C., and Wallenberger, F. T., U.S. Patent 3,538,195 (du Pont de Nemours) (1970); *C.A.* **74**, 32623z (1971).

D63. Ogawa, Y., Notani, K., Yamakawa, Y., and Awata, N., Japanese Patent 72/28,719 (Kanebo Co.) (1972); *C.A.* **78**, 98897f (1973).

D64. Yamashita, Y., and Hane, T., *J. Polym. Sci., Polym. Chem. Ed.* **11**(2), 425 (1973); *C.A.* **78**, 111835f (1973).

D65. Allegrezza, A. E., Jr., Lenz, R. W., Cornibert, J., and Marchessault, R. H., *Polym. Prepr., Am. Chem. Soc., Div. Polym. Chem.* **14**(2), 1232 (1973).

D66. Gobran, R. H., and Osborn, S. W., U.S. Patent 3,504,050 (Thiokol Chemical Corp.) (1970); *C.A.* **72**, 122651v (1970).

D67. Gobran, R. H., and Osborn, S. W., French Patent 1,524,572 (Thiokol Chemical Corp.) (1968); *C.A.* **71**, 13554z (1969).

D68. Korotneva, L. A., Belonovskaya, G. P., Korol, N. A., and Dolgoplosk, B. A., *Dokl. Akad. Nauk SSSR* **178**(5), 1084 (1968); *C.A.* **68**, 96208b (1968).

D69. Boileau, S., and Sigwalt, P., *Makromol. Chem.* **131**, 7 (1970); *C.A.* **72**, 911021 (1970).

D70. Boileau, S., and Sigwalt, P., *Makromol. Chem.* **171**, 11 (1973); *C.A.* **80**, 3843q (1974).

D71. Nevin, R. S., and Pearce, E. M., *J. Polym. Sci., Part B* **3**(6), 487 (1965); *C.A.* **63**, 3069d (1965).

D72. Gourdenne, A., and Sigwalt, P., *Eur. Polym. J.* **3**(3), 481 (1967); *C.A.* **67**, 91142v (1967).

D72a. Gourdenne, A. and Sigwalt, P., *Bull. Soc. Chim. Fr.* No. 7, p. 2249 (1967); *C.A.* **67**, 91147a (1967).

D73. Balcerzyk, E., Pstrocki, H., and Wlodarski, G., *Rocz. Chem.* **44**(7–8), 1583 (1970); *C.A.* **74**, 32225w (1971).

D74. Gourdenne, A., *Makromol. Chem.* **158**, 261 (1972); *C.A.* **77**, 114942d (1972).

D75. Lambert, J. L., and Goethals, E. J., *Makromol. Chem.* **133**, 289 (1970); *C.A.* **73**, 133272y (1970).

D76. Sebenda, J., *J. Macromol. Sci., Chem.* **6**, No. 6, 1145 (1972).

D77. Fasman, G. D., and Tooney, N., *Encycl. Polym. Sci. Technol.* **2**, 837 (1964); *C.A.* **65**, 3965b (1966).

D78. Oya, M., Uno, K., and Iwakura, Y., *Prog. Polym. Sci. Jpn.* **6**, 51 (1973).

D79. Walton, A. G., and Blackwell, J., "Biopolymers," p. 189. Academic Press, New York, 1973.

D80. Itoh, K., and Katabuchi, H., *Biopolymers* **12**(4), 921 (1973); *C.A.* **79**, 19303d (1973).

D81. Katakai, R., Toda, F., Uno, K., Iwakura, Y., and Oya, M., *Chem. Lett.* No. 7, p. 763 (1973); *C.A.* **79**, 105666f (1973).

D82. Reibel, L., and Spach, G., *Bull. Soc. Chim. Fr.* No. 3, p. 1025 (1972); *C.A.* **77**, 20166b (1972).

D83. Itoh, K., and Katabuchi, H., *Biopolymers* **11**(8), 1593 (1972); *C.A.* **77**, 140655g (1972).

D84. Mori, S., *Kobunshi Kagaku* **30**(9), 546 (1973); *C.A.* **80**, 96419u (1974).

D85. Fairweather, R., and Jones, J. H., *J. Chem. Soc., Perkin Trans. 1* No. 19, p. 2475 (1972); *C.A.* **78**, 30246n (1973).

D86. Fairweather, R., and Jones, J. H., *J. Chem. Soc., Perkin Trans. 1* No. 15, p. 1908 (1972); *C.A.* **77**, 114911t (1972).

D87. Paolillo, L., Temussi, P., Trivellone, E., Bradbury, E. M., and Crane-Robinson, C., *Biopolymers* **10**(12), 2555 (1971); *C.A.* **76**, 113729g (1972).

D88. Boni, R., Filippi, B., Ciceri, L., and Peggion, E., *Biopolymers* **9**(12), 1539 (1970); *C.A.* **74**, 76745v (1971).

D89. Szwarc, M., *Adv. Polym. Sci.* **4**, 1 (1965).

D90. Szwarc, M., and Schuerch, C., *Polym. Biol. Syst., Ciba Found. Symp., 1972,* p. 7 (1972); *C.A.* **79**, 5627a (1973).

D91. Ooya, M., Uno, K., and Iwakura, Y., Japanese Patent 70/24,790 (Asahi Chemical Industry Co., Ltd.) (1970); *C.A.* **73**, 110333a (1970).

D92. Imanishi, Y., Sugihara, T., and Higashimura, T., *Biopolymers* **12**(7), 1505 (1973); *C.A.* **79**, 105636w (1973).

D93. Seeney, C. E., and Harwood, H. J., *Polym. Prepr., Am. Chem. Soc., Div. Polym. Chem.* **15**(2), 239 (1974).

D94. Matzner, M., McGrath, J. E., and Noshay, A., U.S. Patent 3,770,849 (Union Carbide Corp.) (1973); *C.A.* **80**, 71354f (1974).

D95. Okazaki, K., and Nakayama, Y., Japanese Patent 68/19,033 (Toyo Rayon Co., Ltd.) (1968); *C.A.* **70**, 69226r (1969).

D96. Litt, M., and Herz, J., *Polym. Prepr., Am. Chem. Soc., Div. Polym. Chem.* **10**(2), 905 (1969).

D97. Oya, M., Uno, K., and Iwakura, Y., *Bull. Chem. Soc. Jpn.* **43**(6), 1788 (1970); *C.A.* **73**, 56475h (1970).

D98. Godfrey, R. A., and Miller, G. W., *J. Polym. Sci., Part A-1* **7**, 2387 (1969).

D99. French Patent 1,598,865 (Farbenfabriken Bayer A.-G.) (1970); *C.A.* **74**, 77234q (1971).

D100. Minoura, Y., Mitoh, M., Tabuse, A., and Yamada, Y., *J. Polym. Sci., Part A-1* **7**(9), 2753 (1969); *C.A.* **71**, 124971c (1969).

D101. Saam, J. C., Gordon, D. J., and Lindsey, S., *Macromolecules* **3**(1), 1 (1970); *C.A.* **72**, 79520n (1970).

D102. Saam, J. C., and Fearon, F. W. G., *Ind. Eng. Chem., Prod. Res. Dev.* **10**, 10 (1971).

D103. Gaines, G. L., Jr., and Bender, G. W., *Macromolecules* **5**(1), 82 (1972); *C.A.* **76**, 141638v (1972).

D104. Saam, J. C., and Fearon, F. W. G., French Patent 2,105,983 (Dow Corning Corp.) (1972); *C.A.* **78**, 16772z (1973).

D105. Juliano, P. C., German Offen. 2,164,469 (General Electric Co.) (1972); *C.A.* **77**, 115360z (1972).

D106. Juliano, P. C., U.S. Patent 3,663,650 (General Electric Co.) (1972); *C.A.* **77**, 127417m (1972).

D107. Minoura, Y., Shundo, M., and Enomoto, Y., *J. Polym. Sci., Part A-1* **6**(4), 979 (1968); *C.A.* **68**, 105726y (1968).

D108. Minoura, Y., Shundo, M., and Enomoto, Y., *Kogyo Kagaku Zasshi* **70**(6), 1025 (1967); *C.A.* **68**, 96250j (1968).

D109. Netherlands Patent Appl. 6,408,970 (Union Carbide Corp.) (1965); *C.A.* **63**, 1963d (1965).

D110. Saam, J. C., German Offen. 2,011,088 (Dow Corning Corp.) (1970); *C.A.* **73**, 121222q (1970).

D111. Saam, J. C., and Fearon, F. W. G., *Polym. Prepr., Am. Chem. Soc., Div. Polym. Chem.* **11**(2), 455 (1970).

D112. Ito, N., and Hosogane, T., Japanese Patent 73/13,705 (Showa Denko K. K.) (1973); *C.A.* **80**, 37859q (1974).

D113. Gourdenne, A., and Sigwalt, P., *Eur. Polym. J.* **3**(3), 481 (1967); *C.A.* **67**, 91142 (1967).

6

A-B-A
Triblock Copolymers

This chapter is comprised of block copolymers containing three segments. Section A describes systems in which all three segments are hydrocarbon in nature, e.g., styrene–butadiene. Section B deals with systems containing one or more acrylic or vinylpyridine segments. Section C describes block copolymers in which one or more of the blocks contains heteroatoms in the backbone.

A. HYDROCARBON BLOCK COPOLYMERS

1. Styrene–Dienes

In this section, the synthesis, characterization, and physical behavior of styrene–diene–styrene A-B-A block copolymers are reviewed. The rejuvenation of interest in block copolymers as a means of achieving new and novel properties may be traced back to the discovery that these particular A-B-A structures are capable of behaving like thermoplastic elastomers (A1). They can be melt processed or solution cast like thermoplastics and yet display rubberlike elasticity in the absence of chemical cross-links. The total market for all thermoplastic elastomer materials in the mid-1980s is predicted to be one billion pounds per year (450,000 metric tons) (A2). The method of choice for preparing these copolymers is the Szwarc technique of "living" anionic polymerization (A3, A4). Styrene–diene polymerizations in hydrocarbon solvents are nonterminating and thus allow the preparation of well-defined structures. The availability of these macromolecules prompted studies of their solution and bulk properties. These findings in turn, have led to a much better understanding of the structure–property relationships in block copolymers in general. In particular, one can now relate molecular weight and composition to morphology, rheology, viscoelastic behavior, solution properties, and mechanical properties. An even greater range of properties has been achieved by chemically modifying one or more of the preformed segments.

One of the larger areas of interest to develop for block copolymers is their utility in modifying the physical properties of other polymers. Such alloys are possible due to the mechanical compatibility of, or phase adhesion between, one segment of a block copolymer and its corresponding homopolymer. Within certain molecular dimensional limits, the homopolymer molecules are indistinguishable from the corresponding block segments. This behavior, of course, is also true for A-B and $\{A\text{-}B\}_n$ structures as well as for A-B-A copolymers. Block copolymers can also, to a certain degree, emulsify two homopolymers. These characteristics have led to a number of applications for styrene–diene–styrene systems.

a. Synthesis

The synthesis technique of anionic block copolymerization is the only approach that produces well-defined styrene–diene A-B-A structures.

The preparation of triblock polymers may be classified by four different approaches, depending on the initiator chosen (A5–A11): (i) Difunctional initiator process—use of a dicarbanion or dicarbanion-forming species (e.g., sodium naphthalene) in either a one-stage process or a two-stage sequential addition process. (ii) Three-stage sequential addition process—use of monofunctional initiators (e.g., alkyllithiums) with three sequential monomer additions. (iii) Coupling process—use of a two-stage process to prepare a diblock polymer that is "coupled" (e.g., with phosgene or alkyl dihalides) to form the A-B-A. (iv) Tapered blocks process—polymerization (partial or complete) of the initial styrene block followed by copolymerization of a mixture of styrene and diene. In this second step, the diene is preferentially polymerized first (in hydrocarbon solvents) to produce a predominantly A-B-A structure directly.

i. *Difunctional Initiator Process.*   The patent literature of 1939 described the discovery by Scott (A12) that sodium metal in various aromatic solvents was capable of polymerizing conjugated hydrocarbon monomers such as styrene and butadiene. In 1956, Paul and coworkers (A13) showed that sodium naphthalene and sodium biphenyl were actually ion radicals that were formed by electron transfer from the sodium atom to the naphthalene molecule (Reaction 6-1). Upon

$$\text{naphthalene} + \text{Na} \xrightarrow{\text{THF}} \left[\text{naphthalene} \right]^{\cdot-} \text{Na}^+ \tag{6-1}$$

addition of an unsaturated monomer such as styrene, Szwarc (A3, A14) proposed an electron transfer to the lowest unoccupied orbital of the monomer to generate a styrene radical anion (Reaction 6-2). Dimeriza-

$$\left[\text{naphthalene} \right]^{\cdot-} \text{Na}^+ + \underset{\text{C}_6\text{H}_5}{\text{CH}_2{=}\text{CH}} \rightleftharpoons \left[\cdot\text{CH}_2{-}\overset{..}{\text{CH}} \right]^- \text{Na}^+ + \text{naphthalene} \tag{6-2}$$

tion of the radical anion produces a relatively stable dianionic moiety capable of growing polymer at both ends (Reaction 6-3). Because of

$$2 \left[\cdot\text{CH}_2{-}\overset{..}{\text{CH}} \right]^- \text{Na}^+ \longrightarrow \text{Na}^+ \; {}^-\text{HC}{-}\text{CH}_2{-}\text{CH}_2{-}\text{CH}^- \; \text{Na}^+ \tag{6-3}$$

the stability of the polymeric dianions (in the absence of water, oxygen, or other impurities), Szwarc described them as "living" polymers. Accordingly, it is possible to polymerize a diene monomer to produce a dianion-terminated macromolecule that is still active and therefore capable of initiating the polymerization of sequentially added styrene monomer (Reaction 6-4). Two other features of the mechanism require

$$\text{Na}^+ \; {}^-\text{Polydiene} \; {}^-\text{Na}^+ + \text{C}_6\text{H}_5\text{CH}{=}\text{CH}_2$$
$$\downarrow$$
$$\text{Polystyrene Polydiene Polystyrene} \tag{6-4}$$

comment. First, assuming macromolecular growth from each anion terminal, the ratio of the monomer concentration to that of the anionic species determines the kinetic number average molecular weight $(\overline{M}_k)$ [Eq. (6-5)]).

$$\overline{M}_k = \tfrac{1}{2}\,(\text{weight of monomer/moles of initiator}) \qquad (6\text{-}5)$$

The value of $\tfrac{1}{2}$ results from the difunctional nature of the initiator. Second, the distribution of molecular weights can be very narrow. A Poisson distribution of molecular weights can be achieved [Eq. (6-6)].

$$\overline{P}_w/\overline{P}_n = 1 + 1/\overline{P}_n \qquad (6\text{-}6)$$

where $\overline{P}_w$ = the weight average chain length and $\overline{P}_n$ = the number average chain length. This mechanism has since been demonstrated many times to be basically true (A4, A15–A17). There are, however, complications, which are largely due to the role of impurities and the often rapid nature of the propagation step. Impurities can deactivate one or both ends of the growing chain. Obviously, this results in diblock polymer and homopolymer formation. One must also consider the situation in which the first monomer is depleted by polymerization before all of the initiator is transformed into polymeric anions (i.e., propagation rate $\gg$ initiation rate). In this case, addition of the second monomer will result in some homopolymerization due to residual initiator.

An important limitation of the electron transfer initiators is that they are generally soluble only in ether or other polar solvents. These media lead to polydiene segments of low 1,4-microstructure content, which results in a sacrifice in the desirable properties of low glass transition temperature. In order to achieve high 1,4-polydiene structure, lithium initiators in predominately hydrocarbon solvents are required. Soluble dilithium initiators have been prepared by reacting lithium with 1,1-diphenyl ethylene (Reaction 6-7) (A18, A76). Small

$$\qquad (6\text{-}7)$$

1,4-Dilithio-1,1,4,4-
tetraphenylbutane

amounts of weakly basic aromatic ethers such as anisole do not significantly alter the polydiene 1,4-microstructure (A18, A19).

   ii. *Three-Stage Sequential Addition Process.*   The three-stage process for synthesizing A-B-A block copolymers comprises initiation of styrene polymerization to form "living" polystyrene anion followed by addition of a diene monomer to form a living diblock and, finally, introduction of a second quantity of styrene to complete the formation of the A-B-A structure. This reaction sequence is shown in Scheme 6-1.

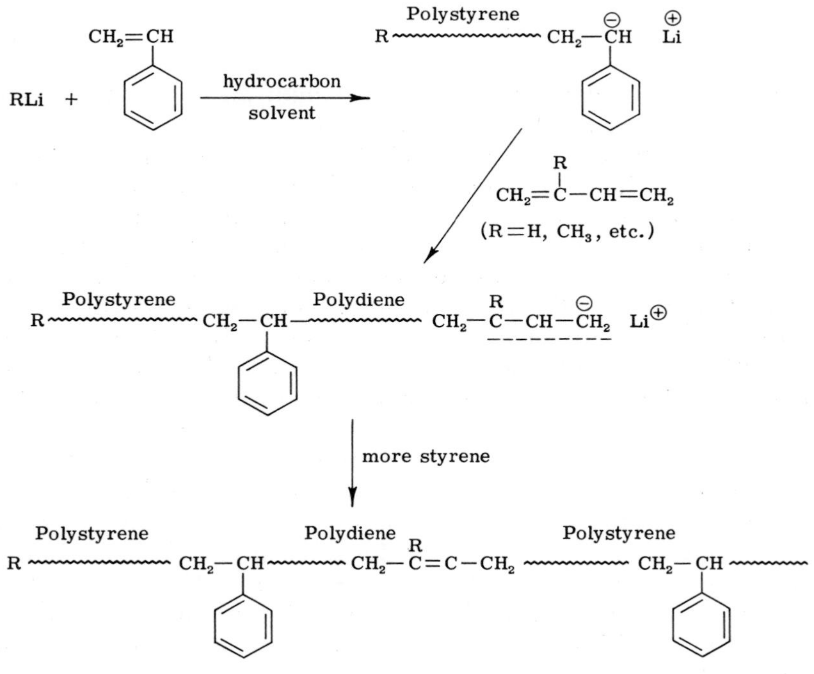

Scheme 6-1

Hydrocarbon soluble organolithium initiators (A23, A24) are most suitable for allowing termination-free polymerizations. The three sequential monomer additions require careful purification (A19–A21) and analytical (A77) procedures to avoid the formation of large amounts of homopolymer and diblock "contaminants." Attention must also be paid to the initiation step in the first styrene polymerization. It must be rapid to assure that all of the alkyllithium is consumed in preparing the relatively low molecular weight (e.g., 10,000–20,000) polystyrene segment. Fast initiation can be achieved with either branched organolithium components (A22, A25) or with primary organolithium compounds promoted with low concentrations of aromatic ethers (A11, A19, A20) or tertiary amines (A16, A21). The crossover reaction of the living polystyrene anion with either butadiene or isoprene is extremely rapid (Scheme 6-2) (A17). The formation of the last

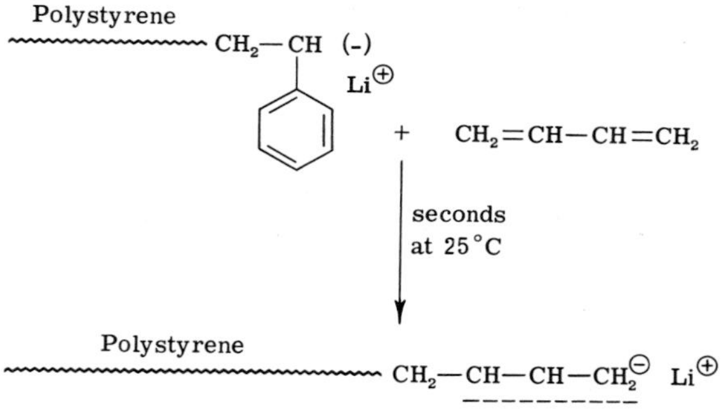

Scheme 6-2

styrene segment via initiation with the polydiene anion is not nearly as rapid as this in pure hydrocarbon media. However, traces of ethers are remarkably effective promoters (A21). That a three-stage process can produce a pure block structure is indicated by the gel permeation chromatographs shown in Fig. 6-1. These block copolymers of styrene–isoprene–styrene and styrene–butadiene–styrene were initiated by ethyllithium in benzene containing a small amount of anisole (A19, A26). The sharpness of the single peaks and the absence of shoulders demonstrate that narrow molecular weight distributions were attained. The number and weight average molecular weights determined independently were in good agreement with the predicted molecular weight and narrow molecular weight distribution.

iii. *Coupling Process.* This technique differs from the previous three-stage approach, since the final step is not addition and polymerization of a second quantity of styrene but rather addition of a difunctional small molecule (such as phosgene), which can, in principle, quantitatively "couple" the diblock chains to form an A-B-A (or perhaps more precisely an A-B–B-A) structure (Scheme 6-3). This approach produces symmetrical end blocks. It also avoids possible termination due to impurities in the third monomer addition step. The principal disadvantage, however, is the requirement for both high efficiency and precision in the stoichiometry of the coupling step (A5). As

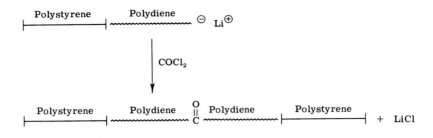

Scheme 6-3

will be discussed later, the mechanical properties are particularly adversely affected by diblock contamination (A7, A8, A26).

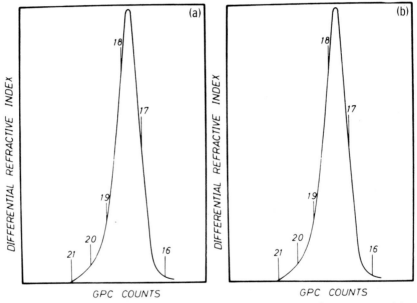

Fig. 6-1. Gel permeation chromatographs of styrene–butadiene–styrene (a) and styrene–isoprene–styrene (b) block copolymers (A5).

iv. *Tapered Blocks Process.* This last method utilizes the fact, discussed in Chapter 5, that diene–styrene monomer mixtures in hydrocarbon media preferentially polymerize the diene first (A16, A17) and then the styrene. This behavior is particularly pronounced for the combination of butadiene and styrene. The isoprene–styrene system forms less pure blocks under similar circumstances (A5, A9). This technique is probably the least sensitive to termination via monomer impurities. For example, knowing the kinetics of lithium alkyl-initiated styrene polymerization, one can allow this reaction to proceed to about 50% conversion and then introduce butadiene. When the butadiene monomer is consumed, the remaining styrene monomer polymerizes to form the third block. Since the preferential butadiene polymerization is not 100% efficient, the center polybutadiene segment contains a few percent (~8%) (A27) of randomly dispersed styrene. The styrene content in the center block increases as the butadiene monomer is depleted, thus producing a "tapered" center block. This merely raises the lower glass transition temperature to a minor extent (A27). The principal contaminant in these products would be expected to be homopolystyrene, formed by termination (by impurities) during the introduction of butadiene. Minor amounts of the homopolymer serve as a filler and do not drastically decrease the mechanical properties. Since there is no second charge of styrene monomer, this process results in a minimal amount of diblock contaminant. Polymerization via a continuous process has been claimed (A100).

For the sake of completeness, several important references pertaining to the synthesis of styrene–diene–styrene A-B-A block copolymers have been tabulated in Tables 6-1 through 6-4 according to the four approaches described above.

TABLE 6-1

Synthesis of Styrene–Diene–Styrene Block Copolymers Via the
Difunctional Initiator Process

Diene	Initiator	Solvent	Reference
Isoprene	Sodium naphthalene	THF	A3, A14, A28
	Cesium naphthalene	THF	A29
	Sodium α-methylstyrene	THF	A30
	Lithium metal	Benzene	A31
	Lithium metal	Dioxane	A32
	Dilithioisoprene	Benzene	A9
	sec-Butyllithium–divinylbenzene	Benzene	A10
Butadiene	Aromatic ketone dilithium complex	Toluene	A33
	Lithium methylnapthalene	Ether–cyclohexane	A34, A35
	1,2-Dilithio-1,2-diphenylethane	Ether, cyclohexane	A36, A37
	Dilithiobutane	Ether	A38
	Sodium α-methylstyrene tetramer	THF	A30
	Dilithioisoprene	Benzene	A9

TABLE 6-2

Synthesis of Styrene–Diene–Styrene Block Copolymers Via the
Three-Stage Sequential Addition Process

Diene	Initiator	Solvent	Reference
Isoprene	n-Butyllithium	Hexane–benzene	
	sec-Butyllithium	Benzene	A22
	Ethyllithium–anisole	Benzene	A11, A19, A26, A39
	n-Butyllithium	Benzene–cyclohexane	A21, A40
	sec-Butyllithium	Benzene–cyclohexane	A41–A46
Butadiene	Ethyllithium–anisole	Benzene	A11, A39
	n-Butyllithium	Benzene–cyclohexane	A21
	Ethyllithium–diphenyl ether	Benzene	A20

b. Characterization of Molecular Weight and Solution Behavior

In Section a, above, it was pointed out that the styrene–diene anionic block copolymerization process leads to predictable composition, microstructure, molecular weights, narrow molecular weight distribution, and a high degree of block purity. A considerable amount of data pertaining to the molecular characterization of these systems is

TABLE 6-3

Synthesis of Styrene–Diene–Styrene Block Copolymers Via the Coupling Process

Diene	Initiator	Solvent	Coupling agent	Reference
Isoprene	sec-Butyllithium	Benzene	Dibromoalkane	A47, A48
	sec-Butyllithium	Benzene	Carbon monoxide	A49
	n-Butyllithium	Benzene	Chlorine, bromine, iodine vapors	A50
	n-Butyllithium	Cyclohexane	Spectral technique to determine living end concentration prior to coupling	A51
Butadiene	sec-Butyllithium	Cyclohexane	Dicarboxylic acid diesters	A52
	Butyllithium	Toluene	Pentaerythrityl bromide or other polyhalogen derivative	A53
	sec-Butyllithium	Benzene	Dibromoalkane	A47, A54–A56
	sec-Butyllithium	Cyclohexane	Vinyl acetate, ethyl acetate	A57

TABLE 6-4

Synthesis of Styrene–Diene–Styrene Block Copolymers
Via the Tapered Block Process

Diene	Initiator	Solvent	Reference
Isoprene	n-Butyllithium	Benzene	A10
	sec-Butyllithium	Benzene	A41–A43, A45
Butadiene	n-Butyllithium	Benzene	A10

available. A variety of physical techniques (A67) have been employed such as spectroscopy, osmometry, viscometry, light scattering, ultracentrifugation, and gel permeation chromatography. The information provided by these various tools is often complimentary and is an essential input to the formulation of a structure–property model.

When an A-B-A triblock is synthesized by any of the procedures previously discussed, it is often possible to isolate and examine the intermediate segments. If synthetic problems occur, it may then be possible to determine the step (e.g., coupling, monomer impurities) that is causing the difficulties (A16, A22, A55). For example, coupling of an A-B block copolymer anion to produce an A-B-A structure should cause the number average molecular weight to double if the coupling efficiency is high.

Characterization data from several investigations are compiled in Tables 6-5 and 6-6. All of the polymers examined were obtained with pure dry reagents under high vacuum conditions. Under these conditions, it was possible to get predictable molecular weights. Light-scattering experiments indicated that narrow molecular weight distributions were achieved in the triblock systems. Light scattering on copolymers is dependent upon compositional heterogenity as shown by Bushak and Benoit (A59). In the case of uniform block copolymer, one might predict that compositional heterogenity is low so that deviations from the "apparent" $\overline{M}_w$ and the "true" value would be very small. This opinion is supported by the work of Angelo and co-workers (A30), who demonstrated that the observed $\overline{M}_w$ was similar regardless of the solvent refractive index. Narrow distributions are, of course, also indicated by the gel permeation chromatography (GPC) curves in Fig. 6-1.

Density-gradient ultracentrifugation appears to be a powerful method for analyzing the purity and molecular weight distribution of block copolymers (A22, A62). Since polystyrene has a higher density than polydienes, the order of density is polystyrene > triblock

TABLE 6-5

**Molecular Characterization of Styrene–Butadiene–Styrene
A-B-A Block Copolymers[a]**

Initiator	Block sequence	Segment $\overline{M}_k \times 10^{-3}$	$\overline{M}_k \times 10^{-3}$	$\overline{M}_n \times 10^{-3}$	$\overline{M}_w \times 10^{-3}$	Reference
n-Butyllithium	SBS	7.5–18–7	32.5	34.8	—	A60, A61
	SBS	10–100–10	120	117	—	A60, A61
	SBS	10–450–9	469	517	—	A60, A61
Ethyllithium– diphenyl ether	SBS	13.7–63.4–13.7	90.7	92.7	122	A11, A20
	SBS	21.2–97.9–21.3	140	147	—	A11, A20
Sodium α-methylstyrene tetramer[b]	BSB	—	—	1.02	124	A30

[a] Prepared via sequential addition of the monomers.

[b] Prepared in THF solvent; all other copolymers referred to in this table were prepared in benzene.

TABLE 6-6

**Molecular Characterization of Styrene–Isoprene–Styrene
A-B-A Block Copolymers**

Initiator	Block sequence	Segment $\overline{M}_k \times 10^{-3}$	$\overline{M}_k \times 10^{-3}$	$\overline{M}_n \times 10^{-3}$	$\overline{M}_w \times 10^{-3}$	Reference
Ethyllithium anisole	SIS	13.7–63.4–13.7	91	89	—	A39
	SIS	8.4–63.4–8.4	82	87	—	A39
	SIS	21.1–63.4–21.1	106	117	—	A39
	SIS	13.5–80–13.5	107	116	103	A19
	SIS	13.5–26–13.5	53	55	—	A19
Sodium α-methylstyrene tetramer[a]	ISI	—	—	111	—	A30
Ethyllithium[b]	SIS	—	122	106	—	A19
			161	142		

[a] Prepared in THF solvent.

[b] Prepared via coupling with CH_2Br_2; all other copolymers referred to in the table were prepared via sequential addition of the monomers.

polymer > diblock polymer, e.g., density is proportional to styrene content. It was estimated that concentrations as low as 1% polystyrene and 99% diblock could be detected by this technique. A typical Schlieren diagram is shown in Fig. 6-2. Earlier characterization by sedimentation was reported by Bresler et al. (A63, A64).

Juliano (A20) investigated the separation of model homopolystyrene, styrene–butadiene diblock, and styrene–butadiene–styrene triblock copolymers by GPC and other techniques. By utilizing rigorous synthesis techniques, it was possible to control all of the styrene and butadiene block molecular weights to 21,000 and 63,000, respectively. Appropriate solution blends could then be made that permitted the evaluation of the GPC separation efficiency. It was concluded (A20) that homopolymer could easily be detected in the presence of either diblock or triblock copolymer. However, detection of diblock in predominantly triblock structures was more difficult. For example, 5% or even 20% of diblock merely broadened and slightly shifted the GPC peak of the triblock copolymer control. Chang (A66, A92) has also investigated the molecular weight analysis of block copolymers by GPC.

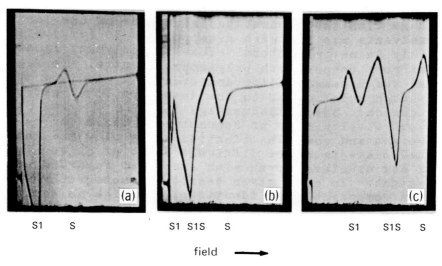

Fig. 6-2. Characterization of an impure styrene–isoprene–styrene block copolymer by density gradient ultracentrifugation. (a) Schlieren diagram of an impure two-block copolymer; the sinusoidal section of curve near center represents polystyrene. Two-block polymers are at left near the meniscus. (b) Schlieren diagram of an impure three-block copolymer. Polystyrene is near the cell center, two and three blocks are at left near the meniscus. (c) Same polymer in a different gradient. A two-block impurity is at left, other components are toward the cell bottom (A22).

The theoretical ramifications of the dilute solution properties of well-defined styrene–butadiene–styrene block copolymers have been discussed in the literature (A20, A60, A61). A two-parameter expression containing short-range and long-range interaction parameters was utilized to analyze the data. Long-range interactions are negligible at the theta temperature of a given homopolymer–solvent combination. Extension of the theta temperature concept to block copolymers is difficult. For homopolymers, the theta temperature may be experimentally defined as the temperature at which the second virial coefficient is equal to zero. However, for a block copolymer, the block interactions do affect the value of the virial coefficient (A20, A60), and the theta temperature therefore includes a factor that compensates for the significant block–block interaction. The true theta temperature could be defined as the point at which all individual long-range interaction parameters vanish identically (A60). Unfortunately, it is not now experimentally possible to determine this "true" theta temperature. Clearly, the size and number of individual blocks, the overall molecular weight, and the chemical nature of the solvent can be important parameters. Nevertheless, Utrachi and co-workers (A60, A61) were able to derive expressions that permitted them to analyze their osmotic and viscometric data. They concluded that the properties of diblock and triblock are similar. Intrinsic viscosities and second virial coefficients were calculated for these block copolymers based on their chemical composition, molecular weight and values of the interaction parameter between styrene and butadiene units. The magnitude of the interaction varied with the solvent. The results were interpreted as meaning that the domains of the polystyrene and polybutadiene blocks were overlapping to a great extent.

The proton spin–lattice relaxation of random and block copolymer has been studied (A65). Local chain motions in the butadiene and styrene domains of butadiene-rich block copolymers were similar to the corresponding motions in the bulk homopolymers.

c. Morphology

i. *Two-Phase Nature of Block Copolymers.* The morphology of block copolymers in general and styrene–diene–styrene systems in particular has been reviewed (A68–A74). The unique morphology observed in block copolymers clearly results from the microphase separation of the dissimilar polymer segments into distinct domains. An important question, then, is to define the structural parameters that determine whether a given system will display this two-phase behavior. One might anticipate that the segment block lengths and the degree of chemical compatibility of the two constituents—as judged, for example, by solubility parameter values or interaction parameters—would be the most important features to be considered. This should be particularly true for the essentially amorphous, nonpolar polymers of styrene and the dienes. In addition, the composition volume ratio, number, and possibly sequence of the blocks could be important.

In the limit of uniform random copolymers where the "block" sequence of either unit is a very small number, it is recognized that one-phase systems are obtained. At the other extreme, as discussed in Chapter 4, two high molecular weight homopolymers are usually highly incompatible with each other. Of course, such blends differ greatly from A-B-A block copolymers, which might be considered to be polymer "blends" in which the segments are chemically linked at two junctions. As a result, the block copolymers can undergo phase separation only on a micro scale or not at all if the block solubility parameters are similar (A75). Nevertheless, on a molecular scale, segment molecular weight is an important parameter governing domain formation. Investigations by Angelo and co-workers (A30) showed that high block molecular weight styrene–diene block copolymers displayed multiple transitions corresponding to each constituent.

The thermodynamics of domain formation for styrene–diene–styrene copolymers have been studied by Meier (A78–A80), Krause (A81, A82, A113), Fedors (A83), McIntyre (A84–A86), Helfand (A87, A88), and others. Utilization of the Flory–Huggins theory, with the assumption that the polymer–polymer interaction parameter is the major parameter, permitted Fedors (A83) to predict the minimum molecular weight levels required to produce microphase separation. For example, by these calculations, a styrene–butadiene–styrene block copolymer should have a minimum molecular weight of 2500 for each styrene segment and 6000 for the butadiene center block in order to develop domains. However, Meier's approach (A78), based on the diffusion equation, predicts that the critical molecular weights necessary for domain formation is 2.5–5 times higher for block copolymers than for simple polymer blends. This is ascribed to the loss in chain configurational entropy at the copolymer junctions. Although derived for diblock copolymers, the theory was stated to be also essentially correct for triblock copolymers. For styrene–butadiene systems, this approach requires a polystyrene block molecular weight of 5000–10,000 when the corresponding polybutadiene value is 50,000. The tensile strength data of Morton (A7, A8) and Holden et al. (A89, A90) show that thermoplastic elastomer properties (which are due to microphase separation) disappear when the styrene block molecular weight is reduced to 6000–7000. These data thus support the Meier theory. The effect on morphology of decreasing diene block molecular weight at a constant styrene block length has not been reported.

Theoretical treatments predict a variation in domain size with molecular weight. At the molecular weights considered, the predicted sizes of 100–300 Å are in general agreement with most of the experimentally determined values (A86). The small domain size relative to the wavelength of light explains why the block copolymers are usually transparent, even though possessing two phases. The shapes of the domains were limited to spheres in the theoretical treatment to simplify the calculations. In fact, however, other morphological structures, such as cylinders and lamellae, are possible at various intermediate compositions as will be discussed below.

ii. *Electron Microscopy.*   Hendus *et al.* (A91) were the first to publish the basic morphological features of styrene–butadiene–styrene block copolymers. They reported that a paracrystalline structure, controlled by the ratio of the constituents, resulted from the intermeshing of the phases and that the periodicity was controlled by the block lengths. Matsuo reported similar observations somewhat later (A93, A94). Spheres, rods, or sheet structures were observed depending on composition. Unfortunately, in both cases the copolymers were not characterized with respect to molecular weight and homopolymer contamination. At about the same time, studies at the University of Akron (A19) on rigorously prepared and well-characterized styrene–isoprene–styrene materials were conducted. At styrene and isoprene block molecular weights of 13,700 and 81,000, respectively, no homopolymer was observed by GPC and $\overline{M}_w/\overline{M}_n$ was <1.1. Electron microscopy investigations showed a highly dispersed polystyrene phase in a continuous polyisoprene matrix (A95). Since this early work, numerous electron microscopic investigations have been reported that have confirmed and further elucidated the morphology. The various types of information reported are summarized in Table 6-7. Most of the work has utilized the OsO_4 staining technique developed by Kato (A96). A review has discussed the morphology of "regular" block copolymers (A69).

The results of Enomoto (A98) and Bradford (A99) clearly showed the effect of block molecular weight on the observed domain dimensions. The effects of synthesis variables, such as molecular weight, composition, and sequential arrangement, on the observed morphological structures have been discussed by a number of authors (e.g., A74, A97,

TABLE 6-7

Electron Microscopy Studies of Styrene–Diene–Styrene Block Copolymer

Study	Reference
Early work	A91, A93–A95, A97, A104
Relationship of molecular weight to domain dimensions	A98, A99, A101, A102
Identification of morphological structures (e.g., spheres, rods, cylinders); theoretical models, etc.	A74, A94, A97, A102, A103, A105–A112, A133
Effect of casting solvent and rate of evaporation	A39, A94, A102, A105, A114–A119, A125
Deformation studies	A103, A105
Effect of added homopolymer	A7, A8, A109, A120
Statistical thermodynamic treatments	A121–A124
Development of lattice models	A86, A105, A114, A126–A130
Liquid crystalline structures	A131, A132

A102, A103, A105, A106, A108–A110). Many structures such as spheres, rods, cylinders, and lamellae have been observed. In general, the most important variable seems to be the volume ratio of the phases. Spheres of one constituent in a continuous matrix of the other can form at a volume fraction of about ≤ 0.25 for the former. Intermediate compositions (volume fractions between 0.25 and 0.75) lead to cylinders, rods, or, frequently, two continuous phases, depending upon film forming rates, type of solvent, etc. The effect of added homopolymer on the phase volume ratio has been discussed (A7, A8, A109, A120).

It has been recognized since the experiments of Merritt on graft copolymers (A134) that the nature of the casting solvent and its interaction with the polymer segments can play a major role in determining the morphology of two-phase systems. The use of selective (or mixed) solvents that are "good" or "poor" for one segment can determine which phase will be the last to come out of solution and thus form the continuous phase. This produces wide variations in morphology, particularly at copolymer compositions at which the two moieties can both be continuous (e.g., 50/50 volume ratio). Various investigators have demonstrated casting solvent effects with styrene–diene–styrene block copolymers (A39, A94, A102, A105, A114–A119, A125).

Another interesting observation (A114) is the effect of solvent evaporation rate on the morphology. In films prepared by very slow solvent evaporation (~24 hours), the polystyrene chains (26 wt%) formed spheres that were monodisperse in size and spacing. Faster evaporation rates gave lamellae, whereas flash evaporation produced a completely disordered but homogeneous mixture of irregularly shaped domains and regions of incomplete phase separation. Compression-molded specimens were similar to the solution-cast film prepared by rapid evaporation. The behavior is shown in Fig. 6-3.

Along analogous lines, polystyrene determined from statistical thermodynamic considerations were found to be in agreement with results from electron microscopy (A123). For a given system, the domain size was inversely related to the temperature of formation. The kinetics of aggregation and the dimensions of the supermolecular structure has also been studied (A118, A135). The development of detailed lattice models for the structure of these block copolymers has now begun (A86, A105, A114, A126–A130). Examination (A114) of

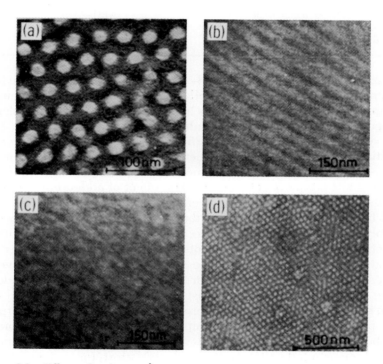

Fig. 6-3. Effect of solvent evaporation rate on the morphology of styrene–butadiene–styrene block copolymer. (a) Film of sample prepared at low rate of evaporation; (b) film of sample prepared at intermediate rate of evaporation; (c) film of sample prepared at high rate of evaporation; (d) film of sample prepared at low rate of evaporation showing grain structure. After Lewis and Price (A114), reproduced by permission of the publishers, IPS Business Press LTD.

the film obtained by slow evaporation (Fig. 6-3) showed that there was a highly ordered array of uniform glassy domains embedded in a rubbery matrix. Whether the domains were spheres, prolate ellipsoids, or cylinders could not be determined from the limited two-dimensional electron micrographs. Line defects similar to those associated with either crystalline lattices or to regions of incomplete phase separation were observed. Electron microscopic evidence of a macroscopic "single crystal" (A129) obtained by extruding styrene–butadiene–styrene has also been reported.

In summary, it has been pointed out (A114) that electron microscopic studies on thin films have produced rather substantial information about the two-phase morphology of these styrene–diene–styrene systems. For example, it reveals long-range order and also the existence of finer details. Second, a number of morphological units such as spheres, cylinders, and rods can be seen. Finally, this technique again points out the importance of the casting solvent and processing conditions (e.g., from the melt or from solution) on the morphology of this class of materials.

iii. *Small-Angle X-Ray Scattering.*   The technique of low-angle X-ray scattering can be used to obtain information about the state of the bulk block copolymer (A84–A86, A114, A136, A138, A139). The data thus complement the visual results obtained by electron microscopy on the surface of thin film. Domain sizes of the order of 100 Å are ideally suited for small-angle X-ray scattering studies (A139). In addition, the average spacing between the domains can be determined as well as the size of the domains themselves. The technique is also amenable to studying the variation of parameters, such as stress, temperature, and pressure.

The theory of small-angle scattering in these systems has been discussed (A84–A86, A114, A139) and will not be repeated here. Methods to calculate the domain size and interdomain spacing have also been reported (A85). It must be emphasized that it is necessary to employ very high-resolution film cameras to obtain quantitative information (A84–A86, A136, A139). The first reports (A91) presented evidence for ordered structures due to interdomain interference, but the higher-order reflections related to the domain radius could not be distinguished. A typical scattering curve for styrene–butadiene–styrene block copolymers having different size styrene end blocks (and hence different compositions) as shown in Fig. 6-4. The domain radius values derived from the experimental small-angle scattering data (A86) are shown in Table 6-8.

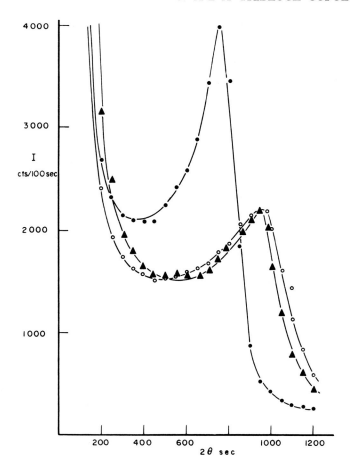

Fig. 6-4. Scattering from styrene–butadiene–styrene polymers having different compositions. Molecular weights: ● = 21,100–63,400–21,100; ▲ = 13,700–63,400–13,700; ○ = 8200–63,400–8200 (A139).

As might be expected, the domain structure was found to be decidedly influenced by the type of casting solvent, the volume fraction of polystyrene, and the presence of impurities (e.g., homopolymer) in the block copolymer. The block purity is also very important (A140). It should be kept in mind that many investigations on commercially available styrene–butadiene–styrene do not take into account the significant amount of nontriblock structure present. The data has been interpreted to mean that the domain boundaries near the interface

TABLE 6-8

Domain Sizes in Different Styrene–Butadiene–Styrene Polymers[a]

Molecular weight (SBS)	Weight percent styrene	Domain radius size (Å)
7,000–35,000–7,000	28.5	93
13,700–63,400–13,700	30	116
21,200–97,900–21,200	30	170
120,000–660,000–120,000	27	207

[a] From Campos-Lopez et al. (A86).

may be somewhat diffuse (A85, A138). Again, however, the effect of casting conditions (e.g., solvent type and evaporation rate) play a major role in the observed degree of phase separation.

Low-angle X-ray studies were carried out on solution-cast and compression-molded test pieces (~26% styrene). The moldings displayed a hexagonal lattice of glassy cylinders embedded in a rubbery matrix (A114). The solvent-cast specimen also exhibited a regular domain structure, but the investigators (A114) did not feel it was possible to make a definite structural assignment. The data were suggestive of a face-centered cubic lattice of spheres. The work of McIntyre and Campos-Lopez (A84) with a well-characterized styrene–butadiene–styrene containing 36% styrene showed spherical domains that were assigned to that of a face-centered orthorhombic lattice. It should be noted that Skoulios (A141) and Krigbaum et al. (A142) have criticized this interpretation of the experimental data. In any event, it would appear that the highly sensitive camera used by McIntyre et al. (A84) may be allowing phenomena to be observed that are not seen with less advanced instrumentation. Further work has also been reported by these investigators (A86) which provides a rebuttal to the comments of Skoulios (A141). They were able to deduce the end-to-end distance, the interfacial region, and the molecular weight dependence of the domain size and spacing. Their data were in good agreement with the theory of Meier. Further evidence of simple cubic geometry in styrene–butadiene–styrene copolymers has been described (A136). Particularly sharp diffraction peaks or spots were observed upon extension to 600%. The fact that spots rather than lines were observed was taken to be strong evidence of an ordered arrangement of spheres rather than rod or sheet morphology. It was also felt that such sharp peaks could not be obtained from a completely random distribution of spheres. The basic structural feature did not change over a temperature range of $-120°C$ to $+130°C$.

Perhaps the most dramatic effects of orientation on morphology have been reported by Keller and his colleagues (A116, A127, A129, A143). Extruded plugs (A69, A144, A145) of commercial styrene–butadiene–styrene ($\sim$25% styrene) were found to exhibit discreet reflections such as are characteristic of an individual single crystal. Mechanical properties have been reported (A146). The "crystal" lattice was hexagonal, with a lateral spacing of 300 Å and an infinite period along the hexagonal axis (i.e., the extrusion direction). No dichroism for either phase could be detected (A147). Annealing at temperatures well above that of polystyrene T_g, (i.e., 150°–225°C) enhanced the "crystal" perfection. The styrene phase was thought to be in the form of cylinders with a diameter of $\sim$150 Å. One might, at first, find the improvement in order at such high temperatures surprising. Apparently, though, this behavior is yet another manifestation of block polymer segmental microphase separation. Of course, the styrene segments in the rubbery state would much more rapidly attain their preferred thermodynamic arrangement. Other investigations (A148) have also shown that similar annealing conditions resulted in increased regularity in the "paracrystalline" arrangements of the domains. Techniques to prepare the anisotropic structures have been patented (A149). Conditions such as high temperature and/or high deformations were reported not to result in gel formation (A148). On the other hand, styrene–isoprene–styrene copolymers had similar morphology before and after annealing at 150°C (A150, A151). Radial coupling has been reported not to seriously alter the lattice morphology in styrene–isoprene–styrene system (A152).

The model that is emerging from the experimental studies described in this section is that of a regular pseudocrystalline lattice held together by entropic forces. The degree of regularity is too high to be accounted for by a more random lateral order. Clearly, however, there are many questions remaining that will require further elucidation.

iv. *Birefringence and Light Scattering.* The application of a stress to a macromolecule causes it to become optically anisotropic and to display properties of a uniaxial crystal with the optical axis parallel to the direction of stress (A153). The initial slope of a plot of birefringence versus stress or strain define, respectively, the stress optical and strain optical coefficients. Birefringence techniques have been quite valuable in interpreting mechanical properties (particularly of single phase rubbers) in terms of molecular structure. Their application to block copolymers is relatively recent (A68) and theoretical treatments are not yet well developed.

It has been shown (A115, A153) that the different morphology obtained as a function of casting solvent is reflected in the birefringence measurements. The strain optical coefficient (SOC) of an methyl ethyl ketone-cast styrene–butadiene–styrene film was less than that of a toluene-cast specimen. Since the elastomeric polybutadiene phase is known to possess a higher SOC than glassy polystyrene, the polybutadiene is believed to be more continuous in the toluene case. Measurements at high speed (A115) show that there is a relaxation of birefringence with time. Furthermore, the MEK samples relax to a greater extent and are much more temperature sensitive. This effect may be seen clearly in Fig. 6-5.

An interpretation (A153) that seems reasonable is that the birefringence arises principally from the orientation of the polybutadiene segments for which the polystyrene domains act as "cross-links" and filler particles (A68). Then the changes with time and temperature are associated with the number of effective network chains.

The birefringence Δ of a styrene–butadiene–styrene block copolymer may be described in terms of the contributions from its components [Eq. (6-8)] (A115, A154):

$$\Delta = \phi_B \Delta_B + \phi_S \Delta_S + \Delta_F \qquad (6-8)$$

where ϕ_B and ϕ_S are the volume fractions of butadiene and styrene, Δ_B and Δ_S are birefringence contributions per unit volume from three microphases, and Δ_F is the form birefringence arising from the anisotropic shape of the phase boundary. The latter term occurs because of the difference in the average refractive index between styrene ($N_S = 1.59$) and butadiene ($N_B = 1.52$) phases. The Δ_F is small for crystalline polymers but can be large for block copolymers; the work of Folkes and Keller demonstrates this clearly (A155). They measured the birefringence of extruded commercial styrene–butadiene–styrene. The sample had been previously shown (A129) to consist of a hexagonal array of styrene cylinders embedded in a butadiene matrix. The samples showed positive birefringence values of about 5×10^{-4} Brewsters. Calculations showed that this would be the value expected (within experimental error) if all of the birefringence was due to the form contribution. This was further demonstrated by Wilkes (A156). The conclusion, then, is that the dispersed phase consists of randomly oriented chains. These results and the findings of Lewis and Price (A114) have also enabled Stein (A157) to propose an explanation for the light-scattering patterns observed for these block copolymers as a

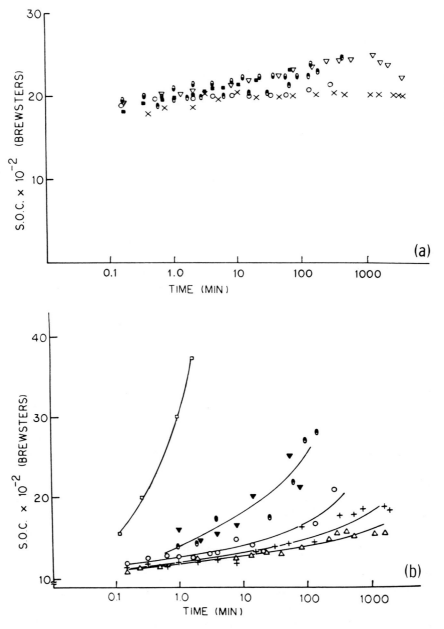

Fig. 6-5. Variation of the stress–optical coefficient with time during relaxation at various temperatures. (a) Toluene-cast film: × = 24°C, ○ = 35°C, ◑ = 41°C, ▽ = 51°C, ◒ = 63°C, ■ = 74°C. (b) MEK-cast film: △ = 31°C, + = 48°C, ○ = 59°C, ◑ = 71°C, ▼ = 75°C; □ = 85°C (A115).

function of strain. It is felt (A157) that the Δ_F should change with strain because of the variation in orientation of the styrene cylinders with elongation of the polymer. The importance of Δ_F has also been discussed by others. This information might also be useful in interpreting the anisotropic light scattering results of Pishareva *et al.* (A159).

d. Rheology

It is, of course, well understood that the A-B-A block copolymers based on styrene and dienes are called thermoplastic elastomers and that they can be melt processed like thermoplastics. However, rheological investigations of these two-phase systems show their melt behavior to be much more complicated than that of their corresponding homopolymers or random copolymers (A160–A166). The molecular structure and melt rheological properties of butadiene–styrene copolymers has been reviewed (A165).

McGrath (A19) noted the effect of styrene molecular weight on compression moldability. Styrene–isoprene–styrene copolymers having molecular weights of 100,000–400,000–100,000 could not be successfully molded. It was speculated that chain entanglements in the hard segment were important. Holden *et al.* (A161) reported that melt viscosity was very sensitive to molecular weight changes, particularly at low shear rates. The behavior was not Newtonian at low shear rates and thus could not be examined by the well-known equation that states that the zero shear viscosity is proportional to molecular weight raised to the power 3.4. At the relatively low shear rate of 10^{-1} seconds, the exponent was approximately 5.5. The melt viscosity of styrene–butadiene–styrene block copolymers is greater than either of the homopolymers at the same molecular weight. It was suggested (A161–A163) that the high viscosities result from the fact that the two-phase structure still exists in the melt. At high temperatures, the polystyrene domains are fluid and flow is possible. The microphase separation, however, requires the polystyrene domains to flow through the polydiene matrix. An additional energy term is needed to describe this feature, which shows up as a higher melt viscosity (A161). The flow behavior is temperature-dependent as shown in Fig. 6-6. Similar dependence on shear rate was noted. Entirely different responses were observed in the low and high shear rate regions.

The effect of block sequence (e.g., styrene–butadiene–styrene or butadiene–styrene–butadiene), molecular weight, and branching on steady flow and dynamic viscosity was investigated (A163). At constant molecular weight and total styrene content, viscosities were greater for polymers terminating in styrene blocks. This effect was more impor-

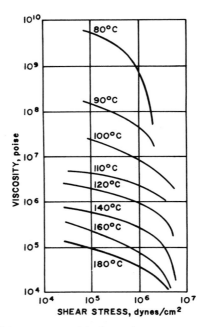

Fig. 6-6. Viscosities of a 10S–52B–10S block copolymer at various temperatures (A161).

tant than the degree of branching. Branching did decrease the viscosity of either butadiene–styrene–butadiene or styrene–butadiene–styrene at a constant $\overline{M}_w$. The length of the terminal block was also shown to be more important than the total molecular weight. Figure 6-7 (A163) shows a schematic drawing of the styrene–butadiene–styrene and butadiene–styrene–butadiene block polymer molecular domain structures in the melt. The styrene domains in butadiene–styrene–butadiene should not be affected to any great extent when the polymer flows. In contrast, the styrene–butadiene–styrene system can not flow without disruption of the styrene domain.

The dynamic viscosities of styrene–butadiene–styrene have also been studied as a function of frequency and temperature using a Weissenberg rheogoniometer (A164). This technique has advantages over the capillary method in the low shear rate range. Also in this investigation (A164), viscous and elastic responses are independent of deformation amplitude and may be analyzed using linear viscoelastic theory. The data confirm a domain-type structure at low frequencies and a thermoplastic behavior in the high frequency range. The unusual nature of styrene–butadiene–styrene may be seen (Fig. 6-8) by compar-

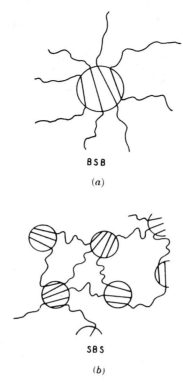

BSB

(a)

SBS

(b)

Fig. 6-7. Schematic representation of domain structures. (a) Polymers terminating in butadiene blocks; (b) polymers terminating in styrene blocks (A163).

ing its behavior at various frequency values and temperatures with that of monodisperse polystyrene (MDPS/97) of a similar molecular weight (97,000). The viscosity of the polystyrene is constant and shows a steady state Newtonian viscosity at low frequencies. Like most thermoplastics, viscosity decreases at higher shear rates. In contrast, it is not possible to obtain a zero shear viscosity value for the styrene–butadiene–styrene. Further experiments by these investigators (A164) showed that the viscoelastic properties of styrene–butadiene–styrene were not amenable to simple thermoplastic characterization. It was proposed that styrene–butadiene–styrene can exist in three distinct states depending upon the rate of deformation. At low frequencies, the molecular network is intact and a very high viscosity is observed. At intermediate regions of shear stress, the domains are disturbed and a viscosity higher than expected is still observed. Finally, at very high shear stresses the domains are largely separated and a typical thermo-

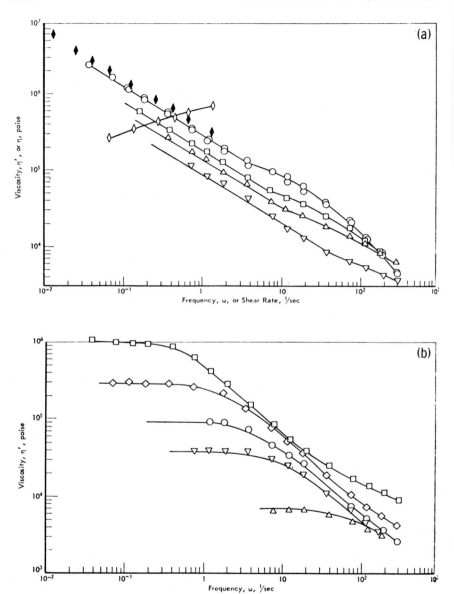

Fig. 6-8. (a) Viscosity of SBS 22-50. Steady shear data (170°C): ◆ = viscosity (poise); ◇ = normal stress (dynes/cm²); dynamic data: ○ = 170°C, □ = 180°C, △ = 190°C, ▽ = 200°C. (b) Dynamic viscosity of MDPS/97: □ = 150°C, ◇ = 160°C, ○ = 170°C, ▽ = 180°C, △ = 200°C (A164).

plastic response is displayed. It is also interesting to note that two vastly different flow activation energies were observed. Above or below ~31 vol% styrene, the values were 38 and 19 kcal/mole, respectively. It was suggested (A164) that the higher value resulted from disrupting a semicontinuous polystyrene phase. These investigators also found that the response in the low-frequency region was much more dependent upon the length of the styrene end block than on the polybutadiene segment.

Studies have also been conducted on the rheology of styrene–butadiene–styrene in dilute (A167) and concentrated (A165, A168) solutions. The non-Newtonian viscosity of concentrated (~16.5–30.6 gm/100 ml) solutions was determined over a shear rate range of 1–10^5 sec^{-1} in five different solvents (A168). All of the block polymer flow curves could be reduced to a single master curve that followed the Graessley theory (A168) and had the same shape exhibited by monodisperse polystyrene. The shift factor parallels the Rouse relaxation time but showed an additional concentration and solvent dependence. It is interesting that the *concentrated* viscosity showed a *minimum* with good solvents (Fig. 6-9) as opposed to a *maximum* (Fig. 6-10) observed for intrinsic viscosity (e.g., infinite dilution).

Steady shear viscosities were measured (A167) in toluene for both random and styrene–butadiene–styrene copolymers. The values for block copolymers fall between values predicted by the Zimm and Rouse theories, while those of the random copolymers were close to the Zimm theory. The results suggested (A167) that the random copolymer had a more compact conformation than the block copolymer in a good solvent, such as toluene.

e. Viscoelastic Behavior

The two-phase nature of diene–styrene block copolymers has been described by Angelo et al. (A30, A169), Kraus et al. (170), Tobolsky and co-workers (A172), and many others (A19, A171). The modulus–temperature behavior has been reviewed (A68). Typical results are shown in Fig. 6-11 (A30). Note that the relatively high transition temperature of the polydiene phase in Fig. 6-11 is due to the high 1,2-enchainment (e.g., synthesized with sodium counter-ion in THF solvent).

The dynamic properties and glass transition temperatures of both random and styrene–butadiene–styrene block copolymers were discussed by Kraus et al. (A170). The moduli of the block polymers above the diene T_g are higher than those of the random copolymers and are proportional to the styrene content. This may be due to a

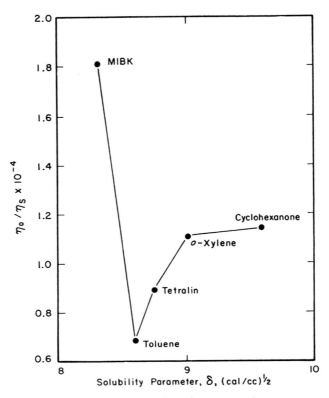

Fig. 6-9. Effect of solvent quality on the relative zero shear viscosity at constant polymer concentration (δ = solvent solubility parameter) (A168).

"filler" effect. It was shown that the *width* as well as the position of the dynamic loss peak was useful in characterizing these systems. Uniformly random copolymers in which the comonomers follow each other in a statistically well-defined way (i.e., composition does not vary along the chain and all molecules have the same composition) showed the narrowest loss peak. More heterogenous (e.g., longer sequence distributions) random copolymers also showed only one loss peak, but it was considerably broadened. The position and width of the block copolymer peaks were dependent upon the block length and compositional purity.

It cannot be overemphasized that morphology governs the viscoelastic and mechanical behavior of these systems. Thus, in addition to block purity, sequence length, and ratio of the components, the method of sample preparation can be critical. For example, as dis-

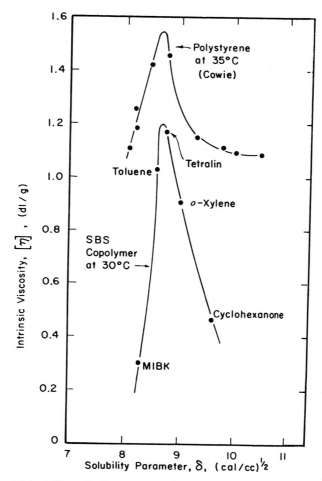

Fig. 6-10. Effect of solvent quality on the intrinsic viscosity (A168).

cussed earlier, the nature of the casting solvent, evaporation rate, drying conditions, and molding times and temperatures can have enormous effects. Of course, additives such as plasticizers and cross-linking agents have an additional bearing on the viscoelastic response. Variation of interfacial morphology on mechanical properties has been discussed by Kaelble (A173, A174).

The variations in the properties of solution-cast films, which are possible merely by changing the nature of the casting solvent, has intrigued a number of investigators. The principal role of the solvent is

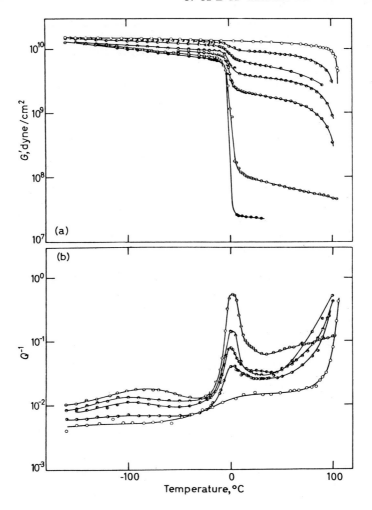

Fig. 6-11. Styrene–butadiene block copolymers; temperature dependence of the torsion modulus G' and internal friction Q^{-1}; $\bigcirc$ = polystyrene, $\pmb{\Phi}$ = S/B$_1$, $\pmb{\mathbb{O}}$ = S/B$_2$, $\pmb{\ominus}$ = S/B$_3$, $\pmb{\ominus}$ = S/B$_4$, $\ominus$ = S/B$_5$, and $\pmb{\bullet}$ = polybutadiene (A30). Reproduced by permission of the publishers, IPS Business Press. Ltd.

often to cause a given phase to become more continuous or more discreet. Alternately, solvents can be chosen to encourage phase separation or phase blending. The mechanism is a development of the technique due to Merrett (A134) wherein "good" or "poor" solvents, respectively, can expand or collapse a chain and predetermine the resultant morphology after solvent evaporation. Fortunately, only three casting systems (as shown in the tabulation below) have been used for much of the experimentation with the styrene–diene–styrene copolymers.

Solvent system	Specific for
THF/MEK	Polystyrene phase
Benzene/heptane	Polyisoprene or polybutadiene phase
Carbon tetrachloride	Both phases

Beecher *et al.* (A105) found that a broad damping peak was observed at 35°C for the CCl₄-cast specimen. Since this transition is intermediate to the peak for the two segments, it was attributed to phase blending. Variation in damping or temperature for the different solvent systems is shown in Fig. 6-12 (A105). Shen and co-workers (A175–A179) performed stress relaxation measurements on styrene–butadiene–styrene over an extended temperature range (−150°C to 100°C). Viscoelastic master curves covering 25 decades of time were constructed. The THF/MEK-cast films had a higher modulus than the benzene-cast films, as expected. The shift factors utilized do not follow the classical Williams-Landell-Ferry equation, which was, of course, derived for single-phase systems. The MEK/THF films are not adequately described by any of the modifications of the WLF equa-

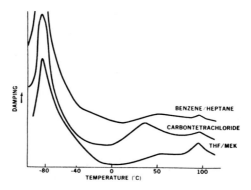

Fig. 6-12. Damping versus temperature for Kraton 101 on samples deposited from different solvent systems (A105).

tion. In additional studies, evidence for an interfacial phase was presented (A180). As discussed earlier in Section c, above, Wilkes and Stein (A115) utilized casting techniques to demonstrate rheooptical relaxation characteristics. Their results have also been discussed in a review (A68).

Block copolymers fabricated from the melt are often vastly different from solution-cast films. The most important reason for the different behavior is the characteristic high melt viscosity. Morphological studies described earlier show that the unoriented melt-fabricated specimens do not develop the same degree of order as carefully prepared cast films. Furthermore, one observes (A20) unusually large effects of molding time and temperature. It was also demonstrated (A171) that the upper transition (styrene) can be quite diffuse, unless the molded film is first submitted to stress relaxation at high elongations. This treatment is believed to improve the degree of phase separation. Several other phenomena that are related to thermal history have been reviewed (A68).

The relaxation rate of molded styrene–butadiene–styrene has been studied (A181). The values were derived from constant extension rate data and were found to be about 8% per decade of time at temperatures from −40°C to +40°C and at extensions from about 20% up to 400%.

An unusual relationship between tensile (E) and shear (G) storage modulus for molded film has been reported (A163, A192). Ratios of $E/3G$ as high as twelve have been observed with 40% styrene materials. By contrast, solution-cast films from toluene had the expected value of about unity. It was suggested (A163) that the disparity resulted from the formation of interconnected polystyrene domains. Further work (A192) confirmed this picture. The loss tangent in tension is heavily weighted by mechanical losses in the polystyrene phase, whereas the corresponding shear value is only moderately affected. A number of other viscoelastic and dielectric studies have been performed and are suggested to the interested reader (A102, A182–A196).

f. Ultimate Properties

The previous sections dealing with synthesis, characterization, morphology, rheology, and viscoelastic behavior have served to lead to this part of the chapter. There can be no doubt that most of the academic and practical interest in this class of two-phase macromolecules is related to the elucidation and utility of their high strengths and elastic recovery characteristics. The ultimate properties are relatively easy to measure, but interpretation of the role of molecular structure in the failure process is most difficult. It is easy to imagine how the presence of impurities in the block structure (e.g., homopolymer, diblock, plasticizer, or residual solvents) could lead to a hopelessly complex situation. The literature on this aspect of styrene–diene–styrene systems falls into two categories: (1) investigations of commercial styrene–butadiene–styrene materials, and (2) studies on laboratory synthesized styrene–butadiene–styrene and styrene–isoprene–styrene copolymers. The ready availability of the commercial copolymer from the Shell Chemical Co. has encouraged numerous physical property investigations. While there is certainly agreement that the commercialization of styrene–butadiene–styrene by anionic block copolymerization was an outstanding technical achievement, it should be noted that the commercial copolymer does contain a substantial quantity of block impurities (A5, A6, A140). Although the ultimate mechanical properties are still good, they are significantly lower (A6) than those that can be obtained with pure systems. Systematic studies, particularly at the University of Akron by Morton and co-workers (A5–A8, A11, A17–A21, A26, A39, A197, A201) since 1965 have been concerned with the physical behavior of very well-defined and characterized styrene–butadiene–styrene and styrene–isoprene–styrene block copolymers of predictable molecular weight. This work has shed considerable light on molecular structure ultimate property relationships. Furthermore, studies with intentionally added impurities have produced a better understanding of the characteristics of the commercial material. Subtle but important distinctions between styrene–butadiene–styrene and styrene–isoprene–styrene have also been noted. In light of the extensive information available from this laboratory, we shall place considerable emphasis on the results of Morton and co-workers.

These studies on high vacuum synthesized model polymers more or less predicted the properties to be expected from less well-defined styrene–butadiene–styrene samples prepared commercially or under less rigorous laboratory conditions.

Early experimental studies on pure styrene–butadiene–styrene and styrene–isoprene–styrene block polymers (A11) were conducted by Juliano (A20) and McGrath (A19), respectively. Differences between the butadiene- and isoprene-containing materials are to be expected. For example, the glass transition temperature of 1,4-polybutadiene is about 30°C lower than that of 1,4-polyisoprene. Thermal or oxidative treatment cross-links polybutadiene but primarily causes chain scission in polyisoprene. Both polydienes have a high (~90%) 1,4 content (with lithium initiator and hydrocarbon solvents), but the polyisoprene is more stereospecific (~80% cis-1,4 configuration) and can, in some situations, crystallize at high extensions.

It was established that there was a relatively narrow molecular weight range ($\overline{M}_n$ of ~10,000–25,000 for the polystyrene and ~30,000–100,000 for the polydiene) in which the desired mechanical properties were observed. The styrene–butadiene–styrene systems were somewhat more narrowly defined than the styrene–isoprene–styrene copolymers. The lower limit is governed by the minimum requirements for the formation of a domainlike two-phase morphology. The upper limit is restricted by the anomolously high melt viscosity of these compositions as discussed above in Section d. The high strength observed for these materials was ascribed to the reinforcing nature of the small (<400 Å) highly dispersed, perfectly bonded polystyrene domains which behave like "filler particles." The high degree of recovery from deformation is due to the physical network formed by the association of the glassy polystyrene segments. The optimum "rubberlike" behavior was noted at between ~20 and 35 wt% styrene. Again, the lower level is believed related to a minimum required concentration of filler and network junctions, whereas the upper limit is due to plastic yielding that results from the development of a semicontinuous polystyrene phase.

High-strength rubbers (250–300 kg/cm² at >1000% elongation) were prepared from both styrene–butadiene–styrene and styrene–isoprene–styrene block copolymers (A19, A20). Some important differences were recorded. The ultimate strengths (at comparable block lengths) of the styrene–butadiene–styrene materials were a function of the styrene content in the elastomeric range of interest (20–40% styrene). However, the tensile strengths of the styrene–isoprene–styrene compositions were independent of the styrene content and less sensitive to molding conditions (A197) over the same compositional range. For example, annealing was required to develop optimum styrene–butadiene–styrene tensile strengths, whereas it had no effect in the styrene–isoprene–styrene system. Typical stress–strain curves illustrating this effect are shown in Figs. 6-13 and 6-14. The possible explanations for these differences were explored by Schwab (A8, A39, A197). It was known that the calculated solubility parameter difference between polystyrene (8.8) and polyisoprene (8.15) was larger than between polystyrene (8.8) and polybutadiene (8.4). Thus, one might argue that a sufficient degree of segment incompatibility for development of the two-phase structure could be achieved in the styrene–isoprene–styrene material at equivalent styrene contents but at lower styrene block lengths. On the other hand, it should be noted

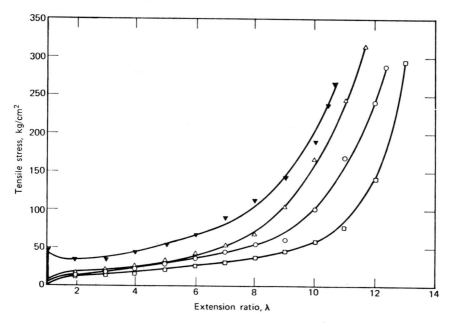

Fig. 6-13. Stress–strain properties of styrene–isoprene–styrene block copolymers (A8).

	Molecular weight ($\times 10^{-3}$)			
	S	I	S	Styrene (%)
▼	13.7	41.1	13.7	40
□	13.7	100.4	13.7	20
○	1:1 blend of 1 and 2			30
△	13.7	63.4	13.7	30

that the melt viscosity values for styrene–butadiene–styrene and styrene–isoprene–styrene, measured by Holden (A90) and communicated by Schwab (A197), show the styrene–butadiene–styrene is higher by nearly an order of magnitude. The corresponding curves are shown in Fig. 6-15. These data help explain why the styrene–butadiene–styrene was observed to be much more sensitive (A20, A197) to molding conditions (e.g., annealing and cooling times).

It is well known that the strength of single-phase elastomers can be related to $T - T_{\mathrm{g}}$, where T is the test temperature and T_{g} is the glass transition temperature (A181). A comparison of styrene–butadiene–

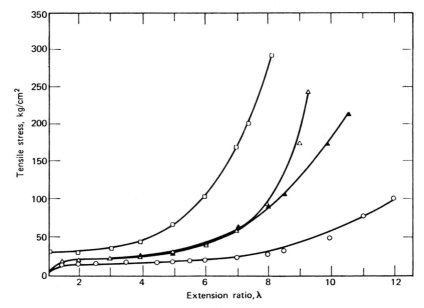

Fig. 6-14. Stress–strain properties of styrene–butadiene–styrene block copolymers (A8).

	Molecular weight ($\times 10^{-3}$)			
	S	B	S	Styrene (%)
○	13.7	100.4	13.7	20
□	13.7	41.2	13.7	40
▲	1:1 blend of 1 and 2			30
△	13.7	63.4	13.7	30

styrene and styrene–isoprene–styrene has been made (A197) at equivalent compositions and block lengths and is shown in Table 6-9.

It may be seen that the styrene–butadiene–styrene (20% styrene) begins to approach the corresponding styrene–isoprene–styrene materials at the $T_g + 60°C$ test conditions. However, at 30 and 40 wt% styrene the styrene–butadiene–styrene has a much higher strength.

Effect of Morphology on Strength. As would be expected from previous discussions, morphology can influence the tensile strength. At 20% styrene, the effect is small, since geometrical considerations prevent spherical domains from becoming continuous at less than ~26

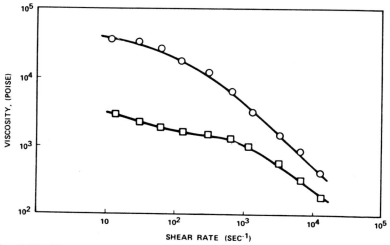

Fig. 6-15. Viscosities of a styrene–butadiene–styrene ($\bigcirc$ = 9.5S–46B–9.5S) and styrene–isoprene–styrene ($\square$ = 8.7S–45I–8.7S) block copolymer (segmental molecular weights $\times 10^{-3}$) (A90).

vol%. Higher styrene compositions (30 or 40%) can be made to show phase inversions as a function of casting solvent. For example, 30% styrene content films cast from benzene show no draw region in their

TABLE 6-9

Tensile Strengths of Styrene–Butadiene–Styrene and Styrene–Isoprene–Styrene at Constant $T - T_g{}^a$

Sample	Weight% styrene	Molecular weight $\times 10^{-3}$	Tensile strength (kg/cm²)[b] at 22°C	at T_g(diene) +60°C
SIS	20	8.4–63.4–8.4	225	340
SBS	20	8.4–63.4–8.4	110	315
SIS	20	13.7–109.4–13.7	300	395
SBS	20	13.7–109.4–13.7	150	285
SIS	30	13.7–63.4–13.7	350	415
SBS	30	13.7–63.4–13.7	360	535
SIS	40	21.1–63.4–21.1	340	435
SBS	40	21.1–63.4–21.1	380	670

[a] From Schwab (A197).

[b] Test temperatures were −5°C and −35°C for the SIS and SBS, respectively.

stress–strain curves, whereas similar specimens from THF/MEK do. This indicates the development of a semicontinuous polystyrene phase in the latter case (A197). Films cast from CCl_4 have both lower modulus and tensile strength than specimens prepared from the other solvents. This is believed due to the relatively high degree of phase blending possible in the CCl_4 system (A105). Cast films of styrene–isoprene–styrene (THF/MEK) give small but significantly higher tensile strength than molded samples. This is probably related to the fact that the cast films can develop a more highly ordered morphology during the slow solvent evaporation, as was discussed earlier (A114).

The effect of polymeric impurities is an important consideration, since no practical anionic polymerization process can completely prevent the termination of growing chains at the A or A-B stage. The effects of added polystyrene, polyisoprene, and diblock polystyrene–polyisoprene (SI) to styrene–isoprene–styrene has been studied (A7, A8, A197). The addition of 1–20 wt% homopolystyrene (molecular weight similar to the block polystyrene) has an insignificant effect on tensile strength. The modulus (at 400% elongation) was enhanced monotonically by the additional "filler." It was concluded that the added polystyrene largely entered the domain of the block polystyrene. Films remained transparent up to 20 wt% added polystyrene. Addition of 5% polyisoprene homopolymer decreased the tensile strength by about 10%. By far the largest effect was noted for the addition of diblock polymer (Table 6-10). Thus, even 2% diblock can have a significant harmful effect. This has been ascribed to the "net-

TABLE 6-10

Effect of Added Diblock on Styrene–Isoprene–Styrene Tensile Strength[a]

Added diblock (%)	Tensile strength (kg/cm^2)
0	319
1	308
2	264
5	244
67	49

[a] From Morton (A8). Styrene–isoprene block molecular weights, 21,100–63,400; styrene–isoprene–styrene block molecular weights, 21,100–63,400–21,100.

work defects" that arise when the diblock chains have one end embedded in polystyrene domains but the other end unattached (A8).

Carefully prepared styrene–butadiene–styrene and styrene–isoprene–styrene block polymers are nearly monodisperse (A19). Therefore, one could expect that the network chains are very uniform. In a sense, broadening the molecular weight distribution (MWD) either by changing the synthetic conditions (A19, A20) or by blending with another triblock (A197) would introduce network impurities. It appears that (A11) there is no more than a small loss of strength upon moderately increasing the MWD. This may be because the total polydiene molecular weight should not be equated to molecular weight between cross-links (A198).

Another type of impurity is one introduced by modification to chemically cross-link the system (A197). Cross-linked styrene–isoprene–styrene block polymer showed a lower tensile strength than its non-vulcanized counterpart. It was reasoned that this decrease in tensile strength was a result of the uneven stresses, which could no longer be removed by slippage of the entanglement cross-links in the polydiene. Such a model has been proposed (A198) (See Fig. 6-16).

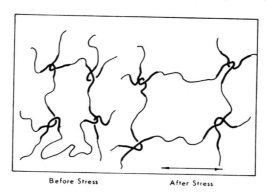

Before Stress   After Stress

Fig. 6-16. Relief of stress in the A-B-A network by entanglement slippage (A198).

The pioneering work of Holden and co-workers (A89, A199, A200) complimented and essentially agreed with studies at Akron University previously cited. Among other accomplishments, Holden *et al.* proposed the schematic phase arrangement shown in Fig. 6-17, that has since been largely confirmed by microscopy, small-angle X-ray, etc. Holden and co-workers were also the first to publish the shape of the stress–strain curves for various compositions (Fig. 6-18).

The 28% styrene content specimen has the most typical elastomeric stress–strain curve. Interest also exists in the high styrene (~60–80

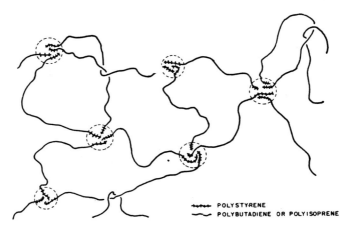

Fig. 6-17. Phase arrangement in styrene–butadiene–styrene and styrene–isoprene–styrene block copolymers (schematic) (A89).

wt%) compositions where there is an interesting combination of impact strength and transparency. Resins containing about 25 wt% polybutadiene have been commercialized. They are believed to be styrene–butadiene–styrene materials possibly of the radial type coupled by polyfunctional agents (see Section 2, below).

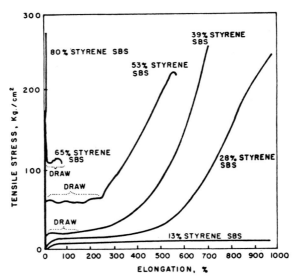

Fig. 6-18. Stress-strain curves (at 2 inches/minute) for styrene–butadiene–styrene polymers of various styrene contents (A89).

Important contributions to the understanding of network structures have been made by a clever study of the swelling behavior of styrene–butadiene–styrene elastomers (A198). Using solvents such as isooctane, which selectively swell the diene segment, it was possible to use techniques that are well known for characterizing chemically cross-linked rubbers. Perhaps the most important conclusion from this work was that the concentration of elastically effective chains is essentially independent of the center block size. Thus, the role of physical entanglements of the rubbery segment, which are trapped between two glassy domains, is of paramount importance. The effective molecular weight between cross-links $(\overline{M}_c)$ is thus associated with an "entanglement" molecular weight $(\overline{M}_e)$ of the polydiene. The swelling results on styrene–butadiene–styrene and styrene–isoprene–styrene give $\overline{M}_e$ values 15,000—16,000 gm/mole, which are in agreement with the expected molecular weights between entanglements for the corresponding diene homopolymers (A5, A6).

The strength and extensibility of elastomers, including styrene–butadiene–styrene has been reviewed (A189). It was concluded that the time and temperature dependence of the high elongation and high tensile strength results primarily from plastic deformation of the polystyrene domains (A181, A202). This is in agreement with electron micrographs of the failure process (A105), which show the polystyrene domains do deform and rupture. It was proposed (A184) that the domain rupture occurs in the vicinity of a slowly developing crack. The domain must be disrupted before the crack's size and the resulting elastically stored energy become sufficiently large to initiate high-speed crack propagation. It should also be noted (A11, A19, A203) that other energy dissipating processes such as stress softening occur in these "filled" elastomers. Typical behavior is shown in Fig. 6-19 and 6-20. For all chemically cross-linked rubbers, a direct relationship exists between W_b, the work required to break, and the energy dissipated, W_d, in stretching to the breaking elongation. In other words, the strongest elastomers are precisely those in which the major part of the energy is dissipated before rupture. Schwab (A197) found that the styrene–diene–styrene block copolymers also fit this criteria quite well.

The above discussion largely summarizes our knowledge of the factors governing the ultimate properties of styrene–butadiene–styrene and styrene–isoprene–styrene. There are, however, several reviews (A68, A184) and other papers and patents recommended that cannot be further discussed here because of space limitations (A83, A204–A217).

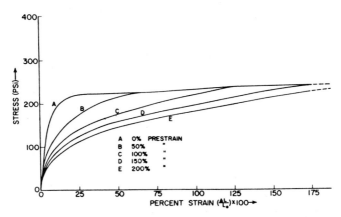

Fig. 6-19. Stress–strain response of the styrene–butadiene–styrene block polymer, Kraton 101, as a function of strain history. Prestrain: (A) 0%, (B) 50%, (C) 100%, (D) 150%, (E) 200% (A203).

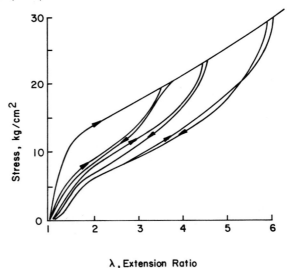

λ, Extension Ratio

Fig. 6-20. Stress-softening behavior of styrene–isoprene–styrene block copolymers (A19).

g. Applications and Alloys

There are many patents and trade journal publications that have reviewed particular application areas. Many alloys have been prepared that are claimed to result in improved product performance. While it is beyond the scope of this book to discuss these products in detail, a limited list of these applications and references are tabulated in Tables 6-11 and 6-12.

Most of the applications have been more or less elastomeric in nature. However, very transparent moderately rigid and ductile butadiene–styrene copolymers have been introduced by Phillips Petroleum as "K resins" (see Section 2, below). Their molecular structure has not been disclosed. Publications and patents (A218) also exist on transparent shock-resistant polystyrenes. Both di- and triblocks were prepared with butadiene contents in the range of 10–30%. Transparency and unnotched impact were superior for the styrene–butadiene–styrene triblock system (A219). It has also been shown that the nature of the continuous phase is important (A222) and that styrene–butadiene–styrene or styrene–butadiene–styrene–butadiene high styrene resins are tough as evidenced by high elongation to break, whereas styrene–butadiene, butadiene–styrene–butadiene, or polyblends of similar molecular weight and composition were brittle. Independently (A223) it was shown that a 79 wt% styrene containing styrene–butadiene–styrene of molecular weight 125,000 had a tensile strength of 217 kg/cm and 126% elongation.

2. Radial Styrene–Dienes

Radial or star-shaped copolymers are a relatively recent interesting development. Accordingly, although they are at an early stage of development, this section will attempt to point out some of the known

TABLE 6-11

Some Applications of Styrene–Diene–Styrene Block Copolymers

Application	Reference	Remarks
Latex film	A224–A226	—
Adhesives	A227–A233, A245	High solubility and strength with no vulcanization
Elastic filaments	A234	Compared with natural rubber
Foam	A235, A236	—
Powders	A237	Carpet backings, etc.
Footwear	A238–A240	—
Packaging films	A241	—
Caulking and sealants	A242–A244	—
Coatings	A246–A248	—
Injection molding or extrusion	A249–A255	Review of properties

TABLE 6-12

Alloys of Styrene–Diene–Styrene Block Polymers with Other Materials

Alloy material	Reference	Remarks
Polyethylene	A256, A270, A271	Improved weather and ozone resistance
Ethylene–vinyl acetate (or ethyl acrylate) copolymer	A257, A258	Increased melt flow and ozone resistance
Poly(2,6-dimethylphenylene oxide)	A259, A261	Transparent compatible system
Oil	A260, A262	Economics processibility
Bitumen	A263–A266, A272	Improved strength of road and building materials
Polystyrene	A267–A269	Improved miscibility
Impact styrene	A90, A273–A275, A285, A286, A304	Improved toughness
Polybutadiene	A276	Good mutual solubility
Polybutadiene or polyisoprene	A277, A278	Improved green strength
Polymers and copolymers of acenaphthalene	A279	Increased strength
Styrene–butadiene random copolymer	A280, A281	Improved green strength
Polyurethane block copolymer	A258	Solvent resistance
ABS grafts	A282, A283	Impact modification
Coumarone–indene	A284	Adhesives

similarities and differences between the "stars" and their linear counterparts.

The concept of a star polymer has been well established for more than a decade (A333). A four-arm star homopolymer may be represented by the structure below.

Such a structure can be prepared by living anionic polymerization followed by an efficient coupling step with a compound such as silicon tetrachloride, e.g., Reaction 6-9.

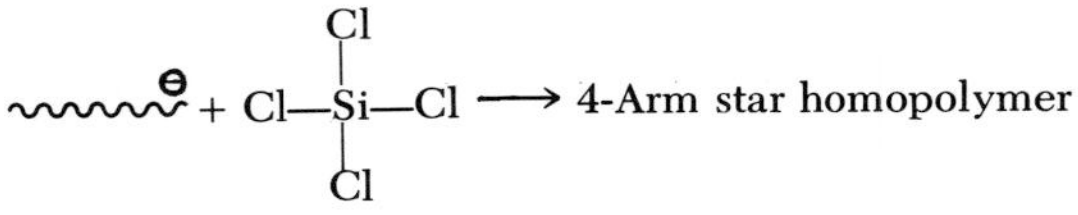

$$\text{\textasciitilde\textasciitilde\textasciitilde}^{\ominus} + \text{Cl—}\underset{\underset{\text{Cl}}{|}}{\overset{\overset{\text{Cl}}{|}}{\text{Si}}}\text{—Cl} \longrightarrow \text{4-Arm star homopolymer} \qquad (6\text{-}9)$$

A number of investigators (A333–A338) have been interested in these systems, often to prepare model branched polymers that could be used to test polymer solution theories (A339, A340). Other linking agents, such as divinylbenzene, were also used (A341–A343).

Thus, it is easy to see how the technique could be extended to the coupling of diblock systems with polyfunctional agents. The coupling technique with divinylbenzene has been refined by Fetters (A344, A345), who has optimized the synthesis to the point that materials with up to twenty-nine or more arms were obtained under very well-defined conditions. Predictable molecular weights and block structures were demonstrated. Other investigators have also used silicon halides (A357–A359) and divinylbenzene (A360) for star block copolymer synthesis.

The radial styrene–butadiene system can also be prepared by coupling the diblock with epoxidized polybutadiene or epoxidized soybean oil (A350), as well as THF-promoted divinylbenzene. The materials must be well stabilized (A350, A352, A353) to prevent cross-linking during processing. Interestingly, the radial copolymers have also been studied as model systems for the characterization of long-chain branched polymers (A351).

A major practical incentive for investigating star-shaped block copolymers appears to be that they display remarkably low melt and solution viscosities, even at very high molecular weights. Typical behavior is shown in Fig. 6-21 and 6-22. Such behavior is important in adhesive applications (A346–A349).

In elastomers and thermoplastics, the desirability of relatively low melt viscosity during processing is obvious. The advantages of com-

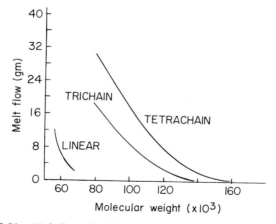

Fig. 6-21. Melt flow of teleblock polymers (30% styrene) (A346).

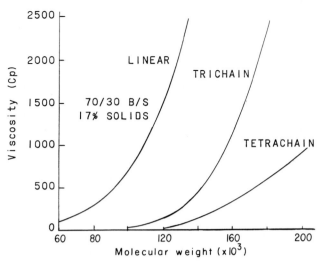

Fig. 6-22. Solution viscosity of teleblock polymers in toluene–naphtha (25:15) (A346).

pounding such radial elastomers have been discussed (A354–A356). It is perhaps less apparent that one may also prepare much higher molecular weight polymers than are easily obtainable with linear systems. Moreover, this can be achieved without excessively high melt viscosity (A361). The morphology can apparently approach "equilibrium" more closely in the star macromolecules than in their linear counterparts. Bi and Fetters (A345) demonstrated that a very well-defined morphology could be developed in their divinylbenzene-linked styrene–isoprene system. A typical electron micrograph is shown in Fig. 6-23 (A345).

In addition to elastomers, rigid star copolymers have been developed as well (A362, A363). An important class of styrene-rich resins has been commercialized under the trade name K Resins (A362, A363). The main attribute of these materials resides in the combination of transparency (due to the small domain size) and satisfactory impact for many packaging applications. Preliminary results with model polymers suggest that the rigid radial structures are more ductile than their linear high molecular weight counterparts at the same weight percent polystyrene (A364).

3. Modified Styrene–Dienes

One can visualize a number of polymer-modifying reactions that could be applied to the styrene–diene–styrene systems. The general

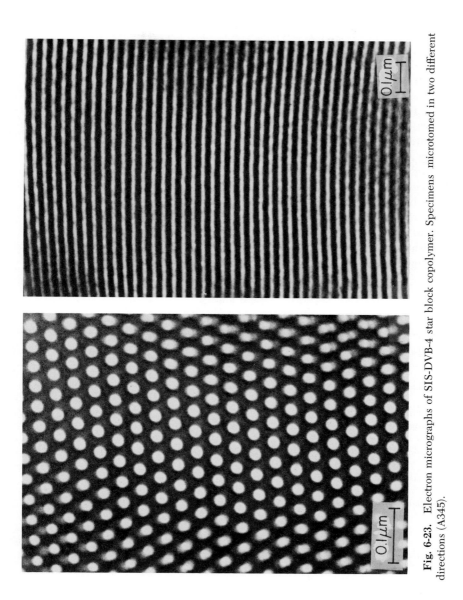

Fig. 6-23. Electron micrographs of SIS-DVB-4 star block copolymer. Specimens microtomed in two different directions (A345).

chemical reactions of polymers containing carbon–carbon double bonds have been reviewed (A287). Table 6-13 shows the polymer modifications that have been reported, primarily in the patent literature. They are essentially reactions that have previously been performed on homopolymers. However, it is impressive that excellent catalysts and mild conditions have been developed that essentially allow quantitative transformations with minimal chain-length degradation.

At least one modification, hydrogenation, has been commercialized. This is the Kraton G family, which is believed to be hydrogenated styrene–butadiene–styrene. The center segment may contain moderate amounts of 1,2-configuration so that, after hydrogenation, one has essentially an ethylene–butene rubbery segment. Of course, hydrogenation of high 1,4-polybutadiene and polyisoprene would lead, respectively, to crystalline polyethylene and rubbery ethylene–propylene blocks. This has actually been done (see Section A,5).

As might be expected, the Kraton G materials have greatly improved thermal and oxidative resistance. For example, Holden (A90) has demonstrated they are stable in the melt at 300°C, as shown by the GPC data in Fig. 6-24.

TABLE 6-13

Modification of Styrene–Diene–Styrene Block Polymers

Type of modification	Reference	Remarks
Sulfonation of SIS or SBS	A288–A292	Allows preparation of desalination membranes
Chlorosulfonation	A293, A294	—
Hydrogenation of SIS and SBS	A295–A321	Improves stability toward ozone, oxygen; reduces creep
Halogenation	A322	Improves adhesion; can be quaternized
Haloalkylation	A323	Improves adhesion; can be quaternized
Maleic anhydride adduct	A324, A365	Improves adhesion; can be quaternized
Reaction with S_2Cl_2	A325	Improves heat resistance
Vulcanization	A326, A327	Solvent resistance; reduced creep
Stabilization	A328, A329	Enhanced resistance to heat degradation
Thermosetting styrene graft	A330	—
Hydroxylation	A331, A332	Membranes; medical applications

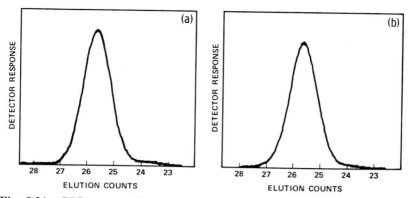

Fig. 6-24. GPC curves of a styrene–ethylene/butyene–styrene block copolymer:
(a) original polymer; (b) after 15 minutes at 300°C or 2000 hours in 300 psig oxygen at
70°C. (A90).

4. Other Alkenylaromatic–Dienes

While the great bulk of the investigations in the literature have
utilized polystyrene as the hard block and either polybutadiene or
polyisoprene as the soft segment, there have been various structural
modifications studied. Many of these are listed in Table 6-14.

It has been clearly demonstrated that higher T_g hard blocks (as compared to polystyrene) can be prepared and utilized. Of course, this
does not overcome the thermal stability limitations ($\leq 200°C$) of the
polydiene soft segment unless hydrogenation is performed. Nevertheless, measurements of tensile strength on well-defined styrene–
isoprene–styrene and α-methylstyrene–isoprene–α-methylstyrene systems of virtually identical composition and molecular weight showed
that the replacement of polystyrene with poly(α-methylstyrene) resulted in both higher room temperature tensile strength and higher
temperature capability (A18). In this study, the novel soluble difunctional initiator 1,4-dilithio-1,1,4,4,-tetraphenylbutane was used to
generate the high 1,4-content polyisoprene prior to the low-
temperature THF-promoted polymerization of the α-methylstyrene
segment. In another reference, a similar copolymer was synthesized
via initiation with α-methylstyrene tetramer and promotion with
hexamethylphosphoramide (A364a).

5. Diene–Dienes

Various studies have been reported on the preparation, properties,
and modification of various types of diene–diene block polymers. These

TABLE 6-14

Miscellaneous A-B-A Diene–Alkenyl Aromatic Block Polymers

Hard block	Soft block	Reference	Remarks
α-Methylstyrene	Isoprene	A18, A36	Improved upper glass transition and mechanical properties relative to SIS
α-Methylstyrene	Isoprene	A366, A367	Higher upper T_g
α-Methylstyrene	Diene	A368	—
α-Methylstyrene, styrene	Butadiene	A369–A373	Improved properties claimed; optionally hydrogenated
2,4-Dimethylstyrene	Isoprene, butadiene	A374	—
4-Vinylbiphenyl	Isoprene	A375	—
1,1-Diphenylethylene–styrene	Butadiene	A376	Improved upper glass transition
Styrene	2-Cyanobutadiene	A391	—
Styrene	Butadiene and ε-caprolactone	A377	—
tert-Butylstyrene	Butadiene, isoprene	A378, A379, A388	—
2,6-Diphenyl-1,6-heptadiene	Butadiene, isoprene	A378	—
1,2-Dihydronaphthalene	Butadiene, isoprene	A378	—
Functionally terminated polystyrene, nylon 66, or polyethylene terephthalate	Hydroxyl-terminated polybutadiene	A382	Attempted condensation polymerization
Vinyltriorganosilanes and styrene or 2,4-dimethylstyrene	Isoprene or butadiene	A380	—
Styrene, and/or acrylonitrile	Butadiene	A383	Phosphorus terminals used to initiate acrylonitrile
Stilbene	Diene	A381, A389	Impact resistant
α-Methylstyrene and styrene	Butadiene	A385	Hydrogenated A-B-C type
α-Methylstyrene and styrene	Butadiene	A387	Hard blocks are compatible
α-Methylstyrene and styrene	Ethylene–butylene	A384	A-B-C type via hydrogenation
α-Methylstyrene–styrene copolymer	Isoprene	A386	Some styrene improves α-methylstyrene rate
α-Methylstyrene–styrene copolymer	Diene	A390	Coupled with divinylbenzene

are listed in Table 6-15. These are, of course, not thermoplastic elastomers per se; however, it is possible to modify the segments to achieve this end. For example, in the hydrogenation of a butadiene–isoprene–butadiene block copolymer, the polybutadiene block is converted to a crystalline "polyethylene" segment, while the polyisoprene block yields a rubberlike "ethylene–propylene copolymer" segment (A392). High 1,2-butadiene center segments become ethylene–butylene, etc. (A392, A393). Other treatments, such as hydrochlorination, can result in crystallinity (A394, A395) and enhanced mechanical properties.

6. Styrene–Alkenylaromatics

The principal alkenylaromatic monomer investigated other than, of course, styrene, has been α-methylstyrene. However, some other sys-

TABLE 6-15

Diene–Diene Block Copolymers

Segment A	Segment B	Reference	Remarks
Polybutadiene	Polyisoprene	A30	Dynamic measurements show compatibility
Polyisoprene	Polybutadiene	A396	Modified by chloral reaction to improved compatibility with PVC
Polyisoprene	Polybutadiene	A394, A395	Selective hydrochlorination gave good film properties
Polyisoprene	Polybutadiene	A397	Calendering was easy, whereas mixture of homopolymers was difficult
Polyisoprene	Polybutadiene	A398	—
cis-1,4-Polybutadiene	trans-1,4-Polybutadiene	A399	Properties better than a random microstructure
1,4-Polybutadiene	1,2-Polybutadiene	A400, A404	—
Polybutadiene	Polyisoprene	A401	—
Polybutadiene	Polyisoprene	A392	Hydrogenated the polybutadiene to polyethylene segments
Polybutadiene	Polyisoprene	A393	Hydrogenated both blocks to yield polyethylene–ethylene–propylene
1,4-Polybutadiene	Polybutadiene (1,4–1,2 mixed)	A402	Hydrogenated to polyethylene–polyethylene–butylene blocks
1,5-trans-Polypentenamer	cis-Polybutadiene	A403	—

tems, such as *tert*-butylstyrene and vinylpyrene have also been discussed and are the subject of this section.

The most novel feature of α-methylstyrene–styrene block copolymers is the fact that they normally form one-phase (one T_g) materials. This is in contrast to the two-phase morphology observed in blends. Experimentally, the first report of this phenomenon is due to Baer (A405). Robeson *et al.* demonstrated (A75) that A-B block copolymers behaved similarly. The best explanation of the intersegment "solubility" is that the segment interaction parameters are very similar. According to Krause's treatment of phase separation (A406), molecular weights in excess of 10^6 are required before phase separation will occur (see also Chapter 5, Section A,3). Block copolymers of α-methylstyrene and styrene were known in the patient literature at least as early as 1959 (A407).

Block copolymers of vinylpyrenes with various monomers were reported by O'Malley and co-workers (A409, A410). However, side reactions involving adduct formation were observed. Also, electron transfer resulted in premature termination. These side reactions were influenced by variables such as solvent polarity, counterion, and, of course, reaction temperature.

tert-Butylstyrene systems have been discussed by Morton *et al.* (A378). More recently, *tert*-butylstyrene–styrene block copolymers have been used by Murray and Schwab (A411–A413) as dispersing agents in the anionic systhesis of divinylbenzene–styrene copolymers. The latter (A413) were reported to be useful as organic reinforcing fillers for rubber.

Interestingly, Lundberg and Makowski prepared styrene–*tert*-butylstyrene–styrene block copolymers, and then selectively plasticized the center segment. The resulting materials reportedly produced tough, transparent, flexible molded products (A414, A415).

B. POLYACRYLIC AND POLYVINYLPYRIDINE BLOCK COPOLYMERS

1. Acrylics

Triblock acrylic-containing block copolymers have usually been prepared with dianionic initiators such as sodium α-methylstyrene tetramer (B1) or electron-transfer initiators such as sodium biphenyl (B2–B4). Low temperatures and other precautions described in Chapter 5 are required to suppress side reactions. The system most frequently studied is methyl methacrylate–styrene–methyl methacrylate

(MSM) and to a lesser extent methyl methacrylate-α-methylstyrene–methyl methacrylate (B5). As discussed in Chapter 5, the living polystyrene must be prepared first and then used to initiate the acrylate monomer polymerization. Furthermore, the styrene anion is preferably capped with 1,1-diphenylethylene to decrease side reactions at the ester carbonyl (B1). Extraction of the crude copolymer with boiling cyclohexane and subsequently with acetonitrile has been suggested for selectively removing the styrene and acrylate homopolymers, respectively (B2–B4), while not affecting the triblock. It has been pointed out (B2) that the A-B copolymer, by contrast, tends to dissolve in boiling cyclohexane.

Compared to the styrene–diene–styrene systems, relatively few characterization investigations have been reported. This may be due, at least in part, to the more difficult synthesis and to the fact that there are as yet no commercially available A-B-A acrylic block copolymers. Nevertheless some excellent investigations concerning the conformational properties of well characterized copolymer in solution have been published. Light-scattering experiments on an MSM copolymer in various solvents over a range of temperatures (18°–65°C) have shown that a conformational transition exists (B1). At lower temperatures, the segments are believed to be segregated and the second virial coefficient is found to be a weighted average of the values for the corresponding homopolymers. At higher temperatures, contacts between the blocks become possible and the additivity rule no longer holds. This change to a "pseudogaussian" conformation also can be seen by specific volume and refractive index increment variations. This behavior was not observed in the case of the diblock SM copolymers. With these materials, the same authors (B1) feel strongly that segregation always occurs. Kotara and co-workers (B2–B4, B6) also studied SM and MSM solution behaviors and concluded that the SM domains are segregated, particularly in selective solvents. By contrast, the behavior of the triblock MSM was critically influenced by the polar nature of the acrylic end segments. A dramatic example is that in p-xylene (at 30°C) (B3), which is a nonsolvent for the acrylic block, the intrinsic viscosity of some MSM copolymers was *smaller* (B8) than that of the precursor polystyrene homopolymer. The MSM was considered to be analogous to a molecular dispersion. The SM diblock, having one soluble and only one collapsed subchain underwent intermolecular association and formed stable micelles. The stability and size of the micelles were, as one might expect, dependent upon molecular weight and composition. The conformational differences of the SM and MSM copolymers were treated theoretically (B3).

Solution studies on methyl methacrylate–$_\alpha$–methylstyrene–methyl methacrylate have also been presented (B4). The block copolymers were soluble in solvents that would not dissolve the corresponding homopolymers. This system is believed (B4) to show some differences from MSM because of the greater compatibility of the α-methylstyrene and methyl methacrylate segments.

The morphology of a methyl methacrylate–α-methylstyrene–methyl methacrylate copolymer has been studied by electron microscopy (B5, B7). Spherical domains of 500–2000 Å were seen and increased in size with the molecular weight of the acrylic block. The micrographs are shown in Fig. 6-25.

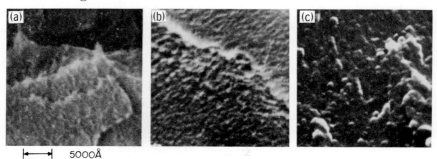

|←——→| 5000Å

Fig. 6-25. Electron micrographs of MMA/MS block copolymers. (a) MMA-1/MS-4; (b) MMA-1/MS-1; (c) MMA-4/MS-1 (B5).

Stress relaxation measurements were made, and it was concluded that the continuous phase was α-methylstyrene. The two segments disturbed each others' motions, as shown by relaxation time and dielectric dispersion measurements.

2. Vinylpyridine

Polyvinylpyridine–polystyrene–polyvinylpyridine (both the 2- and 4-vinyl isomers) block copolymers have been synthesized anionically. Styrene was initiated first, using sodium α-methylstyrene dianion (B9) or sodium biphenyl (B10) difunctional initiators, followed by addition of the vinylpyridine monomer (Scheme 6-4). Structures with the reverse sequence have also been prepared, using a monofunctional initiator, by concluding the synthesis with a coupling step (Scheme 6-5) (B9). The synthesis conditions and properties of these A-B-A block copolymers are similar to those of the A-B block copolymers discussed earlier in Chapter 5, Section B,2.

Noel (B11) discusses the differential thermal analysis performance of

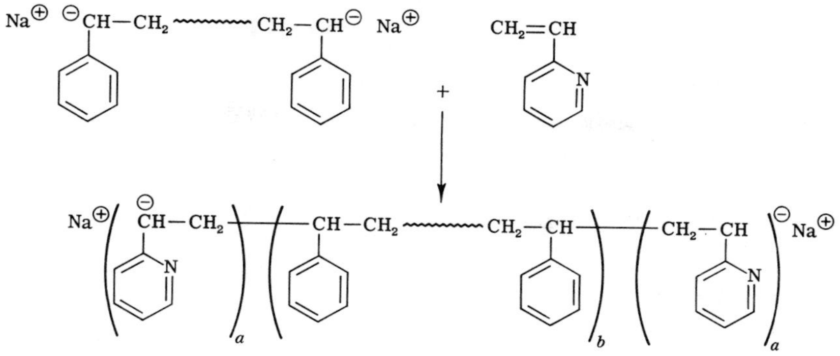

Scheme 6-4

the A-B-A structures. Both segments have T_g values at 105°C. This reference also compares the DTA performance of this block copolymer with that of a similar structure, a poly(p-dimethylaminostyrene)–poly-styrene–poly(p-dimethylaminostyrene) block copolymer. The latter displays two separate T_g values at 105°C and 140°C.

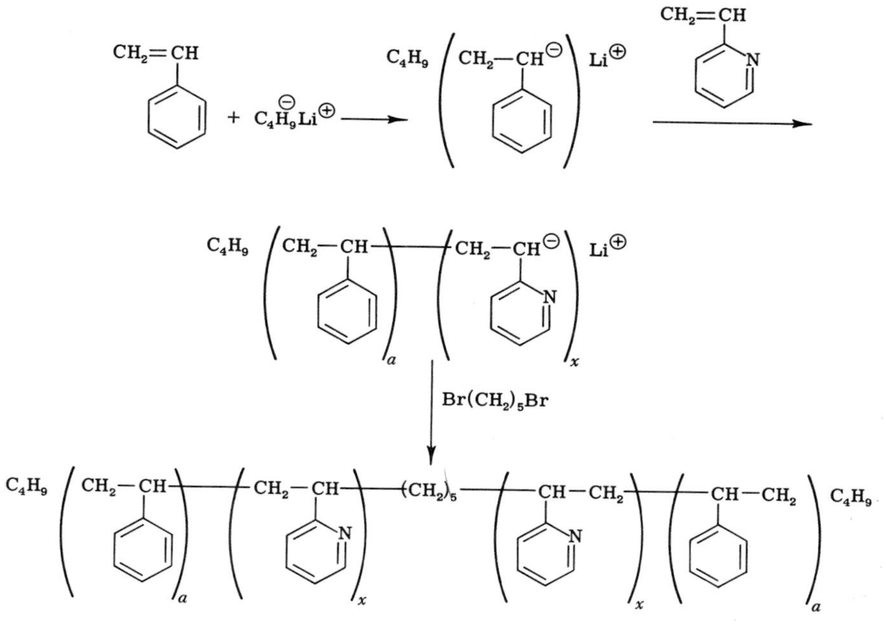

Scheme 6-5

C. HETEROATOM A-B-A BLOCK COPOLYMERS

The A-B-A block copolymers of this section contain one or more segments that have at least one atom other than carbon in the backbone repeat unit. These heteroatoms include oxygen, sulfur, nitrogen, and silicon, and they are found in ether, ester, sulfide, amide, urethane, and siloxane structures. The preparation and properties of these materials are discussed below.

1. Ether–Ether

The block copolymers in this category contain three segments, all of which are polyethers. The segments are derived from epoxide, aldehyde, tetrahydrofuran, or oxetane monomers. Table 6-16 lists the

TABLE 6-16

A-B-A Block Copolymers Containing All-Polyether Segments

A Blocks	Block B	Initiator system	Conditions	Reference
Ethylene oxide	Propylene oxide	$NaOCH_2C(CH_3)HONa$	120°C	C1, C2, C7, C21, C22
Propylene oxide	Ethylene oxide	$NaOCH_2C(CH_3)HONa$	120°C	C1, C8
Ethylene oxide	Propylene oxide–ethylene oxide	NaORONa	50–130°C/1–90 psig	C1, C9
Ethylene oxide–propylene oxide	Propylene oxide–ethylene oxide	NaORONa	50–130°C/1–90 psig	C1, C10
Ethylene oxide	Butylene oxide	NaORONa	135°C/35 psig	C1, C11–C13
	Amylene oxide	NaORONa	135°C/35 psig	C1, C11, C12
	Cyclohexene oxide	NaORONa	135°C/35 psig	C1, C11, C12
	Styrene oxide	NaORONa	135°C/35 psig	C1, C11, C12
Formaldehyde	Tetramethylene oxide	$R_4N^+O^-R'$	25°C	C16, C17
Tetrahydrofuran (A block) and 3,3-bis(chloromethyl) oxacyclobutane (C block)	Tetrahydrofuran–3,3-bis(chloromethyl)oxacyclobutane copolymer	BF_3-epichlorohydrin	0°C	C19, C20, C23

compositions that have been reported along with some of the synthesis conditions used.

a. Olefin Oxide–Olefin Oxide

A-B diblock copolymers containing one poly(ethylene oxide) segment and one poly(propylene oxide) segment were discussed in Chapter 5, Section D. These copolymers were synthesized by initiation with a monofunctional alkali metal alkoxide. A-B-A block copolymers have been prepared (C1, C2) in a similar manner using difunctional initiators to propagate in two directions simultaneously. For example, propylene oxide polymerization was initiated by the disodium salt of propylene glycol at 120°C, after which ethylene oxide was added and polymerized. Neutralization produced the dihydroxyl-terminated A-B-A block copolymer product (Scheme 6-6).

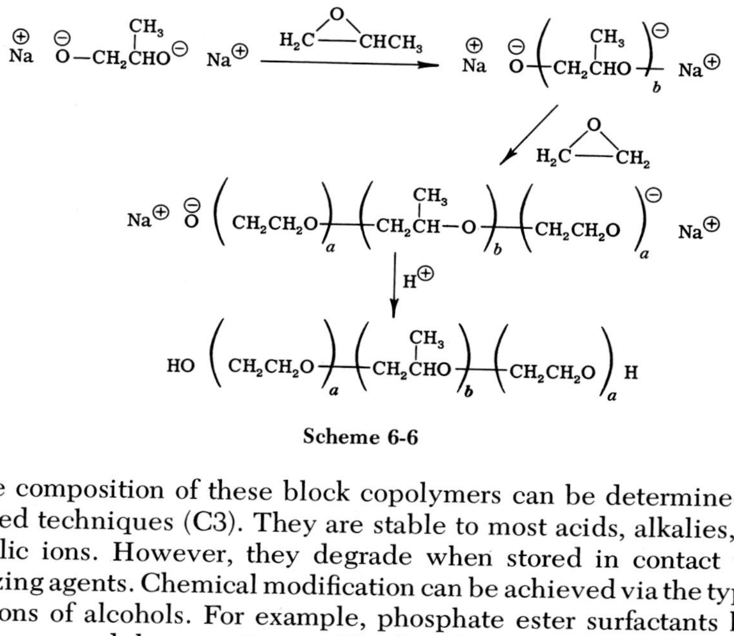

Scheme 6-6

The composition of these block copolymers can be determined by infrared techniques (C3). They are stable to most acids, alkalies, and metallic ions. However, they degrade when stored in contact with oxidizing agents. Chemical modification can be achieved via the typical reactions of alcohols. For example, phosphate ester surfactants have been prepared by reacting a dihydroxyl-terminated A-B-A block copolymer with $POCl_3$ (C1, C4–C5b). As another example, a diquaternary ammonium-terminated composition was prepared by reacting a block copolymer with excess chloroacetic acid followed by reaction of the resulting chloro ester with pyridine (C1, C6, C6a).

Since poly(ethylene oxide) is water soluble and poly(propylene oxide) of ≥ 900 molecular weight is water insoluble, dihydroxyl-terminated A-B-A block copolymers display excellent surface-active properties (C1). These materials contain 20–90% ethylene oxide, have molecular weights of 1000–16,000, and range in appearance from mobile liquids to hard solids. The latter can be flaked and are relatively nonhygroscopic. The copolymers are more soluble in cold water than in hot water due to a greater extent of hydrogen bonding at lower temperatures. They are more soluble in dilute mineral acids than in water because of oxonium ion formation. Other solvents include aromatic hydrocarbons, chlorinated solvents, ketones, and alcohols; ethylene glycol and aliphatic hydrocarbons are nonsolvents. The degree of solubility varies, of course, with the composition of the copolymer. Some of the copolymers form gels in concentrated (>30%) aqueous solution, as a result of hydrate formation and hydrogen bonding in the poly(propylene oxide) segments. (C1, C7).

These block copolymers are low-to-moderate-foaming surfactants, and compositions low in ethylene oxide content are useful as defoaming agents (C1). In 0.01–0.1% solutions at 25°C, they display surface tension values of 36–56 dynes/cm and interfacial tension values of 2–28 dynes/cm (C1). The highest degree of surface activity is observed in copolymers with relatively long poly(propylene oxide) segments and short poly(ethylene oxide) segments. On the other hand, longer poly(ethylene oxide) blocks display superior wetting behavior.

A-B-A block copolymers in which A is poly(propylene oxide) and B is poly(ethylene oxide) have also been prepared by reversing the order of monomer addition, i.e., first ethylene oxide and then propylene oxide (C1, C8). The cloud points and foam heights displayed by these surfactants were found to be significantly lower than those exhibited by ethylene oxide–propylene oxide–ethylene oxide block copolymers of the same weight percent composition.

A-B-A block copolymers have also been synthesized in which A is poly(ethylene oxide) and B is a random copolymer of ethylene oxide–propylene oxide (C1, C9). A mixture of the two epoxides was polymerized first in the presence of a difunctional initiator at 50°–135°C and 1–90 psig. Then the reaction mixture was stripped to remove any unreacted monomer, after which pure ethylene oxide was added to form the end blocks. These compositions also displayed surface-active properties.

This approach is carried one step further in another type of block copolymer in which the A and B segments are both random copolymers of ethylene oxide–propylene oxide, but at different ratios, such that one segment is more hydrophilic than the other (C1, C10).

A-B-A block copolymers have also been reported in which A is poly(ethylene oxide) and B is a higher poly(alkylene oxide) block such as poly(butylene oxide), poly(amylene oxide), poly(cyclohexene oxide), or poly(styrene oxide) (C1, C11, C12). These were prepared by adding the higher alkylene oxide to the potassium salt of a glycol at 135°C and 35 psig followed by the addition of ethylene oxide. These compositions also display detergency, emulsifying, and foam-inhibiting properties (C1, C13). Block copolymers of similar composition but in which poly(ethylene oxide) is the center B segment were found to be lower-foaming surfactants (C1, C13). The surface-active, detergency, and emulsifying properties of the above and related block copolymers are discussed in detail in the literature (C1, C14, C15).

b. Other Polyethers

A-B-A block copolymers have been prepared in which A is polyformaldehyde and B is poly(tetramethylene oxide) (C16, C17). The synthesis was carried out by adding formaldehyde to a solution of preformed, 7000 molecular weight, dihydroxyl-terminated poly(tetramethylene oxide) at 25°C in the presence of a quaternary ammonium hydroxide or acetate catalyst (Reaction 6-10). The hydroxyl end groups

$$CH_2O + HO[(CH_2)_4O]_b H \longrightarrow HO[CH_2O]_a[(CH_2)_4O]_b[CH_2O]_a H$$

$$(6\text{-}10)$$

were acetylated with acetic anhydride and sodium acetate to stabilize the copolymer against degradation by "unzipping." The product (which contained 73% formaldehyde) had a crystallinity value of 40% compared to a formaldehyde homopolymer crystallinity of 65%. The copolymer was superior to the homopolymer in clarity and impact strength but inferior in strength. Similar copolymers were also prepared using amine-terminated poly(tetramethylene glycol.)

The synthesis of triblock copolymers via BF_3-catalyzed cationic polymerization of cyclic oxides has been reported. In one report (C18), a cyclic oxide monomer was added to a preformed dihydroxyl-terminated polyether in the presence of $BF_3 \cdot O(C_2H_5)_2$. In another (C19, C20), A-B-C block copolymers are reported in which A is poly(tetramethylene oxide), C is poly[3,3-bis(chloromethyl)oxacyclobutane], and B, the center block, is a random copolymer of these two. The latter copolymers were prepared by what was called a cationic living polymerization using a BF_3–epichlorohydrin catalyst at 0°C. The three-stage polymerization began with partial polymerization of tetrahydrofuran, followed by the addition of the bicyclobutane monomer. The resulting monomer mixture was polymerized randomly, after which the unreacted tetrahydrofuran was removed (by evaporation) and the remaining bicyclobutane was polymerized to form the third C segment. Termination in the second and third stages was reported to be a problem in this synthesis. The product exhibited thermoplastic elastomeric behavior. X-Ray investigation of stretched samples indicated crystallinity in both terminal blocks. Stress–strain, strain–temperature, and flow rate–temperature properties were comparable to those of a styrene–butadiene–styrene A-B-A block copolymer. Hysteresis, however, was high. These properties were cited as evidence of the block sequence structure of the copolymer.

2. Ether–Vinyl

The block copolymers in this category contain either one polyether segment between two polyvinyl segments or one polyvinyl segment between two polyether segments. The polyether segments are prepared from epoxide, aldehyde, tetrahydrofuran, and fluorinated ketone monomers. The polyvinyl blocks are derived from styrene α-methylstyrene, methyl methacrylate, acrylonitrile, vinylpyridines, butadiene, and isoprene. Table 6-17 presents the compositions reported along with some of the synthesis conditions employed in preparing them.

a. Olefin Oxides

i. *Olefin Oxide–Styrene–Olefin Oxide.*   Szwarc *et al.* (C90) and later other workers (C24, C91) synthesized ethylene oxide–styrene–ethylene oxide block copolymers via a two-step anionic living polymer process. Styrene was polymerized using disodium (or dipotassium) α-methylstyrene tetramer initiators in tetrahydrofuran solution for ~ $\frac{1}{2}$ hour at −80°–0°C. Addition of a small amount of ethylene oxide to the living polystyrene converts the carbanion end groups to alkoxide end groups at the low temperature. At this point, the viscosity of the reaction solution increases substantially, even though the molecular weight does not rise appreciably. This is due to a greater degree of intermolecular association of the sodium alkoxide ion pairs as compared to the sodium carbanion ion pairs as a result of the more localized nature of the negative charge in the alkoxide ion. The newly formed alkoxide end groups are capable of initiating the polymerization of additionally introduced ethylene oxide monomer, but this reaction is slower than that of the first step. Polymerization times of a day or more at 75°C were required to reach high conversions when the sodium form of the initiator was used. Ethylene oxide polymerization rate is proportional to counter ion size, as was illustrated by the observation that the potassium

TABLE 6-17

A-B-A Block Copolymers Containing Polyether and Polyvinyl Segments

Block A	Block B	Initiator system	Conditions	Reference
Ethylene oxide	Styrene	Dianion (Na or K) of α-methylstyrene tetramer	−80° to 0°C, then 25° to 75°C	C24–C26, C90, C91
	Styrene	K biphenyl	—	C57
	Styrene	Poly(ethylene oxide) peroxycarbamate	—	C97
Styrene	Ethylene oxide	End group condensation	—	C91
Ethylene oxide	α-Methylstyrene	Dianion (Na) of α-methylstyrene tetramer	−78°C, then 75°C	C30
Acrylonitrile	Ethylene oxide	$^+M^-O(CH_2Ch_2O)_b^-M^+$	−20° to +10°C	C35, C36, C58
	Ethylene oxide	$Cl(NH_3)_2(SO_4)_3 +$ $HO(CH_2CH_2O)_bH$	25°C	C37–C39
	Ethylene oxide	Peroxycarbamate-terminated poly(ethylene oxide)	—	C97
Methyl methacrylate	Ethylene oxide	Peroxycarbamate-terminated poly(ethylene oxide)	—	C97
Acrylonitrile	Propylene oxide	$^+M^-O(CH_2C(CH_3)HO)_b^-M^+$	−20° to +10°C	C40
	Propylene oxide	$Ce^{4+} +$ $HO(CH_2CH(CH_3))_bOH$	0°C	C41
Methyl methacrylate	Propylene oxide	Peroxycarbamate-terminated poly(propylene oxide)	73°C	C33, C42
Styrene	Propylene oxide	Peroxycarbamate-terminated poly(propylene oxide)	73°C	C31–C33
Styrene	Propylene oxide	Macro-multi-azonitrile of poly(propylene oxide)	85°C	C34
Methyl methacrylate	Propylene oxide	Macro-multi-azonitrile of poly(propylene oxide)	85°C	C34
Vinyl chloride	Propylene oxide	Macro-multi-azonitrile of poly(propylene oxide)	85°C	C34
Vinyl acetate	Propylene oxide	Macro-multi-azonitrile of poly(propylene oxide)	85°C	C34
Styrene	Tetramethylene oxide	Macro-multi-azonitrile of poly(tetramethylene oxide)	85°C	C34
Methyl methacrylate	Tetramethylene oxide	Macro-multi-azonitrile of poly(tetramethylene oxide)	85°C	C34

TABLE 6-17 *(Continued)*

Block A	Block B	Initiator system	Conditions	Reference
Vinyl chloride	Tetramethylene oxide	Macro-multi-azonitrile of poly(tetramethylene oxide)	85°C	C34
Vinyl acetate	Tetramethylene oxide	Macro-multi-azonitrile of poly(tetramethylene oxide)	85°C	C34
Formaldehyde	Styrene	Na naphthalene	$-20°$ to $-60°$C	C43–C47
	Styrene	Na biphenyl	$-80°$C	C93
	α-Methylstyrene	Na naphthalene	$-20°$ to $-60°$C	C43, C44
	Vinylpyridine	Na naphthalene	$-20°$ to $-60°$C	C44
	Butadiene	Na naphthalene	$-20°$ to $-60°$C	C44
	Isoprene	Na naphthalene	$-20°$ to $-60°$C	C44
Acetaldehyde	Styrene	Na naphthalene	$-20°$ to $-60°$C	C48
	α-Methylstyrene	Na naphthalene	$-20°$ to $-60°$C	C48
	2-Methyl-5-vinylpyridine	Na naphthalene	$-20°$ to $-60°$C	C48
	Methyl methacrylate	Na naphthalene	$-20°$ to $-60°$C	C48
	Butadiene	Na naphthalene	$-20°$ to $-60°$C	C48
	Isoprene	Na naphthalene	$-20°$ to $-60°$C	C48
Butyraldehyde	Styrene, 2-methyl-5-vinylpyridine	Na naphthalene	$-20°$ to $-60°$C	C48
Isobutyraldehyde	Styrene, 2-methyl-5-vinylpyridine	Na naphthalene	$-20°$ to $-60°$C	C48
THF	Styrene	Anionic–cationic oligomer condensation	$\sim0°$C	C49
Styrene	THF	Anionic–cationic oligomer condensation	25°C	C53, C54 C92
	THF	THF oligomer + AIBN	50°C/5 days	C54
Methyl methacrylate	THF	THF peroxycarbamate oligomer	25°C	C55
Hexafluoro-acetone	Styrene	Na biphenyl	$-80°$C	C56
	Butadiene	Na biphenyl	$-80°$C	C56
	Methyl methacrylate	Na biphenyl	$-80°$C	C56
Hexafluoroace-tone–thio-carbonyl fluoride	Styrene	Na biphenyl	$-80°$C	C56
Tetrafluoro-ethylene oxide	Styrene	Na biphenyl	$-80°$C	C56
	Isoprene	Na biphenyl	$-80°$C	C56

form of the initiator gave complete reaction of the ethylene oxide in less
than 24 hours at 25°C (C24). The reaction sequence is shown in Scheme
6-7. In some of the reported work, the block copolymer products were

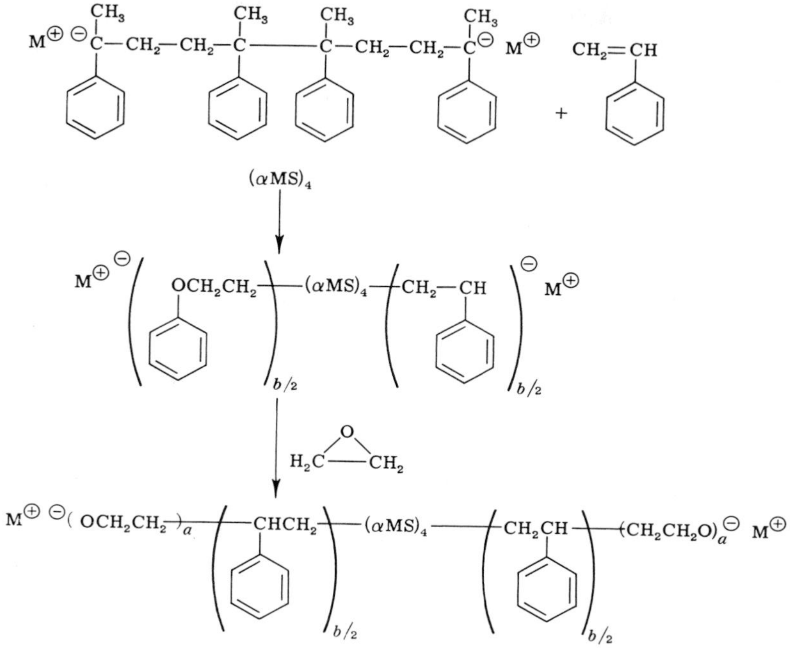

Scheme 6-7

end-capped by reaction with acetic anhydride to enhance thermal
stability.

Solubility behavior (extraction) of the products was cited as proof of
block copolymer structure. Only small amounts (<20%) of homopolymer were found in the product. Addition of a drop of water to
a benzene solution of the block copolymer caused precipitation. This
phenomenon was believed due to intermolecular hydrogen bonding
via hydration of the poly(ethylene oxide) segments (C90). These
copolymers are claimed to be useful in film and sheet applications in
which static electricity charge build-up is undesirable (C25).

Attempts to extend this technique to block copolymers of styrene
with other poly(olefin oxides), e.g., propylene oxide and styrene oxide,
were unsuccessful, reportedly due to premature termination by chain
transfer with monomer (C90).

The thermal transition properties of a series of poly(ethylene
oxide)–polystyrene–poly(ethylene oxide) block copolymers was

studied using differential scanning calorimetry (C24). The molecular weight $(\overline{M}_n)$ of the poly(ethylene oxide) blocks was kept constant at ~10,000, while the polystyrene $\overline{M}_n$ ranged from 5000 to 50,000. The poly(ethylene oxide) segment melting point (T_m) of these copolymers ranged from 53° to 59°C. The higher values were obtained with compositions containing the lower $\overline{M}_n$ polystyrene segments, i.e., higher weight percent poly(ethylene oxide) content. This was said to be due to an increase in crystal perfection with increasing poly(ethylene oxide) content. The glass transition temperature (T_g) of the polystyrene blocks in these copolymers ranged from 75°C to 101°C. The higher values were obtained at higher block molecular weights, as expected from the known molecular weight dependence of homopolystyrene T_g in this $\overline{M}_n$ range. The degree of poly(ethylene oxide) crystallinity in this series of copolymers, as calculated from the differential scanning calorimetry data, decreased from ~90% to ~60% as the polystyrene $\overline{M}_n$ (and therefore, weight percent content) increased. This was observed both for solution-cast and melt-fabricated samples. Comparative work on A-B and A-B-A polystyrene–poly(ethylene oxide) copolymers led these workers to the conclusion that thermal parameters are sensitive to the size of the blocks but relatively insensitive to their placement in the copolymers.

Rheological studies with these block copolymers revealed that they display birefringence even in the melt at temperatures above 250°C (C26). This work also showed that the melts are highly elastic compared to the parent homopolymers, due to intermolecular association in the melt. Shear modulus versus temperature measurements indicated that A-B-A block copolymers [A is poly(ethylene oxide)] were less shear stable than A-B structures. The melt viscosity and melt elasticity of these block copolymers were higher than those of either homopolymer of similar molecular weight or blends of these homopolymers (C27).

As was discussed earlier for A-B structures, mesomorphic phases were observed to form in solutions of ethylene oxide–styrene–ethylene oxide A-B-A block copolymers (C28). At very low concentration (<0.05%) the copolymer exists as a monomolecular micelle, and at high concentrations (up to 30%), it exists as a multimolecular micelle or aggregate (C29).

Poly(ethylene oxide)–poly(α-methylstyrene)–poly(ethylene oxide) copolymers were synthesized (C30) by the same techniques described above for the analogous styrene-containing block copolymers. Isothermal crystallization from dilute solution at 20°–40°C was investigated using a series of block copolymers ranging in poly(ethylene oxide) content from 55 to 86 wt%. The $\overline{M}_n$ of the poly(ethylene oxide) blocks ranged from 8000 to 74,000, and that of the poly(α-methylstyrene) blocks from 16,000 to 82,000. At poly(ethylene oxide) contents >64%, the copolymers crystallized from 0.2% solutions in the form of square tablets, similar to homopoly(ethylene oxide) single crystals. Electron micrographs revealed that poly(α-methylstyrene) chains were excluded from the crystalline part and were deposited as aggregated long loops on the surface of the crystals during chain-folding crystallization. At higher solution concentrations (0.5%), irregularly shaped multilayer crystals containing surface holes formed, suggesting that more tie molecules form as concentration increases. These observations were cited as general evidence for the formation of long loops on the surfaces of lamellar crystals. This results in interlaminar tie molecules that connect lamellae in multilayer crystals during chain folding crystallization.

ii. *Styrene–Olefin Oxide–Styrene.* As described earlier, ethylene oxide polymerization can be initiated with polystyryl dianion to produce ethylene oxide–styrene–ethylene oxide block copolymers. However, it is not possible to use this technique to synthesize copolymers containing a center block of ethylene oxide between two styrene segments, since alkoxide ions are not strong enough to initiate styrene polymerization. However, such a block copolymer has been reportedly prepared via an end-group condensation reaction (C91). Monofunctional polystyrene, prepared by cumylpotassium initiation, was reacted with excess phosgene to produce an acyl chloride-terminated oligomer, which was subsequently reacted with dihydroxyl-terminated poly(ethylene oxide) to form ester linkages between segments as illustrated in Scheme 6-8.

A free radical approach to the synthesis of styrene–ethylene oxide–styrene block copolymers via a peroxycarbamate intermediate was reported which involved the reaction sequence shown in Scheme 6-9 (C97). The product was contaminated with large quantities of homopolystyrene. A preliminary evaluation of the bulk product for use in desalination membrane applications was discouraging.

A similar approach was used to prepare styrene–propylene oxide–styrene block copolymers (C31–C33). Again, considerable quantities of homopolymer were formed. The block copolymer was said to be an effective compatibilizing agent for the homopolymers (C33).

Another radical macroinitiator approach, similar to that described above, is one that employs "macromultiazonitriles" as the initiating species (C34). The authors claimed this to be superior to the

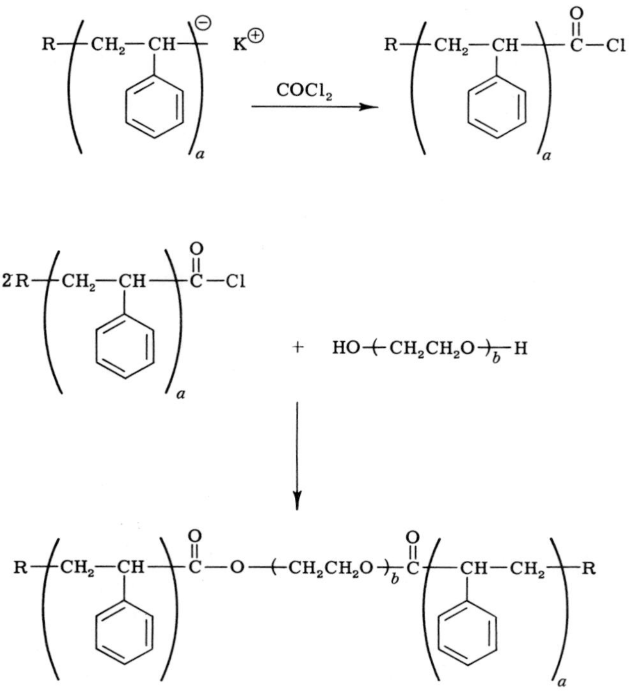

Scheme 6-8

peroxycarbamate approach for the following reasons: (a) The polymeric nature of the initiator leads to macrodiradicals and, therefore, fewer monoradicals (from terminal fragments) that can initiate the formation of homopolymer. (b) Azo initiation is preferable to peroxide initiation, which is accompanied by undesirable side reactions, such as induced decomposition and hydrogen abstraction by resulting oxy radicals. The macroazonitrile initiator was prepared from 2000–4000 molecular weight poly(propylene glycol) [or poly-(tetramethylene oxide)], toluene diisocyanate, and azobis(δ-cyano-*n*-pentanol) (Scheme 6-10). This initiator, which contained six to twenty azo units, was used to polymerize styrene for 7 hours at 85°C. The synthesis was not without side reactions. The products had T_g values characteristic of the blocks. A styrene–tetramethylene oxide–styrene block copolymer made by this technique was said to display dynamic properties similar to that of a styrene–butadiene–styrene thermoplatic elastomer block copolymer. In addition to styrene-containing copoly-

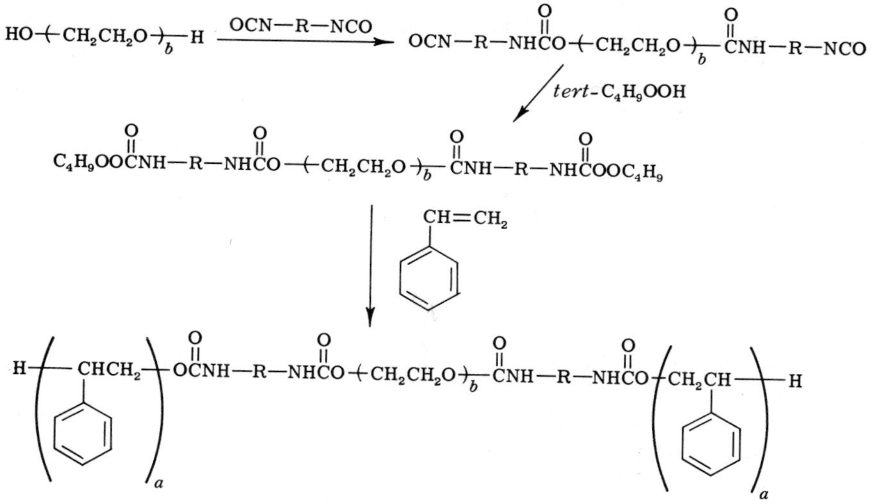

$$HO \text{---} (CH_2CH_2O)_{\overline{b}} \text{---} H \xrightarrow{OCN \text{---} R \text{---} NCO} OCN \text{---} R \text{---} NH\overset{O}{\overset{\|}{C}}O \text{---} (CH_2CH_2O)_{\overline{b}} \overset{O}{\overset{\|}{C}}NH \text{---} R \text{---} NCO$$

$$\downarrow tert\text{-}C_4H_9OOH$$

$$C_4H_9OO\overset{O}{\overset{\|}{C}}NH \text{---} R \text{---} NH\overset{O}{\overset{\|}{C}}O \text{---} (CH_2CH_2O)_{\overline{b}} \overset{O}{\overset{\|}{C}}NH \text{---} R \text{---} NH\overset{O}{\overset{\|}{C}}OOC_4H_9$$

$$\downarrow \quad \underset{\text{(styrene)}}{CH=CH_2}$$

$$H \text{---} \left(\underset{\underset{\text{(phenyl)}}{}}{CHCH_2} \right)_a \text{---} O\overset{O}{\overset{\|}{C}}NH \text{---} R \text{---} NH\overset{O}{\overset{\|}{C}}O \text{---} (CH_2CH_2O)_{\overline{b}} \overset{O}{\overset{\|}{C}}NH \text{---} R \text{---} NH\overset{O}{\overset{\|}{C}}O \left(CH_2CH \right)_a \text{---} H$$

Scheme 6-9

mers, compositions were also prepared by this procedure in which the
A segment was vinyl chloride, vinyl acetate, or methyl methacrylate.

iii. *Acrylic–Olefin Oxide–Acrylic.* The synthesis of acrylonitrile–
ethylene oxide–acrylonitrile block copolymers has reportedly been
accomplished by both anionic and free radical polymerization. Both
routes employed preformed poly(ethylene oxide) oligomers. In the

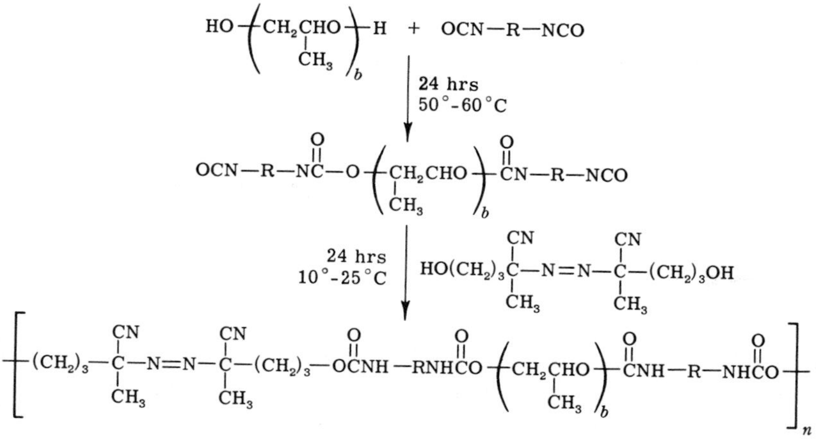

$$HO \text{---} \left(\underset{CH_3}{CH_2CHO} \right)_b \text{---} H \; + \; OCN \text{---} R \text{---} NCO$$

$$\downarrow \begin{array}{l} 24 \text{ hrs} \\ 50°\text{-}60°C \end{array}$$

$$OCN \text{---} R \text{---} N\overset{O}{\overset{\|}{C}} \text{---} O \left(\underset{CH_3}{CH_2CHO} \right)_b \overset{O}{\overset{\|}{C}}N \text{---} R \text{---} NCO$$

$$\downarrow \begin{array}{l} 24 \text{ hrs} \\ 10°\text{-}25°C \end{array} \quad HO(CH_2)_3\underset{CH_3}{\overset{CN}{C}} \text{---} N{=}N \text{---} \underset{CH_3}{\overset{CN}{C}} \text{---} (CH_2)_3OH$$

$$\left[\text{---} (CH_2)_3 \text{---} \underset{CH_3}{\overset{CN}{C}} \text{---} N{=}N \text{---} \underset{CH_3}{\overset{CN}{C}} \text{---} (CH_2)_3 \text{---} O\overset{O}{\overset{\|}{C}}NH \text{---} RNH\overset{O}{\overset{\|}{C}}O \left(\underset{CH_3}{CH_2CHO} \right)_b \overset{O}{\overset{\|}{C}}NH \text{---} R \text{---} NH\overset{O}{\overset{\|}{C}}O \text{---} \right]_n$$

Scheme 6-10

anionic process, the sodium or lithium alcoholate of dihydroxyl-terminated 4000 molecular weight poly(ethylene glycol) was used to initiate the polymerization of acrylonitrile (C35, C36). Polymerization was carried out in dimethylformamide or toluene solution over a period of $\frac{1}{2}-2$ hours. The expected sequence is shown in Reaction 6-11. Acidification with HCl yielded the neutralized form of the prod-

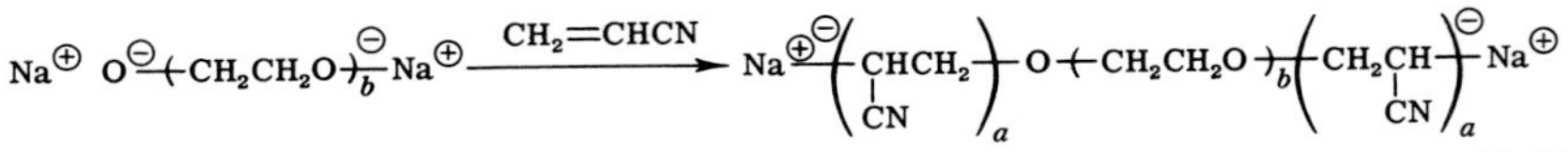

$$Na^{\oplus}\ O^{\ominus}(CH_2CH_2O)_b Na^{\oplus} \xrightarrow{\ CH_2=CHCN\ } Na^{\oplus}\left(\underset{\underset{CN}{|}}{CHCH_2}\right)_a O(CH_2CH_2O)_b\left(\underset{\underset{CN}{|}}{CH_2CH}\right)_a Na^{\oplus}$$

$$(6\text{-}11)$$

uct. The workers concluded from fractionation and light-scattering studies that the yield of block copolymer was quite low due to the inefficiency of the macroinitiator. Consumption of the acrylonitrile monomer was quantitative, yielding a mixture of block copolymers and homopolymers. The polyacrylonitrile blocks were highly branched and polydisperse.

Block copolymers were reportedly synthesized by free radical initiation of acrylonitrile polymerization via a redox mechanism (C37, C38). In this process, $Cl(NH_3)_2(SO_4)_3$ was the oxidizing agent and the hydroxyl end groups of dihydroxyl-terminated poly(ethylene oxide) acted as the reducing agent. The polymerization was carried out for ~2 hours in aqueous solution at room temperature. Biradical initiation of the acrylonitrile polymerization under these conditions was said to occur at the terminal carbon atoms of the poly(ethylene glycol) oligomer, as illustrated in Scheme 6-11. Insolubilization of the product upon reac-

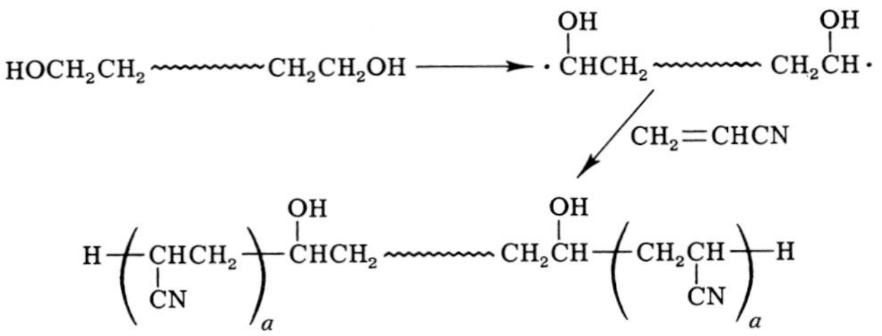

$$HOCH_2CH_2 \sim\sim\sim CH_2CH_2OH \longrightarrow \cdot \overset{OH}{\underset{|}{CHCH_2}} \sim\sim\sim CH_2\overset{OH}{\underset{|}{CH}} \cdot$$

$$\downarrow CH_2=CHCN$$

$$H\left(\underset{\underset{CN}{|}}{CHCH_2}\right)_a \overset{OH}{\underset{|}{CHCH_2}} \sim\sim\sim CH_2\overset{OH}{\underset{|}{CH}}\left(\underset{\underset{CN}{|}}{CH_2CH}\right)_a H$$

Scheme 6-11

tion with a diisocyanate was cited as evidence for the presence of hydroxyl groups in the product and thus the reaction sequence was as depicted. These block copolymers, which contained up to 20% ethylene oxide and had molecular weights as high as 68,000, were claimed to produce solution spun fibers with fatigue and wear resistance properties superior to those of homopolyacrylonitrile fibers and to fibers spun from blends of the two homopolymers (C37, C39).

Another free radical route to acrylonitrile–ethylene oxide–acrylonitrile A-B-A block copolymers is shown in Scheme 6-12 in

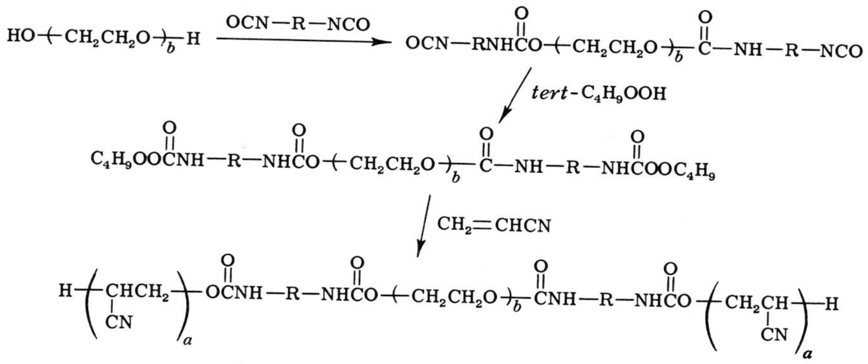

Scheme 6-12

which the use of a peroxycarbamate-terminated oligomer as a macromolecular free radical initiator is involved (C97). The product of this reaction was obviously contaminated with large quantities of homopolyacrylonitrile. A preliminary investigation of the unpurified product in desalination membrane applications was unsuccessful. Similar copolymers in which the A blocks were poly(methyl methacrylate) rather than polyacrylonitrile were also investigated with similar negative results (C97).

Acrylonitrile-containing A-B-A block copolymers in which the B block is poly(propylene oxide) rather than poly(ethylene oxide) have also been reportedly prepared (a) via anionic initiation with a metal alkoxide of poly(propylene oxide) (C40) or (b) via ceric salt redox-initiated free radical polymerization (C41).

Block copolymers in which A is methyl methacrylate and B is propylene oxide were prepared by the peroxycarbamate process described earlier (C33, C42). This involved initiation of methyl methacrylate polymerization with a poly(propylene oxide) oligomer bearing two butyl peroxycarbamate end groups. As discussed above, this process yields products containing large quantities of vinyl homopolymer. Molding of this product at 250°C gave a transparent film with properties intermediate to those of the two homopolymers.

b. Aldehydes

A-B-A block copolymers have been prepared in which the A blocks are polyaldehydes, e.g., formaldehyde or acetaldehyde, and the B block is a vinyl or diene polymer, most frequently styrene. These copolymers are made via anionic polymerization.

Formaldehyde–styrene–formaldehyde block copolymers were prepared by initiating styrene polymerization with sodium naphthalene in tetrahydrofuran solution at $-20°$ to $-60°C$ (C43–C46). Addition of formaldehyde caused the anionic end groups to convert to alkoxide ions which subsequently initiated polymerization of the remaining formaldehyde, as shown in Scheme 6-13. In some cases, the end

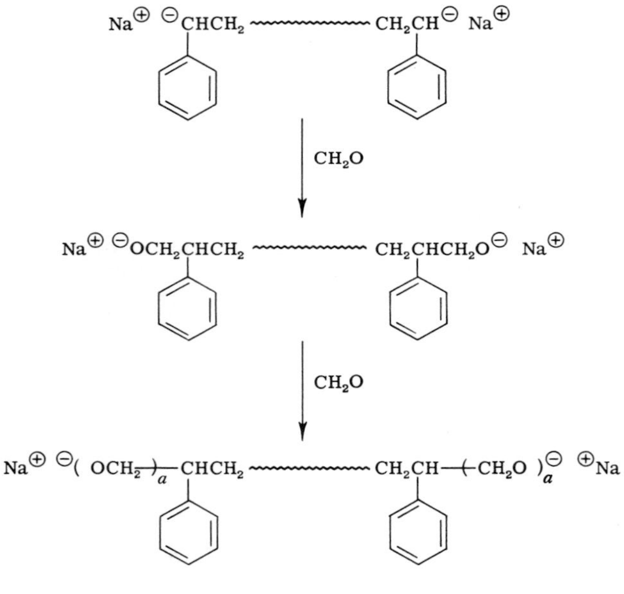

Scheme 6-13

groups were acetylated to improve the stability of the products. Products containing 5–65% styrene were reported with melting points ranging from 160° to 185°C. Solubility and spectroscopic properties were offered as evidence of block structure.

The use of lithium rather than sodium counterion led to an unsuccessful attempt to synthesize a block copolymer (C47). In this work, formate-terminated low molecular weight polystyrene was isolated, leading to the conclusion that the problem was due to a chain terminating Tischenko-type reaction (6-12) between the alcoholate end groups

$$\text{~~~~CH}_2\text{CHCH}_2\text{O}^{\ominus} \quad + \quad CH_2O \quad \longrightarrow \quad \text{~~~~CH}_2\text{CHCH}_2\text{OCH} \qquad (6\text{-}12)$$

and formaldehyde monomer. However, other workers (C43) found that formaldehyde monomer purity is extremely critical if terminating side reactions are to be avoided.

Block copolymers of formaldehyde with other vinyl monomers have also been reported briefly in the literature. Preparative methods similar to the one discussed above were utilized. These compositions are also listed in Table 6-17.

Aldehydes higher than formaldehyde have also been reported to give A-B-A block copolymers with vinyl polymers via anionic polymerization (C48). These compositions are also listed in Table 6-17.

c. Tetramethylene Oxide

A-B-A block copolymers have been synthesized which contain one or more blocks of poly(tetramethylene oxide) [i.e., poly(tetrahydrofuran)] and one or more blocks of a vinyl polymer such as polystyrene or poly(methyl methacrylate). Both ionic and free radical techniques were employed. Table 6-17 lists the compositions reported.

Berger *et al.* (C49) synthesized a THF–styrene–THF block copolymer by coupling a dianionically terminated polystyrene with a cationically terminated poly(tetrahydrofuran) "living" polymer. Several workers have claimed that, under proper conditions, "living" cationic polymers can be prepared, since the rate of self termination is very slow compared to the rates of initiation and propagation (C50–C52). In the Berger *et al.* work, the dianionic polystyrene was prepared in THF solution using α-methylstyrene tetramer dianion as the initiator. The cationic THF polymer was synthesized using Meerwein's catalyst, $(C_2H_5)_3OBF_4$, at ice-bath temperature for 2 hours. Combination of the two oligomers gave the product, presumably via the reaction shown in Scheme 6-14. Solubility behavior different from that of a blend of the two homopolymers was given as evidence of the block structure of the product. It was reported that cationic initiators other than $(C_2H_5)_3OBF_4$ [i.e., $SbCl_5$, $(C_6H_5)_3CSbCl_4$, and BF_3—propylene oxide] can also be used according to this process. Attempts to prepare {A-B}$_n$ block copolymers by this approach using bifunctional epoxides in conjunction

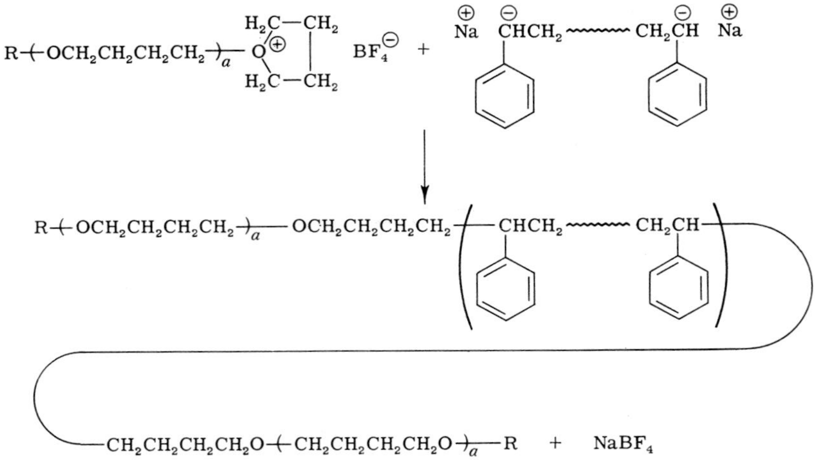

$$R \text{-}(OCH_2CH_2CH_2CH_2\text{-})_a \overset{\overset{H_2C-CH_2}{\underset{H_2C-CH_2}{|}}}{O^{\oplus}} \quad BF_4^{\ominus} \quad + \quad \overset{\oplus}{Na} \overset{\ominus}{CHCH_2} \text{-}\!\!\sim\!\!\text{-} CH_2\overset{\ominus}{CH} \overset{\oplus}{Na}$$

$$R \text{-}(OCH_2CH_2CH_2CH_2\text{-})_a OCH_2CH_2CH_2CH_2 \overset{|}{{}} CHCH_2 \text{-}\!\!\sim\!\!\text{-} CH_2CH \text{-}$$

$$\text{-}CH_2CH_2CH_2CH_2O\text{-}(CH_2CH_2CH_2CH_2O\text{-})_a R \quad + \quad NaBF_4$$

Scheme 6-14

with BF_3 as the cationic initiators were unsuccessful, leading to insoluble, gelled products.

A more interesting block sequence structure than that discussed above is one in which the B block is poly(tetrahydrofuran) and the A blocks are polystyrene. The low melting point (34°C) and T_g (−79°C) of poly(tetrahydrofuran) (C53) should result in interesting thermoplastic elastomeric properties for such a composition. A patent reference (C54) claims the synthesis of such structures by a technique involving a cationic–anionic process similar to the one discussed above. In this process, THF is polymerized via a "living" cationic process using trifluoromethane sulfonic anhydride, $(CF_3SO_2)_2O$, as the initiator for 1–4 hours at 25°C. This was said to produce a poly(tetrahydrofuran) oligomer with two cationic end groups. Combination of this intermediate with a monoanionically terminated polystyrene, prepared via butyllithium initiation, results in the styrene–THF–styrene block copolymer product. Inherent viscosity data (higher for the product than for the intermediates) was cited as evidence of the block structure of the product.

Other workers suggested that similar products can be synthesized using 2,2′-octamethylenebis-1,3-dioxolenium perchlorate as the initiator to form dicationically terminated poly(tetrahydrofuran) oligomers (C53, C92). The chemistry of this coupling reaction is described in Chapter 7, Section B. The A-B-A block copolymer morphology was found to depend both on crystallization and microphase separation.

A styrene–THF–styrene block copolymer was also claimed to have been synthesized by preparing dicationically terminated poly(tetrahydrofuran) using the $(CF_3SO_2)_2O$ initiator, terminating this with H_2S to form a dithiol-terminated oligomer, and polymerizing styrene by free radical initiation with azo bis(isobutyronitrile) in the presence of these chain-transfering sulfhydryl end groups (C54). A single-phase solution of the product in benzene, as compared to a two-phase solution of a blend of the homopolymers, was cited as proof of block structure. However, considerable quantities of homopolymer must have formed in this process.

An A-B-A block copolymer with A being methyl methacrylate and B being THF was also prepared via a free radical process (C55). Dihydroxyl-terminated poly(tetrahydrofuran) was end-capped with tolylene diisocyanate and then reacted with *tert*-butylhydroxymethyl peroxide to form the biscarbamate, which was then used to free radically initiate methyl methacrylate polymerization. The product was a mixture of block copolymer and homopolymers.

d. Perfluorinated Ethers

A patent reference (C56) reports the synthesis of A-B-A block copolymers containing B segments of vinyl polymers such as styrene, butadiene, isoprene, or methyl methacrylate and A segments of perfluoro polyethers such as those derived from hexafluoroacetone or tetrafluoroethylene oxide. Difunctional initiation of the vinyl monomer polymerization was achieved with sodium biphenyl at −80°C in tetrahydrofuran solution, after which the fluoromonomer was added and polymerized over a period of a few minutes. The specific compositions reported are listed in Table 6-17. Infrared and solubility behavior were cited as evidence of block copolymer structure formation. The products were said to be flexible, tough, melt processable, and solvent resistant.

3. Esters

Several A-B-A block copolymers have been reported that contain one or two segments of a polyester, such as polycaprolactone, and one or two polyether or polyvinyl segments. These compositions are listed in Table 6-18.

A caprolactone–ethylene oxide–caprolactone block copolymer was prepared by initiating caprolactone polymerization with the sodium salt of a preformed dihydroxyl-terminated poly(ethylene oxide) oligomer (C59). The reaction, carried out at 60°C in benzene solution,

TABLE 6-18

A-B-A Block Copolymers Containing Polyester Segments

Block A	Block B	Initiator system	Conditions	Reference
Caprolactone	Ethylene oxide	Ethylene oxide oligomer disodium salt	60°C	C59, C65–C68, C94
	Styrene	Na	−78°C	C60
	Butadiene	Difunctional lithium–isoprene adduct	120°–160°C	C95
	Methyl methacrylate	Na α-methylstyrene dianion	—	C63
	Isoprene	Na α-methylstyrene dianion or lithium	—	C63, C69
Pivalolactone	Styrene	Disodium polystyrene dicarboxylate	30°C	C61, C62
	Isoprene	Dilithiopolyisoprene	60°C	C70
Styrene	Alkylene adipate	Peroxycarbamate-terminated poly-(alkylene adipate)	—	C32, C33, C64
Methyl methacrylate	Alkylene adipate	Peroxycarbamate-terminated poly-(alkylene adipate)	—	C32, C33, C64
Acrylonitrile	Alkylene adipate	Peroxycarbamate-terminated poly-(alkylene adipate)	—	C32, C33, C64
Pivalolactone	DL-α-Methyl–α-η-propyl–β-propiolactone	Tetrabutylammonium sebacate	35°C	C96
Styrene	Tetrachloro-bisphenol A adipate	Photodegradation of polyester containing oxime linkages in presence of styrene monomer	20°C	C71
Pivalolactone	Acrylonitrile	Disodium polyacrylonitrile dicarboxylate	30°C	C62

can be illustrated as shown in Scheme 6-15. The product contained significant quantities of caprolactone homopolymer probably due to cleavage of the polyester segments by the strongly basic end groups. X-Ray analysis indicated partial crystallinity in the caprolactone segments.

A caprolactone–styrene–caprolactone block copolymer was prepared by initiating the polymerization of styrene with sodium in THF–benzene solution at −78°C, followed by addition of caprolactone

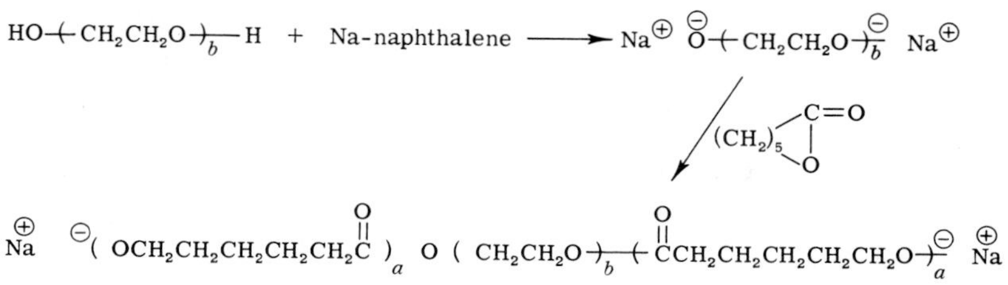

$$HO \xleftarrow{} CH_2CH_2O \xrightarrow{}_b H \ + \ Na\text{-naphthalene} \longrightarrow Na^{\oplus} \ {}^{\ominus}O \xleftarrow{} CH_2CH_2O \xrightarrow{}_b {}^{\ominus} \ Na^{\oplus}$$

$$\underset{Na}{\overset{\oplus}{}} \ {}^{\ominus}(\ OCH_2CH_2CH_2CH_2CH_2\overset{\overset{O}{\|}}{C} \ )_a \ O \ (\ CH_2CH_2O \xrightarrow{}_b \xleftarrow{} \overset{\overset{O}{\|}}{C}CH_2CH_2CH_2CH_2CH_2O \xrightarrow{}_a {}^{\ominus} \ \overset{\oplus}{Na}$$

Scheme 6-15

to the resulting polystyryl dianion (Scheme 6-16) (C60). It was concluded that alcoholate anions are the active species in the propagation
of the caprolactone, since it was observed that caprolactone
homopolymerization was initiated by sodium ethoxide but not by
sodium propionate. The block copolymer products were characterized
via infrared analysis and fractional precipitation behavior.

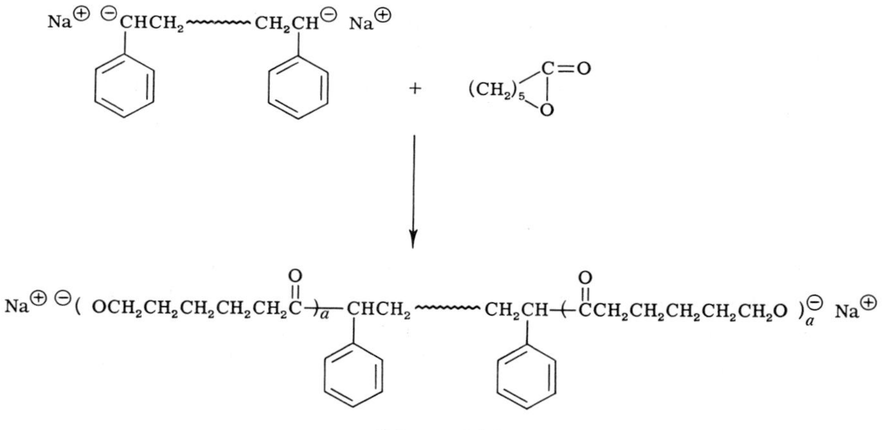

Scheme 6-16

Pivalolactone–styrene–pivalolactone block copolymers were synthesized via a similar route using dicarboxylate terminated polystyrene,
prepared by capping with CO_2, to initiate pivalolactone polymerization
(C61, C62). The product contained unreacted polystyrene dicarboxylate due to the slow rate at which this species initiates lactone
polymerization compared to the propagation rate. Pivalolactone–
acrylonitrile–pivalolactone block copolymers were also prepared in
this way. Caprolactone–methyl methacrylate–caprolactone and
caprolactone–isoprene–caprolactone block copolymers were reportedly prepared by a similar living polymer anionic process (C63).

Block copolymers were prepared from polyalkylene adipate intermediates and vinyl monomers such as styrene, methyl methacrylate, or acrylonitrile (C32, C33, C64). The approach used was to extend a 3000 molecular weight hydroxyl-terminated alkylene adipate polyester oligomer with tolylene diisocyanate, cap the resulting isocyanate end groups with cumene hydroperoxide, and use this diperoxycarbamate-terminated intermediate to initiate the free radical polymerization of the vinyl monomer. Modulus–temperature data for the products, which contain considerable amounts of homopolymer, are discussed in these reports.

4. Sulfides

The A-B-A block copolymers of this category contain at least one poly(alkylene sulfide) segment derived from either an episulfide monomer or a thiocarbonyl monomer. In some of the compositions, both the A and the B blocks are poly(alkylene sulfides). Others contain poly(alkylene sulfide) segments and polyvinyl segments such as those derived from styrene, α-methylstyrene, vinyl pyridine, diene, methyl methacrylate, or acrylonitrile monomers. In most of the polyvinyl-containing copolymers, the alkylene sulfide segments are present as A blocks, but in one case, it was the B block. The various compositions reported are listed in Table 6-19.

a. Olefin Sulfide–Olefin Sulfide

Poly(ethylene sulfide)–poly(propylene sulfide)–poly(ethylene sulfide) block copolymers have been synthesized in tetrahydrofuran solution at 40°C using potassium ethanedithiolate as the anionic initiator (Scheme-6-17) (C72). Since the reaction is a rapid one, the monomers

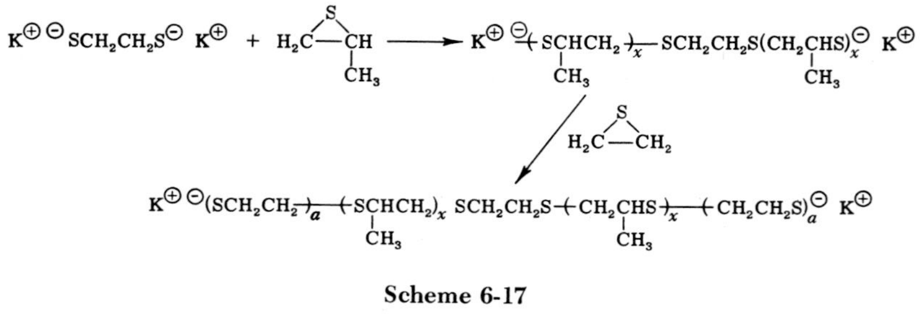

Scheme 6-17

were added slowly to keep the temperature at 40°C. Ethylene sulfide was added $\frac{1}{2}$ hour after completion of the propylene sulfide addition. The product precipitated from solution due to crystallization of the eth-

TABLE 6-19

A-B-A Block Copolymers Containing Polyolefin Sulfide Segments

Block A	Block B	Initiator system	Conditions	Reference
Ethylene sulfide	Propylene sulfide	$^+K^-SCH_2CH_2S^-K^+$	40°C	C72
	Propylene sulfide	Na naphthalene	−70°C	C73
Isobutylene sulfide	Propylene sulfide	Na (or K) naphthalene	−50°C	C74, C75
Propylene sulfide	Isobutylene sulfide	Na carbazyl	−50°C	C75
Ethylene sulfide	Styrene	Na (or K) naphthalene (or biphenyl)	−78°C to +25°C	C76–C80, C87
Propylene sulfide	Styrene	Na (or K) naphthalene (or biphenyl)	−78°C to +25°C	C73, C77
Ethylene sulfide	α-Methylstyrene	Na naphthalene	−78°C to +25°C	C77
Propylene sulfide	α-Methylstyrene	Na naphthalene	−78°C to +25°C	C77
α-Methylstyrene	Propylene sulfide	$C_2H_5Li + COCl_2$ coupling	−78°C to +25°C	C81
Ethylene sulfide	Methyl methacrylate	Na naphthalene	−78°C to +25°C	C77
Propylene sulfide	Methyl methacrylate	Na naphthalene	−78°C to +25°C	C77
Propylene sulfide	2-Vinylpyridine	Na biphenyl or diphenylmethyl sodium	−50°C	C75, C88
Ethylene sulfide	4-Vinylpyridine	Na naphthalene	−78°C	C80
	Butadiene	1,4-Dilithiotetra-phenylbutane	—	C85–C87
	Isoprene	Li naphthalene	−30°C	C83
	Isoprene	1,4-Dilithiotetra-phenylbutane	—	C85–C87
Propylene sulfide	Isoprene	Na naphthalene	−40°C and +25°C	C82, C82a
	Butadiene	Na naphthalene	−40°C and +25°C	C82, C82a
	Butadiene	$Li(CH_2)_4Li$	50°C	C89
Thiocarbonyl-fluoride	Styrene	Na biphenyl	−80°C	C56, C84
	Methyl methacrylate	Na biphenyl	−80°C	C56, C84
	Acrylonitrile	Na biphenyl	−80°C	C56, C84
Tetrafluoroethylene sulfide	Acrylonitrile	Na biphenyl	−80°C	C56, C84
Thiacyclobutane	Isoprene	1,4-Dilithiotetra-phenylbutane	—	C85, C86

ylene sulfide blocks. Product molecular weights were low, probably as a result of premature precipitation and low initiator and solvent purity. Physical properties were poor due to low molecular weight. The products ranged in ethylene sulfide content from 9% to 29% and had softening points of 70°–125°C. The polymers were reported to be tough, rubbery, and solvent resistant, the latter characteristic being due to the ethylene sulfide crystallinity.

An ethylene sulfide–propylene sulfide–ethylene sulfide copolymer was also prepared at −70°C using a sodium naphthalene initiator (C73). The molecular weight of this product was also fairly low, about 10,000–20,000.

A-B-A block copolymers in which A is poly(isobutylene sulfide) and B is poly(propylene sulfide) have been synthesized (C74, C75) via initiation with sodium (or potassium) naphthalene. These copolymers, which contain 96% propylene sulfide, were insoluble in toluene, while an AB block copolymer of the same composition (made by sodium carbazyl initiation) was soluble. Since homopoly(isobutylene sulfide) is crystalline and insoluble and homopoly(propylene sulfide) is amorphous and soluble, this difference in block copolymer solubility was attributed to association of the crystalline poly(isobutylene sulfide) segments in the A-B-A structure. This results in a three-dimensional network via "physical vulcanization." This notion was supported by the solubility behavior of another A-B-A block copolymer with the reverse sequence structure [i.e., A being poly(propylene sulfide) and B being poly(isobutylene sulfide)], which was made by sodium carbazyl initiation and sequential monomer addition (C75). This product, which contained 52% propylene sulfide and had a molecular weight $(\overline{M}_n)$ of 186,000, was soluble in toluene. Presumably this behavior is due to the fact that only one crystalline poly(isobutylene sulfide) segment was present per molecule and, therefore, no opportunity existed for the formation of a "physically cross-linked" system.

b. Olefin Sulfide–Styrene

Ethylene sulfide–styrene–ethylene sulfide block copolymers have been reported by several workers (C76–C80). The general synthetic procedure involved (a) polymerization of styrene in tetrahydrofuran solution at −78°C for about an hour using sodium (or potassium) naphthalene or sodium biphenyl initiator, followed by (b) capping with ethylene sulfide at −78 to −50°C and (c) polymerization of the excess ethylene sulfide at room temperature for about an hour (Scheme 6-18). The products were insoluble in common organic solvents and had $\overline{M}_n$ values of 60,000. They displayed softening points

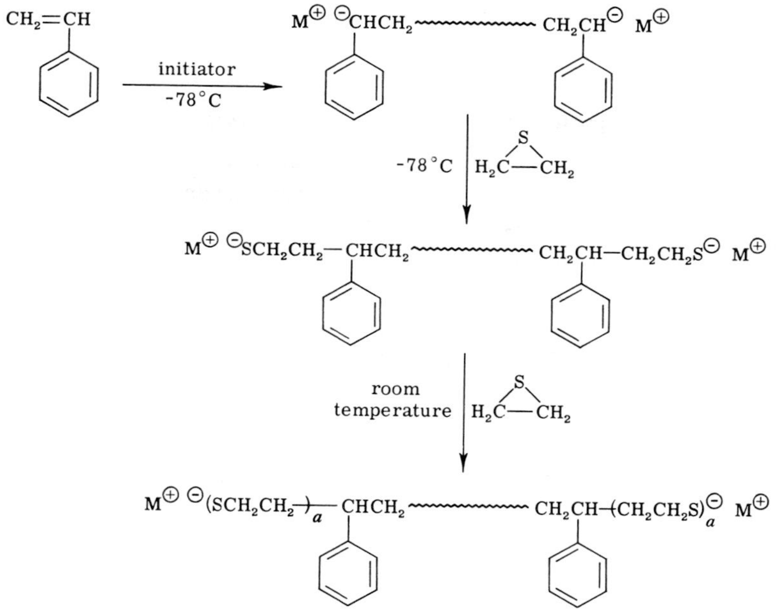

Scheme 6-18

at 200°–227°C, corresponding approximately to the melting point of ethylene sulfide homopolymer (203°–206°C). Extraction studies indicated that the products contained <10% styrene homopolymer. Differential thermal analysis showed the thermal transitions of both segments. The morphology of these block copolymers was studied by electron microscopy (C76). The copolymers were found to be in the form of microplatelet aggregates, which were less distinctly developed than the microhedrite or microspherulite forms observed for ethylene sulfide homopolymers of comparable molecular weight.

Propylene sulfide–styrene–propylene sulfide block copolymers have been prepared via a similar route (C73, C77). Since atactic polypropylene sulfide is amorphous with a T_g at ~−40°C, these block copolymers are more soluble than their crystalline ethylene sulfide counterparts. The copolymers also displayed the room temperature instability characteristic of propylene sulfide homopolymer. One copolymer (C73) was reported to contain only 2% styrene homopolymer and to have a molecular weight $(\overline{M}_n)$ of 230,000. Ethylene sulfide–α-methylstyrene–ethylene sulfide and propylene sulfide–α-methylstyrene–propylene sulfide block copolymers have also been prepared by the above procedure (C77).

A different approach was used by Morton *et al.* (C81) to prepare A-B-A block copolymers. In an effort to achieve thermoplastic elastomer properties, the alkylene sulfide segment was utilized as the center B block, i.e., α-methylstyrene–propylene sulfide–α-methylstyrene. This copolymer was synthesized by (a) initiation of α-methylstyrene polymerization in THF solution at −78°C with ethyllithium, (b) addition of propylene sulfide at −78°C followed by polymerization of the propylene sulfide for 24 hours at 25°C, and finally, (c) coupling of the resulting α-methylstyrene–propylene sulfide anion intermediate by the addition of a stoichiometric quantity of phosgene. The reaction sequence shown in Scheme 6-19, including a novel cross-

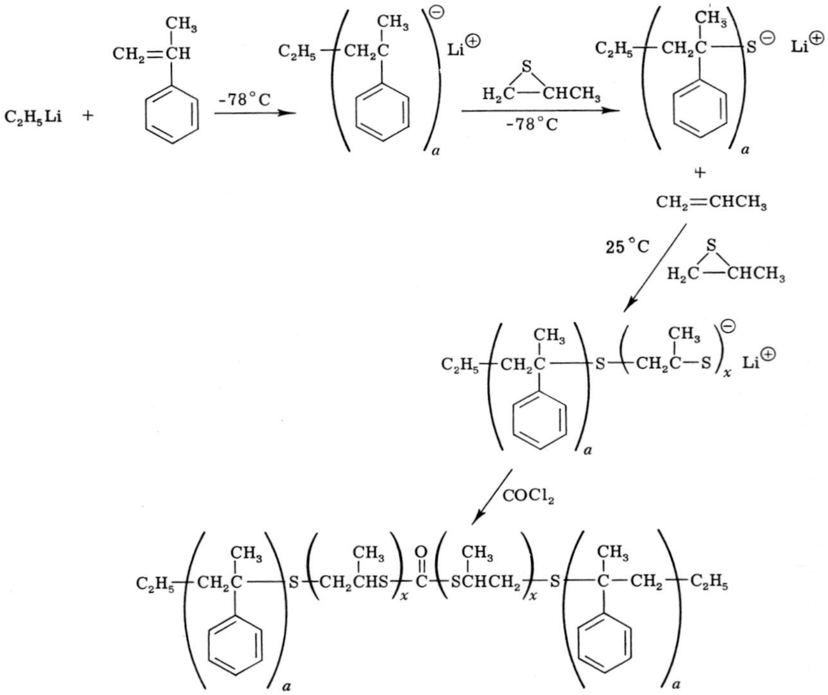

Scheme 6-19

over mechanism, was proposed by these workers (C81) for the propylene sulfide initiation step. Three block copolymers were synthesized in this study, with α-methylstyrene contents varying from 20% to 39%. The data in Table 6-20 illustrates the excellent control exerted on composition and molecular weight. Gel permeation chromatography

TABLE 6-20

Composition and Molecular Weight of
α-Methylstyrene/Propylene Sulfide/α-Methylstyrene Block Copolymers[a]

α-Methylstyrene (wt %)		Molecular weight[c] ($\overline{M}_n \times 10^{-3}$)	
Projected	Found[b]	Segment (theoretical)	Total (found)
20.3	20.2 ± 1.8	8–66–8	84
30.1	29.9 ± 1.0	13–62–13	83
39.0	38.9 ± 0.8	17–55–17	90

[a] From Fetters *et al.* (C81).
[b] Determined by NMR analysis.
[c] Determined by osmometry.

studies showed that the copolymers had narrow molecular weight distributions. This degree of structure control indicates that the crossover reaction was fast and that the coupling reaction was an efficient one. The tensile strength and percent set at break properties of these copolymers were considerably lower than those of styrene–butadiene–styrene, styrene–isoprene–styrene, or α-methylstyrene–isoprene–α-methylstyrene A-B-A thermoplastic elastomer block copolymers. Nevertheless, the stress–strain characteristics were similar.

c. Olefin Sulfide–Vinyl

A-B-A block copolymers in which A is ethylene sulfide or propylene sulfide and B is methyl methacrylate have been reportedly prepared via the anionic living polymer technique described earlier for alkylene sulfide–styrene–alkylene sulfide block copolymers (C77).

A block copolymer of propylene sulfide–2-vinylpyridine–propylene sulfide was similarly prepared using sodium biphenyl to initiate the vinylpyridine polymerization, followed by addition of the propylene sulfide monomer (C75). The molecular weight ($\overline{M}_n$) of this product was 169,000. An ethylene sulfide–4-vinylpyridine–ethylene sulfide block copolymer was similarly synthesized (C80).

A-B-A block copolymers have also been reported in which B is polybutadiene or polyisoprene and A is poly(ethylene sulfide) or poly(propylene sulfide) (C82, C83). Anionic processes were again used in these syntheses, which employed sodium (or lithium) naphthalene initiators. Polymerization of the diene in THF and/or heptane at −30° to −40°C was followed by addition and polymerization of the episulfide monomer at room temperature for several days. Solubility behavior was cited as evidence of block structure. The products displayed elastomeric properties. The specific compositions reported are listed in Table 6-19.

Sulfide–vinyl block copolymers in which A is ethylene sulfide or thiacyclobutane and B is butadiene or isoprene were similarly synthesized using 1,4-dilithiotetraphenylbutane as the initiator (C85, C86). Compositions containing 30% ethylene sulfide were found to display much lower tensile strengths than those of comparable styrene–diene A-B-A block copolymers. Other workers (C87) reported that ethylene sulfide–isoprene block copolymers displayed good thermal stability but poor melt processability. Ethylene sulfide–isoprene–styrene A-B-C block copolymers (C87) displayed unusual flow characteristics and morphology.

d. Perfluorinated Sulfides

Several A-B-A block copolymers are reported in the patent literature (C56, C84) in which the A segments are perfluorinated sulfide polymers derived from thiocarbonyl fluoride

$$(F-\overset{\overset{\displaystyle S}{\|}}{C}-F)$$

or tetrafluoroethylene sulfide

$$(F_2C\overset{\overset{\displaystyle S}{\diagup\ \diagdown}}{\text{------}}CF_2)$$

and the B segments are vinyl polymers such as polystyrene, poly(methyl methacrylate), or polyacrylonitrile (see Table 6-19). These block copolymers were synthesized by anionically polymerizing the vinyl monomer at $-80°C$ using sodium biphenyl initiator, followed by addition and polymerization of the sulfide monomer. The products were claimed to be superior to the vinyl homopolymers in toughness, solvent resistance, and flame resistance.

5. Amides

The block copolymers of this category contain at least one polyamide segment. The other segments include polyamide, polyether, polyester, polyurethane, and polyvinyl blocks.

The initiated anionic polymerization of lactams under controlled conditions offers one of the best approaches for the preparation of an A-B-A polyamide containing block copolymer. Yamashita et al. (C98) have carefully studied the anionic polymerization of α-pyrrolidone and ε-caprolactam (at 30°C and 80°C, respectively) initiated by the bischloroformates of dihydroxy-terminated poly(tetramethylene glycol) and polystyrene. Initiation efficiency was high in both cases. This was judged by the low contents of unreacted polyether or polystyrene found by selective extraction. It was also shown that both the yield and viscosity of the block copolymer increases with polymerization time up to about 50% conversion of the lactam. Such behavior is, of course, expected if the polymerization is essentially "living." The reactions were run in bulk, and thus, high viscosity immobilized the chain ends at high conversion. The data for α-pyrrolidone initiated by poly(tetramethylene glycol) bischloroformate (PTG) are shown in Table 6-21. The catalyst (sodium naphthalene) interacts with the lac-

TABLE 6-21

Anionic Polymerization of α-Pyrrolidone(Pyd) Initiated by
Poly(tetramethylene glycol) Bischloroformate(PTG)[a]

Polymeriza-tion time (hours)	Conversion of Pyd (%)	η_{sp}/c[b]	Block copolymer		Recovered PTG (%)
			Pyd (mole%)		
			From NMR	From conversion	
1	13.9	0.55	68.2	64.9	1.61
3	26.0	0.72	78.3	77.5	1.78
6	33.5	1.25	81.8	80.4	3.21
15	47.1	2.03	83.5	86.0	2.50

[a] From Yamashita et al. (C98). PTG bischloroformate, 2.84×10^{-4} mole; sodium naphthalene, 1.17×10^{-3} mole; Pyd, 0.104 mole; polymerization at room temperature.
[b] Measured in m-cresol at 25°C, $c = 0.5$ gm/dl; η_{sp}/c for PTG alone $= 0.25$.

tam to generate the sodium salt of the lactam. The reaction sequence is shown in Scheme 6-20.

ε-Caprolactam behaved similarly as is shown by the results in Table 6-22. Proton magnetic resonance could be utilized to determine the overall composition of the block copolymers. A typical spectrum is shown in Figure 6-26.

Polystyrene–nylon 6 block copolymers were also efficiently prepared, although it was observed (C98) that the lactam polymerization rate was somewhat slower. Initiation rate was also slow compared to propagation. These slower polymerization rates were attributed to either decreased chain flexibility or solubility of the polystyrene mac-

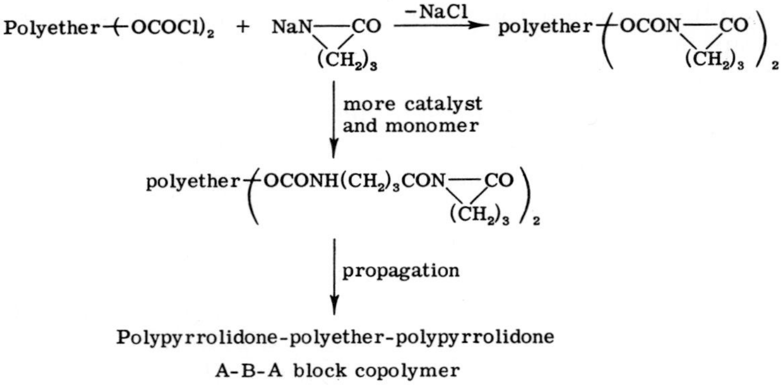

$$\text{Polyether} \{ \text{OCOCl} \}_2 \ + \ \text{NaN} \underset{(CH_2)_3}{\overset{CO}{\diagdown \diagup}} \ \xrightarrow{\ -NaCl\ } \ \text{polyether} \left(\text{OCON} \underset{(CH_2)_3}{\overset{CO}{\diagdown \diagup}} \right)_2$$

$$\Big\downarrow \begin{array}{l} \text{more catalyst} \\ \text{and monomer} \end{array}$$

$$\text{polyether} \left(\text{OCONH(CH}_2)_3 \text{CON} \underset{(CH_2)_3}{\overset{CO}{\diagdown \diagup}} \right)_2$$

$$\Big\downarrow \text{ propagation}$$

Polypyrrolidone-polyether-polypyrrolidone

A-B-A block copolymer

Scheme 6-20

romolecular initiator. It is interesting and important to note that the authors did not observe a side reaction of the monomer anion with the amide link of the chain under the mild polymerization conditions utilized.

Independently, McGrath and co-workers (C99–C104) demonstrated that activated aromatic halides such as 4,4'-dichlorodiphenylsulfone are novel initiators for lactam anionic polymerization (C99, C100) and that activated halide terminated polyethers (C101–104) could be used

TABLE 6-22

Anionic Polymerization of ε-Caprolactam (CL) Initiated by
Poly(tetramethylene glycol) Bischloroformate (PTG)[a]

Polymeriza-tion time (hours)	Conversion of CL (%)		Block copolymer		
				CL (mole%)	Recovered PTG (%)
		η_{sp}/c[b]	From NMR	From conversion	
1	17.1	0.47	66.2	60.2	3.92
3	30.5	0.75	76.5	73.0	2.21
5	22.5	0.79	68.7	66.7	2.54
6	33.6	1.21	75.0	71.7	2.01
8	36.7	1.30	84.2	76.5	2.53
14	48.6	1.73	83.3	81.8	1.98

[a] From Yamashita *et al.* (C98). PTG bischloroformate, 2.86×10^{-4} mole; sodium naphthalene, 1.17×10^{-3} mole; CL, 7.28×10^{-2} mole; polymerization at 80°C.

[b] Measured in *m*-cresol at 30°C; $c = 0.5/\text{gm/dl}$; η_{sp}/c for PTG alone = 0.23.

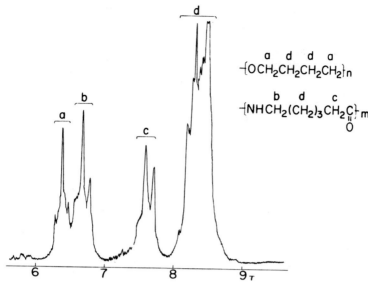

$$\underset{a \quad d \quad d \quad a}{+OCH_2CH_2CH_2CH_2\}_n}$$

$$\underset{b \quad d \quad c}{+NHCH_2(CH_2)_3CH_2\underset{O}{\overset{}{C}}\}_m}$$

Fig. 6-26. Proton NMR spectrum of nylon 6-poly(tetramethylene glycol)–nylon 6 A-B-A block copolymer in formic acid (C98).

to prepare poly(aryl ether)–nylon 6 block copolymers. For example, chlorine-terminated polysulfone

$$Cl-\langle\bigcirc\rangle-SO_2-\langle\bigcirc\rangle-\left[O-\langle\bigcirc\rangle-\underset{CH_3}{\overset{CH_3}{C}}-\langle\bigcirc\rangle-O-\langle\bigcirc\rangle-SO_2-\langle\bigcirc\rangle\right]_n-Cl$$

is readily prepared (C101–C104) and is soluble in liquid caprolactam. Addition of a base catalyst such as sodium hydride results in polymerization from the poly(aryl ether) chain ends. At high temperatures and/or high catalyst levels, a side reaction involving cleavage of the aryl ether link was identified. Nevertheless, extraction with selective polysulfone solvents such as chloroform indicated that a large part of the polyether was chemically attached to the polyamide. Moreover, the mechanical properties of the copolymer, or its alloys, were shown to be attractive (C101–C104) whereas the simple physical blend of the homopolymers displayed poor properties. The attachment of the semicrystalline polyamide segment to the poly(aryl ether) produced a dramatic increase in environmental stress crack resistance of the latter. This effect is illustrated by the effect of nylon 6 content on stress rupture as shown in Figure 6-27. On the other hand, the presence of

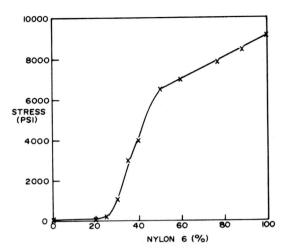

Fig. 6-27. Polysulfone–nylon 6 block copolymers. Stress level required for 1 minute rupture in acetone as a function of nylon 6 composition (C103, C104).

the relatively hydrophobic poly(aryl ether) block decreased the water sensitivity of the polyamide (C103, C104).

Other end groups, such as ester, isocyanate, acyl halide, and terephthaloyllactams, have also been used with varying degrees of success to prepare lactam block copolymers. A summary of this approach is given in Table 6-23.

Polyamide–polyvinyl–polyamide block copolymers have been prepared by the anionic polymerization of alkyl isocyanates (e.g., to form

TABLE 6-23

Polyamide A-B-A Block Copolymers via Anionic Lactam Polymerization

A segment	B segment	Initiating site	Reference
Nylon 6	Poly(tetramethylene glycol)	Chloroformate	C98
	Polystyrene	Chloroformate	C98
	Polystyrene	Isocyanate	C105, C108
Nylon 4	Polystyrene	Chloroformate	C98
Nylon 6	Polysulfone	Chlorophenylsulfone	C101–C104, C109
	Nylon 6/66 copolymer	Bislactam	C106
	Polystyrene	Bisphenol A dichloroformate	C107
	Polycaprolactone	Isocyanate	C110
Nylon 6 or 7	Poly(tetramethylene glycol)	Isocyanate	C115

1-nylons) and vinyl monomers. Isocyanate polymerization was initiated by the dianions of the hydrocarbon center block, which in turn was synthesized by sodium metal- or sodium biphenyl-initiated systems (C111–C114). The polymerizations were conducted at $-78°C$ to minimize the formation of the cyclic trimer of the isocyanate monomer. The polymerizations were carried out, presumably, as shown in Scheme 6-21 for the case of a butyl isocyanate–styrene–butyl isocya-

Scheme 6-21

nate copolymer. Similarly, A-B-A block copolymers were reportedly prepared in which the B segment was polybutadiene, polyisoprene, polyacrylonitrile, or poly(methyl methacrylate) and the A block was derived from methyl isocyanate, decyl isocyanate, or decyl isothiocyanate. It seems doubtful that a high degree of block integrity was achieved in the case of polar monomers such as acrylonitrile.

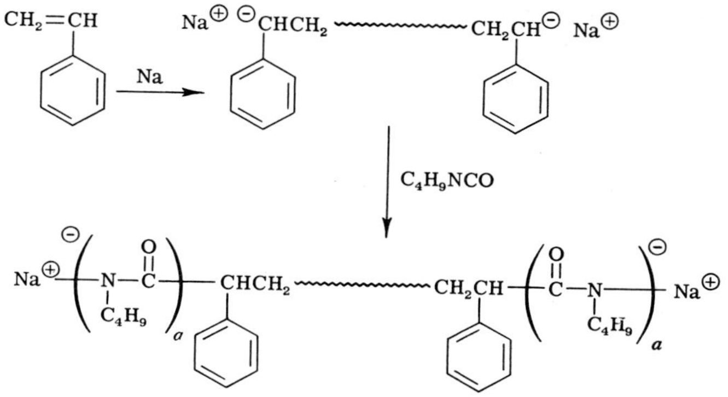

Amine-terminated polyamides have been used to initiate the ring opening polymerization of lactones to form polyester–polyamide–polyester block copolymers (C116–C120). For example, the polymerization of caprolactone by amine-terminated nylon 6 has been studied (C116). It was reported that block copolymer was formed when the reaction temperature was higher than the melting temperature (220°C) of the macromolecular initiator. Also, if the number average molecular weight of the polyamide was greater than 10,000, phase separation took place and the polymerization did not proceed. Formation of block copolymer was confirmed by infrared spectroscopy, differential thermal analysis, solubility tests, and quantitative analysis of amino end groups.

Polyamide A-B-A block copolymers have also been reported to be prepared via the hydrolytic or acid-catalyzed polymerization of lac-

tams. However, it appears that this approach must also generate substantial quantities of homopolymers in addition to the desired segmented structures. Nevertheless, block copolymers of either nylon 12
or nylon 6 prepared in the presence of amine-terminated poly(tet-
ramethyleneglycol) were reported to have improved properties over
the homopolymers (C121–C123).

   Novel biopolymers in which one, two, or three segments are
polypeptides may be expected to assume great importance in the future. Considerable amounts of information concerning these materials
is already in the literature. Invariably, the synthesis route has involved
the amine-initiated polymerization of amino acid N-carboxyanhydride
monomers. Unfortunately, mechanistic unknowns similar to those
cited in Chapter 5, Section B exist. Also, molecular characterization
has been rather limited to date. In principle, the synthesis should be as
indicated in Scheme 6-22.

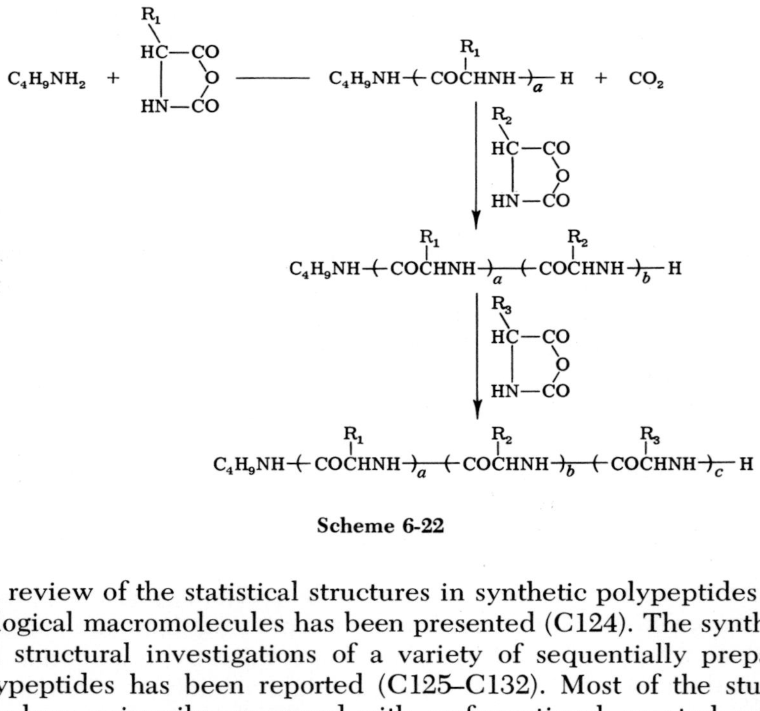

Scheme 6-22

   A review of the statistical structures in synthetic polypeptides and
biological macromolecules has been presented (C124). The synthesis
and structural investigations of a variety of sequentially prepared
polypeptides has been reported (C125–C132). Most of the studies
have been primarily concerned with conformational, spectral, and/or
solubility differences between the block polypeptides and their random counterparts.

6. Siloxanes

The copolymers of this category contain one or more segments of poly(dialkylsiloxane), the alkyl groups being either methyl or phenyl. Some of these copolymers contain only siloxane blocks, while others contain either vinyl polymer blocks or lactam polymer blocks together with polysiloxane blocks. The latter two types display properties characteristic of both the organic and the siloxane polymers. The various siloxane-containing ABA block copolymers reported are listed in Table 6-24.

a. Siloxane–Siloxane

Well-defined A-B-A block copolymers in which A is poly(diphenyl-siloxane) and B is poly(dimethylsiloxane) were first reported by Bostick (C133, C154). The synthesis was carried out by using lithium diphenyl silanolate to initiate the polymerization of hexamethylcyclo-trisiloxane (cyclic trimer of dimethylsiloxane), at room temperature in the presence of tetrahydrofuran followed by the addition of hexaphenylcyclotrisiloxane (Scheme 6-23).

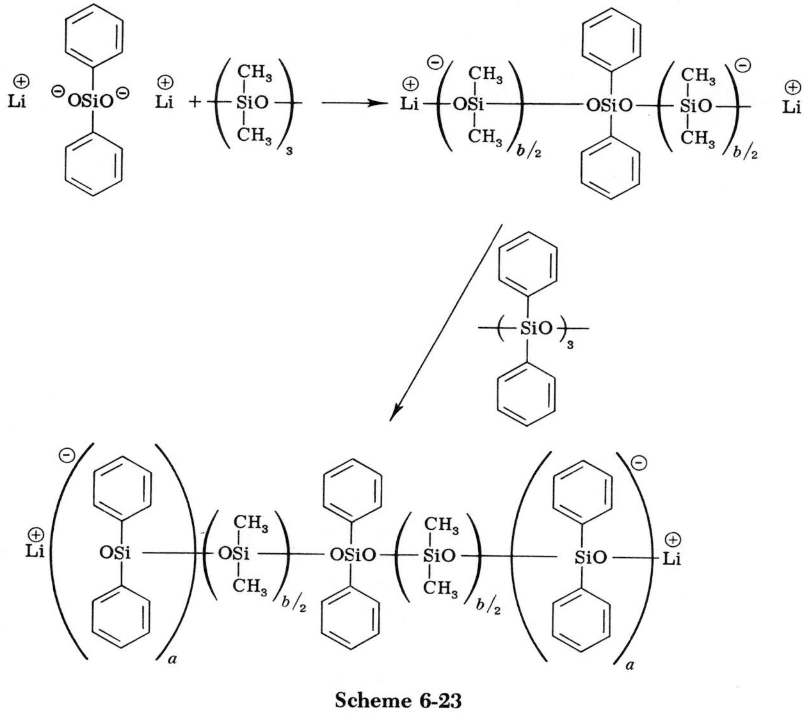

Scheme 6-23

TABLE 6-24

A-B-A Block Copolymers Containing Polysiloxane Segments

Block A	Block B	Initiator system	Conditions	Reference
Diphenylsiloxane	Dimethylsiloxane	LiOSiOLi	25° to 125°C	C133, C135, C151, C154, C155, C158
Dimethylsiloxane	Styrene	Na, K, or Li naphthalene	−80° to 0°C	C136–C139, C152
	Styrene	—CH=CH₂ and —Si—H terminated oligomers	120°C	C141
	Styrene	—Si—OH and —Si—Cl terminated oligomers	—	C141
Styrene	Dimethylsiloxane	Na polystyryl + Cl terminated siloxane oligomer	—	C142
	Dimethylsiloxane	C₄H₉Li + Cl—R—Cl coupling	25°C	C143–C146
α-Methylstyrene	Dimethylsiloxane	C₄H₉Li + Cl—R—Cl coupling	—	C147
Styrene (A block), diphenylsiloxane (C block)	Dimethylsiloxane (B block)	C₄H₉Li	−40° to +200°C	C148, C149
Dimethylsiloxane	α-Methylstyrene	α-Methylstyrene tetramer dianion	25° to 50°C	C140
	Isoprene	K metal or K naphthalene	−80°C	C156, C152
	Methyl methacrylate	Na (or K or Li) naphthalene	0°C	C137
	Acrylonitrile	Na (or K or Li) naphthalene	0°C	C137
Acrylonitrile	Dimethylsiloxane	Na (or K or Li) naphthalene	0°C	C137
2-Vinylpyridine	Dimethylsiloxane	Na (or K or Li) naphthalene	—	C153
Caprolactam	Dimethylsiloxane	Lactam-terminated siloxane oligomer + lactam monomer + LiAlH₄	110°C	C153
Lauryllactam	Dimethylsiloxane	Lactam-terminated siloxane oligomer + lactam monomer + LiAlH₄	110°C	C150

Cyclic trimers are much preferred to cyclic tetramers as monomers in that they are more reactive due to greater ring strain. As a result, the slower redistribution side reactions that can lead to randomization are minimized (C134). Catalysts containing lithium counterion are preferable to sodium or potassium counterions due to the lower catalytic activity of lithium in siloxane redistribution reactions. The presence of aprotic, moderately electron-donating compounds promote the polymerization of cyclic trimer without inducing loss of specificity or randomization. In addition to tetrahydrofuran, other compounds, such as glycol ethers, anisole, triethylamine, and p-dioxane, are also effective in promoting the polymerization rate. The above conditions have been reported to give siloxane polymers of very narrow molecular weight distribution (C134).

The block copolymers were described as tough, white, opaque elastomers that were insoluble in common organic solvents at room temperature but soluble in diphenyl ether at elevated temperatures. Compositions containing 4–27% diphenylsiloxane had dimethylsiloxane segments of 150,000 $\overline{M}_n$ and diphenylsiloxane segments of 9000–75,000 $\overline{M}_n$. They displayed crystalline melting points (measured by differential scanning calorimetry) at $-42°C$ to $-49°C$ for the dimethylsiloxane domains and $210°C$ to $237°C$ for the diphenylsiloxane domains. The dimethylsiloxane segments were amorphous and flexible at room temperature. Nuclear magnetic resonance behavior was cited as evidence for a block sequence structure.

The block copolymer mechanical properties were affected by solvent treatment. Initially elastomeric products were transformed to high-modulus materials by dissolving in hot (230°C) diphenyl ether and coagulating with cold methanol. This behavior was explained on the basis of conversion from a morphological state in which the flexible dimethylsiloxane segments were the continuous phase to one in which the high-modulus diphenylsiloxane segments had become, at least in part, the continuous phase. Another possible interpretation is that the "hard" diphenylsiloxane block had undergone further crystallization during the treatment. However, DSC measurements indicated that the reprecipitation procedure had not resulted in any significant increase in crystallinity.

Block copolymers precipitated from dilute solution were found by electron microscopy to be in the form of lamellae a few hundred angstroms thick and threads about 200 Å in diameter (C155). Solution-cast films contained dispersed crystalline regions of the diphenylsiloxane block. The size of the crystalline regions increased with increasing block length. Annealing the films above 300°C decreased the size and increased the number, and degree of perfection of, the crystalline regions. Unlike styrene–butadiene block copolymers, these siloxane block copolymers did not require staining to achieve sufficient contrast to obtain good quality electron micrographs.

The solution properties of these block copolymers were investigated (C135). Intrinsic viscosity increased with increasing length of the dimethylsiloxane block in polymers of constant degree of polymerization. The effect of solvent solubility parameter (δ) on swelling of the block copolymer showed two maxima, one at 8.0 (due to dimethylsiloxane block) and another at 9.5 (due to the diphenylsiloxane block). The latter was markedly dependent on temperature (more swelling at higher temperatures) while the former was not. This behavior was said to be due to progressive dissolution of the diphenylsiloxane crystallites with increasing temperature.

b. Siloxane–Vinyl

A-B-A block copolymers in which A is poly(dimethylsiloxane) and B is polystyrene have been prepared by using polystyryl dianion to initiate the polymerization of the cyclic trimer (C136) or cyclic tetramer (C137, C138, C156) of dimethylsiloxane in tetrahydrofuran solution at −80° to 0°C. The resulting silanolate anion end groups were terminated with methyl iodide (Scheme 6-24). Products were obtained with 56,000–165,000 $\overline{M}_n$ polystyrene blocks and 80,000–465,000 $\overline{M}_n$ siloxane blocks. The polystyrene segment was reported to have a narrow molecular weight distribution, while the poly(dimethylsiloxane) blocks had a most probable distribution. Precipitation behavior led to the conclusion that some block copolymer was present in the products but that considerable quantities of siloxane homopolymer had also been formed. The presence of the homopolymer is attributed to the siloxane equilibration side reactions. The solution equilibration behavior of block copolymers of this type was reported to be similar to that of siloxane homopolymers (C139).

The kinetics of the anionic polymerization of octamethylcyclotetrasiloxane initiated by α-methylstyrene tetramer dianion was reported in another reference (C140). The data indicated that a slow initiation step is followed by a rapid propagation reaction. The resulting copolymer product has a very short α-methylstyrene B block and two long dimethylsiloxane blocks.

Siloxane–styrene–siloxane A-B-A block copolymers were also reported to have been prepared by the condensation of polysiloxane and polystyrene oligomers bearing reactive functional end groups such as olefin and —Si—H (Scheme 6-25) (C141). Similarly, siloxane oligomers containing two allyl end groups were reacted with a styrene

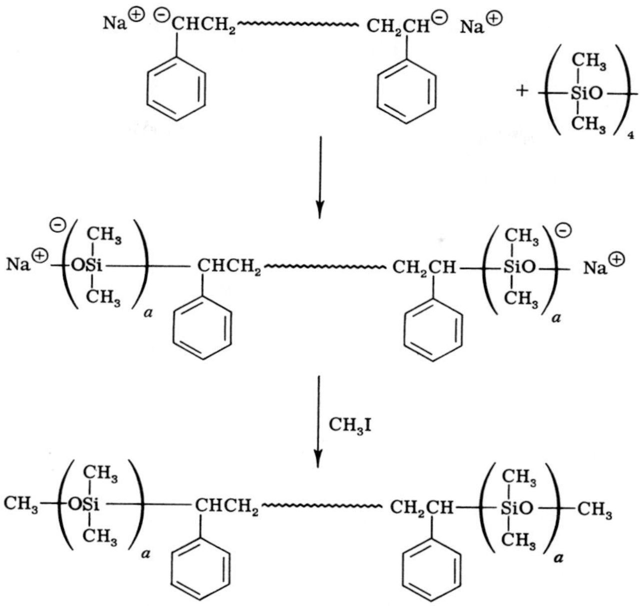

Scheme 6-24

oligomer bearing two —Si—H end groups in a 2 : 1 mole ratio to produce a siloxane–styrene–siloxane block copolymer. Reversal of this mole ratio reportedly gave a styrene–siloxane–styrene block

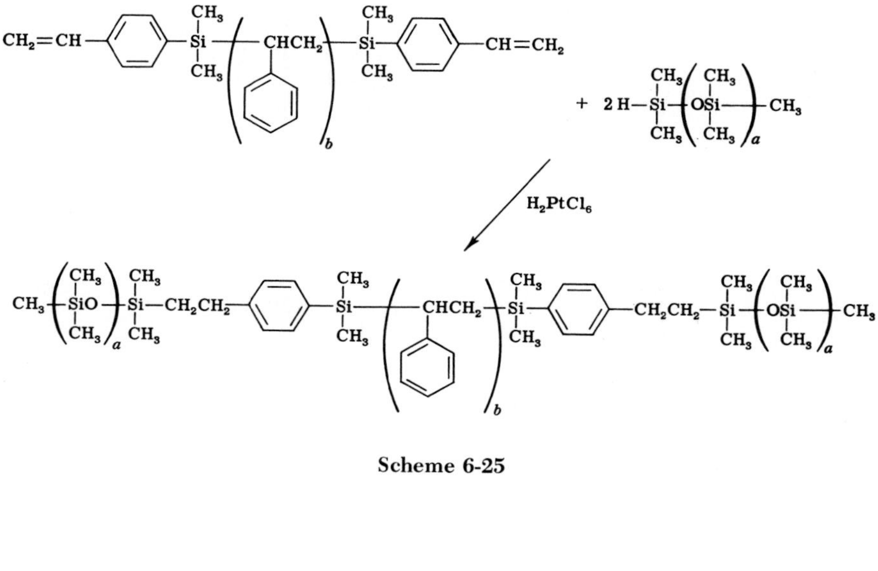

Scheme 6-25

copolymer. Block copolymers were also reportedly obtained by condensing disilanol-terminated polystyrene with dichloro terminated polysiloxane (Scheme 6-26). Again, the reverse sequence structure

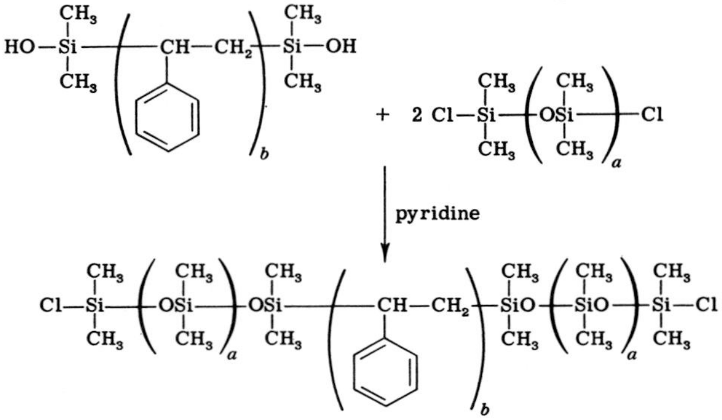

Scheme 6-26

was said to result when two "moles" of the styrene oligomer were condensed with one "mole" of the siloxane oligomer. Obviously, this approach will produce a mixture of homopolymers and copolymers of widely ranging sequence structure due to the statistical nature of the reaction. The use of difunctionally terminated oligomers is better suited to the preparation of $(A-B)_n$ block copolymers, as will be discussed in Chapter 7.

A-B-A block copolymers in which A is polystyrene and B is

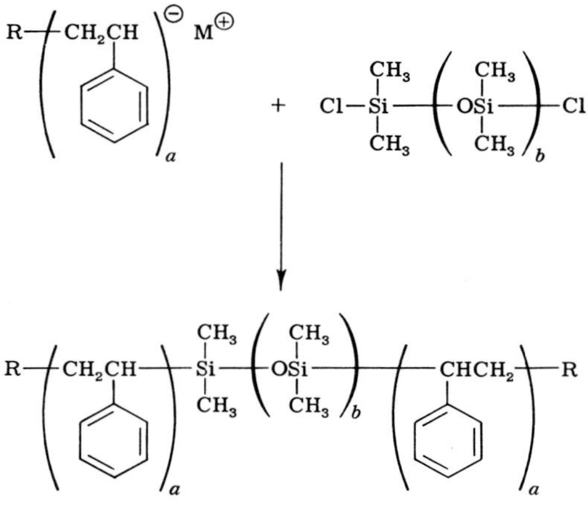

Scheme 6-27

poly(dimethylsiloxane) have reportedly been prepared by condensing polystyryl anions with —Si—Cl-terminated siloxanes (Scheme 6-27) (C142). However, siloxane cleavage by the strongly basic polystyryl anions is a side reaction that can take place to obscure the structure of the product.

Another, more promising route to styrene–siloxane–styrene block copolymers involves a combination of sequential anionic polymerization of styrene and hexamethylcyclotrisiloxane followed by coupling (C143–C145). The coupling reaction is necessitated by the fact that the silanolate anion is much less reactive than the styryl anion and accordingly is not capable of initiating styrene polymerization. The reaction sequence carried out in THF at 25°C is as shown in Scheme 6-28. As discussed earlier, the cyclic trimer of dimethysiloxane is a much more reactive monomer than the cyclic tetramer and results in narrow

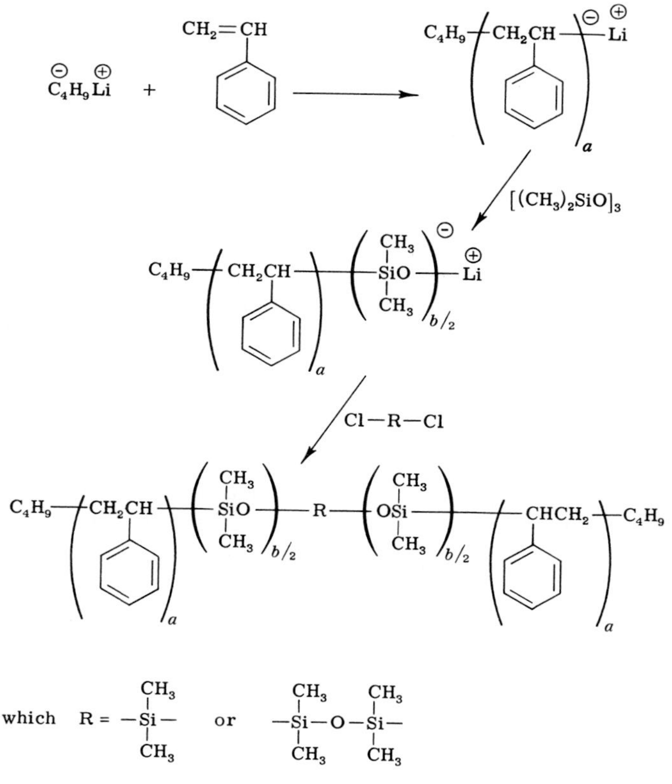

Scheme 6-28

molecular weight distributions due to minimization of the siloxane redistribution side reaction. $\overline{M}_w/\overline{M}_n$ values, as measured by light scattering and osmometry, of 1.1–1.5 were found for products ranging in composition from 15% to 2% polystyrene and in $\overline{M}_n$ from 20,000 to 130,000 (C143, C145). The conformation of these block copolymers in dilute solution was elucidated by investigating their intrinsic viscosity behavior in a good solvent for both blocks (toluene at 25°C) and in theta solvents for the styrene and siloxane blocks (cyclohexane at 34°C and methyl ethyl ketone at 20°C, respectively). Higher intrinsic viscosities were observed than those calculated assuming the copolymers behave like binary mixtures of homopolymers. The data were interpreted as meaning that the copolymers form randomly interpenetrating coils in dilute solution. In good solvents, the coils are expanded by polymer–solvent interactions and A–B interactions are minimum, whereas in a theta solvent for one of the segments, repulsive A–B interactions contribute appreciably to coil expansion. These large repulsive interactions are a result of the incompatibility of the A segments with segment B.

The surface activity of styrene–siloxane–styrene A-B-A block copolymers was studied by measuring the surface tension of solutions of the copolymers in styrene monomer and their effect on the critical surface tension of wetting of solid polystyrene (C146). The data was interpreted in terms of a model for the copolymer in the solvent in which the cross-sectional area of the copolymer is determined by the packing of distorted polystyrene segments.

Styrene–siloxane–styrene block copolymers are claimed to be useful as additives for reducing the critical surface tension of polystyrene molding materials, acrylic surface coatings, and rubber vulcanizates, as well as having useful mold release properties (C144). Their use as impact modifiers for polystyrene has also been claimed (C157).

α-Methylstyrene–siloxane–α-methylstyrene block copolymers have also been synthesized via the coupling of lithio-terminated α-methylstyrene–siloxane A-B structures (C147). These materials were found to exhibit significantly lower tensile strengths than comparable α-methylstyrene–diene–α-methylstyrene A-B-A block copolymers. This behavior was attributed to the low T_g of the siloxane block relative to those of polydienes.

An A-B-C block copolymer containing a polystyrene segment (A), a poly(dimethylsiloxane) segment (B), and a poly(diphenylsiloxane) segment (C) was prepared by initiating the polymerization of styrene with butyllithium and sequentially adding hexamethylcyclotrisiloxane and hexaphenylcyclotrisiloxane (Scheme 6-29) (C148, C149). No randomi-

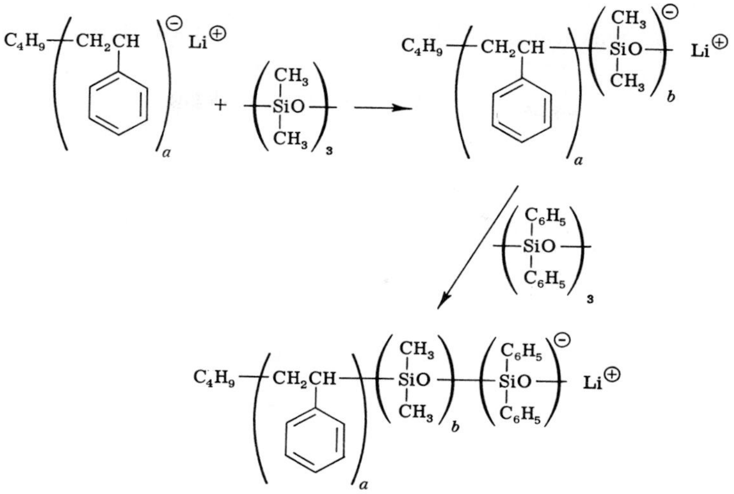

Scheme 6-29

zation (redistribution) was observed using the lithium initiator, but a potassium initiator led to random distribution of the dimethylsiloxy and diphenylsiloxy units.

Dimethylsiloxane–isoprene–dimethylsiloxane block copolymers were synthesized by initiating isoprene polymerization with potassium naphthalene followed by the addition of octamethylcyclotetrasiloxane (C156). Isoprene content ranged from 15 to 73%, and isoprene and dimethylsiloxane block molecular weights were 50,000–300,000 and 25,000–425,000, respectively. Block structure was concluded based on solubility behavior and the fact that vulcanization via a sulfur system yielded an almost completely insoluble product (siloxanes do not cross-link under these conditions). Surprisingly high tensile properties (1200 psi tensile strength/700% elongation) were obtained even in the absence of an added reinforcing filler.

An attempt was made to prepare siloxane–methyl methacrylate–siloxane and siloxane–acrylonitrile–siloxane block copolymers via the anionic living polymer process described above for siloxane–styrene–siloxane copolymers (C137). However, the products contained large quantities of the vinyl homopolymers due to termination and polymer precipitation in the first step.

A copolymeric product (presumably an acrylonitrile–siloxane–acrylonitrile block copolymer) was obtained by adding acrylonitrile to anionically terminated poly(dimethylsiloxane) (C137). However, this product was obtained in low yield, contained only small amounts of siloxane, and was low in molecular weight. This partially successful result, in contrast to the completely negative results obtained with styrene and methyl methacrylate monomers, was probably due to the fact that acrylonitrile polymerization can be initiated by weak bases.

c. Siloxane–Amide

These block copolymers were prepared by initiating the polymerization of hexamethylcyclotrisiloxane with $O[Si(CH_3)_2O^-Li^+]_2$, capping of the resulting dilithiosilanolate-terminated oligomer with

$$Cl(CH_3)_2Si(CH_2)_{10}\overset{\overset{\displaystyle O}{\|}}{C}N(CH_2)_5C=O$$

to form

$$O=\overset{\overset{\displaystyle O}{\|}}{C}(CH_2)_5N\overset{\overset{\displaystyle O}{\|}}{C}(CH_2)_{10}Si(CH_3)_2OSi(CH_3)_2 \mathtt{\sim\!\!\sim} Si(CH_3)_2O(CH_3)Si(CH_2)_{10}\overset{\overset{\displaystyle O}{\|}}{C}N(CH_2)_5C=O$$

and by using this intermediate to initiate the polymerization of caprolactam or lauryllactam in the presence of $LiAlH_4$ catalyst (C150) to form the final lactam–siloxane–lactam block copolymer. The lactam polymerizations were carried out in toluene solution at 110°C. The critical surface tension and coefficient of friction of nylon 6 and of polyethylene were reduced by blending with 2% of the caprolactam and lauryllactam block copolymers, respectively.

REFERENCES

A1. Milkovich, R., S. African Patent 280,712 (Shell) (1963).

A2. Kossoff, R. M., quoted in *Mod. Plast.* 50 (1974).

A3. Szwarc, M., Levy, M., and Milkovich, R., *J. Am. Chem. Soc.* **78**, 2656 (1956).

A4. Szwarc, M., "Carbanions, Living Polymers, and Electron Transfer Processes." Wiley, New York, 1968.

A5. Fetters, L. J., *J. Elastoplast.* **4**, 34 (1972).

A6. Fetters, L. J., *in* "Block and Graft Copolymerization" (R. J. Ceresa, ed.), Vol. 1, p. 99. Wiley, New York, 1973.

A7. Morton, M., *Phys. Chem., Ser. One* **8,**1 (1972); *C.A.* **78**, 30204x (1973).

A8. Morton, M., *Encycl. Polym. Sci. Technol.* **15**, Suppl., 508 (1971); *C.A.* **76**, 100848v (1972).

A9. Cunningham, R. E., Auerbach, M., and Floyd, W. J., *J. Appl. Polym. Sci.* **16**, 163 (1972).

A10. Cunningham, E., and Treiber, M. R., *J. Appl. Polym. Sci.* **12**(1), 23 (1968); *C.A.* **68**, 87981w (1968).

A11. Morton, M., McGrath, J. E., and Juliano, P. C., *J. Polym. Sci., Part C* **26**, 99 (1969); *C.A.* **70**, 107274q (1969).

A12. Scott, N. D., U.S. Patent 2,181,771 (E. I. DuPont de Nemours and Co.) (1939).

A13. Paul, D. E., Lipkin, D., and Weissman, S. I., *J. Am. Chem. Soc.* **78**, 116 (1956).

A14. Szwarc, M., *Nature (London)* **178**, 1168 (1956); *C.A.* **51**, 11815i (1956).

A15. Adams, H. E., Bebb, R. L., Forman, L. E., and Wakefield, L. B., *Rubber Chem. Technol.* **45**, 1252 (1972).

A16. Fetters, L. J., *J. Polym. Sci., Part C* **26**, 1 (1969).

A17. Morton, M., and Fetters, L. J., *Macromol. Rev.* **2**, 71 (1967).

A18. Fetters, L. J., and Morton, M., *Macromolecules* **2**, 453 (1969).

A19. McGrath, J. E., Ph.D. Thesis, University of Akron, Akron, Ohio (1967).

A20. Juliano, P. C., Ph.D. Thesis, University of Akron, Akron, Ohio (1968).

A21. Fetters, L. J., *J. Res. Natl. Bur. Stand., Sect. A* **70**, 421 (1966).

A22. Prud'homme, J., and Bywater, S., *in* "Block Polymers" (S. L. Aggarwal, ed.), p. 11. Plenum, New York, 1970.

A23. Wakefield, B. J., "The Chemistry of Organolithium Compounds." Pergamon, Oxford, 1974.

A24. Hsieh, H. L., and Glaze, W. H., *Rubber Chem. Technol.* **43**(1), 22 (1970); *C.A.* **72**, 121944f (1970).

A25. Hsieh, H. L., *in* "Block and Graft Copolymers" (J. J. Burke and V. Weiss, eds.), p. 51. Syracuse Univ. Press, Syracuse, New York, 1973.

A26. Morton, M., *in* "Block Polymers" (S. L. Aggarwal, ed.), p. 1. Plenum, New York, 1970.

A27. Kraus, G., and Railsback, H. E., *Polym. Prepr., Am. Chem. Soc., Div. Polym. Chem.* **14**(2), 1051 (1973); see also "Recent Advances in Polymer Blends, Grafts and Blocks" (L. H. Sperling, ed.), p. 245. Plenum, New York, 1974.

A28. Schlick, S., and Levy, M., *J. Phys. Chem.* **64**, 883 (1960).

A29. Rembaum, A., Ells, F. R., Morrow, R. C., and Tobolsky, A. V., *J. Polym. Sci.* **61**, 155 (1962).

A30. Angelo, R. J., Ikeda, R. M., and Wallach, M. L., *Polymer* **6**(3), 141 (1965); *C.A.* **63**, 3064b (1965).

A31. Minoux, J., and Leng, M., *C. R. Hebd. Seances Acad. Sci.* **252**, 277 (1961).

A32. French Patent 1,209,992 (Laboratoire de recherche et de controle du Caoutchouc) (1960).

A33. Hsieh, H. L., and Trepka, W. J., U.S. Patent 3,410,836 (Phillips Petroleum Co.) (1968); *C.A.* **70**, 29568x (1969).

A34. Zelinski, R. P., U.S. Patent 3,287,333 (Phillips Petroleum Co.) (1966).

A35. British Patent 964,478 (Phillips Petroleum Co.) (1964); *C.A.* **61**, 10853g (1964).

A36. Zelinski, R. P., and Hsieh, H. L., U.S. Patent 3,078,254 (1963); *C.A.* **58**, 10324d (1963).

A37. Zelinski, R. P., U.S. Patent 3,251,905 (Phillips Petroleum Co.) (1966); *C.A.* **65**, 2370h (1966).

A38. Shiratsuchi, E., Hayashi, S., and Nozaki, C., Japanese Patent 70/01,629 (Japan Synthetic Rubber Co., Ltd.) (1970); *C.A.* **72**, 101624y (1970).

A39. Morton, M., Fetters, L. J., Juliano, P. C., Schwab, F. C., and Strauss, C., "High Temperature Elastomer Networks from Difunctional Block Polymers," Air Force Materials Laboratory Report—Contract AF 33 (615) 5362, Wright Patterson AFB, Ohio 1968.

A40. Zelinski, R. P., Belgian Patent 661,095 (Phillips Petroleum Co.) (1965); *C.A.* **65**, 4080f (1966).

A41. Holden, G., and Milkovich, R., U.S. Patent 3,265,765 (Shell Oil Co.) (1966).

A42. Holden, G., and Milkovich, R., Belgian Patent 627,652 (Shell Internationale Research Maatschappij N.V.) (1963); *C.A.* **60**, 14713f (1964).

A43. Holden, G., and Milkovich, R., U.S. Patent 3,231,635 (Shell Oil Co.) (1966); *C.A.* **64,** 9836f (1966).

A44. British Patent 1,035,873 (Shell Internationale Research Mattschappij N.V.) (1966); *C.A.* **65,** 12389 (1966).

A45. Holden, G., and Milkovich, R., Belgian Patent 627,652 (Shell Internationale Research Maatschappij N.V.) (1963); *C.A.* **60,** 14714f (1964).

A46. Moss, F. D., and Mathews, J. F., U.S. Patent 3,390,207 (Shell Oil Co.) (1968); *C.A.* **69,** 36895j (1968).

A47. Netherlands Patent Appl. 6,405,416 (Shell Internationale Research Maatschappij N.V.) (1964); *C.A.* **62,** 10648e (1965).

A48. Neverthelands Patent Appl. 6,513,888 (Shell Internationale Research Maatschappij N.V.) (1966); *C.A.* **65,** 10692c (1966).

A49. Netherlands Patent Appl. 6,603,846 (Shell Internationale Research Maatschappij N.V.) (1966).

A50. British Patent 1,121,978 (Polymer Corp. Ltd.) (1968); *C.A.* **69,** 78286s (1968).

A51. Netherlands Appl. Patent 69/09,189 (Shell Internationale Research Mattschappij N.V.) (1969); *C.A.* **72,** 91154d (1970).

A52. De la Mare, H. E., and Bullard, E. F., German Offen. 1,905,422 (Shell Internationale Research Maatschappij N.V.) (1969); *C.A.* **71,** 92470v (1969).

A53. French Patent 1,534,150 (Badische Anilin-und Soda-Fabrik A.-G.) (1968); *C.A.* **71,** 13978r (1969).

A54. Belgian Patent 671,460 (Shell Internationale Research Maatshappij N.V.) (1966).

A55. British Patent 1,014,999 (Shell Internationale Research Maatschappij N.V.) (1966).

A56. French Patent 1,418,831 (Shell Internationale Research Maatschappij N.V.) (1965).

A57. De la Mare, H. E., and Wilcoxen, C. H., Jr., German Offen. 1,940,278 (Shell Internationale Research Maatschappij N.V.) (1970); *C.A.* **72,** 122643u (1970).

A58. Pavelich, W. A., and Livigni, R. A., *J. Polym. Sci., Part C* **21,** 215 (1967); *C.A.* **68,** 96154f (1968).

A59. Bushuk, W., and Benoit, H., *Can. J. Chem.* **36,** 1616 (1958).

A60. Utracki, L. A., Simha, R., and Fetters, L. J., *J. Polym. Sci., Part A-2* **6,** 2051 (1968).

A61. Utracki, L. A., and Simha, R., *Polym. Prepr., Am. Chem. Soc., Div. Polym. Chem.* **9**(1), 742 (1968); *C.A.* **71,** 125083b (1969).

A62. Bresler, S. E., Pyrkov, L. M., and Frenkel, S. Ya., *Vysokomol. Soedin.* **2**(2), 216 (1960).

A63. Bresler, S. Ye., Korotkov, A. A., Mosebitskii, M. I., and Poddubnyl, I. Ya., *Zh. Tekh. Fiz.* **28,** 114 (1958).

A64. Bresler, S. E., Pyrkov, L. M., Frenkel, S. Ya., Laius, L. A., and Klenin, S. I. *Vysokomol. Soedin.* **4,** 250 (1962).

A65. Anderson, J. E., and Liu, K. J., *Macromolecules* **4**(2), 260 (1971); *C.A.* **75,** 21599r (1971).

A66. Chang, F. S. C., *Polym. Prepr., Am. Chem. Soc., Div. Polym. Chem.* **12**(2), 835 (1971).

A67. Molau, G. E., *N.A.S.—N.R.C., Publ.* **1573,** 245 (1968).

A68. Estes, G. M., Cooper, S. L., and Tobolsky, A. V., *J. Macromol. Sci., Rev. Macromol. Chem.* **4**(2), 313 (1970).

A69. Folkes, M. J., and Keller, A., *in* "The Physics of Glassy Polymers" (R. N. Haward, ed.), Chapter 10. Wiley, New York, 1973.

A70. Molau, G. E., ed., "Colloidal and Morphological Behavior of Block and Graft Copolymers." Plenum, New York, 1971.

A71. Bever, M. B., and Shen, M., *Mater. Sci. Eng.* **15,** 145 (1974).

A72. Bradford, E. B., and McKeever, L. D., *Prog. Polym. Sci.* 3, 109 (1971).

A73. Dawkins, J. V., *in* "Block Copolymers" (D. C. Allport and W. H. Janes, eds.), p. 363. Wiley, New York, 1973.

A74. Laflair, R. J., *IUPAC Meet.* 8, 195 (1971).

A75. Robeson, L. M., Matzner, M., Fetters, L. J., and McGrath, J. E., *in* "Recent Advances in Polymer Blends, Grafts and Blocks" (L. H. Sperling, ed.), p. 281. Plenum, New York, 1974.

A76. Morton, M., and Fetters, L. J., German Offen. 2,042,624 (Goodyear Tire and Rubber Co.) (1971); *C.A.* **75**, 7103t (1971).

A77. Uraneck, C. A., and Smith, R. L., U.S. Patent 3,755,273 (Phillips Petroleum Co.) (1973); *C.A.* **80**, 4093g (1974).

A78. Meier, D. J., *J. Polym. Sci., Part C* **26**, 81 (1967); *C.A.* **70**, 107049v (1969).

A79. Meier, D. J., *Polym. Prepr., Am. Chem. Soc., Div. Polym. Chem.* **11**(2), 400 (1970).

A80. Meier, D. J., *in* "Block and Graft Copolymers" (J. J. Burke and V. Weiss, eds.), p. 105. Syracuse Univ. Press, Syracuse, New York, 1973.

A81. Krause, S., *Polym. Prepr., Am. Chem. Soc., Div. Polym. Chem.* **11**(2), 568 (1970).

A82. Krause, S., *in* "Block and Graft Copolymers" (J. J. Burke and V. Weiss, eds.), p. 143. Syracuse Univ. Press, Syracuse, New York, 1973.

A83. Fedors, R. F., *J. Polym. Sci., Part C* **26**, 189 (1969).

A84. McIntyre, D., Rounds, N., and Campos-Lopez, E. *Polym. Prepr., Am. Chem. Soc., Div. Polym. Chem.* **10**(2) 531 (1969).

A85. McIntyre, D., and Campos-Lopez, E., *Macromolecules* 3(3), 322 (1970); *C.A.* **73**, 46406x (1970).

A86. Campos-Lopez, E., McIntyre, D., and Fetters, L. J., *Macromolecules* 6(3), 415 (1973); *C.A.* **79**, 43186p (1973).

A87. Helfand, E., *Macromolecules* 5(3), 301 (1971); *C.A.* **77**, 75631e (1972).

A88. Helfand, E., *in* "Recent Advances in Polymer Blends, Grafts and Blocks" (L. H. Sperling, ed.), p. 141. Plenum, New York, 1974.

A89. Holden, G., Bishop, E. T., and Legge, N. R., *J. Polym. Sci., Part C* **26**, 37 (1969).

A90. Holden, G., *in* "Block and Graft Copolymerization" (R. J. Ceresa, ed.), Vol. 1, p. 133. Wiley, New York, 1973.

A91. Hendus, H., Illers, K. D., and Ropte, E., *Kolloid Z. & Z. Polym.* **216/217**, 110 (1967); *C.A.* **67**, 44211g (1967).

A92. Chang, F. S. C., *J. Chromatogr.* 55(1), 67 (1971); *C.A.* 74, 126198s (1971).

A93. Matsuo, M., Ueno, T., Horino, H., Chujyo, S., and Asai, H., *Polymer* 9(8), 425 (1968); *C.A.* **69**, 87820p (1968).

A94. Matsuo, M., Sagae, S., and Jyo, Y., *J. Electron Microsc.* 17(4), 309 (1968); *C.A.* **71**, 30940q (1969).

A95. Morton, M., McGrath, J. E., and Juliano, P. C., Presented at the *Am. Chem. Soc., Rubber Div. Meet.*, 1967.

A96. Kato, K., *Polym. Lett.* 4, 35 (1966).

A97. Matsuo, M., Ueno, T., Horino, H., Chujyo, S., and Asai, H., *Polymer* 9, 425 (1968).

A98. Enomoto, S., and Wada, H., *Kobunshi Kagaku* 26(294), 673 (1969); *C.A.* 72, 67307k (1970).

A99. Bradford, E. B., and Vanzo, E., *J. Polym. Sci., Part A-1* 6(6), 1661 (1968); *C.A.* **69**, 3264z (1968).

A100. Fahrbach, G., Bronstert, K., Illers, K. H., Ladenberger, V., Simak, P., and Wittmer, P., German Offen. 2,111,966 (Badische Anilin- und Soda-Fabrik A.-G.) (1972); *C.A.* **78**, 4755h (1973).

A101. Price, C., Watson, A. G., and Chow, M. T., *Polymer* 13, 333 (1972).

A102. Uchida, T., Soen, T., Inoue, T., and Kawai, H., *J. Polym. Sci., Part A-2* **10**, 101 (1972).

A103. Fischer, E., *J. Macromol. Sci., Chem.* **2**(6), 1285 (1968); *C.A.* **70**, 20791g (1969).

A104. Moacanin, J., Holden, G., and Tschoegl, N. W., eds., "Block Copolymers," Wiley (Interscience), New York, 1969.

A105. Beecher, J. F., Marker, L., Bradford, R. D., and Aggarwal, S. L., *Polym. Prepr., Am. Chem. Soc., Div. Polym. Chem.* **8**(2), 1532 (1967); *C.A.* **70**, 107298a (1969).

A106. Matsuo, M., Nozaki, C., and Iyo, Y., *J. Electron Microsc.* **17**(1), 7 (1968); *C.A.* **69**, 36876d (1968).

A107. Marker, L., *Polym. Prepr., Am. Chem. Soc., Div. Polym. Chem.* **10**(2), 524 (1969).

A108. Matsuo, M., Sagae, S., and Asai, H., *Polymer* **10**(2), 79 (1969); *C.A.* **71**, 30941r (1969).

A109. Molau, G. E., *Polym. Prepr., Am. Chem. Soc., Div. Polym. Chem.* **10**(2), 700 (1969).

A110. Kaempf, G., Hoffmann, M., and Kroemer, H., *Ber. Bunsenges. Phys. Chem.* **74**(8-9), 851 (1970); *C.A.* **74**, 4393f (1971).

A111. Douy, A., and Gallot, B., *Makromol. Chem.* **156**, 81 (1972); *C.A.* **77**, 76361d (1972).

A112. Gallot, B., and Douy, A., *Semin. Chim. Etat Solide* **5**, 13 (1971); *C.A.* **77**, 49020k (1972).

A113. Krause, S., *Macrmolecules* **3**(1), 84 (1970); *C.A.* **72**, 79512m (1970).

A114. Lewis, P. R., and Price, C., *Polymer* **12**, 258 (1971).

A115. Wilkes, G. L., and Stein, R. S., *J. Polym. Sci., Part A-2* **7**(9), 1525 (1969); *C.A.* **71**, 125296y (1969).

A116. Miyamoto, T., Kodama, K., and Shibayama, K., *J. Polym. Sci., Part A-2* **8**, 2095 (1970).

A117. Lewis, P. R., and Price, C., *Nature (London)* **223**(5205), 494 (1969); *C.A.* **71**, 92009b (1969).

A118. Hoffmann, M., Kampf, G., Kromer, H., and Pampus, G., *Adv. Chem. Ser.* **99**, 351 (1971); *C.A.* **75**, 152725b (1971).

A119. Schwab, F. C., Ph.D. Thesis, University of Akron, Akron, Ohio (1970).

A120. Bradford, E. B., *Polym. Prepr., Am. Chem. Soc., Div. Polym. Chem.*, **11**(2), 392 (1970).

A121. Bianchi, U., Pedemonte, E., and Turturro, A., *J. Polym. Sci., Part B* **7**(11), 785 (1969); *C.A.* **72**, 4158s (1970).

A122. Leary, D. F., and Williams, M. C., *J. Polym. Sci., Part B* **8**(5), 335 (1970); *C.A.* **72**, 88359q (1970).

A123. Bianchi, U., Pedemonte, E., and Turturro, A., *Polymer* **11**(5), 268 (1970); *C.A.* **73**, 35988n (1970).

A124. Leary, D. F., and Williams, M. C., *J. Polym. Sci., Polym. Phys. Ed.* **11**(2), 345 (1973); *C.A.* **78**, 125035g (1973).

A125. Fielding-Russell, G. S., *Rubber Chem. Technol.* **45**(1), 252 (1972); *C.A.* **77**, 63066m (1972).

A126. Dlugosz, J., Folkes, M. J., and Keller, A., *J. Polym. Sci., Polymer Phys. Ed.* **11**, 929 (1973); also see *Polymer* **12**, 793 (1971).

A127. Keller, A., Pedemonte, E., and Willmouth, F. M., *Kolloid Z. & Z. Polym.* **238**(1-2), 385 (1970); *C.A.* **73**, 66983m (1970).

A128. Price, C., and Lewis, P. R., *Lab. Pract.* **19**(6), 599 (1970); *C.A.* **73**, 78296c (1970).

A129. Dlugosz, J., Keller, A., and Pedemonte, E., *Kolloid Z. & Z. Polym.* **242**, 1125 (1970).

A130. Pedemonte, E., Turturro, A., Bianchi, U., and Devetta, P., *Polymer* **14**, 145 (1973).

A131. Douy, A., and Gallot, B. R., *Mol. Cryst. Liq. Cryst.* **14**(3-4), 191 (1971); *C.A.* **76**, 15103d (1972).

A132. Douy, A., and Gallot, B., *C.R. Hebd. Seances Acad. Sci., Ser. C* **272**(17), 1478 (1971); *C.A.* **75**, 36870w (1971).

A133. Kroemer, H., Hoffmann, M., and Kaempf, G., *Ber. Bunsenges. Phys. Chem.* **74**(8–9), 859 (1970); *C.A.* **74**, 4435w (1971).

A134. Merrett, F. M., *Trans. Faraday Soc.* **50**, 759 (1954).

A135. Kaempf, G., Kroemer, H., and Hoffmann, M., *Kolloid Z. & Z. Polym.* **247**(102), 820 (1971); *C.A.* **76**, 113698w (1972).

A136. Brown, D. S., Fulcher, K. U., and Wetton, R. E., *Polym. Lett.* **8**, 659 (1970).

A137. Kim, H.-G., *Macromolecules* **5**, 594 (1972).

A138. LeGrand, D. G., *J. Polym. Sci., Part B* **8**(3), 195 (1970); *C.A.* **73**, 15604f (1970).

A139. McIntyre, D., and Campos-Lopez, E., *in* "Block Polymers" (S. L. Aggarwal, ed.), p. 19. Plenum, New York, 1970.

A140. Fetters, L. J., Meyer, B. H., and McIntyre, D., *J. Appl. Polym. Sci.* **16**, 2079 (1972).

A141. Skoulios, A., *Macromolecules* **4**, 268 (1971).

A142. Krigbaum, W. R., Yazgan, S., and Tolbert, W. R., *J. Polym. Sci. Polym. Phys. Ed.* **11**, 511 (1973).

A143. Keller, A., Pedemonte, E., and Willmouth, F. M., *Nature (London)* **225**, 538 (1970); *C.A.* **72**, 101611s (1970).

A144. Folkes, M. J., Keller, A., and Scalisi, F. P., *Kolloid Z. & Z. Polym.* **251**, 1 (1973).

A145. Keller, A., Dlugosz, J., Folkes, M. J., Pedemonte, E., Scalisi, F. P., and Willmouth, F. M., *J. Phys. (Paris), Colloq.* (**5**), 295 (1971); *C.A.* **77**, 49703k (1972).

A146. Arridge, R. G. C., and Folkes, M. J., *J. Phys. D* **5**(2), 344 (1972); *C.A.* **76**, 142081v (1972).

A147. Folkes, M. J., Keller, A., and Scalisi, F. P., *Polymer* **12**(12), 793 (1971); *C.A.* **76**, 113716a (1972).

A148. Harpell, G. A., and Wilkes, C. E., *in* "Block Polymers" (S. L. Aggarwal, ed.), p. 31. Plenum, New York, 1970.

A149. Hoffmann, M., Kaempf, G., Kroemer, H., and Pampus, G., German Offen. 2,112,142 (Farbenfabriken Bayer A.-G.) (1972); *C.A.* **78**, 59394v (1973).

A150. Pedemonte, E., Turturro, A., Bianchi, U., and Devetta, P., *Polymer* **14**(4), 145 (1973); *C.A.* **78**, 160407b (1973).

A151. Pedemonte, E., Turturro, A., Bianchi, U., and Devetta, P., *Chim. Ind. (Milan)* **54**(8), 689 (1972); *C.A.* **77**, 141170g (1972).

A152. Price, C., Watson, A. G., and Chow, M. T., *Polymer* **13**(7), 333 (1972); *C.A.* **78**, 73329j (1973).

A153. Fischer, E., and Henderson, J. F., *Rubber Chem. Technol.* **40**, 1373 (1967); *C.A.* **68**, 69965q (1968).

A154. Wilkes, G. L., Fukuda, M., and Stein, R. S., *Polym. Prepr., Am. Chem. Soc., Div. Polym. Chem. Polymer* **10**(2), 694 (1969).

A155. Folkes, M. J., and Keller, A., *Polymer* **12**, 222 (1971).

A156. Wilkes, G. L., *J. Polym. Sci., Part A-2* **10**, 767 (1972).

A157. Stein, R. S., *Polym. Lett.* **9**, 747 (1971).

A158. Nishioka, A., Furukawa, J., Yamashita, S., and Kotani, T., *J. Appl. Polym. Sci* **14**(3), 799 (1970); *C.A.* **72**, 122607k (1970).

A159. Pishareva, E. P., Erenburg, E. G., and Poddubnyi, I. Y., *Dokl. Akad. Nauk SSSR* **189**(3), 571 (1969); *C.A.* **72**, 79582j (1970).

A160. Kraus, G., and Gruver, J. T., *J. Appl. Polym. Sci.* **11**(11), 2121 (1967); *C.A.* **68**, 40715x (1968).

A161. Holden, G., Bishop, E. T., and Legge, N. R., *J. Polym. Sci., Part C* **26**, 37 (1969); *C.A.* **70**, 116020f (1969).

A162. Holden, G., Bishop, E. T., and Legge, N. R., *Proc. Int. Rubber Conf., 5th, 1967* (1968).

A163. Kraus, G., Naylor, F. E., and Rollmann, K. W., *J. Polym. Sci., Part A-2* **9**, 1839 (1971).

A164. Arnold, K. R., and Meier, D. J., *J. Appl. Polym. Sci.* **14**(2), 427 (1970); *C.A.* **72**, 79736n (1970).

A165. Kotaka, T., and White, J. L., *Trans. Soc. Rheol.* **17**(4), 587 (1973); *C.A.* **80**, 108972j (1974).

A166. Minor, H. B., Shaw, A. W., and Wilcoxen, C. H., Jr., U.S. Patent 3,507,934 (Shell Oil Co.) (1970).

A167. Sakanishi, A., and Tanaka, H., *J. Phys. Soc. Jpn.* **24**(1), 222 (1968); *C.A.* **68**, 79365h (1968).

A168. Paul, D. R., St. Lawrence, J. E., and Troell, J. H., *Polym. Eng. Sci.* **10**(2), 70 (1970); *C.A.* **72**, 122584a (1970).

A169. Ikeda, R. M., Wallach, M. L., and Angelo, R. J., *in* "Block Polymers" (S. L. Aggarwal, ed.), p. 43. Plenum, New York, 1970.

A170. Kraus, G., Childers, C. W., and Gruver, J. T., *J. Appl. Polym. Sci.* **11**, 1581 (1967).

A171. Canter, N. H., *J. Polym. Sci., Part A-2* **6**(1), 155 (1968); *C.A.* **68**, 79347d (1968).

A172. Cooper, S. L., and Tobolsky, A. V., *Text. Res. J.* Sept., 1966, p. 802.

A173. Kaelble, D. H., and Cirlin, E. H., *J. Polym. Sci., Polym. Symp.* **43**, 131 (1973).

A174. Kaelble, D. H., *Trans. Soc. Rheol.* **15**(2), 235 (1971).

A175. Shen, M., Cirlin, E. H., and Kaelble, D. H., *Polym. Prepr., Am. Chem. Soc., Div. Polym. Chem.* **11**(2), 686 (1970).

A176. Kaelble, D. H., Cirlin, E. H., and Shen, M., *Polym. Prepr., Am. Chem. Soc., Div. Polym. Chem.* **11**(2), 676 (1970).

A177. Kaelble, D. H., Cirlin, E. H., and Shen, M., *in* "Colloidal and Morphological Behavior of Block and Graft Copolymers" (G. E. Molau, ed.), p. 295. Plenum, New York, 1971.

A178. Kaelble, D. H., Cirlin, E. H., and Shen, M., *in* "Colloidal and Morphological Behavior of Block and Graft Copolymers" (G. E. Molau, ed.), p. 307. Plenum, New York, 1971.

A179. Kaya, A., Choi, G., and Shen, M., *U.S. N.T.I.S., AD Rep.* **AD-748213** (1972); *C.A.* **78**, 58897z (1973).

A180. Shen, M. C., and Kaelble, D. H., *J. Polym. Sci., Part B* **8**(3), 149 (1970); *C.A.* **73**, 15993g (1970).

A181. Smith, T. L., and Dickie, R. A., *J. Polym. Sci., Part C* **26**, 163 (1967); *C.A.* **70**, 107272n (1969).

A182. Nishioka, A., Furukawa, J., and Yamashita, S., *Kogyo Kagaku Zasshi* **72**(11), 2440 (1969); *C.A.* **72**, 122608m (1970).

A183. Fesko, D. G., and Tschoegl, N. W., *Int. J. Polym. Mater.* **3**, 51 (1974).

A184. Harpell, G. A., and Thrasher, D. B., *Polym. Prepr., Am. Chem. Soc., Div. Polym. Chem.* **11**(2), 888 (1970).

A185. Morton, M., Tschoegl, N. W., and Froelich, D., *U.S. C.F.S.T.I., AD Rep.* **AD-701736**; from *U.S. Gov. Res. & Dev. Rep.* **70**(8), 114 (1970); *C.A.* **73**, 56654r (1970).

A186. Murakami, K., Tamura, S., Shiina, K., and Ueno, T., *Kogyo Kagaku Zasshi* **73**(7), 1602 (1970); *C.A.* **73**, 110269j (1970).

A187. Pezzin, G., and Ajroldi, G., *Atti Congr. Int. Mater. Plast. Elastomeriche, 1968* 07-10 p. 31 (1969); *C.A.* **74**, 3877e (1971).

A188. Uematsu, I., Takagi, K., Moriguchi, T., Sukeiima, H., and Fukuzawa, T., Asahi, *Garasu Kogyo Gijutsu Shoreikai Kenkyu Hokoku* **15**, 593 (1969); *C.A.* **73**, 88309y (1970).

A189. Smith, T. L., *in* "Rheology: Theory and Applications" (F. R. Eirich, ed.), Vol. 5, p. 127. Academic Press, New York, 1970.

A190. Shibayama, K., *Kogyo Kagaku Zasshi* **73**(7), 1268 (1970); *C.A.* **73**, 99209s (1970).

A191. Cooper, S. L., *Diss. Abstr. B* **28**(5), 1894 (1967); *C.A.* **68**, 69960 (1968).

A192. Kraus, G., Rollmann, K. W., and Gardner, J. O., *J. Polym. Sci., Polym. Phys. Ed.* **10**, 2061 (1972).

A193. Turturro, A., Bianchi, U., Pedemonte, E., and Ravetta, P., *Chim. Ind. (Milan)* **54**(9), 782 (1972); *C.A.* **78**, 17304k (1973).

A194. Sokolova, N. F., Kondrat'ev, A. N., Safonova, V. P., and Novichikhina, A. B., *Kauch. Rezina* **2**, 20 (1974); *C.A.* **81**, 4609a (1974).

A195. Choi, G., Kaya, A., and Shen, M., *Polym. Eng. Sci.* **13**(3), 231 (1973); *C.A.* **79**, 54601w (1973).

A196. Shen, M., Kaniskin, V. A., Biliyar, K., and Boyd, R. H., *J. Polym. Sci., Polym. Phys. Ed.* **11**(11), 2261 (1973); *C.A.* **80**, 48625m (1974).

A197. Schwab, F. C., Ph.D. Thesis, University of Akron, Akron, Ohio (1970).

A198. Bishop, E. T., and Davison, S., *J. Polym. Sci., Part C* **26**, 59 (1969); *C.A.* **70**, 116024k (1969).

A199. Luftglass, M. A., Hendricks, W. R., Holden, G., and Bailey, J. T., *Annu. Tech. Conf., Soc. Plast. Eng.* (1966).

A200. Bailey, J. T., Bishop, E. T., Hendricks, W. R., Holden, G., and Legge, N. R., *Rubber Age* **98**, 69 (1966).

A201. Morton, M., Fetters, L. J., Schwab, F. C., Strauss, C. R., and Kammereck, R., *Int. Synth. Rubber Symp., Lect., 4th, 1969* Vol. 3, p. 70 (1969); *C.A.* **73**, 131791p (1970).

A202. Smith, T. L., *Polym. Prepr., Am. Chem. Soc., Div. Polym. Chem.* **10**(2), 670 (1969).

A203. Cooper, S. L., Hugh, D. S., and Morris, W. J., *Ind. Eng. Chem., Prod. Res. Dev.* **7**(4), 248 (1968); *C.A.* **70**, 20809u (1969).

A204. Netherlands Patent Appl. 6,600,172 (Shell Internationale Research Maatschappij N.V.) (1966); *C.A.* **65**, 20335b (1966).

A205. Cooper, S. L., and Tobolsky, A. V., *J. Appl. Polym. Sci.* **11**, 1361 (1967).

A206. Henderson, J. F., Grundy, K. H., and Fischer, E., *J. Polym. Sci., Part C* **16**, Pt. 6, 3121 (1968); *C.A.* **70**, 4761s (1969).

A207. Kovalev, N. F., Korotkov, A. A., Petrov, G. N., Reikh, V. N., Sidorovich, A. E., and Eventova, L. A., *Kauch. Rezina* **26**(12), 6 (1967); *C.A.* **68**, 50791h (1968).

A208. Childers, C. W., and Kraus, G., *Rubber Chem. Technol.* **40**(4), 1183 (1967).

A209. Minekawa, S., Yamaguchi, K., Tomoto, K., Fujimoto, E., and Suzuki, T., U.S. Patent 3,449,179 (Ashi Chemical Industry Co., Ltd.) (1966); *C.A.* **71**, 62686c (1969).

A210. Robinson, R. A., and White, E. F. T., *Polym. Prepr., Am. Chem. Soc., Div. Polym. Chem.* **10**(2), 662 (1969).

A211. Vlig, M., *Kautsch. Gummi, Kunstst.* **22**(4), 183 (1969); *C.A.* **71**, 13957h (1969).

A212. Fischer, E., and Henderson, J. F., *J. Polym. Sci., Part C* **26**, 149 (1969); *C.A.* **70**, 107273p (1969).

A213. Kusamizu, S., *Nippon Gomu Kyokaishi* **42**(2), 117 (1969); *C.A.* **70**, 107304z (1969).

A214. Brunwin, D. M., Fischer, E., and Henderson, J. F., *J. Polym. Sci., Part C* **26**, 135 (1967); *C.A.* **70**, 107286v (1969).

A215. Gent, A. N., and Hirakawa, H., *J. Polym. Sci., Part A-2* **6**, 1481 (1968); *C.A.* **69**, 4462f (1968).

A216. Ambrose, R. J., *Polym. Prepr., Am. Chem. Soc., Div. Polym. Chem.* **11**(2), 880 (1970).

A217. Montiel, R., Kuo, C., and McIntyre, D., *Polym. Prepr., Am. Chem. Soc., Div. Polym. Chem.* **15**(2), 169 (1974).

A218. Smith, R. C., French Demande 2,175,782 (Firestone Tire and Rubber Co.) (1973); *C.A.* **80**, 121731z (1974).

A219. Trukenbrod, K., and Weber, G., *Chim. & Ind., Genie Chim.* **104**(10), 1271 (1971); *C.A.* **75**, 99133d (1971).

A220. Trukenbrod, K., Weber, H., and Dasch, J., German Offen. 2,026,308 (Chemische Werke Huels A.-G.) (1971); *C.A.* **76**, 114136y (1972).

A221. Horiie, S., Kurematsu, S., and Asai, S., German Offen. 2,120,232 (Electro Chemical Industrial Co., Ltd.) (1971); *C.A.* **76**, 114502q (1972).

A222. Angelo, R. J., Ikeda, R. M., and Wallach, M. L., *Am. Chem. Soc., Div. Org. Coat. Plast. Chem., Preprints* **34**(2), 315 (1974).

A223. Fetters, L. J., and McGrath, J. E., unpublished results.

A224. British Patent 1,028,357 (Shell Int. Res. Mij. N.V.) (1966).

A225. Bailey, J. T., and Nyberg, D. D., Belgian Patent 637,822 (Shell Internationale Research Maatschappij N.V.) (1964); *C.A.* **62**, 7976a (1965).

A226. Netherlands Patent Appl. 6,412,809 (Shell Internationale Research Maatschappij N.V. (1965); *C.A.* **63**, 16578d (1965).

A227. Pritchard, J., and Hammersley, I. C., S. African Patent 67/05,966 (Dunlop Co. Ltd.) (1968); *C.A.* **70**, 29696n (1969).

A228. Pritchard, J., and Hammersley, I. C., S. African Patent 67/05,967 (Dunlop Co. Ltd.) (1968); *C.A.* **70**, 58705v (1969).

A229. Harlan, J. T., Jr., U.S. Patent 3,239,478 (Shell Oil Co.) (1966).

A230. Davis. F. C., Luther, W. B., and Martinson, D. L., British Patent 1,116,426 (Shell Internationale Research Maatschappij N.V.) (1968); *C.A.* **69**, 36811d (1968).

A231. Osborne, A. P., British Patent 1,145,861 (Dunlop Co. Ltd.) (1969); *C.A.* **70**, 97822a (1969).

A232. Korpman, R., French Demande 2,006,161 (Johnson and Johnson) (1969); *C.A.* **73**, 4768m (1970).

A233. Russell, T. E., German Offen. 1,931,562 (Flintkote Co.) (1970); *C.A.* **73**, 4513z (1970).

A234. French Patent 1,536,608 (Dunlop Co.) (1968); *C.A.* **71**, 31302y (1969).

A235. Netherlands Patent Appl. 68,09,136 (Shell Internationale Research Maatschappij N.V.) (1968); *C.A.* **70**, 97812x (1969).

A236. Belgian Patent 664,284 (Shell Internationale Research Maatschappij N.V.) (1965).

A237. Netherlands Patent Appl. 6,600,913 (Shell Internationale Research Maatschappij N.V.) (1966); *C.A.* **66**, 11627y (1967).

A238. Netherlands Patent Appl. 6,600,173 (Shell Internationale Research Maatschappij N.V.) (1966); *C.A.* **66**, 11675n (1967).

A239. Martin, J. W., U.S. Patent 3,477,148 (Shell Oil Co.) (1969); *C.A.* **72**, 32922e (1970).

A240. Holden, G., and Luther, W. B., U.S. Patent 3,503,143 (Shell Oil Co.) (1970); *C.A.* **72**, 112449w (1970).

A241. British Patent 1,112,836 (Polymer Corp. Ltd.) (1968); *C.A.* **69**, 11150z (1968).

A242. Grasley, M. H., U.S. Patent 3,325,430 (Shell Oil Co.) (1967); *C.A.* **67**, 55236v (1967).

A243. Berry, P. M., and Kavalir, J. J., S. African Patent 68/01,873 (Uniroyal, Inc.) (1968); *C.A.* **70**, 69186c (1969).

A244. Rot, A. N., and Kavalir, J. J., French Demande 2,002,860 (Uniroyal, Inc.) (1969); *C.A.* **72**, 80140v (1970).

A245. Bailey, J. T., *J. Elastoplast.* **1**, 2 (1969); *C.A.* **71**, 22747s (1969).

A246. British Patent 1,097,150 (Shell Interntionale Research Maatschappij N.V.) (1967); *C.A.* **68**, 41091c (1968).

A247. Henten, K. U., and De Jager, D., British Patent 1,347,655 (Shell Internationale Research Maatschappij B.V.) (1974); *C.A.* **81**, 4784d (1974).

A248. Harlan, J. T., Jr., U.S. Patent 3,792,005 (Shell Oil Co.) (1974); *C.A.* **80**, 147066m (1974).

A249. Van der Bie, G. J., and Vlig, M., *Rev. Plast. Mod.* **20**(156), 483 (1969); *C.A.* **71**, 92447t (1969).

A250. Baumann, G. F., *Adv. Chem. Ser.* **96**, 30 (1969); *C.A.* **72**, 3946k (1970).

A251. Kimura, H., *Kagaku Kogyo* **20**(9), 1004 (1969); *C.A.* **72**, 4146m (1970).

A252. Vrazel, M., *Kozarstvi* **18**(4), 107 (1968); *C.A.* **69**, 20161f (1968).

A253. Van Breen, A. W., and Vlig, M., *Rubber Plast. Age* **47**(10), 1070 (1966). *C.A.* **65**, 20329f (1966).

A254. Van Breen, A. W., and Vlig, M., *Rev. Belge Matieres Plast.* **9**(7), 459 (1968); *C.A.* **70**, 20774d (1969).

A255. Giacchero, A., and Goretti, G., *Ind. Gomma* **11**(9), 39 (1967); *C.A.* **68**, 22594c (1968).

A256. Netherlands Patent Appl. 6,613,391 (Shell) (1967); *C.A.* **67**, 109499d (1967).

A257. Landau, L., German Offen. 1,902,228 (Phillips Patents Ltd.) (1969); *C.A.* **71**, 102932p (1969).

A258. Holden, G., German Offen. 1,913,683 (Shell Internationale Research Maatschappij N.V.) (1969); *C.A.* **71**, 125761w (1969).

A259. Toyomoto, K., Ibata, J., Suzuoki, K., and Uda, B., Japanese Patent (Kokai) 73/97,945 (Asahi Chemical Industry Co., Ltd.) (1973); *C.A.* **80**, 122160z (1974).

A260. Netherlands Patent Appl. 6,406,862 (Shell Internationale Research Maatschappij N.V.) (1964).

A261. Kambour, R. P., German Offen. 2,000,118 (General Electric Co.) (1970); *C.A.* **73**, 88592k (1970).

A262. Netherlands Patent Appl. 68/10,062 (Shell Internationale Research Maatschappij N.V.) (1969); *C.A.* **70**, 97802u (1969).

A263. Netherlands Patent Appl. 68/01,747 (Shell Internationale Research Maatschappij N.V.) (1968); *C.A.* **70**, 12381h (1969).

A264. Netherlands Patent Appl. 67/13,344 (Shell Internationle Research Maatschappij N.V.) (1968); *C.A.* **69**, 20517b (1968).

A265. Netherlands Patent Appl. 67/06, 408 (Shell Internationale Research Maatschappij N.V.) (1968); *C.A.* **70**, 88557r (1969).

A266. Van Breen, A. W., and Vlig, M., British Patent 1,166,799 (Shell Internationale Research Maatschappij N.V.) (1969); *C.A.* **71**, 125753v (1969).

A267. Ko, K., *Kobunshi* **15**(169), 340 (1966); *C.A.* **69**, 67940z (1968).

A268. Satake, K., and Maeda, K., *Nippon Gomu Kyokaishi* **42**(7), 506 (1969); *C.A.* **72**, 56447e (1970).

A269. Cohen, R. E., and Tschoegl, N. W., *Int. J. Polym. Mater.* **3**, 3 (1974).

A270. Luftglass, M. A., and Hendricks, W. R., U.S. Patent 3,459,831 (Shell Oil Co.) (1969).

A271. Legge, N. R., and Snyder, J. L., U.S. Patent 3,459,830 (Shell Oil Co.) (1969).

A272. Van Beem, E. J., and Knibbe, D. E., German Offen. 1,944,337 (Shell Internationale Research Maatschappij N.V.) (1970); *C.A.* **73**, 16461u (1970).

A273. Durst, R. R., Griffith, R. M., Urbanic, A. J., and Van Essen, W. J., *Am. Chem. Soc., Div. Org. Coat. Plast. Chem.*, **34**(2), 320 (1974).

A274. Netherlands Patent Appl. 6,500,632 (Shell Internationale Research Maatschappij N.V.) (1965); *C.A.* **63**, 18368e (1965).

A275. Childers, C. W., Kraus, G., Gruver, J. T., and Clark, E., *Polym. Prepr., Am. Chem. Soc., Div. Polym. Chem.* **11**(2), 553 (1970).

A276. Satake, K., and Maeda, K., *Nippon Gomu Kyokaishi* **42**(7), 500 (1969); *C.A.* **72**, 56449g (1970).

A277. British Patent 1,033,113 (Shell Internationale Research Maatschappij N.V.) (1966).

A278. Holden, G., and Milkovich, R., U.S. Patent 3,322,856 (Shell Oil Co.) (1967); *C.A.* **67**, 33637z (1967).

A279. LaFlair, R. T., and Henderson, J. F., Canadian Patent 786,864 (Polymer Corp. Ltd.) (1968); *C.A.* **69**, 44484q (1968).

A280. Harrison, S. R., Cooper, W., and Bennett, J. V., British Patent 1,147,650 (Dunlop Co. Ltd.) (1969); *C.A.* **70**, 107677c (1969).

A281. Netherlands Patent Appl. 6,601,741 (Shell Oil Co.) (1967); *C.A.* **67**, 109474s (1967).

A282. Darcy, J., and Palmer, P. G., French Patent 1,545,402 (Polymer Corp. Ltd.) (1968); *C.A.* **71**, 22570d (1969).

A283. Skendrovich, S. G., and Bauer, J. N., U.S. Patent 3,536,784 (Koppers Co., Inc.) (1970); *C.A.* **74**, 4209a (1971).

A284. Netherlands Patent Appl. 6,407,182 (Shell Internationale Research Maatschappij N.V.) (1964); *C.A.* **63**, 5874e (1965).

A285. Durst, R. R., German Offen. 2,219,756 (General Tire and Rubber Co.) (1973); *C.A.* **80**, 48698n (1974).

A286. Kraton Product Literature, Shell Chem. Co., Houston, Texas (1975).

A287. Golub, M. A., *in* "The Chemistry of Alkenes" (J. Zabicky, ed.), Vol. 2, Chapter 9, p. 411. Wiley (Interscience), New York, 1970.

A288. Mann, R. H., and Bailey, J. T., U.S. Patent 3,471,431 (Shell Oil Co.) (1969); *C.A.* **71**, 125736s (1969).

A289. Winkler, D. E., U.S. Patent 3,577,357 (Shell Oil Co.) (1971); *C.A.* **75**, 37250f (1971).

A290. Lopatin, G., Newey, H. A., Bishop, E. T., O'Neill, W. P., and Krewinghaus, A. B., *U.S. Off. Saline Water, Res. Dev. Prog. Rep.* **690** (1971); *C.A.* **76**, 34961y (1972).

A291. O'Neill, W. P., and Turner, W. V., Jr., U.S. Patent 3,642,953 (U.S. Dept. of Health, Education and Welfare (1972); *C.A.* **76**, 155271g (1972).

A292. Lopatin, G., Newey, H. A., Bishop, E. T., O'Neill, W. P., and Krewinghaus, A. B., *U.S. Off. Saline Water, Res. Dev. Prog. Rep.* **694** (1971).

A293. Winkler, D. E., U.S. Patent 3,607,979 (Shell Oil Co.) (1971); *C.A.* **76**, 15363p (1972).

A294. Shaw, A. W., U.S. Patent 3,694,521 (Shell Oil Co.) (1972); *C.A.* **78**, 30737y (1973).

A295. Johnson, O., Belgian Patent 660,829 (Shell Internationale Research Maatschappij N.V.) (1965); *C.A.* **64**, 11392c (1966).

A296. Netherlands Patent Appl. 6,404,532 (Shell Internationale Research Maatschappij N.V.) (1964); *C.A.* **62**, 6660d (1965).

A297. Netherlands Patent Appl. 6,500,703 (Shell Internationale Research Maatschappij N.V.) (1965); *C.A.* **63**, 18428e (1965).

A298. Porter, L. M., British Patent 1,053,596 (Shell Internationale Research Maatschappij N.V.) (1967); *C.A.* **66**, 66110e (1967).

A299. British Patent 1,028,388 (Boehme Fettchemie G.m.b.H.) (1966).

A300. French Patent 1,428,012 (Shell Internationale Research Maatschappij N.V.) (1966).

A301. British Patent 1,020,709 (Shell Internationale Research Maatschappij N.V.) (1966).

A302. Haefele, W. R., Dallas, C. A., and Deisz, M. A., U.S. Patent 3,485,787 (Shell Oil Co.) (1969); *C.A.* **72**, 80115r (1970).

A303. Lunk, H. E., French Patent 1,483,215 (Shell Internationale Research Maatschappij N.V.) (1967); *C.A.* **68**, 13724t (1968).

A304. Childers, C. W., and Gruver, J. T., U.S. Patent 3,499,949 (Phillips Petroleum Co.) (1967).

A305. Falk, J. C., *J. Polym. Sci., Part A-1* **9**, 2617 (1971).

A306. Hassell, H. L., German Offen. 2,242,190 (Shell Internationale Research Maatschappij N.V.) (1973); *C.A.* **79**, 5881d (1973).

A307. Hassell, H. L., Watson, J. C., and Shaw, B. W., German Offen. 2,244,682 (Shell Internationale Research Maatschappij N.V.) (1973); *C.A.* **78**, 160333z (1973).

A308. Falk, J. C., *Makromol. Chem.* **160**, 291 (1972); *C.A.* **78**, 4727a (1973).

A309. Hassell, H. L., German Offen. 2,045,621 (Shell Internationale Research Maatschappij N.V.) (1971); *C.A.* **74**, 142672x (1971).

A310. Sequeira, R. M., U.S. Patent 3,686,365 (1972); *C.A.* **77**, 153523n (1972).

A311. Hawkins, J. R., and Locke, J. M., German Offen. 2,145,948 (International Synthetic Rubber Co., Ltd.) (1972); *C.A.* **77**, 21181q (1972).

A312. Duck, E. W., Hawkins, J. R., and Locke, J. M., *J. IRI* **6**(1), 19 (1972); *C.A.* **77**, 21133a (1972).

A313. De la Mare, H. E., and Shaw, A. W., U.S. Patent 3,670,054 (Shell Oil Co.) (1972); *C.A.* **77**, 75801k (1972).

A314. De Vault, A. N., U.S. Patent 3,696,088 (Phillips Petroleum Co.) (1972); *C.A.* **78**, 44503f (1973).

A315. Wald, M. M., and Quam, M. G., U.S. Patent 3,700,633 (Shell Oil Co.) (1972); *C.A.* **78**, 17353a (1973).

A316. Hoeg, D. F., Goldberg, E. P., and Pendleton, J. F., U.S. Patent 3,598,886 (Borg-Warner Corp.) (1971).

A317. Bishop, E. T., Haefele, W. R., and Hendricks, W. R., U.S. Patent 3,639,163 (Shell Oil Co.) (1972); *C.A.* **76**, 142009c (1972).

A318. Pendleton, J. F., Hoeg, D. F., and Goldberg, E. P., *Adv. Chem. Ser.* **129**, 27 (1973); *C.A.* **80**, 108976p (1974).

A319. Crossland, R. K., and Holden, G., German Offen. 2,314,136 (Shell Internationale Research Maatschappij N.V.) (1973); *C.A.* **80**, 84505f (1974).

A320. Pendleton, J. F., Hoeg, D. F., and Goldberg, E. P., *Polym. Prepr., Am. Chem. Soc., Div. Polym. Chem.* **13**(1), 427 (1972); *C.A.* **80**, 84054h (1974).

A321. Gillies, G. A., U.S. Patent 3,792,127 (Shell Oil Co.) (1974); *C.A.* **80**, 109589b (1974).

A322. Winkler, D. E., U.S. Patent 3,576,912 (Shell Oil Co.) (1971); *C.A.* **75**, 22163f (1971).

A323. Petit, D., Blackwell, F. B., Carter, A. R. K., and Harlan, J. T., Jr., German Offen. 1,942,462 (Shell Internationale Research Maatschappij N.V.) (1970); *C.A.* **72**, 112547b (1970).

A324. Taylor, G. L., Cotton, J. W., and Winkler, D. E., German Offen. 1,945,846 (Shell Internationale Research Maatschappij N.V.) (1970); *C.A.* **72**, 122400n (1970).

A325. Muller, R. H., and Archer, J. M., British Patent 1,187,358 (Franklin, J. G. and Sons Ltd.) (1970); *C.A.* **73**, 26412t (1970).

A326. Kraus, G., and Short, J. N., U.S. Patent 3,113,912 (Phillips Petroleum Co.) (1963); *C.A.* **60**, 13422f (1964).

A327. Wei, Y.K., Canadian Patent 786,865 (Polymer Corp. Ltd.) (1968); *C.A.* **69**, 4458j (1968).

A328. Hecker, A. C., and Abramoff, C., U.S. Patent 3,520,952 (Argus Chemical Corp.) (1970).

A329. Martin, J. W., U.S. Patent 3,432,578 (Shell Oil Co.) (1969); *C.A.* **70**, 116077e (1969).

A330. Richards, H. F., U.S. Patent 3,459,700 (Shell Oil Co.) (1969); *C.A.* **71**, 71702g (1969).

A331. Winkler, D. E., and Shaw, A. W., U.S. Patent 3,607,982 (Shell Oil Co.) (1971); *C.A.* **76**, 4740h (1972).

A332. Kray, W. C., and Winkler, D. E., U.S. Patent 3,663,659 (U.S. Dept. of Health, Education and Welfare) (1972); *C.A.* **77**, 49825b (1972).

A333. Morton, M., Helminiak, T. E., Gadakary, S. D., and Bueche, F., *J. Polym. Sci.* **57**, 471 (1962).

A334. Roovers, J. E. L., and Bywater, S., *Macromolecules* **5**, 384 (1972).

A335. Roovers, J. E. L., and Bywater, S., *Macromolecules* **7**, 443 (1974).

A336. Gervasi, J. A., and Gosnell, A. B., *J. Polym. Sci., Part A-1* **4**, 1391 (1966).

A337. Gosnell, A. B., Gervasi, J. A., and Schindler, A., *J. Polym. Sci., Part A-1* **4**, 1401 (1966).

A338. Berry, G. C., *J. Polym. Sci., Part A-2* **9**, 687 (1971).

A339. Zimm, B. H., and Kilb, R. W., *J. Polym. Sci.* **37**, 19 (1959).

A340. Stockmayer, W. H., and Fixman, M., *Ann. N.Y. Acad. Sci.* **57**, 334 (1953).

A341. Decker-Freyss, D., and Rempp, P., *C.R. Hebd. Seances Acad. Sci.* **261**, 1977 (1965).

A342. Worsfold, D. J., Zilliox, J. G., and Rempp, P., *Can. J. Chem.* **47**, 3379 (1969).

A343. Kohles, A., Polacek, J., Koessler, L., Zilliox, J. G., and Rempp, P., *Eur. Polym. J.* **8**, 627 (1972).

A344. Bi, K. L., Fetters, L. J., and Morton, M., *Polym. Prepr., Am. Chem. Soc., Div. Polym. Chem.* **15**(2), 157 (1974).

A345. Bi, L. K., and Fetters, L. J., *Macromolecules* **8**(1), 98 (1975); *C.A.* **82**, 112551e (1975); see also, *Macromolecules,* in press (1976).

A346. Marrs, O. L., Zelinski, R. P., and Doss, R. C., presented at the *Am. Chem. Soc., Rubber Div. Meet., 1973.*

A347. Marrs, O. L., Naylor, F. E., and Edmonds, L. O., *J. Adhes.* **4**, 211 (1972).

A348. Marrs, O. L., and Edmonds, L. O., *Adhes. Age* **14**(12), 15 (1971); *C.A.* **76**, 100569e (1972).

A349. Marrs, O. L., U.S. Patent 3,753,936 (Phillips Petroleum Co.) (1973); *C.A.* **79**, 127240q (1973).

A350. Fodor, L. M., U.S. Patent 3,859,250 (Phillips Petroleum Co.) (1975); *C.A.* **82**, 99349y (1975).

A351. Kraus, G., and Stacy, C. J., *J. Polym. Sci., Polym. Symp.* **43**, 329 (1973).

A352. Farrar, R. C., and Rothlisberger, A. C., U.S. Patent 3,855,189 (Phillips Petroleum Co.) (1974); *C.A.* **82**, 87405a (1975).

A353. Drake, W. O., German Offen. 2,019,140 (Phillips Petroleum Co.) (1970); *C.A.* **74**, 23515p (1971).

A354. Haws, J. R., and Middlebrook, T. C., *Rubber World* **167**(4), 27 (1973); *C.A.* **78**, 112434m (1973).

A355. Walker, J. H., and Bailey, F. W., U.S. Patent 3,766,114 (Phillips Petroleum Co.) (1973); *C.A.* **80**, 37809y (1974).

A356. Farrar, R. C., S. African Patent 72/06,179 (Phillips Petroleum Co.) (1973); *C.A.* **80**, 84470r (1974).

A357. Gomoll, M., and Seide, H., *Plaste Kautsch.* **16**(6), 412 (1969); *C.A.* **71**, 82338u (1969).

A358. Hirshfield, S. M., U.S. Patent 3,755,283 (NASA) (1973); *C.A.* **80**, 4104m (1974).

A359. Price, C., and Woods, D., *Eur. Polym. J.* **9**(8), 827 (1973); *C.A.* **80**, 37545j (1974).

A360. Eschwey, H., Hallensleben, M., and Burchard, W., *Makromol. Chem.* **173**, 235 (1973); *C.A.* **80**, 133882n (1974).

A361. Haws, J. R., *Am. Chem. Soc., Div. Org. Coat. Plast. Chem., Preprints* **34**(1), 114 (1974).

A362. Fodor, L. M., Kitchen, A. G., and Biard, C. C., *Am. Chem. Soc., Div. Org. Coat. Plast. Chem., Preprints,* **34**(1), 130 (1974).

A363. Kitchen, A. G., and Szalla, F. J., U.S. Patent 3,639,517 (Phillips Petroleum Co.) (1972).

A364. Fetters, L. J., and McGrath, J. E., unpublished results.

A364a. Karoly, G. *In* "Block Polymers," (S. Aggarwal, ed.), p. 153. Plenum, 1970.

A365. Meyer, G., Widmaier, J. M., and Toussaint, J. M., *Bull. Soc. Chim. Fr.* Nos. 1-2, Part 2, 286 (1975); *C.A.* **82**, 156948u (1975).

A366. Karoly, G., *Polym. Prepr., Am. Chem. Soc., Div. Polym. Chem.* **10**(2), 837 (1969).

A367. Shimomura, T., Nagata, H., and Murakami, Y., German Offen. 2,224,616 (Sumitomo Chemical Co., Ltd.) (1972); *C.A.* **78**, 98759n (1973).

A368. Shatalov, V. P., Grigor'eva, L. A., and Kistereva, A. E., U.S.S.R. Patent 254,773 (1968); *C.A.* **72**, 122639x (1970).

A369. French Patent 1,573,989 (Polymer Corp., Ltd.) (1969); *C.A.* **72**, 56482n (1970).

A370. Neumann, F. E., German Offen. 2,015,249 (Shell Internationale Research Maatschappij N.V.) (1970); *C.A.* **74**, 13992c (1971).

A371. Neumann, F. E., German Offen. 2,015,250 (Shell International Research Maatschappij N.V.) (1970); *C.A.* **74**, 23231t (1971).

A372. Shimomura, T., Nagata, H., and Murakami, Y., Japanese Patent (Kokai) 73/00,790 (Sumitomo Chemical Co., Ltd.) (1973); *C.A.* **78**, 112535y (1973).

A373. Elgert, K. F., Seiler, E., Puschendort, G., and Kantow, H. J., *Makromol. Chem.* **165**, 245 (1973); *C.A.* **78**, 160668n (1973).

A374. Beilin, S. I., Vollershtein, E. L., Teterina, M. P., Shvarts, M. N., and Dolgoplosk, B. A., *Vysokomol. Soedin., Ser. B* **11**(10), 733 (1969); *C.A.* **72**, 32267v (1970).

A375. Heller, J., Schimscheimer, J. F., Pasternak, R. A., Kingsley, C. B., and Moacanin, J., *J. Polym. Sci., Part A-1* **7**(1), 73 (1969); *C.A.* **71**, 3699m (1969).

A376. Trepka, W. J., *J. Polym. Sci., Part B* **8**(7), 499 (1970); *C.A.* **74**, 4396j (1971).

A377. Wright, R. F., U.S. Patent 3,730,934 (Phillips Petroleum Co.) (1973).

A378. Morton, M., Fetters, L. J., Kammereck, R. F., Mikesell, S. L., and Strauss, C., "High Temperature Elastomer Networks from Difunctional Block Polymers," Tech. Rep. AFML-TR-67-368, Part IV. Wright Patterson AFB, Ohio, 1970.

A379. McKeever, L. D., and Cheng, W. J., U.S. Patent 3,728,193 (Dow Chemical Co.) (1973); *C.A.* **78**, 160553w (1973).

A380. Nametkin, N. S., Durgar'yan, S. G., Solov'ev, E. V., and Piryatinskii, V. M., German Offen. 1,917,863 (Topchiev, A. V., Institute of Petrochemical Synthesis) (1969); *C.A.* **72**, 33016z (1970).

A381. Balas, J., U.S. Patent 3,448,176 (Shell Oil Co.) (1969); *C.A.* **71**, 51058y (1969).

A382. French Patent 1,521,223 (International Synthetic Rubber Co., Ltd.) (1968); *C.A.* **70**, 116075c (1969).

A383. De la Mare, H. E., and Naumann, F. E., German Offen. 1,938,289 (Shell Internationale Research Maatschappij N.V.) (1970); *C.A.* **72**, 90999w (1970).

A384. Wald, M. M., and Quam, M. G., U.S. Patent 3,706,817 (Shell Oil Co.) (1972); *C.A.* **78**, 59575e (1973).

A385. Shatalov, V. P., Grigor'eva, L. A., Sokolova, N. F., Naidenov, S. T., Kistereva, A. E., and Alekhin, V. D., *Prom. Sin. Kauch., Ref. Sb.* **3**, 7 (1972); *C.A.* **78**, 16069p (1973).

A386. Roest, B. C., and Schepers, H. A. J., Netherlands Patent Appl. 71/04,949 (Stamicarbon N.V.) (1972); *C.A.* **78**, 11943q (1973).

A387. Fielding-Russell, G. S., and Pillai, P. S., *Polymer* **15**, 97 (1974).

A388. Fiere, E. M., *Diss. Abstr. Int. B* **34**(4), 1459 (1973) (Univ. Microfilms, Ann Arbor, Michigan, Order No. 73-22, 1973); *C.A.* **80**, 109527e (1974).

A389. Horiya, S., Asai, S., and Saito, C., Japanese Patent 73/20,038 (Denki Kagaku Kogyo K.K.) (1973); *C.A.* **80**, 134193a (1974).

A390. Roest, B. C., and Pijpers, E. M. J., German Offen. 2,261,214 (Stamicarbon N.V.) (1973); *C.A.* **79**, 127214j (1973).

A391. Mueller, E., Mayer-Mader, R., and Dinges, K., *Nuova Chim.* **49**(2), 66 (1973); *C.A.* **79**, 5661g (1973).

A392. Falk, J. C., and Schlott, R. J., *Angew. Makromol. Chem.* **21**, 17 (1972); *C.A.* **76**, 128392r (1972).

A393. Falk, J. C., and Schlott, R. J., *Macromolecules* **4**(2), 152 (1971); *C.A.* **75**, 7052a (1971).

A394. Winkler, D. E., and Shaw, A. W., U.S. Patent 3,452,118 (Shell Oil Co.) (1969); *C.A.* **71**, 62164f (1969).

A395. Henderson, J.F., and Darcy, J., French Patent 1,522,823 (Polymer Corp., Ltd.) (1968); *C.A.* **70**, 116065z (1969).

A396. Mann, R. H., U.S. Patent 3,458,600 (Shell Oil Co.) (1969); *C.A.* **71**, 71414q (1969).

A397. Kovalev, N. F., Korotkov, A. A., Petrov, G. N., Reikh, V. N., Lisochkin, G. F., Dugina, L. V., and Eventova, L. A., *Kauch. Rezina* **25**(12), 2 (1966); *C.A.* **66**, 95949r (1967).

A398. Gourdenne, A., and Sigwalt, P., *Bull. Soc. Chim. Fr.* **10**, 3678 (1967); *C.A.* **68**, 13445c (1968).

A399. Netherlands Patent Appl. 6,605,985 (Badische Anilin- & Soda-Fabrik A.-G.) (1966); *C.A.* **66**, 96003q (1967).

A400. Kuntz, I., U.S. Patent 3,140,278 (Esso Research and Engineering Co.) (1964).

A401. Netherlands Patent Appl. 6,514,631 (Shell Internationale Research Maatschappij N.V.) (1966); *C.A.* **66**, 11635z (1967).

A402. Shaw, A. W., German Offen. 2,045,622 (Shell Internationale Research Maatschappij N.V.) (1971); *C.A.* **75**, 7117a (1971).

A403. Canadian Patent 949,692 (Bayer) (1974).

A404. Sommer, N., and Nordsiek, K. H., U.S. Patent 3,829,409 (Chemische Werke Huls, A.G.) (1972).

A405. Baer, M., *J. Polym. Sci., Part A-2* **2**, 417 (1964).

A406. Krause, S., *Macromolecules* **3**(1), 84 (1970); *C.A.* **72**, 79512m (1970).

A407. Kastning, E. G., and Bronstert, K., U.S. Patent 3,070,574 (Badische Anilin- & Soda-Fabrik) (1962).

A408. Kennedy, J. P., and Melby, E. G., *Polym. Prepr., Am. Chem. Soc., Div. Polym. Chem.* **15**(2), 175 (1974).

A409. O'Malley, J. J., Yanus, J. F., and Pearson, J. M., *Macromolecules* **5**(2), 158 (1972); *C.A.* **77**, 35020y (1972).

A410. O'Malley, J. J., German Offen. 2,136,721 (Xerox Corp.) (1972); *C.A.* **77**, 6086u (1972).

A411. Schwab, F. C., U.S. Patent 3,770,712 (Mobil Oil Corp.) (1973); *C.A.* **80**, 83917e (1974).

A412. Schwab, F. C., German Offen. 2,222,957 (Mobil Oil Corp.) (1972); *C.A.* **78**, 44252y (1973).

A413. Murray, J. G., and Schwab, F. C., German Offen. 2,429,249 (Mobil Oil Corp.) (1975); *C.A.* **82**, 157583h (1975).

A414. Lundberg, R. D., and Makowski, H. S., German Offen. 2,308,492 (Esso Research and Engineering Co.) (1973); *C.A.* **80**, 4499u (1974).

A415. Lundberg, R. D., and Makowski, H. S., French Demande 2,160,555 (Esso Research and Engineering Co.) (1973); *C.A.* **80**, 96767f (1974).

B1. Dondos, A., Rempp, P., and Benoit, H. C., *Polymer* **13**(3), 97 (1972); *C.A.* **77**, 35134p (1972).

B2. Kotaka, T., Tanaka, T., and Inagaki, H., *Polym. J.* **3**(3), 327 (1972).

B3. Tanaka, T., Kotaka, T., and Inagaki, H., *Polym. J.* **3**(3), 338 (1972).

B4. Kotaka, T., Ohnuma, H., and Inagaki, H., *Polym. Prepr., Am. Chem. Soc., Div. Polym. Chem.* **11**(2), 660 (1970); *C.A.* **77**, 5932e (1972).

B5. Hsiue, G., Yasukawa, T., and Murakami, K., *Makromol. Chem.* **139**, 285 (1970); *C.A.* **74**, 32009d (1971).

B6. Kotaka, T., Ohnuma, H., and Inagaki, H., *Bull. Inst. Chem. Res., Kyoto Univ.* **46**(2), 107 (1966).

B7. Yasukawa, T., Murakami, K., and Hsiue, G., *Tohoku Daigaku Hisuiyoeki Kagaku Kenkyusho Hokoku* **21**(1), 75 (1971); *C.A.* **76**, 86251m (1972).

B8. Endo, H., and Yamamoto, M., *Rep. Prog. Polym. Phys. Jpn.* **15**, 1 (1972).

B9. Schindler, A., and Williams, J. L., *Polym. Prepr., Am. Chem. Soc., Div. Polym. Chem.* **10**(2), 832 (1969).

B10. Fontanille, M., and Sigwalt, P., *Bull. Soc. Chim. Fr.* **1**, 4095 (1967); *C.A.* **68**, 69393b (1967).

B11. Noel, C., *Bull. Soc. Chim. Fr.* **1**, 3733 (1967); *C.A.* **68**, 22313k (1968).

C1. Schmolka, I. R., *in* "Nonionic Surfactants" (M. J. Schick, ed.), Chapter 10, p. 300. Dekker, New York, 1967.

C2. Lunsted, L. G., U.S. Patent 2,674,619 (Wyandotte Chem. Co.) (1954).

C3. Weis, G., *Fette, Seifen, Anstrichm.* **70**(5), 355 (1968); *C.A.* **69**, 19712e (1968).

C4. British Patent 918,430 (General Aniline and Film Corp.) (1963).

C5. British Patent 951,445 (Farbwerke Hoechst A.G.) (1964).

C5a. Mansfield, R. C., Canadian Patent 716,029 (Rohm and Haas Co.) (1965).

C5b. British Patent 994,136 (General Aniline and Film Corp.) (1965).

C6. DeGroote, M., and Keiser, B., U.S. Patent 2,771,450 (to Petrolite Corp.) (1956).

C6a. DeGroote, M., and Keiser, B., U.S. Patent 2,792,400 (to Petrolite Corp.) (1957).

C7. Schmolka, I. R., and Bacon, L. R., *J. Am. Oil Chem. Soc.* **44**(10), 559 (1967); *C.A.* **67**, 118450h (1967).

C8. Jackson, D. R., and Lundsted, L. G., U.S. Patent 3,036,118 (Wyandotte Chemicals Corp.) (1962).

C9. Lundsted, L. G., U.S. Patent 3,022,335 (Wyandotte Chemicals Corp.) (1962).

C10. Patton, J. T., U.S. Patent 3,101,374 (Wyandotte Chemicals Corp.) (1963).

C11. Spriggs, J. S., U.S. Patent 2,828,345 (Dow Chemical Co.) (1958).

C12. Lunsted, L. G., British Patent 722,746 (Wyandotte Chemicals Corp.) (1955).

C13. Teot, A. S., Canadian Patent 698,568 (Dow Chemical Co.) (1964).

C14. Morehouse, E. L., U.S. Patent 3,489,698 (Union Carbide Corp.) (1970); *C.A.* **72**, 67752b (1970).

C15. Kucharski, S., and Chlebicki, J., *J. Colloid Interface Sci.* **46**(3), 518 (1974).

C16. Cline, E. T., German Patent 1,139,974 (E. I. du Pont de Nemours and Co.) (1962); *C.A.* **58**, 5804c (1963).

C17. British Patent 807,589 (E. I. du Pont de Nemours and Co.) (1959); *C.A.* **55**, 2199c (1961).

C18. Hammond, J. M., Hooper, J. F., and Robertson, W. G. P., *Weapons Res. Estab. Rep.* **WRE-TN-CPD-146** (1968); *C.A.* **71**, 81761q (1969).

C19. Saegusa, T., Matsumoto, S., and Hashimoto, Y., *Macromolecules* 3(4), 377 (1970); *C.A.* **73**, 77644c (1970).

C20. Saegusa, T., Matsumoto, S., and Hashimoto, Y., *in* "Block Polymers" (S. L. Aggarwal, ed.), p. 293. Plenum, New York, 1970.

C21. Vaughn, T. H., Jackson, D. R., and Lundsted, L. G., *J. Am. Oil Chem. Soc.* **29**, 240 (1952).

C22. Canadian Patent 939,868 (Gen. Motors Corp.) (1970).

C23. Saegusa, T., and Matsumoto, S., U.S. Patent 3,636,132 (1972); *C.A.* **76**, 127778j (1972).

C24. O'Malley, J. J., Crystal, R. G., and Erhardt, P. F., *Polym. Prepr., Am. Chem. Soc., Div. Polym. Chem.* **10**(2), 796 (1969).

C25. Szwarc, M., U.S. Patent 3,050,511 (Dow Chemical Co.) (1962); *C.A.* **58**, 1558e (1963).

C26. Erhardt, P. F., O'Malley, J. J., and Crystal, R. G., *Polym. Prepr., Am. Chem. Soc., Div. Polym. Chem.* **10**(2), 812 (1969).

C27. Erhardt, P. F., O'Malley, J. J., and Crystal, R. G., *in* "Block Polymers" (S. L. Aggarwal, ed.), p. 195. Plenum, New York, 1970; *C.A.* **75**, 6487j (1971).

C28. Skoulios, A., Finaz, G., and Parrod, J., *C.R. Hebd. Seances Acad. Sci.* **251**, 739 (1960).

C29. Sadron, C., *Rev. Gen. Caoutch. Plast., Ed. Plast.* **2**(2), 112 (1965); *C.A.* **66**, 29241c (1967).

C30. Kawai, T., Shiozaki, S., Sonoda, S., Nakagawa, H., Matsumoto, T., and Maeda, H., *Makromol. Chem.* **128**, 252 (1969); *C.A.* **71**, 125110h (1969).

C31. Tobolsky, A. V., and Rembaum, A., U.S. Patent 3,257,476 (1966); *C.A.* **65**, 10780d (1966).

C32. Tobolsky, A. V., U.S. Patent 3,291,859 (1966); *C.A.* **66**, 38447b (1967).

C33. Tobolsky, A. V., and Rembaum, A., *J. Appl. Polym. Sci.* **8**(1), 307 (1964).

C34. Furukawa, J., Takamori, S., and Yamashita, S., *Agnew. Makromol. Chem.* **1**(1), 92 (1967); *C.A.* **68**, 96448e (1968).

C35. Furukawa, J., Saegusa, T., and Mise, N., *Makromol. Chem.* **38**, 244 (1960).

C36. Galin, J. C., *Makromol. Chem.* **124**, 118 (1969); *C.A.* **71**, 30901c (1969).

C37. Novitskaya, M. A., and Konkin, A. A., *Karbotsepnye Volokna* p. 173 (1966); *C.A.* **68**, 88053g (1968).

C38. Novitskaya, M. A., and Konkin, A. A., *Vysokomol. Soedin.* **7**(10), 1719 (1965); *C.A.* **64**, 3710b (1966).

C39. Novitskaya, M. A., and Konkin, A. A., *Karbotsepnye Volokna* p. 246 (1966); *C.A.* **67**, 100890s (1966).

C40. Japanese Patent 70/20300-R (Asahi Chemical Co. Ltd.) (1964).

C41. Novitskaya, M. A., *Tr. Sib. Tekhnol. Inst.* **38**, 203 (1966); *C.A.* **68**, 96224d (1968).

C42. Tobolsky, A. V., and Rembaum, A., Belgian Patent 622,032 (FMC Corp.) (1972); *C.A.* **59**, 7731a (1963).

C43. Smith, W. E., Galiano, F. R., Rankin, D., and Mantell, G. J., *J. Appl. Polym. Sci.* **10**(11), 1659 (1966); *C.A.* **66**, 2838j (1967).

C44. Noro, K., Kawazura, H., Moriyama, T., and Yoshioka, S., *Makromol. Chem.* **83**, 35 (1965); *C.A.* **63**, 3069b (1965).

C45. Kirkland, E. V., and Roberts, W. J., Belgian Patent 636,370 (Celanese Corp. of America) (1963); *C.A.* **62**, 6590f (1965).

C46. Kirkland, E. V., and Roberts, W. J., U.S. Patent 3,219,725 (Celanese Corp. of America) (1965); *C.A.* **62**, 10547f (1965).

C47. Carter, J. H., and Michelotti, F. W., *Polym. Prepr., Am. Chem. Soc., Div. Polym. Chem.* **5**(2), 614 (1964); *C.A.* **64**, 12806g (1966).

C48. Takida, H., and Noro, K., *Kobunshi Kagaku* **21**(234), 459 (1964); *C.A.* **62**, 10537e (1965).

C49. Berger, G., Levy, M., and Vofsi, D., *J. Polym. Sci., Part B* **4**(3), 183 (1966); *C.A.* **64**, 14288a (1966).

C50. Vofsi, D., and Tobolsky, A. V., *J. Polym. Sci., Part A* **3**, 3261 (1965).

C51. Dreyfus, M. P., and Dreyfus, P., *Polymer* **6**, 93 (1965).

C52. Bawn, C. E., Bell, R. M., and Ledwith, A., *Polymer* **6**, 95 (1965).

C53. Yamashita, Y., Hirota, M., Matsui, H., Hirao, A., and Nobutoki, K., *Polym. J.* **2**(1), 43 (1971).

C54. Zimmerman, D. D., Smith, S., and Hubin, A. J., U.S. Patent 3,523,144 (Minnesota Mining and Manufacturing Co.) (1970); *C.A.* **73**, 78032p (1970).

C55. Erdy, N. Z., Ferraro, C. F., and Tobolsky, A. V., *J. Polym. Sci., Part A-1* **8**(3), 763 (1970); *C.A.* **72**, 11874u (1970).

C56. Baker, W. P., Jr., French Patent 1,357,933 (E. I. du Pont de Nemours and Co.) (1964); *C.A.* **62**, 4184h (1965).

C57. Zgonnik, V. N., Shibaev, L. A., and Nikolaev, N. I., *Kinet. Mech. Polyreactions, Int. Symp. Macromol. Chem., Prepr., 1969* Vol. 4, p. 319 (1969); *C.A.* **75**, 64363b (1971).

C58. Galin, J. C., *Peint., Pigm., Vernis* **43**(8), 531 (1967); *C.A.* **67**, 117620b (1967).

C59. Perret, R., and Skoulios, A., *C.R. Hebd. Seances Acad. Sci., Ser. C* **268**(3), 230 (1969); *C.A.* **70**, 78632x (1969).

C60. Nobutoki, K., and Sumitomo, H., *Bull. Chem. Soc. Jpn.* **40**(8), 1741 (1967); *C.A.* **67**, 82452u (1967).

C61. Yamashita, Y., Japanese Patent (Kokai) 73/00,792 (Mitsubishi Chemical Industries Co., Ltd.) (1973); *C.A.* **78**, 11979f (1973).

C62. Yamashita, Y., Nakamura, Y., and Kojima, S., *J. Polym. Sci., Polym. Chem. Ed.* **11**, 823 (1973).

C63. Tabuchi, T., Nobutoki, K., and Sumitomo, H., *Kogyo Kagaku Zasshi* **71**(11), 1926 (1968); *C.A.* **70**, 58488b (1969).

C64. Zaganiaris, E., and Tobolsky, A., *J. Appl. Polym. Sci.* **14**(8), 1997 (1970); *C.A.* **73**, 110641z (1970).

C65. Koleske, J. V., Roberts, R. M. J., and DelGiudice, F. P., U.S. Patent 3,670,045 (Union Carbide Corp.) (1972); *C.A.* **77**, 76043b (1972).

C66. Perret, R., and Skoulios, A., *Makromol. Chem.* **156**, 143 (1972); *C.A.* **77**, 75511r (1972).

C67. Perret, R., and Skoulios, A., *Makromol. Chem.* **156**, 157 (1972); *C.A.* **77**, 88958s (1972).

C68. Perret, R., and Skoulios, A., *Makromol. Chem.* **162**, 163 (1972); *C.A.* **78**, 44152r (1973).

C69. Sharkey, W. H., German Offen. 2,013,553 (E. I. du Pont de Nemours and Co.) (1970); *C.A.* **74**, 13980x (1971).

C70. Foss, R., German Offen. 2,307,740 (E. I. du Pont de Nemours and Co.) (1973); *C.A.* **80**, 97171u (1974).

C71. Lanza, E., Berghmans, H., and Smets, G., *J. Polym. Sci., Polym. Phys. Ed.* **11**(1), 95 (1973); *C.A.* **78**, 136715e (1973).

C72. MacKillop, D. A., *J. Polym. Sci., Part B* **8**(3), 199 (1970); *C.A.* **73**, 15956x (1970).

C73. Boileau, S., and Sigwalt, P., *C.R. Hebd. Seances Acad. Sci.* **261**(1)(7), 132 (1965); *C.A.* **63**, 13422b (1965).

C74. Boileau, S., and Sigwalt, P., *Makromol. Chem.* **131**, 7 (1970); *C.A.* **72**, 91102k (1970).

C75. Gourdenne, A., *Polym. Prepr., Am. Chem. Soc., Div. Polym. Chem.* **10**(2), 826 (1969).

C76. Balcerzyk, E., Pstrocki, H., and Włodarski, G., *J. Appl. Polym. Sci.* **11**(7), 1179 (1967); *C.A.* **67**, 54519c (1967).

C77. Nevin, R. S., and Pearce, E. M., *J. Polym. Sci., Part B* **3**(6), 487 (1965); *C.A.* **63**, 3069d (1965).

C78. Baker, W. P., Jr., French Patent 1,357,934 (E. I. du Pont de Nemours and Co.) (1964); *C.A.* **62**, 1821c (1965).

C79. Pstrocki, H., Polish Patent 53,771 (Instytut Wlokien Sztucznych i 'Syntetycznych) (1967); *C.A.* **68**, 87743v (1968).

C80. Pstrocki, H., *Rocz. Chem.* **38**, 899 (1964); *C.A.* **66**, 2882y (1967).

C81. Fetters, L. J., Morton, M., and Kammereck, R. F., *Macromolecules* **4**(1), 11 (1971); *C.A.* **74**, 112941s (1971).

C82. Gourdenne, A., and Sigwalt, P., *Eur. Polym. J.* **3**(3), 481 (1967); *C.A.* **67**, 91142v (1967).

C82a. Gourdenne, A., and Sigwalt, P., *Bull. Soc. Chim. Fr.* No. 7, p. 2249 (1967); *C.A.* **67**, 91147a (1967).

C83. French Patent 1,566,887 (Dunlop Co., Ltd.) (1969); *C.A.* **72**, 4219n (1970).

C84. Baker, W. P., Jr., U.S. Patent 3,274,295 (E. I. du Pont de Nemours and Co.) (1966); *C.A.* **66**, 19138k (1967).

C85. Morton, M., and Mikesell, S. L., *Polym. Prepr., Am. Chem. Soc., Div. Polym. Chem.* **13**(1), 61 (1971); *C.A.* **80**, 84419f (1974).

C86. Morton, M., and Mikesell, S. L., *J. Macromol. Sci., Chem.* **7**(7), 1391 (1973); *C.A.* **80**, 133873k (1974).

C87. Cooper, W., Hale, P. T., and Walker, J. S., *Polymer* **15**, 175 (1974).

C88. Gourdenne, A., *Makromol. Chem.* **158**, 271 (1972); *C.A.* **77**, 114943e (1972).

C89. Furukawa, J., Yamashita, S., Harada, N., and Arai, Y., Japanese Patent 72/08,641 (Japan Synthetic Rubber Co., Ltd.) (1972); *C.A.* **77**, 63094u (1972).

C90. Richards, D. H., and Szwarc, M., *Trans. Faraday Soc.* **55**, 1644 (1959).

C91. Finaz, G., Rempp, P., and Parrod, J., *Bull. Soc. Chim. Fr.* p. 262 (1962); *C.A.* **57**, 1055f (1962).

C92. Takahashi, A., and Yamashita, Y., *Am. Chem. Soc., Div. Polym. Chem. Polym. Prepr.*, **15**(1), 184 (1974).

C93. Baker, W. P., Jr., French Patent 1,357,936 (E. I. du Pont de Nemours and Co.) (1964); *C.A.* **62**, 1817b (1965).

C94. Perret, R., and Skoulios A., *Makromol. Chem.* **162**, 147 (1972).

C95. Mueller, F. X., Jr., and Hsieh, H. L., U.S. Patent 3,585,257 (Phillips Petroleum Co.) (1971).

C96. Allegrezza, A. E., Jr., Lenz, R. W., Cornibert, J., and Marchessault, R. H., *Polym. Prepr., Am. Chem. Soc., Div. Polym. Chem.* **14**(2), 7 (1973).

C97. Brooks, T. W., and Daffin, C. L., *Polym. Prepr., Am. Chem. Soc., Div. Polym. Chem.*, **10**(2), 1174 (1969).

C98. Yamashita, Y., Matsui, H., and Ito, K., *J. Polym. Sci., Polym. Chem. Ed.* **10**, 3577 (1972).

C99. Matzner, M., Robeson, L. M., Greff, R. J., and McGrath, J. E., *Angew. Makromol. Chem.* **26**, 137 (1972).

C100. McGrath, J. E., and Matzner, M., U.S. Patent 3,770,849 (Union Carbide) (1974).

C101. Matzner, M., and McGrath, J. E., U.S. Patent 3,657,385 (Union Carbide Corp.) (1972); *C.A.* **77**, 35521n (1972).

C102. McGrath, J. E., and Matzner, M., U.S. Patent 3,655,822 (Union Carbide Corp.) (1972); *C.A.* **77**, 49451b (1972).

C103. McGrath, J. E., Robeson, L. M., and Matzner, M., *Polym. Prepr., Am. Chem. Soc., Div. Polym. Chem.* **14**(2), 1032 (1973).

C104. McGrath, J. E., Robeson, L. M., and Matzner, M., *Polym. Sci. Technol.* **4**, 195 (1974).

C105. Ambrose, R., and Hergenrother, W. L., *J. Polymer. Sci.* **12**(11), 2613 (1974).

C106. Lincoln, J., British Patent 1,150,725 (Courtaulds Ltd.) (1969); *C.A.* **71**, 3910y (1969).

C107. Matzner, M., Noshay, A., and McGrath, J. E., U.S. Patent 3,770,849 (Union Carbide Corp.) (1973).

C108. Matzner, M., Barclay, R., Jr., and McGrath, J. E., unpublished results.

C109. Gaskin, S., and Nield, E., German Offen. 2,210,348 (Imperial Chemical Industries Ltd.) (1972); *C.A.* **77**, 165536g (1972).

C110. Werner, B. H., and Hayes, R. A., U.S. Patent 3,758,631 (Firestone Tire and Rubber Co.) (1973); *C.A.* **80**, 28175m (1974).

C111. Godfrey, R. A., and Miller, G. W., *J. Polym. Sci., Part A-1* **7**(8), 2387 (1969); *C.A.* **71**, 102300f (1969).

C112. Baker, W. P., Jr., U.S. Patent 3,225,119 (DuPont) (1965).

C113. Fetters, L. J., *J. Res. Natl. Bur. Stand., Sect. A* **70**, 421 (1966).

C114. Furukawa, J., Yamashita, S., Harada, T., and Arai, Y., Japanese Patent 72/08,618 (Japan Synthetic Rubber Co., Ltd.) (1972); *C.A.* **77**, 49809z (1972).

C115. Schaeffer, P. R., and Steely, N. E., U.S. Patent 3,511,893 (Polymer Corp.) (1970); *C.A.* **73**, 4740w (1970).

C116. Goto, Y., and Miwa, S., *Kobunshi Kagaku* **25**(281), 595 (1968); *C.A.* **70**, 38166s (1969).

C117. Japanese Patent 69/16,023 (Asahi Chem. Ind. Co., Ltd.) (1969).

C118. Japanese Patent 7,023,954 (Toyo Rayon Co. Ltd.) (1966).

C119. Saotome K., Japanese Patent 69/16,023 (Asahi Chemical Industry Co., Ltd.) (1969); *C.A.* **71**, 125253g (1969).

C120. Saotome, K., and Kodaira, Y., Japanese Patent 19,265 (Asahi Chemical Industry Co., Ltd.) (1967); *C.A.* **68**, 13569w (1968).

C121. Shalaby, S. W., Reimschuessel, H. K., and Pearce, E. M., *Polym. Eng. Sci.* **13**(2), 88 (1973); *C.A.* **78**, 136743n (1973).

C122. Shalaby, S. W., Pearce, E. M., and Reimschuessel, H. K., *Ind. Eng. Chem., Prod. Res. Dev.* **12**(2), 128 (1973).

C123. Shalaby, S. W., Reimschuessel, H. K., and Pearce, E. M., *Polym. Eng. Sci.* **13**(2), 98 (1973).

C124. Perly, B., Douy, A., and Gallot, B., *C.R. Hebd. Seances Acad. Sci., Ser. C* **279**(26), 1109 (1974); *C.A.* **82**, 125638p (1975).

C125. Storey, H. T., Thompson, R. C., Hardy, P. M., and Rydon, H. N., *Polymer* **15**(11), 690 (1974); *C.A.* **82**, 125689f (1975).

C126. Trudelle, Y., *J. Chem. Soc., Perkin Trans. 1* **10**, 1001 (1973); *C.A.* **79**, 42873y (1973).

C127. Bradbury, E. M., Cary, P. D., and Crane-Robinson, C., *Macromolecules* **5**(5), 581 (1972); *C.A.* **78**, 16619e (1973).

C128. Spach, G., Reibel, L., Loucheux, M. H., and Parrod, J., *J. Polym. Sci., Part C* **16**(8), 4705 (1969); *C.A.* **71**, 22383v (1969).

C129. Carita, M. M., and D'Alagni, M., *Polymer* **13**(11), 515 (1972); *C.A.* **78**, 16703c (1973).

C130. Lewis, A., and Scheraga, H. A., *Macromolecules* **4**(5), 539 (1971); *C.A.* **76**, 34678m (1972).

C131. Oya, M., Uno, K., and Iwakura, Y., *Bull. Chem. Soc. Jpn.* **43**(6), 1788 (1970); *C.A.* **73**, 56475h (1970).

C132. Iio, T., *Biopolymers* **10**(9), 1583 (1971); *C.A.* **76**, 34648b (1972).

C133. Bostick, E. E., and Fessler, W. A., German Offen. 2,049,547 (General Electric Co.) (1971); *C.A.* **75**, 21478a (1971).

C134. Lee, C. L., Frye, C. L., and Johannson, O. K., *Polym. Prepr., Am. Chem. Soc., Div. Polym. Chem.* **10**(2), 1361 (1969).

C135. Fritsche, A. K., and Price, F. P., *Polym. Prepr., Am. Chem. Soc., Div. Polym. Chem.* **11**(2), 462 (1970).

C136. French Patent 2,134,884 (General Electric Co.) (1973); *C.A.* **79**, 54055w (1973).

C137. Minoura, Y., Mitoh, M., Tabuse, A., and Yamada, Y., *J. Polym. Sci., Part A-1* **7**(9), 2753 (1969); *C.A.* **71**, 124971c (1969).

C138. Dean, J. W., German Offen. 2,116,837 (General Electric Co.) (1971); *C.A.* **76**, 73475n (1972).

C139. Jones, F. R., *Eur. Polym. J.* **10**, 249 (1974).

C140. Lee, C. L., and Johannson, O. K., *J. Polym. Sci., Part A-1* **4**(12), 3013 (1966); *C.A.* **66**, 38271q (1967).

C141. Greber, G., and Balciunas, A., *Makromol. Chem.* **79**, 149 (1964); *C.A.* **62**, 1756f (1965).

C142. Greber, G., Reese, E., and Balciunas, A., *Farbe U. Lack* **70**, 249 (1964); *C.A.* **61**, 5800a (1964).

C143. Davies, W. G., and Jones, D. P., *Polym. Prepr., Am. Chem. Soc., Div. Polym. Chem.* **11**(2), 477 (1970).

C144. Kendrick, T. C., and Owen, M. J., German Offen. 1,915,789 (Midland Silicones Ltd.) (1969); *C.A.* **71**, 125476g (1969).

C145. Davies, W. G., and Jones, D. P., *Ind. Eng. Chem., Prod. Res. Dev.* **10**(2), 168 (1971); *C.A.* **75**, 49823h (1971).

C146. Owen, M. J., and Kendrick, T. C., *Macromolecules* **3**(4), 458 (1970); *C.A.* **73**, 77757s (1970).

C147. Morton, M., Kesten, Y., and Fetters, L. J., *Polym. Prepr., Am. Chem. Soc., Div. Polym. Chem.* **15**(2), 175 (1974).

C148. Bostick, E. E., U.S. Patent 3,483,270 (General Electric Co.) (1969); *C.A.* **72**, 44357n (1970).

C149. Bostick, E. E., U.S. Patent 3,483,270 (General Electric Co.) (1969).

C150. Owen, M. J., and Thompson, J., *Br. Polym. J.* **4**(4), 297 (1972); *C.A.* **78**, 125243y (1973).

C151. Bostick, E. E., and Zdaniewski, J. J., German Offen. 2,048,914 (General Electric Co.) (1971); *C.A.* **75**, 21496e (1971).

C152. Morton, M., and Rembaum, A., U.S. Patent 3,051,684 (University of Akron) (1962); *C.A.* **57**, 16890 (1962).

C153. Dean, J. W., U.S. Patent 3,673,272 (General Electric Co.) (1972); *C.A.* **77**, 89111x (1972).

C154. Bostick, E. E., *Polym. Prepr., Am. Chem. Soc., Div. Polym. Chem.* **10**(2), 877 (1969).

C155. Fritsche, A. K., and Price, F. P., *Polym. Prepr., Am. Chem. Soc., Div. Polym. Chem.* **10**(2), 893 (1969).

C156. Morton, M., Rembaum, A. A., and Bostick, E. E., *J. Appl. Polym. Sci.* **8**(6), 2707 (1964); **62**, 2839h (1965).

C157. French Patent 2,134,885 (General Electric Co.) (1973); *C.A.* **79**, 54374z (1973).

C158. Bostick, E. E., U.S. Patent 3,337,497 (General Electric Co.) (1967); *C.A.* **67**, 82684w (1967).

7

{A-B}_n
Multiblock Copolymers

This chapter deals with copolymers that contain many A and B segments, i.e., $\{A\text{-}B\}_n$ structures in which $n > 2$. The preceding A-B and A-B-A chapters, 5 and 6, are subdivided into separate sections discussing copolymers containing at least one heteroatom segment and copolymers containing only carbon backbones. However, this chapter deals only with structures containing heteroatom segments. The reason for this is that almost all $\{A\text{-}B\}_n$ block copolymers are synthesized by condensation reactions of one type or another, and such reactions result in structures containing heteroatoms. This generalization, however, is not without exceptions. But the number of references uncovered that deal with all-carbon-backbone $\{A\text{-}B\}_n$ block copolymers was so small that these were incorporated into the corresponding A-B or A-B-A chapters. For example, the coupling of dilithio-terminated styrene–butadiene–styrene A-B-A block copolymers with phosgene produces an $\{A\text{-}B\}_n$ structure. This is discussed in the styrene–butadiene section of the A-B-A Chapter 6.

The $\{A\text{-}B\}_n$ block copolymers of this chapter contain one or more segments that have at least one atom other than carbon in the backbone repeat unit. These heteroatoms include oxygen, sulfur, nitrogen, and silicon and are found in ether, ester, sulfide, amide, imide, ester, carbonate, urethane, and siloxane structures. The synthesis and properties of these block copolymers are discussed below.

A. ETHER–ETHER

These multiblock copolymers contain segments of two different polyethers, including those derived from alkylene oxides, formaldehyde, dioxolane, and arylene oxides (i.e., polysulfone).

As was discussed in Chapter 6, ethylene oxide–propylene oxide block copolymers are very useful materials in surfactant applications. Multiblock copolymers with more than three segments have been prepared by the same techniques described earlier by using multifunctional initiators in polymerizing the olefin oxide monomers (A1–A3). Trifunctional initiators such as glycerol produce copolymers with three polyethylene oxide segments and three polypropylene oxide segments via initiation at each of the three hydroxyls of the glycerol. Similarly, copolymers with four segments of each type are produced by using a tetrafunctional initiator such as ethylenediamine. Even higher degrees of branching are obtained by using pentafunctional (e.g., diethylenetriamine), hexafunctional (e.g., sorbitol), or octafunctional (e.g., sucrose) initiators. The technology of the surfactant applications of these materials is discussed in a review article (A1).

A patent reference (A4) claims the synthesis of block copolymers containing segments of polyethylene oxide and segments of an ethylene oxide–formaldehyde random copolymer by coupling the two performed hydroxyl-terminated prepolymers with a diisocyanate. This product will, of course, contain the segments in random rather than strictly alternating sequence, due to the approximately equivalent reactivity of the end groups of the prepolymers. Polyformaldehyde–polydioxolane block copolymers were claimed (A5) to be prepared by an interchange reaction between the two homopolymers in the presence of Lewis acid catalysts.

Another patent reference (A6) claims the preparation of a rubbery block copolymer containing poly(tetramethylene oxide) segments and polysulfone arylene ether segments (Reaction 7-1). This reaction involved the interaction of a chloroformate-terminated tetramethylene oxide oligomer with a hydroxyl-terminated polysulfone oligomer.

McGrath et al. (A7) synthesized block copolymers containing seg-

(7-1)

ments of polyhydroxyether and segments of polysulfone by the approach shown in Reaction 7-2. In an alternative technique, the bisphenol A and epichlorohydrin were replaced with bisphenol A diglycidyl ether. Although physical blends of the two respective homopolymers were opaque two-phase systems, the block copolymers were transparent and displayed single-phase morphology (see Figs. 7-1 and 7-2). This phenomenon was attributed to the chemical similarity of the polymers as indicated by the small difference in their solubility parameters ($\Delta = 0.7$). The presence of the chemical bond joining the two segments in the block copolymer apparently overcomes the weak physical forces causing phase separation, resulting in single-phase morphology.

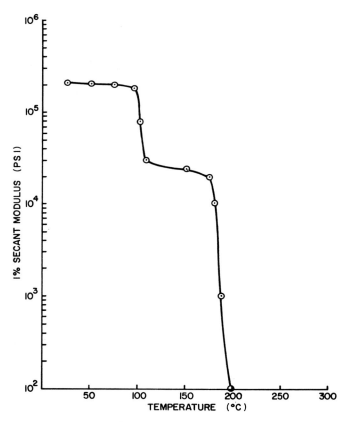

Fig. 7-1. Dynamic mechanical behavior of a 50 : 50 polysulfone–polyhydroxyether blend (A7).

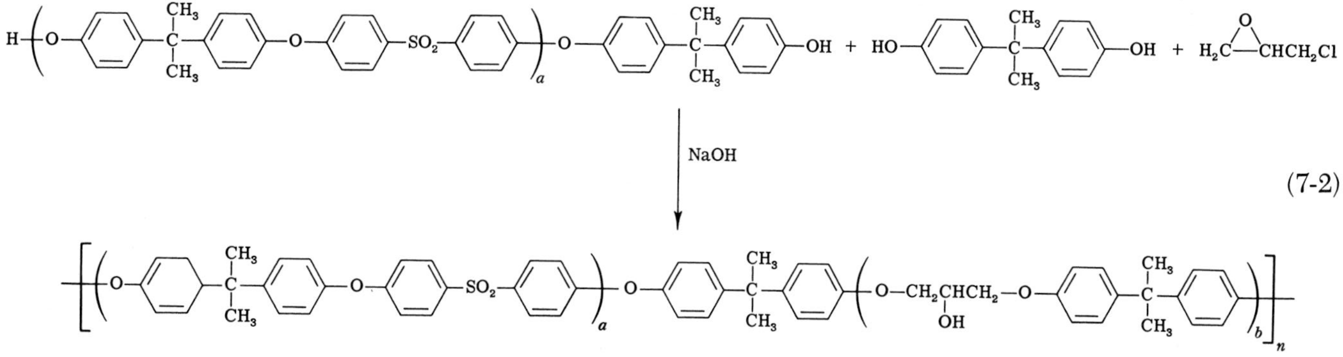

$$\text{HO} - \!\!\!\!\bigcirc\!\!\!\! - \overset{\underset{\displaystyle CH_3}{|}}{\underset{\underset{\displaystyle CH_3}{|}}{C}} - \!\!\!\!\bigcirc\!\!\!\! - OH \;+\; H_2C\!\!-\!\!\overset{O}{\overbrace{}}\!\!-CHCH_2Cl$$

CH₃ C CH₃ — SO₂ — O — OH

$$\xrightarrow{\text{NaOH}}$$

(7-2)

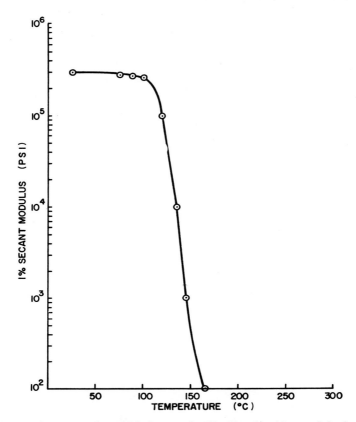

Fig. 7-2. Dynamic mechanical behavior of a 50 : 50 polysulfone–polyhydroxyether block copolymer (A7).

B. ETHER–VINYL

The multiblock copolymers of this category contain polyether segments, such as poly(alkylene oxides) [e.g., poly(ethylene oxide) and poly(tetramethylene oxide)] or polyarylene ethers (e.g., polysulfone) together with vinyl polymer segments such as polystyrene and poly(α-methylstyrene).

The first attempt to synthesize multiblock copolymers of polystyrene and poly(ethylene oxide) was reported by Richards and Szwarc (B1). Sodium alkoxide-terminated ethylene oxide–styrene–ethylene oxide A-B-A block copolymers were reported to increase in molecular weight by a factor of 5 when coupled with phosgene or adipyl chloride. Finaz *et al.* (B2) prepared similar copolymers by converting a dicarbanion-terminated polystyrene to a diacyl chloride-terminated oligomer with excess phosgene, followed by condensation of this intermediate with dihydroxyl-terminated poly(ethylene oxide) (Reaction 7-3).

$$(7\text{-}3)$$

The reverse procedure has also been used, i.e., preparation of a bischloroformate-terminated polyether [e.g., poly(tetramethylene oxide)] followed by condensation of this intermediate with dicarbanion-terminated polystyrene or poly(α-methylstyrene) (B3).

Shimura *et al.* (B4) converted a dicarbanion-terminated polystyrene to a dihydroxyl-terminated oligomer by capping with styrene oxide. Addition of 2 moles of a diisocyanate to this, followed by addition of dihydroxyl-terminated poly(ethylene oxide) produced styrene–ethylene oxide (A-$B)_n$ block copolymers with urethane connecting linkages. Block $\overline{M}_n$ values ranged from 400 to 6000, and the $\overline{M}_n$ values for the block copolymer products (by membrane osmometry) were found to be in the vicinity of 50,000. The molecular weight distributions (by gel permeation chromatography) of the products were relatively broad. The copolymers displayed two differential scanning calorimetry peaks. One was observed at 55°–70°C due to the ethylene oxide T_m and another at 90°C due to the styrene T_g. It was concluded from the differential scanning calorimetry studies that crystallization and fusion of the poly(ethylene oxide) blocks is disturbed by the presence of the polystyrene blocks to a greater extent than was observed in A-B-A structures.

Yamashita *et al.* (B5–B7) synthesized styrene–tetramethylene oxide (A-$B)_n$ block copolymers by preparing a difunctionally terminated poly(tetramethylene oxide) oligomer by a reportedly living cationic polymerization process and coupling this with a dicarboxylate-terminated polystyrene oligomer. The polyether dication was synthesized by initiating the polymerization of tetrahydrofuran (THF) with the difunctional initiator 2,2'-octamethylenebis-1,3-dioxolenium perchlorate. An increase in molecular weight with increasing conversion was offered as support of the living nature of the system. The molecular weight distribution of this oligomer was relatively broad due to a slow rate of initiation compared to the rate of propagation. Coupling of this with the dicarboxylate-terminated polystyrene (used since the carbanion form caused cleavage due to high basicity) yielded {A-B}ₙ block copolymers with more than thirty repeat units of each segment. Scheme 7-1 illustrates the expected reaction sequence.

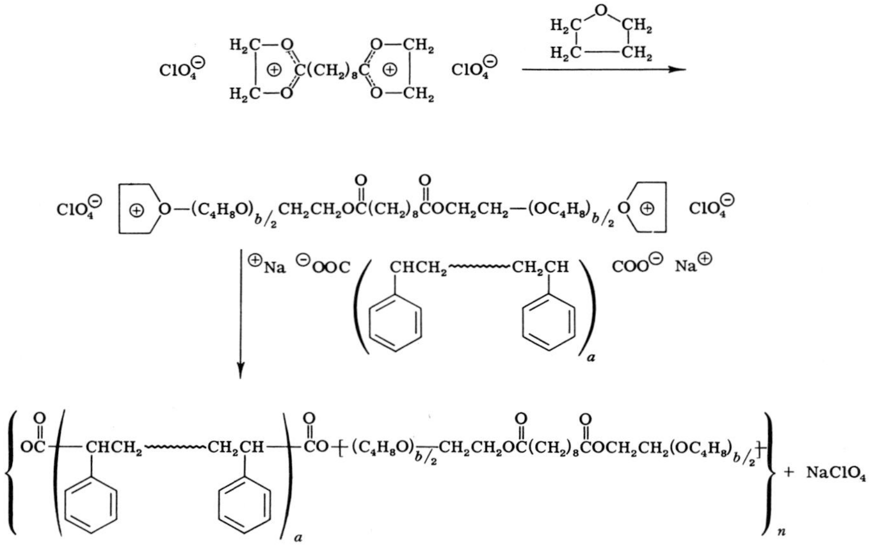

Scheme 7-1

Copolymers containing polystyrene blocks and blocks of a poly (arylene ether) (polysulfone) were synthesized by McGrath *et al.* (B8). This was accomplished by coupling the preformed dihydroxyl-terminated oligomers via phosgenation. The polystyrene oligomers were prepared by anionic polymerization (via sodium biphenyl initiation) followed by end capping with ethylene oxide and acidification, or by free radical polymerization (via benzoyl peroxide initiation) followed by base-catalyzed hydrolysis of the resulting benzoate terminated oligomer. The coupling reaction was carried out as shown in

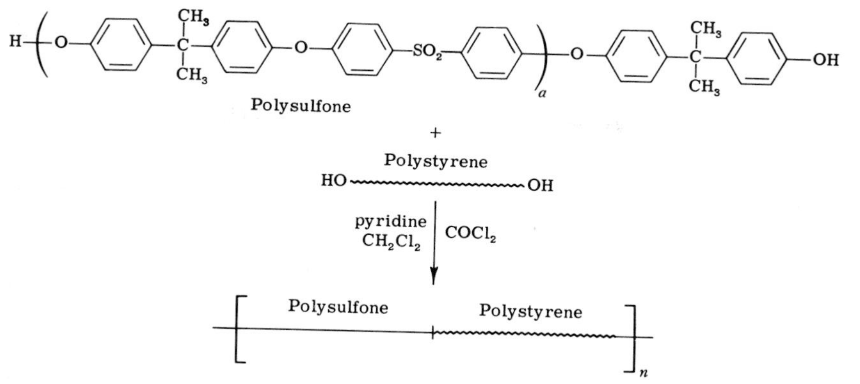

+

Polystyrene

HO ⁓⁓⁓⁓⁓⁓⁓⁓⁓⁓ OH

pyridine
CH₂Cl₂ | COCl₂

| Polysulfone | Polystyrene |

Scheme 7-2

Scheme 7-2. The morphology of these block copolymers was shown to be dependent upon block molecular weight. A composition in which the molecular weights of the polystyrene and polysulfone segments were ~4000 and ~1000, respectively, approached single-phase morphology. Higher molecular weight segments produced two phase systems (see Fig. 7-3).

C. ETHER–ESTER

This subclass of multiblock copolymers is composed of polyether segments such as poly(alkylene oxides) or poly(arylene oxides) [i.e., polysulfone] and polyester segments such as poly(alkylene terephthalates), poly(arylene terephthalates), poly(alkylene adipates), and polycaprolactone.

Several of these materials are attractive for use in elastomeric fiber and thermoplastic elastomer applications similar to those in which thermoplastic polyurethanes have been used. The polyether soft blocks provide the flexibility by virtue of their low glass transition temperatures, and the polyester hard blocks provide the physical cross-linking effect due to their crystallizability. The various compositions reported in the literature are listed in Table 7-1.

1. Alkylene Terephthalates

Block copolymers containing poly(ethylene terephthalate) segments and poly(ethylene oxide) segments were synthesized by Coleman (C1, C2) by the melt polycondensation of dimethyl terephthalate, ethylene

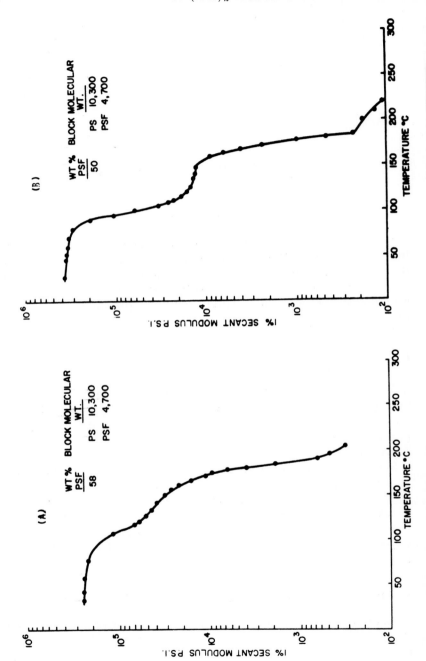

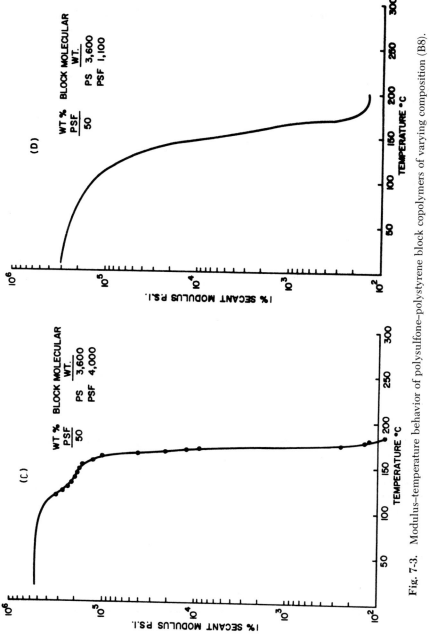

Fig. 7-3. Modulus–temperature behavior of polysulfone–polystyrene block copolymers of varying composition (B8).

TABLE 7-1

{A-B}$_n$ Block Copolymers Containing Polyether and Polyester Segments

Block A	Block B	Synthetic method	References
Ethylene terephthalate	Ethylene oxide	Melt condensation of dialkyl terephthalate + diol + HO-terminated polyether	C1–C9, C55, C56
	Ethylene oxide	Melt oligomer condensation	C57
	Tetramethylene oxide (THF)	Melt oligomer condensation	C10–C12, C58–C60
	Ethylene oxide–propylene oxide	Melt oligomer condensation	C13
	Ethylene oxide–tetramethylene oxide	Melt oligomer condensation	C14, C15
Tetramethylene-2,6-naphthalene dicarboxylate	Tetramethylene oxide	Melt oligomer condensation	C61
Xylylene-2,6-naphthalene dicarboxylate	Ethylene oxide–propylene oxide	Melt oligomer condensation	C16
Tetramethylene terephthalate	Ethylene oxide	Dialkylterephthalate + diol + polyether	C17, C18
Tetramethylene terephthalate	Tetramethylene oxide	Dialkylterephthalate + diol + polyether	C19–C33, C62–C65
	Propylene oxide	Dialkylterephthalate + diol + polyether	C66
Tetramethylene Iso-terephthalate	Tetramethylene oxide	Dialkylterephthalate + diol + polyether	C37, C38
Tetramethylene–glyceryl terephthalate	Tetramethylene oxide	Dialkylterephthalate + diol + polyether	C39
Hexamethylene terephthalate	Tetramethylene oxide	Dialkylterephthalate + diol + polyether	C40
1,4-Dimethylolcyclohexane terephthalate	Propylene oxide	Dialkylterephthalate + diol + polyether	C41, C70
p-Benzenediethyl terephthalate	Tetramethylene oxide	Dialkylterephthalate + diol + polyether	C71
1,4-Dimethylolcyclohexane terephthalate	Tetramethylene oxide	Coupling of ClCOO–polyether + HO–polyester	C42

TABLE 7-1 (*Continued*)

Block A	Block B	Synthetic method	References
Bisphenol A terephthalate	Tetramethylene oxide	Coupling of ClCOO– polyether + HO– polyester	C42
	Ethylene oxide	Coupling of ClCOO– polyether + HO– polyester	C42, C67
	Ethylene oxide	Acid chloride + diol + polyol	C43, C44
Ethylenebis(4-Carboxy-phenoxyethane)	Ethylene oxide or tetra-methylene oxide	Ester interchange	C46–C51, C68
Ethylene adipate	Ethylene oxide	Oligomer condensation	C52
4,4'-(2-Norbornylidene)di-phenol terephthalate	Propylene oxide	Oligomer condensation	C69
4,4'-(Hexahydro-4,7-methanoindan-5-ylidene) diphenol terephthalate or isophthalate	Tetramethylene oxide	Acid chloride + diol + polyol	C54
Ethylene terephthalate	Polysulfone	Acid chloride + diol + polyol	C53
1,4-Butylene terephthalate	Polysulfone	Acid chloride + diol + polyol	C53

glycol, and hydroxyl-terminated poly(ethylene oxide) in the presence of lead oxide (Reaction 7-4). Other workers (C3) prepared these compositions using manganese, antimony, tin, or magnesium based catalysts. Poly(ethylene oxide) oligomers of 1000–6000 molecular

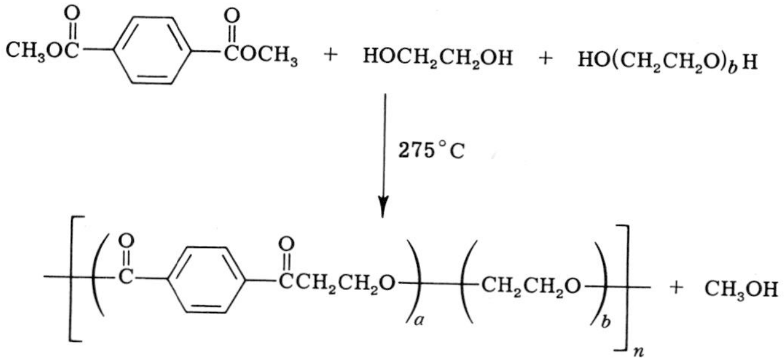

$$\begin{array}{c} \text{CH}_3\text{OC}-\!\!\!\left\langle\;\;\right\rangle\!\!\!-\text{COCH}_3 \;+\; \text{HOCH}_2\text{CH}_2\text{OH} \;+\; \text{HO(CH}_2\text{CH}_2\text{O)}_b\,\text{H} \\[2mm] \downarrow \;\; 275\,^{\circ}\text{C} \\[2mm] \left[\left(\!\overset{O}{\overset{\|}{C}}-\!\!\!\left\langle\;\;\right\rangle\!\!\!-\overset{O}{\overset{\|}{C}}\text{CH}_2\text{CH}_2\text{O}\!\right)_{\!a}\!\!\left(\text{CH}_2\text{CH}_2\text{O}\right)_{\!b}\right]_{\!n} \;+\; \text{CH}_3\text{OH} \end{array}$$

(7-4)

weight were employed. The melting points decreased from 263°C to 241°C as the poly(ethylene oxide) oligomer molecular weight decreased from 6000 to 1000 (at a constant 10 wt% poly(ethylene oxide) content). The glass transition temperature (measured by the penetrometer method) dropped from 70°C to 20°C as poly(ethylene oxide) content rose from 5 wt% to 25 wt%. X-Ray photographs of copolymers containing 15 wt% poly(ethylene oxide) were similar to those of ethylene terephthalate homopolymer. Fibers spun from these block copolymers were not stable to ultraviolet light, and dyed fibers displayed poor light fastness. Another report (C4) claims that these compositions display improved dyeing properties with acetate and vat dyes.

Similar block copolymers were prepared by Brooks *et al.* and were investigated for reverse osmosis desalination behavior (C5). The selectivity of the membranes increased with decreasing poly(ethylene oxide) content (70–50 mole%) and block length (DP = 102 to 15). A composition containing 70 mole% poly(ethylene oxide) of 15 DP block length gave the optimum results and appeared similar to cellulose (2.5) acetate in desalination behavior. The compaction that occurred as these membranes were exposed to increasing operating pressures was said to be completely reversible due to the elasticity of the compositions. These block copolymers were also found by Lyman (C6) to display encouraging hemodialysis membrane properties.

Ethylene terephthalate–ethylene oxide block copolymers were also claimed to have been prepared via transesterification by treating the polyester homopolymer with hydroxyl-terminated poly(ethylene oxide) at high temperatures (e.g., 275°C) in the presence of antimony or calcium catalysts (C7, C8). The products were claimed to be superior to poly(ethylene terephthalate) in antistatic properties (electrical resistance of 10° ohm-cm) and in dyeability.

A study of the ultraviolet degradation of ethylene terephthalate–ethylene oxide block copolymers indicated that chain scission followed first-order kinetics and that the activation energy of the photodegradation was 14 kcal/mole (C9).

Copolymers containing poly(ethylene terephthalate) segments and poly(tetramethylene oxide) segments have been reported (C10–C12). These compositions were prepared by the techniques described above for ethylene oxide-containing copolymers and were claimed to be useful in fiber and molding applications.

Similarly, elastic fibers were reportedly prepared from copolymers containing poly(ethylene terephthalate) blocks and blocks of ethylene oxide–propylene oxide (C13) or ethylene oxide–tetramethylene oxide (C14, C15) random copolymers. Another copolymer containing an ethylene oxide–propylene oxide random copolymer soft segment and a hard segment of poly(xylylene-2,6-naphthalene dicarboxylate) was reported to produce an interesting elastic fiber (C16).

Tetramethylene terephthalate–ethylene oxide block copolymers were reportedly prepared by the interaction of butanediol, dimethyl terephthalate, and poly(ethylene oxide) at 250°C in the presence of phenylenediamine and Ti(OBu)$_4$–MG(OAc)$_2$ (C17, C18). The products were elastomeric.

Nishimura (C19) prepared block copolymers containing tetramethylene terephthalate segments and tetramethylene oxide segments by the melt condensation route at 250°C using Mg[HTi(OC$_4$H$_9$)$_6$]$_2$ as the catalyst. The poly(tetramethylene oxide) segments had $\overline{M}_n$ values of 2000–3000, and the mole ratio of hard block to soft block ranged from 2 to 5. Elastic fibers were prepared from these block copolymers. Strength increased with increasing block length. The films displayed good elastic recovery (99%) from 200% extension at 25°C due to the physical cross-linking effect exerted by the crystalline hard blocks. Recovery at 100°C was poorer, but this property was improved by the use of small amounts of triols in the synthesis to produce chemical cross-links.

Subsequent to Nishimura's work, Witsiepe (C20–C24) reported the synthesis and properties of tetramethylene terephthalate–tetramethylene oxide block copolymers. The synthesis, which involved the use of 1000–4000 molecular weight poly(tetramethylene oxide), was carried out via melt polymerization at 250°C. Higher temperatures resulted in degradation. Improved thermal stability was obtained by carrying out the polymerization in the presence of an antioxidant [sym-di(β-naphthyl)-p-phenylenediamine] using Ti(OC$_4$H$_9$)$_4$–Mg(OOCCH$_3$)$_2$ as the catalyst (C25, C26). Hydrolytic stability was improved by blending with a polycarbodimide (C27, C28).

The products were insoluble in most organic solvents, but soluble in m-cresol. In addition to good solvent resistance, the materials displayed good abrasion and tear resistance. Ultimate properties, such as tensile strength and tear strength, were proportional to copolymer molecular weight (see Fig. 7-4).

X-Ray studies showed that the tetramethylene terephthalate segments displayed the same crystal spacing as that of the homopolymer. The melting point transitions of both the ester and the ether phases were lower than those of the respective homopolymers, presumably due to a combination of the effects of low block molecular weight and some degree of phase blending. The block copolymer melting points (polyester phase) increased with increasing mole fraction of polyester in the copolymer (see Fig. 7-5). Increasing the polyester content also

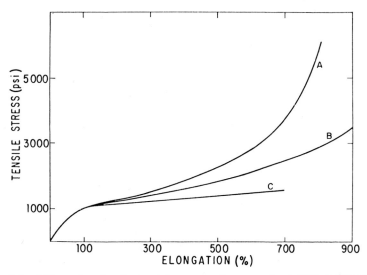

Fig. 7-4. Effect of molecular weight on tensile strength of 33% 4GT/PTMEG-T copolymer. (A) η_{inh} = 1.97; (B) η_{inh} = 1.40; (C) η_{inh} = 0.99 (C22, C23).

changed the copolymers from soft elastomers to impact-resistant rigid systems.

Upon molding, the block copolymer displayed very fast crystallization rates, even when quenched. Therefore, annealing is not necessary to achieve maximum properties, and they were not affected appreciably by annealing. Recovery after compression loading at elevated temperatures was poor, probably due to a combination of creep and recrystallization of the sample while in the stressed condition. A composition containing 33% tetramethylene terephthalate and 67% tetramethylene oxide displayed the following properties: T_m = 176°C; T_g = −78°C; stress at 300% elongation = 1600 psi; tensile strength = 5700 psi; elongation at break = 800%.

Buck and Cella (C29, C30) and Cooper *et al.* (C31) investigated the morphology of these "random segmented block copolymers" via electron microscopy (using phosphotungstic acid as a staining agent), and via X-ray, calorimetry, infrared dichroism, and small-angle light scattering. The ester segments were found to exist in lamellar structures ~100 Å in thickness and several thousands of angstrom units in length. The crystalline regions dispersed in the amorphous polyether phase were highly interconnected rather than discrete and formed the intermolecular tie points responsible for the elastomeric properties of the copolymers (see Fig. 7-6). This is in contrast to the discrete nature of

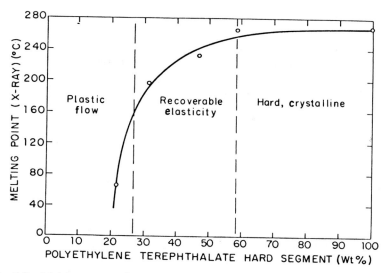

Fig. 7-5. Melting points of segmented polyesters (4000 $\overline{M}_n$ polyether soft segment) (C22, C23).

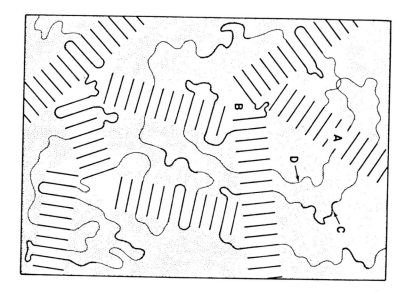

Fig. 7-6. Schematic diagram of the proposed morphology of copolyester polymers. (A) crystalline domain; (B) junction area of crystalline lamella; (C) polymer soft segment; (D) polymer hard segment that has not crystallized (C22, C23).

the amorphous polystyrene phases in styrene–diene block copolymers. This morphological model is consistent with the stress–strain
characteristics of the material. When stress is applied, it is transmitted
initially through the crystalline phase by the tie points. As the crystallites become oriented, the stress is gradually transferred to the elastomeric portions of the polymer network. Finally, the elastomeric domains support all of the stress with the polymer responding like a
typical cross-linked elastomer.

DuPont has recently introduced a new family of polyester elastomer
commercial products, called Hytrel (C28, C32, C33), which are based
on the tetramethylene terephthalate–tetramethylene oxide block
copolymer structure described above. The product is claimed to display a combination of good melt-flow properties, low melt viscosity,
and good melt stability and thus to be useful in rotational molding,
powder coating, injection molding, and casting applications (C28,
C34–C36). The properties of two grades of Hytrel in comparison to
those of a thermoplastic urethane and a plasticized nylon 11 are shown
in Table 7-2. Since Hytrel polymers are based on aromatic polyester
hard blocks, they are claimed to be inherently more stable to hydrolysis than are thermoplastic polyurethanes based on aliphatic
polyester soft blocks (C28). Thermal stability at 150°C was said to be
comparable to that of cured heat resistant rubbers such as neoprene,
nitrile, and chlorosulfonated polyethylene. Chemical resistance, flex
fatigue resistance, tear resistance, abrasion resistance, and creep resistance are also claimed to be good.

Poly(tetramethylene oxide)-containing block copolymers similar to
the above were prepared with hard blocks that were copolymers of
hexamethylene isophthalate and terephthalate (C37, C38). In addition, branched terephthalate structures were synthesized by replacing

TABLE 7-2

Properties of Hytrels as Compared to Polyurethane and Nylon 11[a]

Property	Hytrel 53D	Hytrel 63D	Thermo-plastic ester urethane	Plasticized nylon 11
Tensile at break	5030	6250	5200	8400
Elongation at break	490	605	435	375
100% modulus (psi)	2475	2160	2000	3810
300% modulus (psi)	3530	2680	3680	6125

[a] From Brown and Witsiepe (C32).

a small portion of the difunctional butanediol with trifunctional glycerol (C39). The latter compositions were said to have improved high temperature properties.

Hexamethylene terephthalate–tetramethylene oxide block copolymers of varying block length were synthesized and investigated by wide-angle X-ray diffraction (C40). Below 2000 molecular weight, the polyether segment was wholly amorphous. However, higher molecular weight segments were long enough to be crystallizable, yielding block copolymers that displayed two crystalline phases. The relative rate of crystallization of the polyester and polyether components is important in determining the morphology of the block copolymer. The main features are determined by the faster crystallizing species, while the other component follows the pattern laid down by the first.

Elastomeric block copolymers were also prepared containing 1,4-dimethylolcyclohexane terephthalate segments together with tetramethylene oxide segments (C41) and with ethylene oxide segments (C42). The former were prepared by the usual condensation process described above. However, the latter compositions were synthesized by Riches (C42) from preformed oligomers bearing mutually reactive end groups (Scheme 7-3). Block molecular weights ranged from 400 to

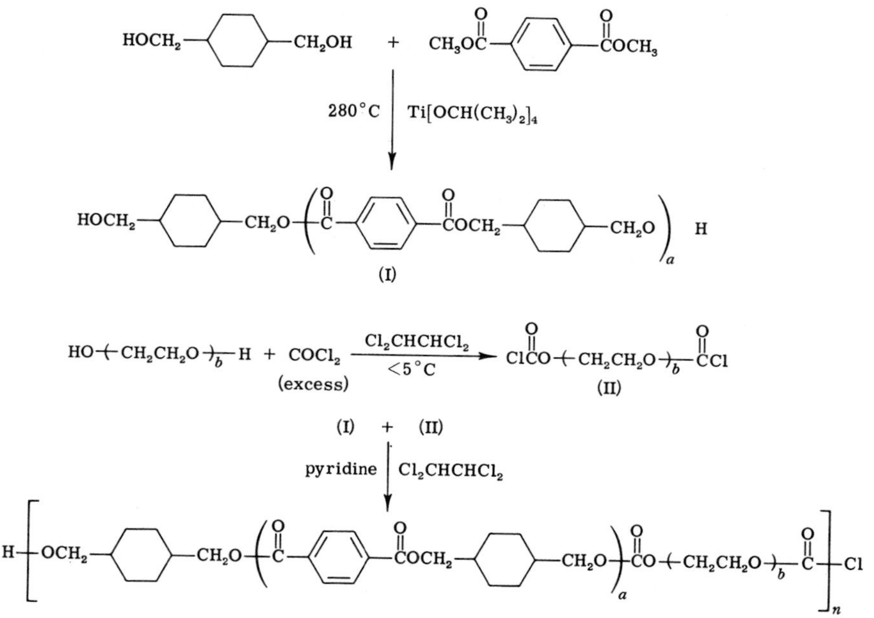

Scheme 7-3

1000 for the poly(ethylene oxide) and from 500 to 3500 for the terephthalate. It was concluded that stress–strain properties were determined primarily by the proportion of the two blocks present and that block length was a minor factor over the range studied. This is illustrated by Fig. 7-7. X-Ray analysis of solution cast films indicated that the polyester segments were more highly crystalline than the corresponding terephthalate homopolymer. There was no evidence of crystallinity in the ethylene oxide blocks except for those copolymers in which the stoichiometry of the reaction had been controlled to produce ethylene oxide end blocks.

Rigid block copolymers containing segments of poly(butylene terephthalate) and segments of polysulfone were synthesized (C53) by the interaction of 1,4-butanediol with terephthaloyl chloride in the presence of terephthaloyl end-capped polysulfone oligomer (Reaction 7-5). Attempts to synthesize good-quality block copolymers with ethylene terephthalate rather than butylene terephthalate segments

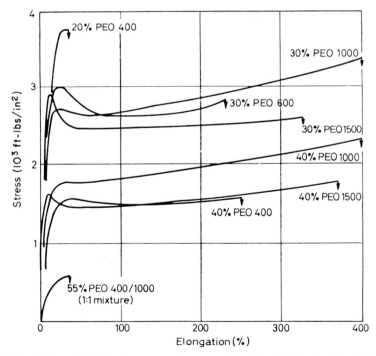

Fig. 7-7. Effect of block size on tensile properties of poly(dimethylolcyclohexane terephthalate)–PEO block polymers (C42). [Reproduced by permission of the publishers, IPS Business Press Ltd.]

(7-5)

were not successful. The butylene terephthalate compositions displayed a T_g transition and a melting transition. The T_g was observed at 139°C, which is intermediate to the T_g values of polysulfone and amorphous butylene terephthalate, indicating that the amorphous fraction of the polyester and the amorphous polysulfone were mutually soluble. The crystalline transition (at 200°C) was attributed to the crystalline polyester phase, thereby demonstrating the importance of crystallization as a driving force for phase separation in an otherwise completely amorphous single-phase system.

2. Arylene Terephthalates

Riches (C42) prepared copolymers containing segments of bisphenol A terephthalate and segments of ethylene oxide or tetramethylene oxide. Block $\overline{M}_n$ ranged from 800 to 2700. The copolymers were prepared from chloroformate-terminated polyether oligomers and hydroxyl-terminated polyester oligomers, as was described in the previous section for Riches' work with dimethylolcyclohexane terephthalate-containing block copolymers. In contrast to the latter copolymers, the bisphenol A based materials showed a marked tendency for the stress to rise after the onset of yielding (see Fig. 7-8). This behavior was interpreted to be due to crystallization of these blocks during stretching.

Copolymers were also prepared from bisphenol A, terephthaloyl chloride, and hydroxyl terminated polyethylene oxide (C43, C44). Bisphenols other than bisphenol A (e.g., hydroquinone, resorcinol) were used to prepare block copolymers by a similar technique. The ethylene oxide blocks were said to increase the solubility of the polyesters while maintaining relatively high softening points. Miscellaneous other aryl terephthalate–ethylene oxide block copolymers have also been reported by this technique (C45).

3. Other Esters

Elastomeric block copolymers with soft blocks of ethylene oxide or tetramethylene oxide and hard blocks derived from ethylene glycol and the dimethyl ester of bis(4-carboxyphenoxy)ethane have been synthesized by ester interchange (C46–C51) (Reaction 7-6). These materials, which were prepared with ethylene oxide block lengths of 1500–4000, had softening points in the range of 137°–190°C. Fibers extruded at 200–260°C were claimed to display interesting properties. Compared to ethylene terephthalate–ethylene oxide block copolymers, they displayed a greater degree of crystallinity but lower soften-

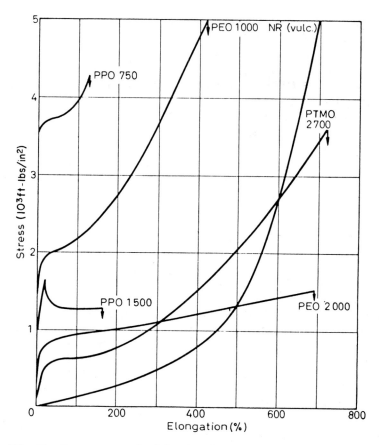

Fig. 7-8. Tensile properties of poly(bisphenol A terephthalate)–poly(alkylene oxide) block copolymers. PPO = poly(propylene oxide); PEO = poly(ethylene oxide); PTMO = poly(tetramethylene oxide); NR = natural rubber (C42). [Reproduced by permission of the publishers, IPS Business Press Ltd.]

$$CH_3O\overset{O}{\overset{\|}{C}}-\!\!\!\bigcirc\!\!\!-OCH_2CH_2O-\!\!\!\bigcirc\!\!\!-\overset{O}{\overset{\|}{C}}OCH_3 + HOCH_2CH_2OH + HO(CH_2CH_2O)_b H$$

$$\downarrow \quad 275\,^\circ C \quad Zn(O\overset{O}{\overset{\|}{C}}CH_3)_2 \tag{7-6}$$

$$\left[\left(-OCH_2CH_2O\overset{O}{\overset{\|}{C}}-\!\!\!\bigcirc\!\!\!-OCH_2CH_2O-\!\!\!\bigcirc\!\!\!-\overset{O}{\overset{\|}{C}}\right)_a O\left(CH_2CH_2O\right)_b\right]_n + CH_3OH$$

ing points and elastic recovery. Crimped fibers were obtained by melt-spinning a blend of the block copolymer and the corresponding polyester homopolymer followed by annealing at 100°C. The crimping was said to be due to the difference in crystallization time between the block copolymer and the homopolymer. The crimping was claimed to be retained over a long period of time, in contrast to the behavior displayed by mechanically crimped fibers.

Ethylene adipate–ethylene oxide block copolymers were synthesized by condensing dihydroxyl-terminated polyethylene oxide oligomers with diisocyanate-capped ethylene adipate oligomers (C52). These compositions were investigated as model systems for polyurethanes.

D. ESTER–ESTER

The segments of the block copolymers of this category are both polyesters. The various compositions reported are listed in Table 7-3.

In 1957, Coffey and Meyrick (D1) reported the preparation of block copolymers in which both blocks were aliphatic polyesters such as ethylene sebacate, ethylene succinate, ethylene adipate, and propylene adipate. These compositions were synthesized by coupling hydroxyl- or carboxyl-terminated polyesters with hexamethylene diisocyanate. For example, the synthesis of ethylene adipate–ethylene sebacate block copolymers from hydroxyl-terminated oligomers is shown in Reaction 7-7. However, the blocks are probably arranged randomly, rather than in perfectly alternating sequence as depicted here, since all four of the end groups on the two polyester oligomers are equally reactive with the diisocyanate. The segments of this type of block copolymer are connected by a urethane linkage. This type of copolymer should not be confused with those in which one of the segments contains a large number of urethane repeat units. The latter structures, which are discussed in Section H, display properties that reflect the higher concentration of urethane groups. The much lower concentration of urethane linkages in the presently discussed polyester–polyester block copolymers exerts much less of an influence on properties. However, it has been reported that such urethane linkages can distort the polyester structure to some extent and thus modify its crystallizability (D27).

The melting points of a series of ethylene sebacate–ethylene adipate block copolymers were measured by a penetrometer technique. The "melting points" were constant (72°C) from 100% to 40% sebacate

$$
\mathrm{HOCH_2CH_2O}\!\left(\!\overset{\overset{\displaystyle O}{\|}}{C}\!-\![CH_2]_4\!-\!COCH_2CH_2O\!\right)_{\!a}\!\!H
$$

$$
+
$$

$$
\mathrm{HOCH_2CH_2O}\!\left(\!\overset{\overset{\displaystyle O}{\|}}{C}\!-\![CH_2]_6\!-\!COCH_2CH_2O\!\right)_{\!b}\!\!H
$$

$$
+\ \mathrm{OCN(CH_2)_6NCO}\ \longrightarrow
$$

$$
\left[\!+OCH_2CH_2O\!\left(\!\overset{O}{\overset{\|}{C}}\!+CH_2\!\,\right]_{\!4}\!COCH_2CH_2O\!\Big)_{\!a}\!CNH(CH_2)_6\!-\!NHC\!-\!OCH_2CH_2O\!\left(\!\overset{O}{\overset{\|}{C}}\!+CH_2\!\,\right]_{\!6}\!COCH_2CH_2O\!\Big)_{\!b}\!CNH(CH_2)_4NHC\right]_n
$$

$$
(7\text{-}7)
$$

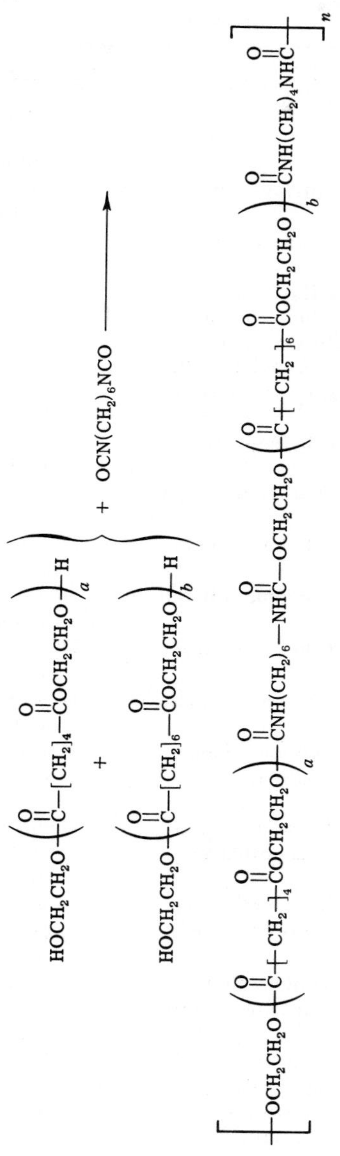

TABLE 7-3

{A-B}$_n$ Block Copolymers Containing Only Polyester Segments

Block A	Block B	Synthetic method	Reference
Ethylene sebacate	Ethylene adipate	Diisocyanate coupling of oligomers	D1
	Propylene adipate	Diisocyanate coupling of oligomers	D1
Ethylene succinate	Ethylene adipate	Diisocyanate coupling of oligomers	D1
Diethylene succinate	Diethylene sebacate	Diisocyanate coupling of oligomers	D2
Ethylene adipate	β-Hydroxyethoxy-benzene adipate	—	D14
Hexamethylene sebacate	2-Ethyl,2-methyl-propylene adipate	—	D28
Ethylene terephthalate	Ethylene succinate	Diisocyanate coupling of oligomers	D3
	Ethylene adipate	Diisocyanate coupling of oligomers	D3
	Diethylene adipate	Diisocyanate coupling of oligomers	D3
	Ethylene azelate	Diisocyanate coupling of oligomers	D3
	Ethylene sebacate	Diisocyanate coupling of oligomers	D3
	Ethylene phthalate	Diisocyanate coupling of oligomers	D3
	Ethylene isophthalate	Diisocyanate coupling of oligomers	D3
Ethylene isophthalate and pentamethylene isophthalate	Tetramethylene adipate	Diisocyanate coupling of oligomers	D15
Decamethylene terephthalate	Decamethylene isophthalate	Coupling of HO- and ClCO-terminated oligomers	D4
Tetramethylene terephthalate	Tetramethylene isophthalate	Coupling of HO- and ClCO-terminated oligomers	D4
Tetramethylene terephthalate	Pentamethylene dithioterephthalate	Coupling of HO- and ClCO-terminated oligomers	D4
Bis A isophthalate	Ethylene terephthalate	Oligomer coupling	D5
Bis A iso–terephthalate	Neopentyl adipate	Oligomer coupling	D6
Bis A hexahydro-terephthalate	2,2-Diethyl-1,3-propylene sebacate	Oligomer coupling	D7

TABLE 7-3 (*Continued*)

Block A	Block B	Synthetic method	Refer-ence
Bis A iso–terephthalate	Diethyleneglycol phthalate–maleate	Oligomer coupling	D8, D9
Bis A iso–terephtha-late–biphenyl-sulfonate	Hexamethylene succinate	Oligomer coupling	D10
Ethylene terephthalate	Ethylene sebacate	Transesterification	D16
	Diethyleneglycol terephthalate	Transesterification	D17, D18
	ε-Caprolactone	Transesterification	D19
Ethylene terephthalate	Ethylene-2,5-di-methylterephthalate	Transesterification	D20
	Hydroxypivalate	Transesterification	D21, D22
	Dimethylolbicyclo-heptane terephthalate	Transesterification	D23
	—SO$_3$Na-substituted neohexyl isophthalate	Transesterification	D24
	—SO$_3$H-substituted ethylene maleate	Transesterification + postsulfonation	D25
Dimethylbis A terephthalate	Hexamethylene terephthalate	Sequential esterification	D26
Dichlorobis A terephthalate	Hexamethylene terephthalate	Sequential esterification	D26
Trans-1,4-Cyclo-hexylene–dimethy-lene terephthalate	*Cis*-1,4-Cyclo-hexylene–dimethy-lene terephthalate	Crystallization-in-duced ester inter-change of random copolymers	D12
Ethylene terephthalate	Ethylene-2-methylsuccinate	Crystallization-in-duced ester inter-change of random copolymers	D13

content, but dropped to 48°C at 30% and lower sebacate concentra-tions. Similar behavior was noted for physical blends of the two homopolymers. Although not stated by the authors, this behavior was undoubtedly due to conversion from a crystalline sebacate continuous phase to a crystalline adipate continuous phase at the 30–40% seba-cate composition level. Copolymers containing crystalline ethylene sebacate and amorphous propylene adipate blocks at the 15–30% propylene adipate level displayed excellent impact strength. The authors attributed this to a "built-in plasticizer effect."

The hexamethylene diisocyanate coupling technique has also been used to prepare diethylene succinate–diethylene sebacate block copolymers (D2). Oligomers of ~2000 molecular weight were reacted at a low (60°C) temperature (to minimize randomizing ester interchange reactions) to produce block copolymers of ~20,000 molecular weight. The glass transition temperature behavior of these block copolymers was more similar to that of random copolymers than that of physical blends. This may have been due to the low molecular weight of the blocks, resulting in a low degree of phase separation of the structurally similar segments.

Copolymers containing aromatic ester blocks and aliphatic ester blocks have been prepared by a similar technique from 2000–3000 molecular weight hydroxyl-terminated polyesters and tetramethylene diisocyanate (D3). The aromatic ester segment was ethylene terephthalate and the aliphatic ester segments were either ethylene succinate, ethylene adipate, diethylene adipate, ethylene azelate, or ethylene sebacate. As mentioned above, a random incorporation of the blocks is likely, since the end groups of the two oligomers are similarly reactive with the isocyanate coupling agent. Under the synthesis conditions used (170°C in nitrobenzene solution), it was reported that little of the undesirable ester interchange side reaction took place. The block copolymers did, however, undergo ester interchange at elevated temperatures (e.g., 250°C) leading to randomization. This was illustrated by the decreasing melting point observed (from 250°C to 120°C) when a 50/50 ethylene terephthalate–ethylene adipate block copolymer was heated at 250°C for 2 hours.

The melting points of these ethylene terephthalate-containing block copolymers were dependent upon the average length of the crystallizing block rather than on the composition, as is the case with random copolymers. The crystalline melting points were ~250°C for compositions containing 30–100% ethylene terephthalate. At concentrations <10%, the melting points observed were characteristic of the other polyester segment. There was a sharp rise in melting point between 10% and 30%. In contrast to these melting point observations, the glass transition temperatures of these compositions were proportional to composition. This behavior, which is typical of random copolymers, is undoubtedly due to the low molecular weight of the segments, which results in segment compatibility in the amorphous regions.

Block copolymers containing 15–35% ethylene terephthalate and 85–65% ethylene adipate displayed elastomeric properties (D3). This was said to be due to joining of the terephthalate crystallites to the flexible adipate chains to form a network similar to that obtained by cross-linking.

The isocyanate-coupling technique has also been used to prepare block terpolymers from various mixtures of ethylene isophthalate, pentamethylene isophthalate, and tetramethylene adipate oligomers of 600–2000 $\overline{M}_n$ (D15). The interrelationship of block molecular weight, composition, and properties are tabulated in the article by Cusano *et al.* (D15).

Flory (D4), in 1954, prepared decamethylene terephthalate–decamethylene isophthalate block copolymers by the condensation of preformed acyl chloride-terminated and hydroxyl-terminated oligomers (Reaction 7-8). The oligomers had degrees of polymerization of

$$(7\text{-}8)$$

about 25. Unlike the isocyanate-coupling technique, this procedure gives alternating rather than random placement of the segments, since the oligomers can react with each other but not with themselves. Block copolymers containing 70–90 mole% of the isophthalate segment displayed a crystalline melting pointy at ~125°C. In contrast to this, random copolymers of the same compositions were soft and rubbery and only slightly crystalline. Tetramethylene terephthalate–tetramethylene isophthalate and tetramethylene terephthalate–pentamethylene dithioterephthalate block copolymers, also prepared by this technique, displayed melting points of 215°–230°C.

Techniques similar to the above, utilizing acyl chloride-, carboxy-, ester-, and hydroxyl-terminated oligomers, have been employed to prepare block copolymers containing bisphenol A isophthalate or terephthalate segments (D5–D9). These compositions are listed in Table 7-3.

A block copolymer has been reported (D10) that contains segments of a random copolymer of bisphenol A (isophthalate/terephthalate/biphenylsulfonate) and segments of hexamethylene succinate. These sulfonate-containing polymers, which were made by the interfacial copolycondensation of the phthaloyl chlorides, the sulfonyl chloride, and bisphenol A in the presence of the acyl chloride-terminated succinate oligomer, were claimed to display good hydrolysis resistance and high heat distortion temperatures.

Several references report the preparation of polyester block copolymers by means of the limited transesterification of a blend of two polyester homopolymers at high temperatures (+280°C) in the melt. These reactions are sometimes carried out in an extruder. At first, ester interchange takes place to produce large segments, but as the reaction proceeds, the interchange continues and the segments become shorter and shorter, until finally, at equilibrium, a random copolymer is obtained. Obviously, the structure of block copolymers prepared by this technique is very difficult to control. The products are usually blends of homopolymers and copolymers of very high polydispersity. Some of the products made by this technique are listed in Table 7-3. Some of these products are reportedly useful in textile fiber applications (D11).

Lenz *et al.* (D12, D13) have explored an unusual and interesting route to preparing "blocky" ester–ester copolymers via crystallization-induced transesterification reactions. The technique involves heat-treating a random copolyester near the melting point in the presence of an active ester-interchange catalyst. The copolyesters investigated included poly(*cis,trans*-1,4-cyclohexylene)dimethylene terephthalate and poly(ethylene terephthalate/ethylene 2-methyl succinate). Under these conditions, crystallization-induced chemical reorganization processes occurred that converted equilibrated random copolymers to nonequilibrium blocky structures. The proposed driving force for this anti-entropic chemical process was said to be the placement of additional units on the ends of crystallizable blocks and their subsequent removal from the reaction zone by crystallization, thereby making them inaccessible for further ester interchange. Melting point, crystallinity, and NMR data supported the notion that blockiness had increased.

E. ESTER–VINYL

The block copolymers of this subclass are composed of alkylene terephthalate segments and polyvinyl segments, such as polystyrene or polybutadiene, or copolymers thereof.

One report (E1) claims the synthesis of hydroxyl-terminated polystyrene oligomers by the polymerization of styrene in the presence of $AlCl_3$ and $SOCl_2$ followed by KOH hydrolysis of the resulting chlorine-terminated intermediate. Styrene–ethylene terephthalate block copolymers were then prepared by the interaction of these oligomers with dimethyl terephthalate and ethylene glycol at 280°C in the presence of Sb_2O_3 and $Cr(OCCH_3)_2 \cdot 2H_2O$. These copolymers were claimed to be useful in plastic and fiber applications.

Polydiene–poly(tetramethylene terephthalate) block copolymers were claimed to be useful in elastomeric fiber applications (E2). Other investigators claim the preparation of ethylene terephthalate polymers containing segments of polybutadiene (E3, E4) or segments of styrene–butadiene random copolymers (E5) by the transesterification of the hydroxyl-terminated vinyl oligomer with ethylene terephthalate homopolymer at 270°C. The products, which contained 5% of the vinyl segments, were claimed to be superior to ethylene terephthalate homopolymer in impact strength.

F. CARBONATES

The block copolymers of this category contain at least one polycarbonate segment. The other segment is either (a) another polycarbonate of different structure, (b) a polyether, (c) polystyrene, (d) a polyester, (e) a polyurethane or (f) a polysulfone. The various structures reported are presented in Table 7-4.

1. Carbonate–Carbonate

These block copolymers have been prepared by the phosgenation of pairs of bisphenols of varying structure. The syntheses were carried out via the technique of interfacial polycondensation. In a typical reaction, a bisphenol was converted to the sodium salt by reaction with aqueous NaOH, followed by the addition of a methylene chloride solution of phosgene at 15°C to form the bisphenol polycarbonate oligomer. Simultaneously, a second bisphenol was similarly converted to an oligomer in a separate flask, after which the two oligomer solutions were combined in the presence of a tertiary amine [i.e., $(C_4H_9)_3N$] and allowed to interact to form the block copolymer. The bisphenols used include bisphenol A, chlorinated bisphenol A, aryl-substituted bis(hydroxyphenyl)ethanes, and bis(hydroxyphenyl)sulfone. The various compositions are shown in Table 7-4.

The block copolymers were reported to be more thermally stable than alternating random copolymers of similar composition (F1). The systems were said to display heat distortion temperatures in the vicinity of 200°–250°C (F2, F3), and some were claimed to be useful as photographic film base materials (F2, F3).

TABLE 7-4

⟨A-B⟩ₙ Block Copolymers Containing Polycarbonate Segments

Block A	Block B	Synthetic method	References
Bisphenol A carbonate	3,3',5,5'-Tetrachloro-bisphenol A carbonate	Phosgenation via interfacial polycondensation	F1, F27–F29
	3,3,5,5-Tetramethyl-bisphenol A carbonate	Phosgenation via interfacial polycondensation	F30
	1,2-bis(4-hydroxy-phenyl)ethane carbonate	Phosgenation via interfacial polycondensation	F27
	Bis(4-hydroxy-phenyl)sulfone carbonate	Phosgenation via interfacial polycondensation	F27
	1,1-Bis(4-hydroxy-phenyl)-1-(3,4-dichlorophenyl)-ethane carbonate	Phosgenation via interfacial polycondensation	F3
1,1-Bis(4-hydroxyphenyl)-1-phenylethane carbonate	1,1-Bis(4-hydroxy-phenyl)-1-(3,4-dichlorophenyl)-ethane carbonate	Phosgenation via interfacial polycondensation	F31
Bis(4-hydroxyphenyl)naphthylmethane carbonate	3,3',5,5'-Tetrachloro-bisphenol A carbonate	Phosgenation via interfacial polycondensation	F2

Bisphenol A carbonate	Ethylene oxide	Interfacial condensation of preformed oligomers	F4, F6
	Ethylene oxide	Solution phosgenation of Bis A and polyether	F7
	Ethylene oxide–propylene oxide block copolymers	Solution phosgenation of Bis A and polyether	F7
	Ethylene oxide–tetramethylene oxide	Solution phosgenation of Bis A and polyether	F8–F10
	Tetramethylene oxide	Solution phosgenation of bisphenol and polyether	F11
	Tetramethylene oxide	Solution phosgenation of bisphenol and polyether	F11
	Tetramethylene oxide	Solution phosgenation of bisphenol and polyether	F11

TABLE 7-4 (*Continued*)

Block A	Block B	Synthetic method	References
[chemical structure: tetrahydronaphthalene bisphenol carbonate]	Tetramethylene oxide (diisocyanate-extended)	Solution phosgenation of bisphenol and polyether	F12
[chemical structure: naphthalene diol malonate carbonate, —OOCCH₂CH₂COO—]	Tetramethylene oxide	Solution phosgenation of bisphenol and polyether	F33
[chemical structure: biphenyl diol malonate carbonate, —OOCCH₂CH₂COO—]	Tetramethylene oxide	Solution phosgenation of bisphenol and polyether	F13
[chemical structure: phenylene malonate carbonate, —COOCH₂CH₂OOC—]	Tetramethylene oxide	Solution phosgenation of bisphenol and polyether	F14
Bisphenol A Carbonate	Polystyrene	Coupling of end-capped living polystyrene	F15
	Polystyrene	Interfacial condensation of capped, peroxide-initiated polystyrene	F4

Carbonate	Polymer	Method	Reference
Bisphenol A carbonate	Polycaprolactone	Solution phosgenation	F16
Hydroquinone carbonate	Polycaprolactone	Solution phosgenation	F16
Dihydroxybenzophenone carbonate	Polycaprolactone	Solution phosgenation	F16
Bisphenol A carbonate	Polycaprolactone extended diol	Solution phosgenation	F16
Hydroquinone carbonate	Polycaprolactone extended diol	Solution phosgenation	F16
2,2,4-Tetramethylcyclobutylene carbonate	Polycaprolactone extended diol	Solution phosgenation	F17
Bisphenol F carbonate	Polycaprolactone	Solution phosgenation	F17
Chlorinated bisphenol A carbonate	Polycaprolactone	Solution phosgenation	F18
4,4'-Dihydroxydiphenylsulfone carbonate	Polycaprolactone	Solution phosgenation	F19
(see structure below)	Polycaprolactone	Solution phosgenation	F20
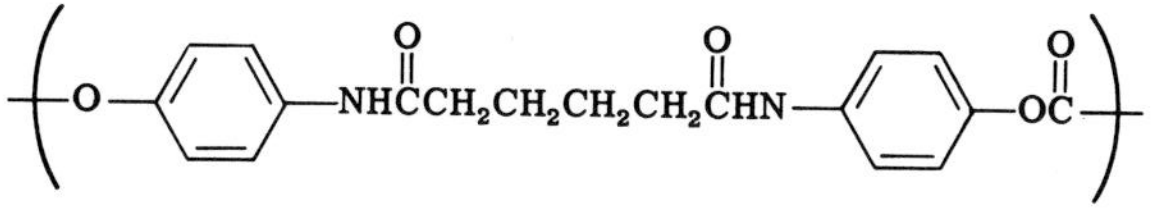 Bisphenol A carbonate	Polycaprolactone	Solution phosgenation	F21, F22
Bisphenol A carbonate	Ethylene adipate	Interfacial condensation of preformed oligomers	F4, F32
	Ethylene terephthalate	Ester interchange	F23
	Bisphenol A carbonate–bisphenol A adipate copolymer	Solution phosgenation of bisphenol A + adipic acid	F24
Bisphenol A carbonate–bisphenol A adipate copolymer	Ethylene oxide	Solution phosgenation of poly(ethylene oxide) + bisphenol A + bisphenol A + dicarboxylic acid	F7

Structure:

$$\left(-O-\underset{}{\bigcirc}-NH\overset{O}{\overset{\|}{C}}CH_2CH_2CH_2CH_2CHN-\underset{}{\bigcirc}-O\overset{O}{\overset{\|}{C}}-\right)$$

TABLE 7-4 (Continued)

Block A	Block B	Synthetic method	References
Bisphenol A carbonate–bisphenol A isophthalate copolymer	Ethylene oxide	Solution phosgenation of poly(ethylene oxide) + bisphenol A + dicarboxylic acid	F7
Bisphenol A carbonate–bisphenol A terephthalate copolymer	Ethylene oxide	Solution phosgenation of poly(ethylene oxide) + bisphenol A + dicarboxylic acid	F7
Bisphenol A carbonate	Bisphenol A–4,4'-Diisocyanatodiphenylmethane urethane	Phosgenation in presence of urethane oligomer	F25
	Bisphenol A chloroformate–piperazine urethane	Phosgenation in presence of urethane oligomer	F25
	Polysulfone	Phosgenation in presence of polysulfone oligomer	F26

2. Carbonate–Ether

These block copolymers contain segments of a polycarbonate derived from bisphenols of varying structure and segments of a polyether such as poly(ethylene oxide), poly(tetramethylene oxide), or an aromatic polyether. The various structures reported are listed in Table 7-4.

In 1961, Merrill (F4) prepared block copolymers containing bisphenol A polycarbonate segments and segments of poly(ethylene oxide) or poly(tetramethylene oxide). The technique employed was the interfacial condensation of preformed hydroxyl-terminated polycarbonate oligomers and chloroformate-terminated polyether oligomers, e.g., Reaction 7-9. The polycarbonate oligomers were pre-

$$(7\text{-}9)$$

pared by the interfacial condensation of bisphenol A with a deficiency of phosgene, and the polyether oligomers were synthesized by end-capping hydroxyl terminated polyethers with excess liquid phosgene. The molecular weights $(\overline{M}_n)$ of both the carbonate and ether oligomers were in the 1000–3000 range. The use of gel permeation chromatography for measuring the $\overline{M}_n$ and molecular weight distribution of bisphenol carbonate oligomers has been described by Robertson et al. (F5).

Traces of amine in the isolated products caused hydrolytic degradation. Long-term exposure to air also resulted in degradation presumably due to peroxide formation. This type of degradation was inhibited by the presence of traces of amine.

Bisphenol A carbonate crystallization occurred more readily in the block copolymers than the corresponding carbonate homopolymer. Thermodynamically, the crystalline state is favored in poly(bisphenol A carbonate) based on symmetry considerations. However, crystallization in the homopolymer is slow since the kinetics are unfavorable. The flexibilizing effect of the polyether segments in the block copolymer enhance the crystallization kinetics of the carbonate block. Crystalline melting points were observed in the 185°–205°C range. X-Ray studies disclosed crystalline polycarbonate patterns with an estimated degree of crystallinity of ~25%, but no polyether crystallinity was observed. The copolymer containing 1600 $\overline{M}_n$ polycarbonate and 3200 $\overline{M}_n$ poly(tetramethylene oxide) segments was reported to display elastomeric properties without vulcanization. This behavior was attributed to the high melting point and the low (<0°C) glass transition temperatures of the composition.

Merrill and Petrie (F6) investigated the effect of composition and block length on the thermal and mechanical properties of bisphenol A polycarbonate–poly(ethylene oxide) block copolymers. Oligomer molecular weights ranged from 2000 to 4700 for the polycarbonate and from 200 to 3000 for the poly(ethylene oxide). The properties of these compositions are presented in Table 7-5. About 20–30 wt% of the ethylene oxide block was necessary to flexibilize the carbonate segments sufficiently to allow thermal crystallization. The melting point of the crystalline carbonate segments was not greatly affected by the polyether blocks. However, the T_g of amorphous compositions decreased with increasing polyether content and with increasing polyether block length. Modulus and tensile strength decreased and elongation increased with increasing polyether content. Two crystalline phases were observed in copolymers containing long (3000 $\overline{M}_n$) polyether blocks. This behavior is consistent with expected morphological effects. The short polyether blocks are compatible with the amorphous carbonate fraction and "plasticizes" it, thereby depressing

TABLE 7-5

Properties of Bisphenol A Carbonate–Ethylene Oxide Block Copolymers[a]

Oligomer $\overline{M}_n$		Young's modulus (kg/cm² × 10⁻⁴)	Tensile strength (kg/cm²)	Elonga- tion (%)	T_g (°C)	Melt- ing point (°C)
Bisphenol A carbonate	Ethylene oxide					
2000	200	2.91	620	3	105	—
2000	1050	0.07	103	>200	−15	188
2000	3000	0.05	47	>200	—	36; 176
3000	620	2.30	460	3	55	225
3000	1340	0.37	150	—	17	231
3000	3000	0.33	32	10	—	41; 220
4700	1050	1.90	420	6–58	42	227
4700	3000	0.97	240	67–95	—	36; 230

[a] Taken from Merrill and Petrie (F6).

it's T_g. The less compatible longer polyether segment enhances phase separation, resulting in the lower T_g levels that are more characteristic of the polyether. The longest polyether blocks are crystallizable and exhibit the melting point of crystalline poly(ethylene oxide).

Goldberg (F7) synthesized bisphenol A carbonate–ethylene oxide block copolymers by phosgenating pyridine solutions of bisphenol A and hydroxyl-terminated poly(ethylene oxide) oligomer mixtures. This technique is more convenient than the preformed oligomer interfacial condensation method employed by Merrill, but it gives less control of block length, since the polyether oligomers can condense with themselves as well as with bisphenol A carbonate sequences. The ethylene oxide oligomers ranged in $\overline{M}_n$ from 1000 to 20,000, and the copolymers ranged in composition from 5 to 70 wt% poly(ethylene oxide) content. Copolymers containing 50 wt% poly(ethylene oxide) (>3 mole%) displayed thermoplastic elastomer properties. These materials exhibited softening temperatures >180°C, tensile strengths of >7000 psi, elongations of ~700%, and snappy elasticity (>90% immediate recovery). For copolymers prepared with a 4000 $\overline{M}_n$ poly (ethylene oxide) oligomer, T_g decreased appreciably from 140°C to 44°C, as the ethylene oxide content increased from 0 to 30 wt%. In contrast, the softening point, due to the polycarbonate T_m, decreased only from 220°C to 180°C over the ethylene oxide concentration range of 0–65 wt%. Figures 7-9 and 7-10 illustrate the effect of composition on T_g, softening temperature, and recovery.

Goldberg (F7) also prepared similar block copolymers using ethylene oxide–propylene oxide copolymers as the soft segment rather than homopoly(ethylene oxide). These compositions were somewhat lower melting but displayed similar elastic recovery properties. Attempts to prepare block copolymers from propylene oxide, butylene oxide, and styrene oxide oligomers resulted in low molecular weight products due to poor end group integrity.

Other workers (F8–F10) used a similar technique to prepare block copolymers containing bisphenol A carbonate segments and segments of ethylene oxide–tetramethylene oxide random copolymers for use as oil-resistant elastomers.

Copolymers containing carbonate segments derived from bisphenols other than bisphenol A together with segments of poly(tetramethylene oxide) have also been prepared. These compositions, which are enumerated in Table 7-4, were made by the pyridine solution phosgenation technique from norbornylidene-containing bisphenols (F11), tetrahydronaphthylenediols (F12), bishydroxyaryl adipates (F13, F33), and bishydroxybenzoates (F14). The norbornylidene-containing copolymers were said to display attractive

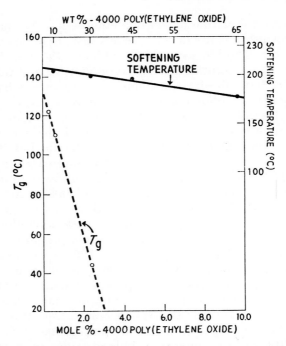

Fig. 7-9. Effect of composition upon the softening temperature and glass transition of 4000 $\overline{M}_n$ poly(ethylene oxide)–bisphenol A copolycarbonates (F7).

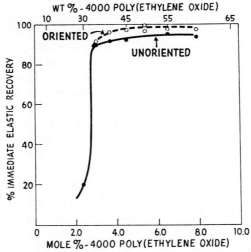

Fig. 7-10. Change in elastic recovery as a function of 4000 $\overline{M}_n$ poly(ethylene oxide)–bisphenol A copolycarbonate composition (F7).

elastomeric properties useful in spandex fiber-type applications. This behavior was attributed to the bulky nature of the bicyclic carbonate segments that "makes movement of the hard segment difficult when the elastomer is stretched and therefore behaves as a tie-down point" (F11). While this steric effect may be a contributing factor, the primary reason for the elastomeric properties must be the two-phase morphology of these systems, as it is with other thermoplastic elastomeric block copolymers.

3. Carbonate–Styrene

Berger *et al.* (F15) synthesized this type of block copolymer by phosgenating disodio-terminated living polystyrene, reacting the resulting acyl chloride end groups with bisphenol A, and then phosgenating a mixture of this oligomer and bisphenol A (Scheme 7-4).

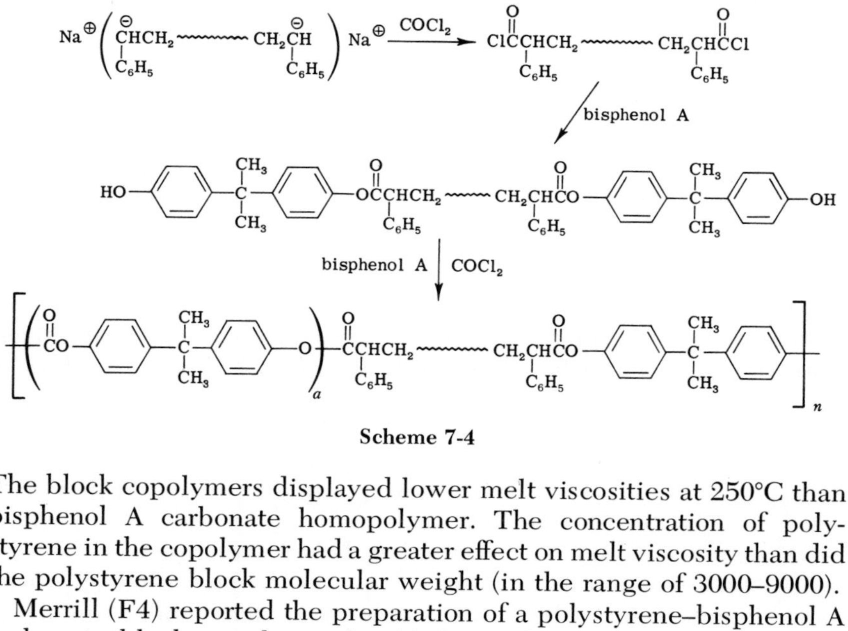

Scheme 7-4

The block copolymers displayed lower melt viscosities at 250°C than bisphenol A carbonate homopolymer. The concentration of polystyrene in the copolymer had a greater effect on melt viscosity than did the polystyrene block molecular weight (in the range of 3000–9000).

Merrill (F4) reported the preparation of a polystyrene–bisphenol A carbonate block copolymer by (a) free-radical initiation of styrene polymerization with benzoyl peroxide followed by (b) saponification of the resulting benzoate-terminated oligomer to form hydroxyl end groups, and then (c) condensation of this 3000 $\overline{M}_n$ intermediate with a 3000 $\overline{M}_n$ bisphenol A carbonate oligomer via interfacial polymerization. This product contained a considerable quantity of styrene homopolymer, as would be expected, but it had a reasonably high molecular weight indicating a fair degree of end-group integrity in the polystyrene oligomer. The mechanical properties were intermediate between those of the two homopolymers.

4. Carbonate–Ester

These copolymers contain segments of polycarbonates derived from bisphenols or from cyclobutylene glycols and segments of lactone or phthalate-based polyesters.

Matzner (F16) synthesized polybisphenol A carbonate–poly-caprolactone block copolymers by the solution phosgenation of mixtures of bisphenol A and hydroxyl-terminated polycaprolactone oligomers of 500–2000 $\overline{M}_n$. The latter were prepared by the base-catalyzed ring-opening polymerization of ε-caprolactone using diethylene glycol as the initiator (Reaction 7-10). These block copolymers,

$$HOCH_2CH_2OCH_2CH_2OH \ + \ (CH_2)_5 \overset{\displaystyle C\overset{O}{\diagup}}{\underset{O}{\diagdown}}$$

$$\Big\downarrow \ \text{base} \qquad\qquad (7\text{-}10)$$

$$H \overline{\Big(O(CH_2)_5\overset{O}{\overset{\|}{C}}\Big)}_b OCH_2CH_2OCH_2CH_2O \overline{\Big(\overset{O}{\overset{\|}{C}}(CH_2)_5O\Big)}_b H$$

at polycaprolactone concentrations of 40–60 wt%, displayed attractive thermoplastic elastomer properties. However, it was necessary to anneal molded compositions (molded at 120°–170°C) for 1–3 days at 100°–130°C in order to achieve ultimate high-temperature properties. Annealing enhances crystallization of the bisphenol A carbonate block. The T_g of these compositions (~25°C) was not significantly affected by annealing, but the melting point (T_m) increased from 40°–100°C to 160°–190°C upon annealing.

Oligomers referred to as "extended diols," preformed by the controlled phosgenation of mixtures of low molecular weight caprolactone oligomers and other various low molecular weight dihydroxy compounds and polyethers, were also employed. These extended copolymer carbonate diols were liquids even though they were higher in $\overline{M}_n$ (~15,000) than the caprolactone homopolymers. The use of these higher $\overline{M}_n$ diols, as a consequence of reaction stoichiometry considerations, resulted in block copolymers containing longer bisphenol A carbonate segments. This, in turn, led to an increased rate of crystallization of the carbonate block and therefore a reduced annealing time requirement (e.g., 3–8 hours rather than 1–3 days). The use of ~15,000 $\overline{M}_n$ caprolactone homopolymers to achieve this end is unsatisfactory due to undesirable crystallization of the highly regular long polycaprolactone chains. This detracts from elastomeric properties. In contrast, the extended diol segments are not crystallizable due to their irregular nature.

In this reference Matzner (F16) also describes block copolymers based on hydroquinone carbonate and dihydroxybenzophenone carbonate segments. It contains considerable data on compositions containing various hard blocks and extended diol segments.

Matzner (F17) has also reported the synthesis and properties of block copolymers containing segments of 2,2,4,4-tetramethyl-

cyclobutanediol carbonate (~50% trans isomer) together with segments of polycaprolactone or of polycaprolactone "extended diols," as described above. These copolymers were synthesized by a two-step process of (a) preparing the cyclobutane diol dichloroformate and (b) reacting this with a mixture of additional cyclobutane diol and hydroxyl-terminated polycaprolactone (or extended diol). Compositions containing 45–60 wt% of 2000 $\overline{m}_n$ polycaprolactone displayed attractive thermoplastic elastomer properties. Unlike the bisphenol A carbonate-based copolymers discussed above, these materials did not require any annealing step, since the cyclobutylene carbonate segments crystallized directly upon molding at 160°C. The T_g was ~25°C and the T_m ranged from 125°–200°C (increasing with decreasing polycaprolactone content). The materials were claimed to have excellent ultraviolet stability due to the cyclobutylene carbonate block.

Other workers have reported block copolymers containing 500–5000 $\overline{M}_n$ segments of polycaprolactone together with segments of bisphenol F carbonate (F18), chlorinated bisphenol A carbonate (F19), aryl sulfone carbonate (F20), and amide carbonates (F21, F22). These were prepared by direct phosgenation in pyridine solution, and they displayed attractive fiber properties. Melting points of 200°–260°C were reported for these materials. The specific compositions are presented in Table 7-4.

Block copolymers have also been reported that contain ethylene adipate segments together with bisphenol A carbonate segments. One composition was synthesized by the interfacial condensation of a hydroxyl-terminated 6000 $\bar{M}_n$ ethylene adipate oligomer and a chloroformate-terminated 1500 $\bar{M}_n$ bisphenol A carbonate oligomer (F4). The X-ray crystallinity patterns for the two blocks of this copolymer were superimposed. Differential thermal analysis disclosed two T_m transitions, one at 47°C (due to the adipate block) and another at 210°C (due to the carbonate block).

A patent reference (F23) claims the preparation of ethylene terephthalate–bisphenol A carbonate block copolymers by refluxing a nitrobenzene solution of the two homopolymers in the presence of $Zn(OOCCH_3)_2 \cdot 2H_2O$, presumably via ester interchange.

Goldberg et al. have reported an elegant, novel technique for synthesizing ester- and carbonate-containing block copolymers in which one of the segments is bisphenol A carbonate and the other segment is a random copolymer of bisphenol A carbonate–bisphenol A adipate (F24). In this procedure, phosgene is added to a solution of bisphenol A in pyridine at room temperature to build up a short bisphenol A carbonate block. Phosgenation is stopped and adipic acid (possibly together with additional bisphenol A) is added, after which phosgene addition is resumed. This leads to the formation of random copolymer segments of bisphenol A carbonate–bisphenol A adipate via the reaction of bisphenol A chloroformate with adipic acid to produce the adipate ester and carbon dioxide (Reaction 7-11). The overall reaction

$$
2 \left[-\!\!\bigcirc\!\!-\overset{\text{O}}{\underset{\|}{\text{OC}}}\!\!-\text{Cl} \right] \quad + \quad \text{HOOC}-(\text{CH}_2)_4\text{COOH}
$$

$$
\left[-\!\!\bigcirc\!\!-\text{OC}-\text{O}-\text{C(CH}_2)_4\text{C}-\text{O}-\text{CO}\!\!-\bigcirc\!\!- \right]
$$

$$
\left[-\!\!\bigcirc\!\!-\text{OC(CH}_2)\text{CO}\!\!-\bigcirc\!\!- \right] + \quad 2\,\text{CO}_2
$$

(7-11)

sequence presumably proceeds as shown in Scheme 7-5. The block

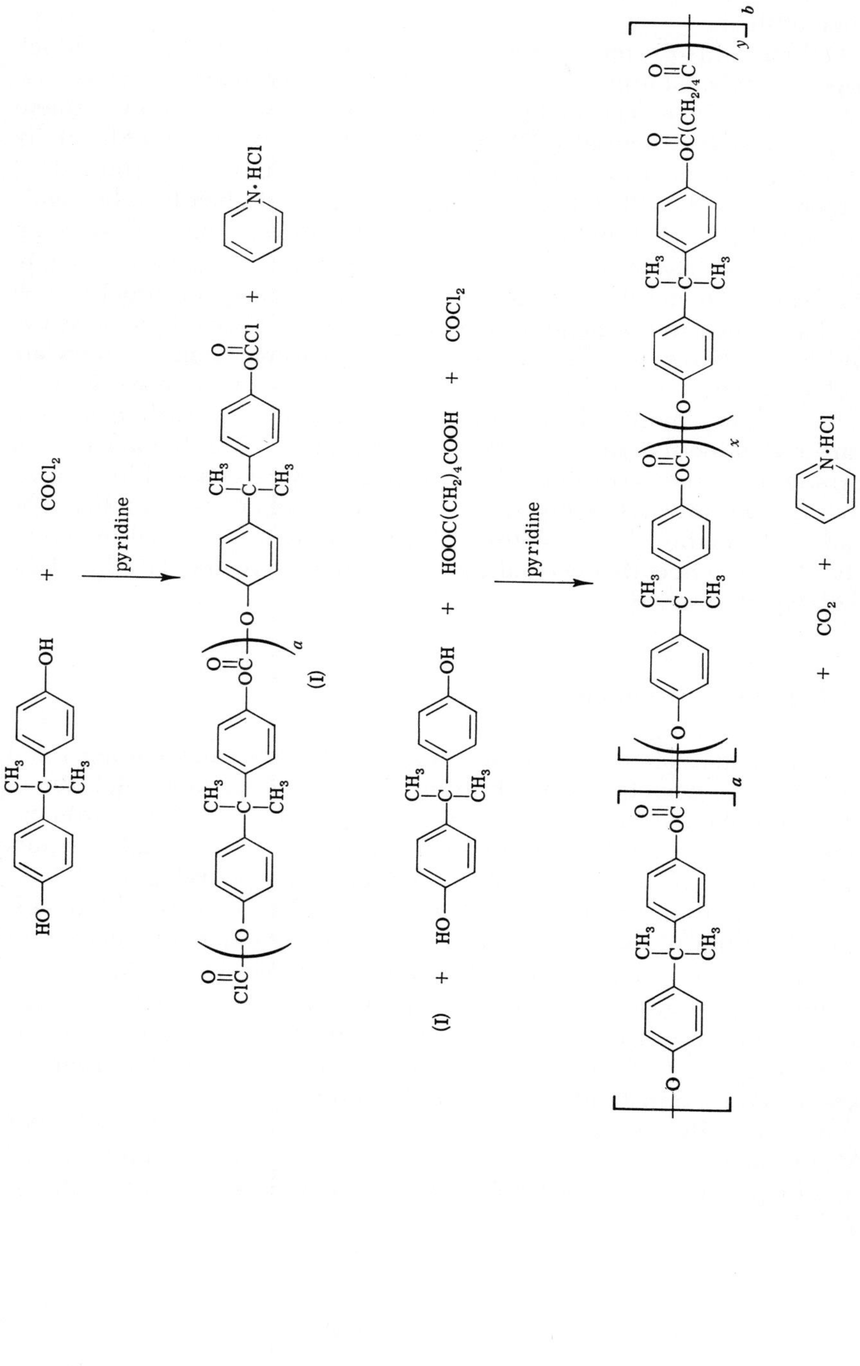

Scheme 7-5

copolymers had higher softening points (155°–220°C) than totally random carbonate–adipate copolymers (100°–175°C) of comparable composition, presumably due to the T_m transition of the bisphenol A carbonate block.

Goldberg also used a variation of this technique to prepare block copolymers containing various bisphenol A carbonate–carboxylate random copolymers as the first segment together with another segment of poly(ethylene oxide) (F7). These syntheses were carried out by phosgenating mixtures of hydroxyl-terminated poly(ethylene oxide) oligomers (4000–20,000 $\overline{M}_n$), bisphenol A, and dicarboxylic acids (isophthalic, terephthalic, or adipic) in pyridine solution at room temperature. These resulting complex structures presumably contain random bisphenol A carbonate–bisphenol A carboxylate blocks, such as that shown in the Scheme 7-5, together with poly(ethylene oxide) blocks, the two blocks being joined both by carbonate and carboxylate linkages. These materials displayed softening points that were much lower than those displayed by bisphenol A carbonate (*no* carboxylate)–ethylene oxide block copolymers. This behavior is attributable to the disruption of the bisphenol A carbonate block structure by the carboxylate units, resulting in loss of crystallizability. The carbonate–carboxylate structures displayed greater thermal oxidative stability, an effect that was said to be due to the greater stability of the carboxylate ester group.

5. Carbonate–Urethane

Block copolymers containing bisphenol A carbonate segments and urethane segments derived from bisphenol A and 4,4'-diisocyanatodiphenyl methane have been reported (F25). The urethane block was formed first in refluxing acetone solution, after which additional bisphenol A, a small amount of *tert*-butyl phenol (end capper), methylene chloride, and aqueous NaOH was added, followed by the introduction of phosgene. The reaction sequence presumably occurred as shown in Scheme 7-6. These products were claimed to be useful in photographic film and recording tape base applications requiring a high degree of dimensional stability. However, structures of this type would be expected to suffer from the known thermal instability of aryl polyurethanes.

A reportedly tough block copolymer containing <5 wt% of a 3000 $\overline{M}_n$ urethane block based on bisphenol A and piperazine and >95 wt% of a bisphenol A carbonate block was prepared as shown in Scheme 7-7 (F25).

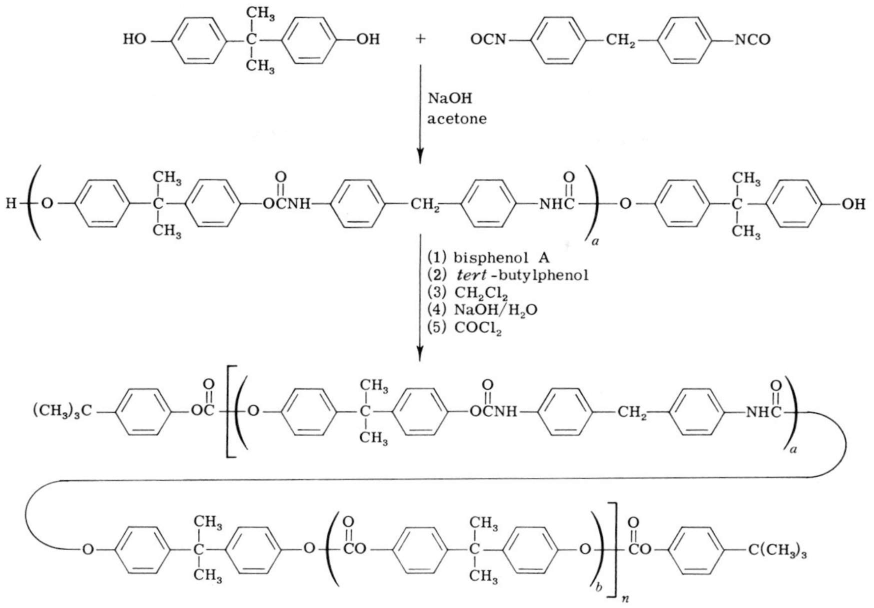

Scheme 7-6

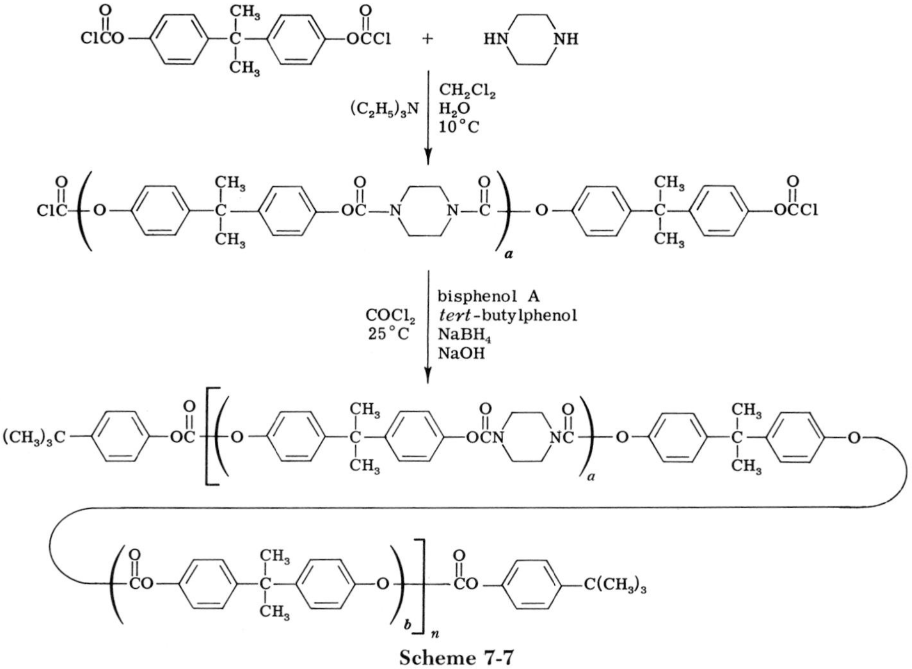

Scheme 7-7

(7-12)

6. Carbonate–Polysulfone

Block copolymers containing segments of polysulfone and segments of bisphenol A polycarbonate were synthesized by McGrath et al. (F26) via the phosgenation reaction (7-12).

The mechanical properties of these compositions were good. Interestingly, they were found to be transparent, single-phase systems, in contrast to the opaque two-phase system obtained by physically blending the two respective homopolymers (see Figs. 7-11 and 7-12). This phenomenon was explained to be due to the chemical similarity of the two polymer structures, as indicated by a small difference in their solubility parameters ($\Delta = 0.7$). The intersegment linkage connecting the two segments is apparently influential enough to make the

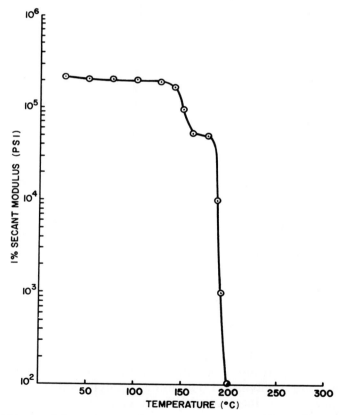

Fig. 7-11. Modulus–temperature behavior of a 50 : 50 wt% polysulfone–polycarbonate blend (F26).

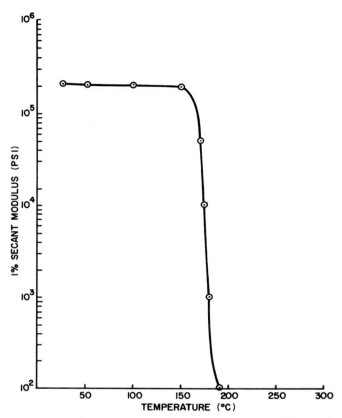

Fig. 7-12. Dynamic mechanical behavior of a 50 : 50 polysulfone–polycarbonate block copolymer (F26).

chemically similar structures mutually soluble, thereby producing a single-phase system.

G. AMIDES

The copolymers discussed in this section contain at least one polyamide segment. The companion segment in these compositions may be another polyamide or a different block, such as a polyether, a polyester, a vinyl polymer, a polyimide, a polythioether, or a polycarbonate. Some miscellaneous polyimide-containing systems are also briefly described. Many of the compositions reported are compiled in Tables 7-6, 7-7, and 7-8, in addition to being cited in the text.

TABLE 7-6

TABLE 7-6

Typical Syntheses of {A-B}$_n$ Polyamide–Polyamide
Block Copolymers

Method	Reference
Amide interchange	G10–G15
Oligomer step-growth or coupling	G14, G16–G19
Interfacial	G20–G22
Solution	G23, G24

1. Amide–Amide

Polyamides, or nylons, such as nylon 6 are one of the most

$$\{NH\text{-}(CH_2)_5CO\}_n$$
Nylon 6

thoroughly studied structures in synthetic macromolecules. Accordingly, it is not surprising that polyamide block copolymers were prepared at least thirty years ago (G1) via amide interchange processes. A number of synthesis routes to polyamides are known (G2, G3), and many can be adapted to the synthesis of block copolymers. In this section, the salient features of polyamide {A-B}$_n$ block copolymer synthesis, characterization, and mechanical behavior will be reviewed. However, no attempt will be made to include all peripheral aspects of the voluminous polyamide literature.

Since the polyamides are generally crystalline materials, soluble only in acidic solvents (phenol, trifluoroethanol, etc.), the segmented systems have often not been subjected to detailed molecular characterization. Nevertheless, it is possible to assess molecular weight (G3) parameters such as end group (e.g., amine, carboxyl) concentrations quite precisely via techniques such as potentiometric titrations (G2). Furthermore, bonding sequences in these copolymers can be determined, at least to some extent by observing variations in the NMR chemical shifts (G4).

The most frequently reported characterization is the examination of the crystalline melting point (T_m) depression as a function of the second constituent. Most investigators (G2) have utilized Flory's equation, which relates the T_m of *random* copolymers to their composition [Eq. (7-13)].

$$(1/T_m) - (1/T_m{}^\circ) = (-R/H\mu) \ln X_A \qquad (7\text{-}13)$$

where T_m and $T_m°$ are, respectively, the melting points of the copolymer and homopolymer, R is the gas constant, $H\mu$ is the homopolymer latent heat of fusion, and X_A is the molar fraction of the major constituent. The equation correctly predicts the T_m to pass through a minimum for the random systems. By contrast, block copolymers possessing sufficiently long segment lengths do not display a similarly depressed crystalline melting point. Therefore, crystallizable block copolymers such as the polyamides can retain high melting points. On the other hand, glass transition temperatures (T_g) are well known to be related to the *weight* fraction of the constituents (particularly the amorphous contents). Thus, in some polyamide and polyester segmented systems one finds only one T_g, i.e., the amorphous phases are compatible. Therefore, T_g may sometimes be varied without significantly affecting T_m, as discussed by Kenney (G5). Thermal analysis, such as differential scanning calorimetry (DSC) or differential thermal analysis (DTA) are powerful tools for characterizing crystalline polyamide block copolymers.

There are several reviews (G2, G5, G6) dealing with the preparation and thermal stability (G7, G8) of amide–amide systems. Some possible synthesis routes are illustrated in Table 7-6. Perhaps the most popular and experimentally simplest is the transamidation or amide interchange approach. Basically, this is the reaction of one amide linkage with another to form two new linkages such as is shown in Reaction 7-14 (G9). In 1944, this reaction was performed in the melt with poly-

$$R^1CONHR^2 + R^3CONHR^4 \rightarrow R^1CONHR^4 + R^3CONHR^2 \qquad (7\text{-}14)$$

amides such as nylon 6 and nylon 66 (G1). In this process, (G73) two homopolymers are transformed into long block copolymers and eventually (if the reaction is continued long enough) into random copolymers. Charch and Shivers (G25) and more recently Kenney (G5) have described the process as shown schematically in Fig. 7-13. The same sequence is followed for other copolymers containing labile groups such as polyesters.

A second way to visualize the process is shown in Fig. 7-14. As indicated earlier, the melting characteristics. of block and random copolyamides are markedly different. This is further illustrated in Fig. 7-15 (G6). The vertical line on the curve indicates that about 8 hours at 285°C would be required to randomize the subject material. Since fabrication processes such as melt spinning, injection molding, or extrusion do not require such long times, it is possible to utilize these block copolymers in practical fabrication operations. However, one can imagine the difficulty of a detailed molecular characterization on

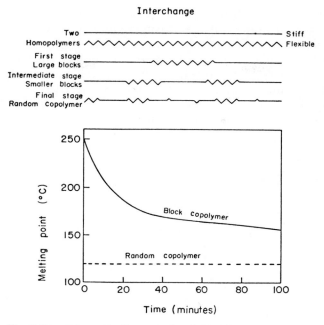

Fig. 7-13. Schematic diagram of melt blending process (G5).

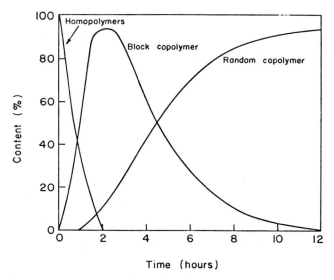

Fig. 7-14. Variation in the composition of nylon 6,6–nylon 6 mixture during a melt blending process (G5).

such a system. Interestingly, it was reported some time ago (G9) that randomization of amide linkages during nylon 66 synthesis would rule out the preparation of nonrandom (presumably aliphatic) polyamides from low molecular weight oligomers. However, use of partially aromatic segments and carefully controlled reaction conditions can apparently permit such a synthesis to take place (G9).

One of the practical incentives for using amide–amide block copolymers is the possibility of eliminating "flat spotting" in nylon tire

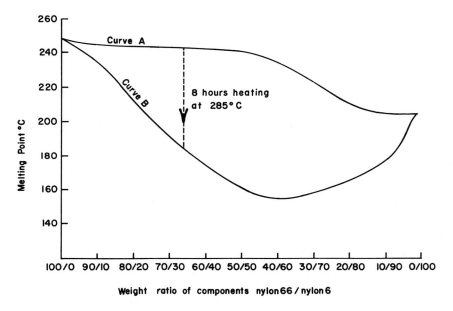

Fig. 7-15. Melting behavior of nylon 6–nylon 6,6 segmented (curve A) and random (curve B) copolymers (G6).

yarns. For example, nylon 66 and other linear aliphatic polyamides have glass transitions in the range of room temperature to 80°C (G11). Melt blending (and block copolymerization) with aromatic–aliphatic polyamides raises the modulus and reduces creep. The effect of a series of aromatic structures (G10) on the set characteristics of nylon 6 or nylon 66 have been reviewed by Sweeny (G2) and are shown in Table 7-7. It is clear that the set value decreases as the aromatic content increases. Similar data on flatspot reduction has also be reported (G5).

TABLE 7-7

Tension Set Values of Block Copolyamides[a]

Aromatic structure	Weight %	Nylon 6,6 (weight %)	Nylon 6 (weight %)	Set[b] (%)	Tenacity gm/ denier
—NH–⟨C6H4⟩–NH–C(O)–(CH₂)₄–C(O)—	—	100	—	1.7	8.5
	30	70	—	0.56	8.7
—NH–⟨C6H4⟩–NH–C(O)–(CH₂)₄–C(O)—	20	80	—	1.04	7.6
—NH–⟨C6H4⟩–NH–C(O)–(CH₂)₄–C(O)—	20[c]	80[c]	—	2.3[b]	3.1
—C(O)–⟨C6H4⟩–C(O)–NH–(CH₂)₆–NH—	30	70	—	0.86	5.4
—C(O)–⟨C6H4⟩–C(O)–NH–(CH₂)₅–NH—	20	80	—	0.64	4.7
	—	—	100	2.1	—
—NH–⟨C6H4⟩–NH–C(O)–(CH₂)₄–C(O)—	30	—	70	0.79	—
—C(O)–⟨C6H4⟩–C(O)–NH–(CH₂)₆–NH—	35	65	65	1.07	—

[a] From Sweeny and Zimmerman (G2).
[b] Determined at a relative humidity of less than 10%.
[c] Random copolymer.

2. Amide–Ether

The preparation of polyamide–polyether $\{A-B\}_n$ block copolymers (Table 7-8) generally utilizes the interaction of an amino or isocyanate terminated polyether polyol with a carboxyl or amino terminated

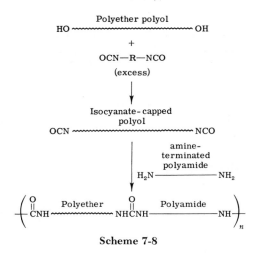

Scheme 7-8

polyamide (Scheme 7-8) (G26). An alternative approach is shown in Scheme 7-9 (G27).

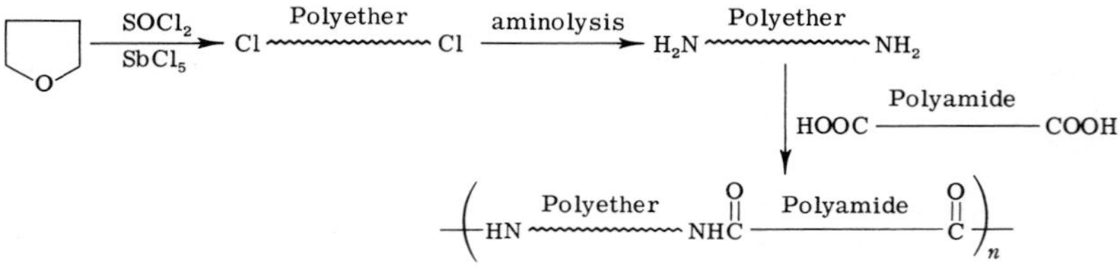

Scheme 7-9

Blocklike structures may also form directly from monomers under some conditions. It was reported (G28) that variation in the choice of solvent, acid acceptor, and mixing sequence determined whether or not blocks were formed from bis(*p*-aminophenyl)ether, bis(*p*-aminophenyl)sulfone and isophthaloyl chloride. The sequence distribution could be determined by NMR spectra in *N,N*-dimethylacetamide containing lithium chloride. The results showed that block copolymers tended to form in sulfolane containing pyridine, and alternating polymers were formed if propylene oxide was the acid acceptor.

Polyamide–polyether materials may be of use in thermoplastic elastomers or elastic fibers (G26, G27) or as additives (G28, G29, G31, G32, G34).

TABLE 7-8

Polyamide–Polyether (A-B)ₙ Block Copolymers

Polyamide	Polyether	Method	Reference
Nylon 66	THF/ethylene oxide oligomer	Step-growth between amine-terminated polyether and acid-functional polyamide	G27
Nylon 66, nylon 6	Poly(alkylene oxide)	Step-growth; hydrolytic polymerization of ε-caprolactam onto amine-terminated polyether	G28–G34
Nylon 66, nylon 6	Poly(alkylene oxide)	Step-growth of isocyanate-capped polyol with amine-terminated polyamide	G26, G35

3. Amide–Ester

Amide–ester (A-B)ₙ block copolymers can, in principle, be prepared via interaction of amino-, carboxyl-, or ester terminated polyamides with suitably terminated (e.g., hydroxyl, isocyanate, acid halide) polyesters of the appropriate molecular weight. Crystallinity and/or end group thermal instability are major limiting factors to the synthesis of well-defined systems. Numerous investigators have attempted melt interchange reactions between preformed polyesters and polyamides. Generally speaking, this approach results in products that are extremely complex and very difficult to characterize. Some exceptions to this situation may exist. For example, it was reported (G36) that the solid state copolymerization (for 6 hours under vacuum) of an equimolar mixture of carboxy-terminated nylon 66 oligomer and hydroxy-terminated poly(ethylene terephthalate) oligomer gave the expected block copolymer. Evidence based on measurements of intrinsic viscosity, NMR sequence distribution, and differential thermal analysis were cited.

As in the case of other step growth systems, it is sometimes possible to prepare blocklike polyamide–polyester structures directly via nonequilibrium polymerizations. Some of the problems encountered in nonequilibrium copolycondensation have been reviewed (G37).

A compilation of the polyamide-polyester systems reported is given in Table 7-9. Many of the copolymers are oriented toward fiber type applications.·

TABLE 7-9

Polyamide–Polyester Block Copolymers

Polyamide	Polyester	Method	Reference
Aromatic polyamides	Polycaprolactone	Cap hydroxyl-terminated polyester with isocyanate; urea link then formed with amine terminated polyamide	G38
Nylon 6 or nylon 6,10	Poly(ethylene terephthalate)	Ester link formed between carboxyl-terminated polymide and the polyester diol	G36, G39–G41
Nylon 6,10	Poly(ethylene terephthalate),	Amide link formed between acid chloride-terminated polyester and amine end groups in polyamide	G42–G46
Nylon 6; nylon 6,6; or nylon 6,10	Poly(ethylene terephthalate)	"Interchange" between ester and amide	G47–G57
Nylon 6,10	Poly(2,2-dimethyl-1,3-propanediol sebacate)	Step-growth; amide link formed via amine–ester reaction	G58
Amino-substituted triazine amides	Polyester	Chloroformate-capped polyester plus amine-terminated polyamide forms an amide link	G59
Polyamides	Aliphatic polyester	Oligomers may be coupled with bisacyllactams	G60

4. Amide–Vinyl

Multiblock copolymers of polyamides with vinyl polymers such as polystyrene and polydienes are in a rather early stage of development. The insolubility of the polyamides in common organic solvents is a formidable obstacle to success in this area.

Shimura and Ikeda (G61) were able to prepare the well-characterized model compound bis-ϵ-aminocaproylaminocaproylhexamethylenediamine (which is an analog of nylon 6 pentamer) $H\{NH(CH_2)_5CO\}_2HN(CH_2)_6NH\{CO(CH_2)_5NH\}_2H$ by an elegant polypeptide synthesis. Attempts to couple the pentamer (which was soluble in aprotic dipolar solvents) with hydroxyl-terminated polystyrene via 4,4-diphenylmethane diisocyanate (MDI) led only to heterogeneous products of low molecular weight. Coupling of the pentamer with α,ω-diepoxypolystyrene was successful and yielded a material having an $\overline{M}_n$ of 18,000.

Polyamide (nylon 6)–polybutadiene block copolymers were reportedly prepared in nitrobenzene via addition of MDI to polybutadiene diol and amino-terminated nylon 6 (G62).

5. Miscellaneous Amides or Imides

Some types of $\{A\text{-}B\}_n$ amide- and imide-containing block copolymers have been prepared that are not incorporated into the previous classifications listed in this section. Nevertheless, several novel systems have been studied and are briefly discussed below.

Crivello and Juliano (G63) have prepared novel polyimido–thioether–polysulfide block copolymers. The polyimidothioether hard segments were prepared by the base-catalyzed condensation of bis-maleimides and hydrogen sulfide (Reaction 7-15). Polysulfides, such

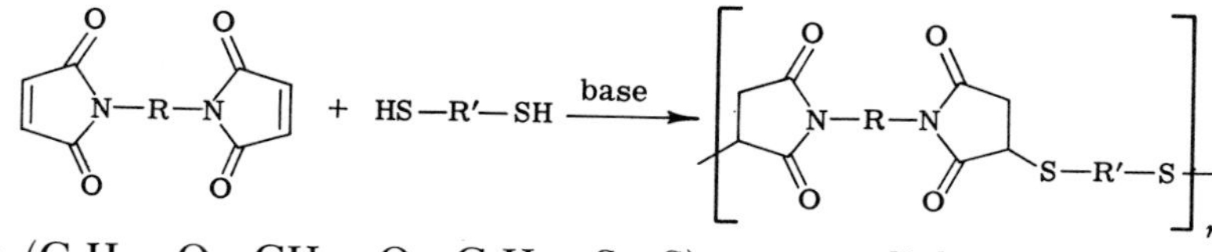

$$(7\text{-}15)$$

as $\{C_2H_4\text{—}O\text{—}CH_2\text{—}O\text{—}C_2H_4\text{—}S\text{—}S\}_n$, are well known soft blocks. Therefore, it was logical to expect multiblock systems containing these two segments to show interesting elastoplastic behavior. The general synthesis of polyimidothioether–polysulfide block polymers is shown in Scheme 7-10.

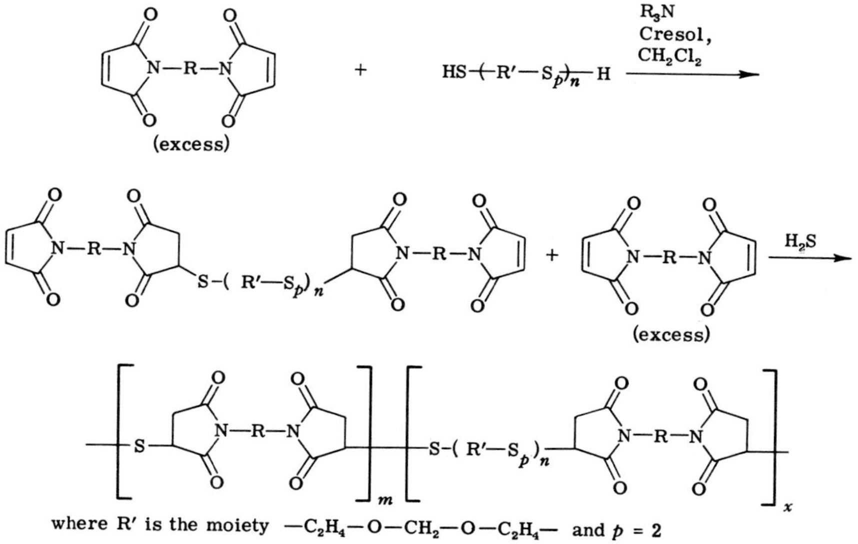

where R' is the moiety $-C_2H_4-O-CH_2-O-C_2H_4-$ and $p = 2$

Scheme 7-10

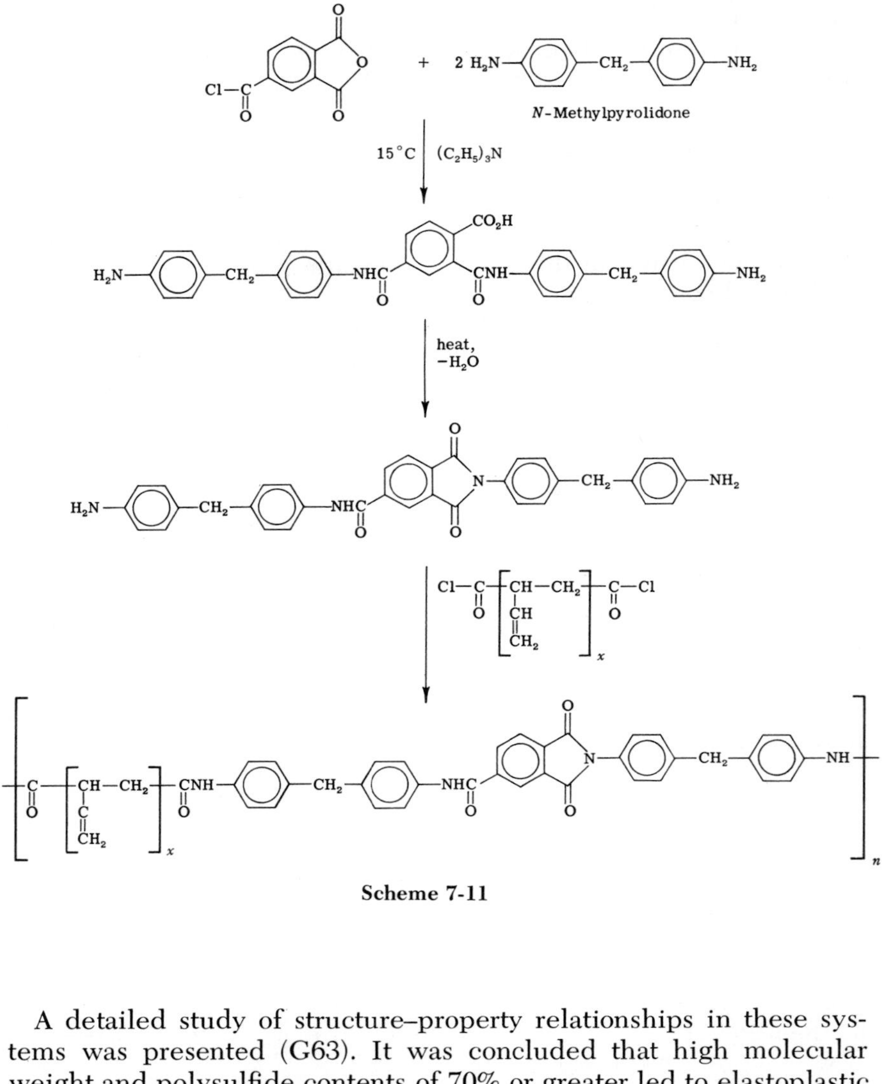

Scheme 7-11

A detailed study of structure–property relationships in these systems was presented (G63). It was concluded that high molecular weight and polysulfide contents of 70% or greater led to elastoplastic materials that could be milled or compression molded.

Polyarylsulfone–polyamic acid block copolymers have been reported that were converted, on heating at 260°C, into the corresponding polyimide system (G64). Aromatic polyamide–polyimide and polybenzoxazoleimide block copolymers have been prepared (G65). Polyamide–imide Schiff bases (G66), polyimide elastomers (G67), polyimide–polyurethanes (G68), and polyamide copolymers containing azo groups (G69) have also been investigated.

Multiblock copolymers of polyimide and 1,2-polybutadiene were also prepared as shown in Scheme 7-11 (G70). Cast films were elastic and flexible. Presumably, they were also readily cross-linked. The structure was confirmed via NMR studies.

Polyamide–polycarbonate block copolymers have been reported (G71). Enantiomorphic block copolymers of γ-methylglutamate have been prepared from the *N,N*-dimethylethylene diamine initiated sequential polymerization of the corresponding L- and D-forms of *N*-carboxyanhydrides (G72). Physical characteristics, such as lack of cholesteric color, and anisotropic properties were quite different for the block copolymer relative to the homopolymer.

H. URETHANES

The polyurethanes are a large family of very important materials that are probably the oldest (H1) and most used block copolymers. New applications as biomaterials (H2) and for interfacing with blood (H3, H4) may be anticipated. Because of their importance, numerous excellent articles, reviews, and monographs have appeared during the past ten years (H5–H26). It is therefore not our goal or within the scope of this section to exhaustively review such topics as the historical development, raw materials, polymerization catalysis, side reactions, and application areas of polyurethanes. Instead, it is our intention to focus on the linear thermoplastic "segmented" polyurethanes (H22, H27–H29) from the point of view of recent developments in structural characterization, structure–property relationships, morphology, thermal analysis, and dynamic mechanical behavior. However, a large number of recent articles on the other urethane areas have been compiled, and these are cited where appropriate and useful. We will also not discuss the use of a small amount of diisocyanate to link dissimilar hydroxyl-terminated polymers, since those materials were described earlier in this chapter.

It is instructive to note the rapid growth of urethane products as shown in Table 7-10. Foamed materials clearly are the major type of urethane product. However, the foaming process per se is not restricted to urethane block copolymers, and therefore, the interested reader is referred to the reviews previously cited for further details. It

TABLE 7-10

Urethane Production[a]

Urethane product	Production (millions of pounds)					
	1955	1960	1965	1970	1972	1975
Flexible foam	1	95	250	673	780	1125
Rigid foam	—	12	71	224	288	645
Noncellular products	1	22	71	106	141	201

[a] From Bedoit (H30).

should also be noted that structural characterization of foamed thermosetting polyurethane systems, is of course, rather difficult. Thermoplastic polyurethane foams are known (H31) but are still relatively insignificant.

1. Structural Types

a. Carbamates (Diol-Based)

The term "polyurethane" as used in elastomer technology includes both carbamate and urea hard segments. The polyurethane block copolymers usually combine a low glass transition (T_g) amorphous or low-melting "soft" block with a rigid "hard" segment that often has a crystalline melting point well above room temperature. The soft block is usually derived from a polyol such as a hydroxyl-terminated polyester [e.g., poly(ethylene adipate), poly(ϵ-caprolactone)] or polyether [e.g., poly(propylene oxide), poly(tetramethylene glycol)] of about 1000–3000 $\overline{M}_n$. Other specialty polyols (H32) can be employed such as hydroxyl-terminated polybutadiene (H33–H39), butadiene copolymers polyisobutylene (H40), fluorinated polyethers (H41–H43), polydimethylsiloxane (H44), and phosphorus-containing oligomers (H45, H46).

The urethane block copolymer may be prepared by "capping" the polyol with an excess of difunctional isocyanate and then chain-extending the macrodiisocyanate with a glycol (e.g., diol-based carbamate hard block) so that the final —OH/—NCO ratio is ≥ 1.0. This step is also often termed the "curing" step. This is a phrase taken from cross-linked rubber technology, which is, of course, not appropriate for the essentially linear thermoplastic elastomers. The idealized process involving a glycol may be written as shown in Scheme 7-12.

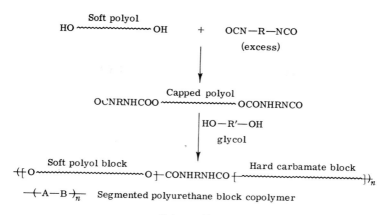

Soft polyol

HO ~~~~~~~~~~~ OH   +   OCN—R—NCO

(excess)

Capped polyol

OCNRNHCOO ~~~~~~~~~~~ OCONHRNCO

HO—R'—OH

glycol

Soft polyol block   Hard carbamate block

+(O ~~~~~~~~~~~~~~~ O}—CONHRNHCO {~~~~~~~~~~~~~~}_n

—(A—B)_n—   Segmented polyurethane block copolymer

Scheme 7-12

Stoichiometry considerations (H47) are quite critical if one wishes to prepare a soluble, linear high molecular weight structure. The excess diisocyanate first reacts with the polyol to form an isocyanate-terminated prepolymer, which then interacts with the residual diisocyanate and the glycol to simultaneously build up the hard carbamate segment and form the $(A-B)_n$ block copolymer structure. There are many variations possible. The reaction of the glycol with the high molecular weight and low molecular weight diisocyanates is statistical. Therefore, the final block copolymer structure actually formed is not perfectly alternating as shown in the above equation, and the blocks themselves are quite polydisperse. An additional assumption is that the isocyanate/hydroxyl ratio is sufficiently less than one so as to encourage the formation of the relatively more chemically and thermally stable hydroxyl end groups.

One can also react the polyol, glycol, and diisocyanate directly, which leads to an even more polydisperse block copolymer. The reaction, however, is straightforward and has been performed in an extruder (H48). In principle, it is possible to prepare a "purer" version of the block copolymer by preforming oligomers and coupling the "hard" and "soft" oligomers through their predesigned hydroxyl and isocyanate functionalities.

b. Ureas (Diamine Based)

Similar reactions can be postulated for the synthesis of block copolymers using a diamine rather than a glycol as the "extender." However, the reactivity of diamines with diisocyanates is much higher than that of glycols, and one normally must be even more concerned with stoichiometry, reaction conditions, etc. Melt stability is much less satisfactory for these urea segments than in the case of the carbamate blocks (i.e., those derived from the "glycol" extenders). In addition to linear chain growth steps, other reactions (H5, H6) between the urea or carbamate group in the backbone and the isocyanate group can occur. For instance, biuret and allophonate groups are formed at 90°C and ~140°C, respectively.

The interested reader is referred to the reviews mentioned earlier for detailed discussions of synthesis conditions (e.g., catalysis and reagent purity).

One additional synthesis route avoids the use of diisocyanates entirely (H49). It involves the use of phosgene or an excess of a bis-chloroformate to form a chloroformate-terminated polyol. The latter can be reacted with a secondary diamine such as piperazine plus more chloroformate or phosgene to produce the block copolymer shown in its idealized form in Scheme 7-13. Again, one assumes a slight excess

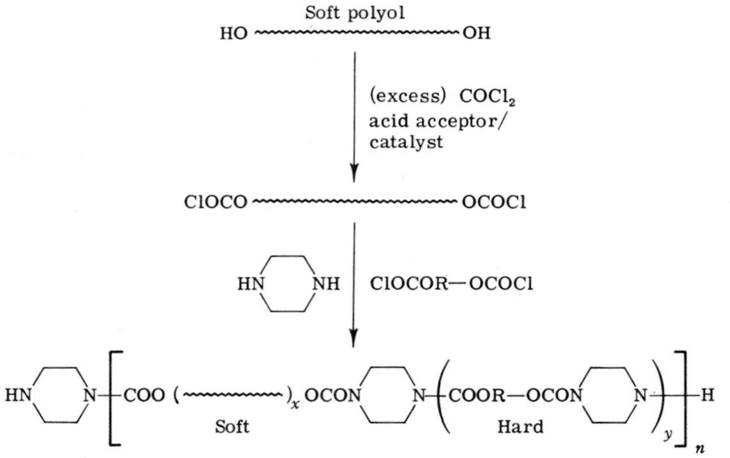

Scheme 7-13

of the piperazine in order to avoid the presence of the highly reactive and thermally unstable chloroformate terminal. The final block copolymer structure prepared by this route has the same "imperfections" described earlier. For example, the average size and the size distributions of the hard segments are a function of the degree of polymerization, y. Furthermore, some of the piperazine can react exclusively with the soft block chloroformate to alter its size distribution. An elegant synthesis procedure was developed by Harrell (H50–H52) to prepare well-defined monodisperse amine-terminated hard segments that could be coupled with chloroformate-terminated poly(tetrahydrofuran) to yield perfectly alternating $(A\text{-}B)_n$ blocks. Further discussion of this work will be deferred to Section H,5,b.

2. Synthesis

The basic chemistry involved in the preparation of linear polyurethanes possessing hard carbamate or urea segments and soft polyol blocks was essentially described in the previous section. However, many variations in the synthesis are possible. Many of these are discussed in the references listed in Table 7-11.

TABLE 7-11

Synthesis of Polyurethanes

Subject	Reference	Subject	Reference
Segmented polyurethanes containing monodisperse hard segments	H50	Polyurethanes soluble in methyl ethyl ketone	H69
Nondiisocyanate-based polyurethanes	H49, H88	Polyether–polyester mixed polyols	H70
One-stage or two-stage routes	H53	Telechelic polymers	H71
Reaction mechanisms	H57	Diene–urea copolymers	H73
Use of piperazine derivatives	H58	Powdered polyurethanes	H74
Linear oligoethylene urethanes	H59	Transparent polyurethane carbonates	H75
Catalysis by tributyltin chloride	H60	Transparent polyurethane laminates	H76
Hydrogen-bonding effects on kinetics	H61	Polyurethanes soluble in polar solvents	H77
Chlorine- or fluorine-containing polyols	H55, H62	Thermoplastic moldable polyurethanes	H27–H29, H78–H80, H85
Effective functionality and intramolecular reactions	H47		
Preparation in a screw extruder	H48, H72	Linear polyurethanes with good recovery	H81
Chemistry and characteristics	H63	Triazine systems	H82
Low-temperature and oil-resistant elastomers	H56	Linear polybiurets	H83
		Water-swellable polyurethanes	H84
Sulfuric acid catalysis	H64		
Polyurethane ionomers	H54	Water-soluble polyurethanes	H86
Kinetics and mechanism	H65	Cycloaliphatic systems	H87
Dioxazolone systems	H66		
Biodegradable polyurethanes	H67, H68		

The stability of the synthesized polyurethane segments is of considerable importance (H89–H101, H107, H108, H139–H141). In general, polyurethanes based on polyether polyols display better hydrolysis resistance and poorer oxidative stability than those containing polyester soft blocks. Polyester hydrolysis is claimed to be improved with a polyol made from 1,6-hexanediol, adipic acid, and diphenyl carbonate (H102–H104). Studies have also shown (H105) that although the polyester structure influenced the hydrolytic stability of polyester-derived urethanes, the dominant factor was the "degree of acidity" of the polyester. Polycarbodiimides have been reported to have a pronounced stabilizing effect, which was attributed (H105) to a "mending" or recoupling action on the broken polyester urethane chains. Other stabilizers have also been reported (H109–H111).

Ultraviolet or, more generally, light instability, has long been noted for most aromatic urethanes (H6, H106, H112–H115). Evidence has been presented (H106) that discoloration and mechanical property loss can be related to the ultraviolet-initiated autoxidation of urethane linkages to quinone imide structures. Polyurethanes derived from diisocyanates structurally incapable of forming such species, even though aromatic, were shown to retain excellent physical properties after accelerated aging (Scheme 7-14) (H106). Alternatively (H115),

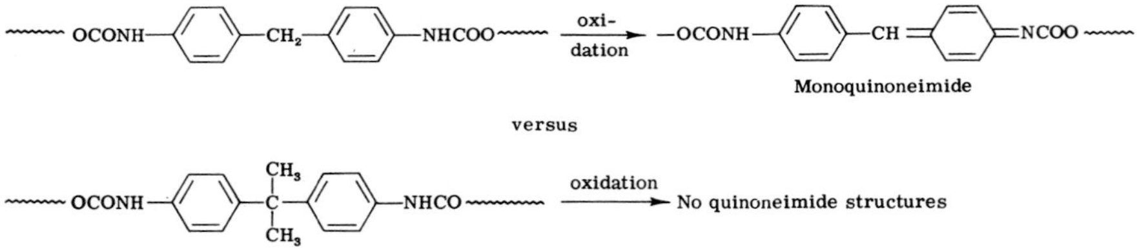

Scheme 7-14

model compound studies indicated that exposure of polyurethanes, based on aromatic diisocyanates and diols, to light and oxygen at room temperature led to formation of primary aromatic amine and hydroperoxy groups. Color formation was attributed to oxidation of the amino group and radical reactions involving the hydroperoxide.

3. Characterization

The analytical chemistry of polyurethanes has been reviewed in a book (H116) that covered the literature through the year 1965. It compliments the earlier books by Saunders and Frisch (H20, H21). In addition, the chemical and compositional characterization of linear polyurethanes (H117) and isocyanate terminated prepolymers has been discussed (H118). However, in comparison to the styrene–diene block copolymer systems, very little work (H119, H120) has been reported on parameters such as molecular weight, molecular weight distribution, and dilute solution properties (chain dimensions, etc.) of linear urethane block copolymers. The molecular weight of the oligomers is usually determined by titrating the hydroxyl terminals (H116). Molecular weight data obtained by titration (e.g., with butylamine) of isocyanate-terminated oligomers and by vapor pressure osmometry have also been reported (H118, H121). The overall number average molecular weights of the Estane thermoplastic polyurethanes has been stated to be in the range of 35,000–50,000.

Considerable useful information on the detailed structure of polyurethanes and polyols (H122, H123) has resulted from NMR, infrared spectroscopic, and pyrolytic–gas chromatography (H124). The number average molecular weight (up to 50,000) can be determined from the ratio of the NMR absorption intensity of the polymeric chain segments to that of the endgroup (H125, H126). The chemical shift characteristics of urea and urethane model compounds were used to identify every component of the linear polyurethanes. Sequence distribution of both poly(ester–urethane), and poly(ether–urethane) systems has been reported (H127–H129). It was possible to distinguish between different degrees of blockiness at equivalent compositions (H128). If the polymer was prepared in a homogenous solution, the sequence distribution was related to the isocyanate reactivity (H129). In the absence of a solvent the hard segment length was longer than expected from reactivity considerations. This was attributed to the heterogeneous polymerization conditions. A complimentary technique employed was a perchloric acid depolymerization (H130, H131). It was reported that this approach could selectively degrade the blocks and allow the molecular weight of the urea segments to be measured. The values deviated from those expected from the equal reactivity of the two isocyanate groups in 4,4'-diphenylmethane diisocyanate (MDI). NMR has also been used to study the ester–urethane exchange reaction (H132).

The rheology of viscous polyester urethane solutions has been studied (H133), but very little basic information seems to be available.

4. Morphology

The morphology of urethane block copolymers is complicated not only by the two-phase structure, as in the styrene–diene systems, but

also by the possibilities of other physical phenomena such as crystallization and hydrogen bonding in both segments. Side reactions resulting in, for example, biuret or allophanate bonds can further confuse
the situation. The problem has nevertheless intrigued many investigators who have utilized a variety of physical techniques to elucidate
the detailed micro and superstructure. In this section, a review of
these results will be presented.

a. Thermal Analysis and X-Ray Diffraction

One might consider the simplest polyol–urethane system to be the
one in which only a stoichiometric quantity of diisocyanate is reacted
with the macroglycol (e.g., without a diol or amine "extender"). A
study of the melting and transition phenomena of poly(ethylene adipate) diols with molecular weights from 1800 to 15,000 and coupled
with toluene diisocyanate was made (H134). Even the relatively minor
amount of connecting urethane linkages apparently can distort the
neighboring crystallizable units to the extent that the melting points,
T_m, are much lower than predicted by random copolymer theory. Additional information is available on the thermal behavior of "simple"
polyurethanes derived from diols and diisocyanates (H135–H138). In
particular, MacKnight and co-workers (H138) prepared the model
series of polyurethanes shown below in which x varied from 2 to 10:

Type A

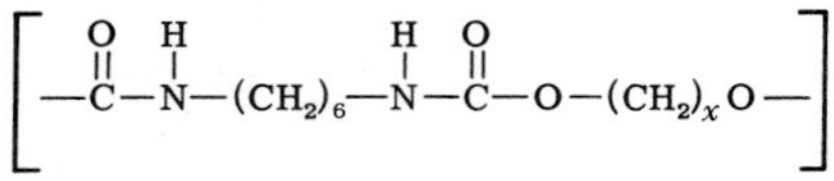

$$\left[-\overset{\overset{\displaystyle O}{\|}}{C}-\overset{\overset{\displaystyle H}{|}}{N}-(CH_2)_6-\overset{\overset{\displaystyle H}{|}}{N}-\overset{\overset{\displaystyle O}{\|}}{C}-O-(CH_2)_x O- \right]$$

Type B

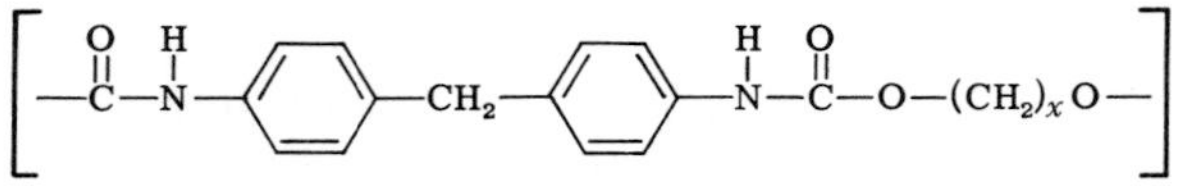

$$\left[-\overset{\overset{\displaystyle O}{\|}}{C}-\overset{\overset{\displaystyle H}{|}}{N}-\bigcirc-CH_2-\bigcirc-\overset{\overset{\displaystyle H}{|}}{N}-\overset{\overset{\displaystyle O}{\|}}{C}-O-(CH_2)_x O- \right]$$

Type C

$$\left[-\overset{\overset{\displaystyle O}{\|}}{C}-\overset{\overset{\displaystyle H}{|}}{N}-\bigcirc-\overset{\overset{\displaystyle H}{|}}{N}-\overset{\overset{\displaystyle O}{\|}}{C}-O-(CH_2)_x O- \right]$$
$$CH_3$$

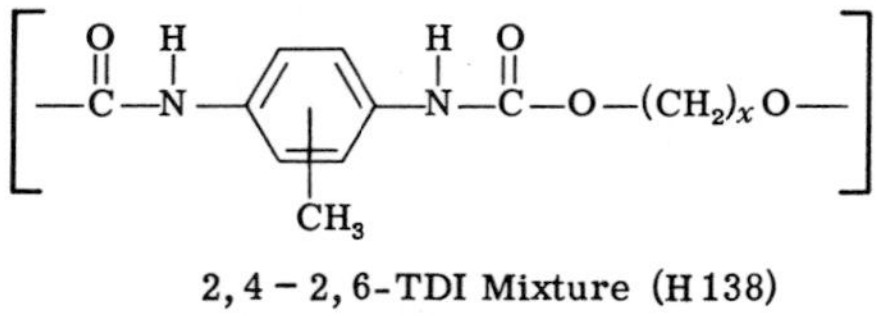

2,4 – 2,6-TDI Mixture (H138)

The glass transition and crystalline melting points for these high
molecular weight "homopolymers" were determined by differential
scanning calorimetry (DSC) and are shown in Tables 7-12 and 7-13,
respectively. As expected, type B, which contains the largest amount
of aromatic moiety, has the highest T_g values. The T_g values in this

TABLE 7-12

Glass Transition Temperatures for
Urethane Homopolymers[a]

X	T_g (°C)		
	Type A	Type B	Type C
2	56	139	52
3	55	119	72
4	59	109	42
5	58	95	52
6	59	91	32
7	55	84	61
8	58	79	64
9	58	72	62
10	55	72	18

[a] From MacKnight et al. (H138).

polymer is more sensitive to the number of methylene units in the diol than the all aliphatic type A series derived from hexamethylene diisocyanate (HDI). In this connection, it is of interest to note that an

TABLE 7-13

Melting Points of
Urethane Homopolymers[a]

X	T_m (°C)	
	Type A	Type B
2	166	—[b]
3	163	241
4	182	248
5	157	192
6	171	200
7	151	198
8	162	201
9	154	194
10	161	194

[a] From MacKnight et al. (H138).
[b] Data not available due to decomposition overlapping the melting point.

additivity scheme for predicting T_g values has been reported (H142). Of the polymers studied by MacKnight (H138), only the polymers derived from the symmetrical methylenediphenyl isocyanate (MDI) or HDI were observed to be crystalline.

The cited results are important to a discussion of the block polyurethanes transition behavior. They allow one to suggest that the hard segments prepared from MDI and butanediol derive their integrity at elevated temperatures from their crystalline or microcrystalline character and not simply from hydrogen bonding. Because of shorter segment lengths and statistical coupling in these block copolymers the corresponding transitions will, of course, be observed at somewhat lower temperatures and is difficult to detect. The effect of the hard block length on the melting transition in a well-defined system (H50) is graphically shown in Figure 7-16. Furthermore, in some less well-defined polydisperse systems that also feature a relatively low concentration of crystallizing segments dispersed in the rubbery soft block matrix, it is difficult to establish the presence of crystallinity. In retrospect, then, it is not surprising that considerable controversy has existed in the literature over the nature of the polyurethane hard segments (H53, H143–H149).

The transitions in block polyurethanes have been discussed by

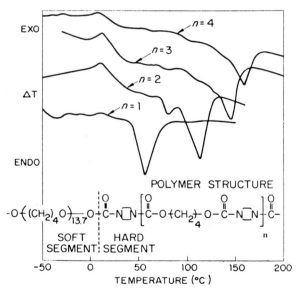

Fig. 7-16. Differential calorimetric scans of polymers with monodisperse hard segment sizes (H50).

numerous authors. The effects of both block length and structural variations on phase separation have been studied but are certainly not completely elucidated (H22).

In attempting to analyze and rationalize the results of many investigators, it is, of course, essential to determine how well the macromolecular structure was defined and to note what type of physical characterization techniques were utilized. It is therefore very helpful to focus attention on the results of Harrell (H50). This work, which was briefly discussed earlier, is probably the most well-characterized study in the entire literature of linear polyurethanes. In this work, narrow distribution perfectly alternating $\{A\text{-}B\}_n$ polyurethanes were prepared. They showed unequivocal crystalline transitions without hydrogen bonding, since they were devoid of N–H groups. There seems little doubt in these very well-defined systems that the thermoplastic elastomeric behavior of these materials is due to the presence of the soft elastomeric phase and to the reinforcing and physical cross-link characteristics of the microcrystalline and glassy (at room temperature) urethane hard blocks. Further detailed characterization of these systems has been reported by Wilkes and co-workers (H150–H152). These results are important, since many of the papers in the literature that report physical measurements on presumably less well-defined block polyurethanes do not present as simple a picture.

A series of papers by Schneider and co-workers (H149, H153–H155) has shed considerable light on the structural organization and transitions in several polyester and polyether urethanes. The soft segments were poly(butylene adipate) and poly(tetramethylene glycol) respectively, while the hard segment was derived from butanediol and MDI. Several transitions were detected via thermal analysis. The low-temperature glass transition was identified for the soft block. The transition at 60°–80°C was ascribed (H153) to urethane–polyester or urethane–polyether "hydrogen bond disruption." An alternate explanation could be that this is the glass transition of the hard phase for the particular block length employed. A transition at 205°C occurred in the polyether based systems. This coincides with the melting of birefringent regions and is believed to be the polyurethane crystalline block T_m. Such a transition was not seen in the polyester urethane. The latter material also was much more transparent, in agreement with smaller crystallite and/or a much lower degree of crystallinity. Note, however, that wide-angle X-ray crystallinity could not be detected in either unstressed copolymer. On the other hand, it was possible to observe stress-induced crystallization of *both* soft blocks. It was concluded (H153, H155) that segregation of the polyol and urethane segments into domain-like structures occurs in both copolymers, but that it exists to a greater extent in the polyether material. The results of a small-angle X-ray study (H154) agreed with this conclusion. It seems probable that phase separation could occur more easily and at lower block molecular weights for the polyether as opposed to the polyester soft phase because of larger differential solubility parameter considerations (H12). Along related lines, it has been reported (H156) that increasing the molecular weight of poly(tetramethylene glycol) from 1000 to 1500 in a 1,4-butanediol (BDO)–MDI urethane caused the soft block glass transition to *decrease* from −43°C to −55°C. This may be interpreted to be due to an increase in phase separation in this molecular weight range. This same paper (H156) concluded that thermal transitions depended on polydispersity as well as on the distribution of the urethane groups as suggested earlier in this section.

Several other studies of thermal or X-ray behavior have also been published (H148, H157–H166, H247). Swelling has been shown to enhance the crystalline development of segmented polyurethanes (H167). Recent study of differential scanning endotherms in block polyurethane elastomers allowed the conclusion that the observed transitions resulted from short- and long-range ordering rather than from hydrogen bond dissociation. Study of the thermal transition behavior (H168) of block polyurethanes prepared from pure 2,4-TDI and pure 2,6-TDI is most instructive. In commerce, the isomers are usually

2, 4-TDI   2, 6-TDI

not separated, and an 80/20 2,4/2,6 mixture is often employed. Schneider and co-workers (H168) showed that the symmetrical 2,6-TDI led to regular ordered hard blocks and observable high-temperature transitions, whereas the polyurethane derived from 2,4-TDI, which can undergo both head-to-head and head-to-tail placement with respect to the methyl group, displayed no transitions above T_g.

b. Spectroscopy

The infrared technique, including dichroism studies, has been used extensively by Cooper and co-workers (H169–H172) and others (H173–H175, H181, H187). MacKnight has studied some model polyurethanes (H178). An analysis of hydrogen bonding was made

(H169). Investigation of the urethane N—H stretching indicated almost all of these groups are hydrogen bonded, whereas about 60% of the urethane carbonyls are similarly associated. Studies of orientation are also possible using infrared dichroism (H171, H182, H183, H281). These studies showed that the soft segments may be readily oriented by an applied stress and yet return to the unoriented state when the stress is removed. By contrast, the hard blocks orient to much the same extent as the soft domains but tend to remain oriented after removal of the applied stress. Thus, this phenomenon provides a possible molecular interpretation for stress hysteresis in these copolymers.

c. Electron Microscopy and Light Scattering

Transmission electron microscopy was employed by Cooper et al. (H176) to study polyester and polyether segmented polyurethanes. Iodine was used as a staining agent for, presumably, the hard phase. Domain structures were observed for both systems as shown in Fig. 7-17 and 7-18.

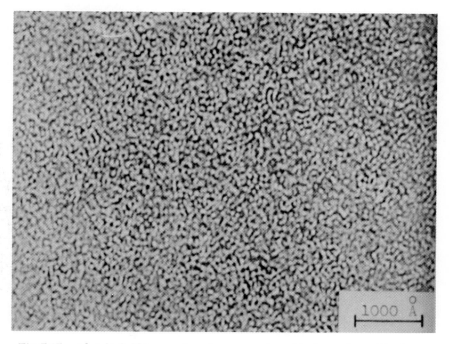

Fig. 7-17. A bright-field image of a polyester–urethane block copolymer film that was solvent etched and stained by iodine. The dark domains correspond to about 30–100 Å in width. Size variances of the domains were observed throughout some of the samples (H176).

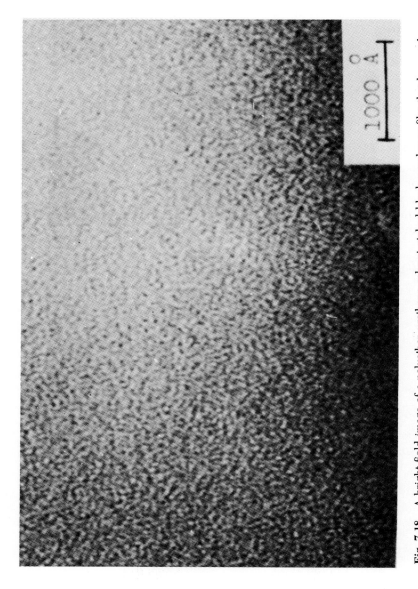

Fig. 7-18. A bright-field image of a polyether–urethane solvent etched block copolymer film that has not been stained. Again, a domain structure is evident with the dark regions of 50–100 Å in size corresponding to the electron dense regions of the sample (H176).

It was concluded that the hard domains were from 30 to 100 Å for both the polyester and polyether urethanes. This is somewhat smaller than for the styrene–diene block copolymers. However, it should be pointed out that the block lengths are considerably shorter in the urethane systems.

The superstructure in the segmented urethanes prepared by Harrell (H50) has been studied by Wilkes and co-workers by photographic light scattering (H177). It was shown that patterns characteristic of spherulitic structure could be observed as shown in Fig. 7-19. The "spherulite" radius was proportional to N, where N is the DP of the hard segment. A particularly well-developed superstructure is shown by scanning electron microscopy when $N = 4$ (Fig. 7-20).

Wilkes and Samuels have also recently reviewed the use of rheo-optical studies for the characterization of block copolymers (H179, H180). Cooper and co-workers (H185) have also discussed stress and birefringence relaxation. Optical anisotropy in segmented poly-urethanes has also been studied (H184).

5. Mechanical Properties

a. Dynamic Mechanical Behavior

The dynamic mechanical behavior of segmented polyurethanes has been studied by a number of investigators (H22, H149, H153, H163, H180, H186, H188–H194, H246). There appears to be general agreement on the temperatures and/or frequency dependence of the transitions and on the fact that they can be markedly influenced by such parameters as block length, composition, and thermal history. There is also agreement that hydrogen bonding plays a very significant role in these systems. Due to the structural complexity of these materials, assignment of a molecular motion to a particular transition is difficult.

Ferguson et al. (H190) studied a series of nine copolymers with a systematically varied hard block to soft block ratio. The soft segment was poly(tetramethylene glycol) and the hard block was the polyurea from 1,3-diaminopropane and MDI. A very low transition ($-130°C$), a soft segment T_g ($-20°C$) and a higher transition could be observed. As the hard segment content increased, spherulites were detected and crystalline melting could be measured. In addition, modulus decreases at around 70°C were noted that were interpreted as being due to breakdown of hydrogen bonding, in agreement with Clough and Schneider (H153). Results of Huh and Cooper (H163) agree experimentally with these findings. However, it was suggested, quite logi-

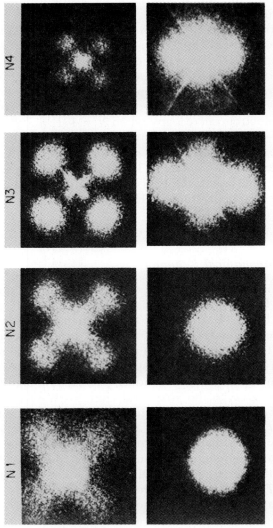

Fig. 7-19. Effect of hard block length on scattering patterns. H_v and V_v scattering patterns for undeformed films on N1, N2, N3, and N4 (H177).

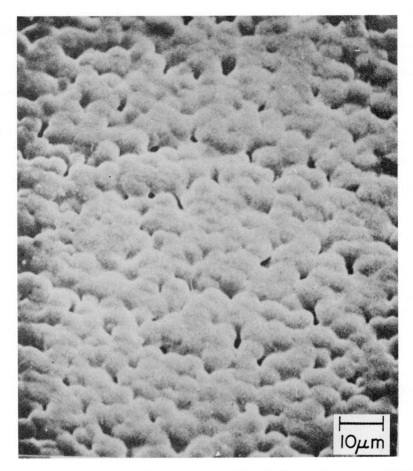

Fig. 7-20. Scanning electron photomicrograph of the N4 block polyurethane (H177).

cally, that at least part of this relaxation is due to the amorphous phase contribution of the hard domain, e.g., T_g for the butanediol–MDI segment. The high temperature transition in the block polymers containing longer segments was again attributed to the melting of the microcrystalline portions of the hard block (H163).

b. Structure–Property Relationships

A number of distinctions have already been made between the polyether and polyester urethanes. Several excellent studies [e.g., molecular weight–property relations (H211), permeability (H241,

H242)] have been published on the effect of diol, diisocyanate, and polyol type on various features of mechanical behavior (H5–H7, H16–H18, H24, H193, H195–H211, H244–H246, H277). A number of useful papers describing the processability and properties of polyurethanes can also be cited (H212–H218, H221). An attempt to illustrate these effects through the use of selected papers is given below.

   i. *Effect of Diol Type.* Koleske, Critchfield and co-workers (H193, H244) have investigated the influence of aliphatic glycol "extenders" ranging in methylene content from two to twelve. The polyol employed was 2000 $\overline{M}_n$ polycaprolactone and the diisocyanate was MDI. A constant mole ratio (1 : 2 : 1) of polyol to diisocyanate to diol was used. The mechanical properties obtained in this study were uniformly excellent, as shown in Table 7-14. However, it was observed in these systems that the modulus values show a dependence on the glycol type. It was suggested that the elastomers containing 1,6-hexanediol and 1,7-heptanediol may have fewer "pseudo" or physical corss-links than the other elastomers. The modulus values pass through a minimum, as seen in Fig. 7-21. The effect of the presence of a urethane hard block segment between polyol segments as compared to that of a single carbamate linkage present in a simple coupled polyol is illustrated in Fig. 7-22. There is a well-defined plateau about the T_g of polycaprolactone in the segmented polyurethane system. In contrast, the modulus–temperature behavior of the coupled polyol is similar to that of high molecular weight polycaprolactone ($T_m \sim 50°C$), i.e., the carbamate linkage is not noticed.

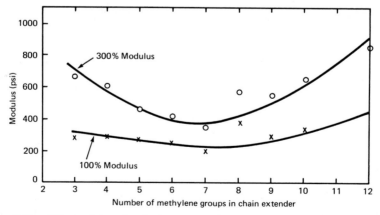

Fig. 7-21. Effect of length of chain extender on 100% and 300% modulus of polycaprolactone–MDI polyurethanes (H193).

TABLE 7-14

Urethane Thermoplastic Elastomers Prepared by Reacting 2000 $\overline{M}_n$ Polycaprolactone, MDI, and a Diol at a 1/2/1 Molar Ratio with the Diols Differing in Methylene Group Concentration[a]

Property	None	\% Lactone in Elastomer / Diol										Test procedure
	88.6	77.6	77.2	76.7	76.3	75.9	75.5	75.1	74.9	74.3	73.5	
	None	Ethylene glycol	1,3-Propane-diol	1,4-Bu-tane-diol	1,5-Pen-tane-diol	1,6-Hex-ane-diol	1,7-Hep-tane-diol	1,8-Oc-tane-diol	1,9-Non-ane-diol	1,10-Dec-ane-diol	1,12-Dodec-ane-diol	
Hardness, shore A	63[b]	63[c]	65	64	65	67	58	65	67	65	72	ASTM D2240
100% Modulus, psi	770	245	275	290	280	255	200	365	290	340	425	ASTM D412
300% Modulus, psi	835	435	665	610	460	420	350	585	550	650	850	ASTM D412
Tensile strength, psi	4200	4400	4500	4400	3400	4200	3600	4100	4300	4200	900	ASTM D412
Elongation, %	835	685	590	705	650	705	760	700	665	610	645	ASTM D412
Graves tear, psi	445	340	300	320	275	235	225	285	255	270	300	ASTM 624C
B Compression set, %	93	63	68	73	70	85	94	68	71	72	68	ASTM D395B
Zwick rebound	58	66	70	69	65	69	63	65	62	70	66	
Tg (G'), 30°C	-40	-38	-38	-40	-40	-42	-42	-42	-40	-43	-43	DIN 53-512
nsp/C, 30°C, DMF, 0.2 gm/dl.	0.89	0.85	1.19	1.05	0.93	1.26	0.89	0.88	1.08	1.01	0.99	

[a] From Critchfield et al. (H193).
[b] Crystallized in less than one week. Properties are for the partially crystalline polymer.
[c] Crystallized on long standing.

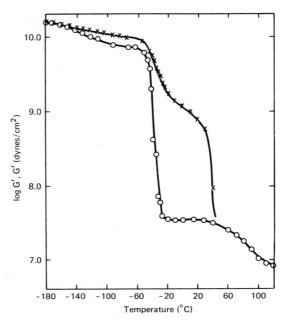

Fig. 7-22. Real component of the complex shear modulus for (×) plastic from NIAX polyol D560/MDI and (○) elastomer from NIAX polyol D560/MDI/1,4-butanediol. Specimens were aged about 1 week at room temperature after molding (H193).

Complimentary studies on the effects of glycol type are shown in Table 7-15. It is noteworthy that the aliphatic glycol containing a partially aromatic structure (polymer E) has the highest hardness and modulus.

ii. *Effect of Polyol Type.* The polyol contributes low-temperature properties and extensibility to the segmented polyurethane. Therefore, it is obvious that features such as its T_g will be highly important. Furthermore, crystallinity (if any), T_m, and ability to crystallize under deformation will certainly influence the ultimate mechanical properties. It has been reported that stress induced crystallization can improve tear resistance and tensile strength (H166, H219, H220), while at the same time diminishing the recovery characteristics (H219). It is well known that conventional cross-linked rubbers such as natural rubber have high strength for similar reasons. However, natural rubber shows good recovery because its crystallinity is largely lost when the deformation is removed. Some effects of polyol type on mechanical properties are shown in Table 7-16. Minor differences are noted in

TABLE 7-15

Effect of Glycol Structure on the Properties of Some Thermoplastic Polyurethanes[a]

Poly- mer	Glycol	Dilute solution viscosity	Hard- ness (Shore A)	Tens. strength (psi)	Elong. (%)	300% modulus (psi)	T_2[b] (°C)
A	Ethylene glycol	0.851	80	6,500	500	1,000	161
B	Trimethylene glycol	—	80	5,900	575	1,200	158
C	Tetramethylene glycol	0.904	88	7,800	530	1,300	160
D	Hexamethylene glycol	0.751	87	5,700	580	1,100	139
E	1,4-Bis(β-hydroxy-ethoxy) benzene	—	93	3,700	550	1,900	128

[a] From Schollenberger (H205). *Components:* diphenylmethane p,p'-diisocyanate (2.00 mols), poly(tetramethylene adipate)glycol, molecular weight about 1000 (1.00 mol), glycol (1.00 mol).

[b] Processing temperature.

TABLE 7-16

Effect of Polyol Structure on the Properties of Some Thermoplastic Polyurethanes[a]

Poly- mer	Polyol/ molecular weight	DSV[b]	Shore hard- ness	Tensile strength (psi)	Elon- gation (%)	300% modulus (psi)	T_2 (°C)
A	Poly(ethylene adipate)glycol/980	0.824	86 (A)	7400	655	900	136
B	Poly(tetramethylene adipate)glycol/989	0.904	88 (A)	7800	530	1300	160
C	Poly(hexamethy-lene adipate)gly-col/986	1.058	82 (A)	8600	560	1200	147
D	Poly(1,4-cyclohexyl-dimethylene adipate)glycol/1190	0.697	60 (D)	5600	355	4800	142
E	Poly(tetramethyl-ene glycol)/974	0.935	90 (A)	5300	725	1000	130
F	Poly(propylene glycol)/1005	0.874	76 (A)	4200	800	640	146

[a] From Schollenberger (H205). *Components:* diphenylmethane-p,p'-diisocyanate (2.00 moles), macroglycol (1.00 mol), 1,4-butanediol (1.00 mol).

[b] DSV = dilute solution viscosity.

polymers A, B, and C (Table 7-16). A dramatic increase in hardness and stiffness can be seen with the bulky cycloaliphatic structure D. The polyethers E and F show significant differences in hardness, modulus, and tensile strength. These data reflect the ability of the regular poly(tetramethylene glycol) to crystallize under deformation, whereas the atactic stereoisomers present in the poly(propylene glycol) prevent crystallization in this case.

iii. *Effect of Diisocyanate Type.*   The structure of the diisocyanate can exert a critical influence on the properties of a thermoplastic polyurethane. Investigations by Schollenberger (H205) and others can be cited. The effects can be illustrated by the data in Table 7-17. Factors such as high symmetry and rigidity in p-phenylene diisocyanate lead to the fact that polymer A has a high modulus. Also, the intermolecular forces are strong enough that the block copolymer is not soluble in solvents such as dimethyl formamide. The additional flexibility in MDI due to the methylene linkage is reflected in the mechanical behavior of Polymer B. Nevertheless, the symmetry of MDI is sufficient to allow the preparation of semicrystalline hard blocks. The meta structures in Polymers C and D, and particularly the additional methyl substituent in the latter case seriously disrupt symmetry and potential crystallizability. It is felt that the upper use temperature of polymers such as polymer D is related to the glass transition of the hard segment. Other structural variations not shown above include the use of aliphatic [e.g., hexamethylene diisocyanate (HDI)] or cycloaliphatic [e.g., hydrogenated MDI (HMDI)] diisocyanates, often for the purpose of enhanced weatherability. The use of HDI leads to readily crystallizable block polymers. However, HMDI usually is employed as a mixture of isomers, and the derived polyurethanes do not possess as high a use temperature as those from MDI. Such behavior would, in fact, seem to be another good argument that the high temperature transition in polyurethane hard segments are of a crystalline nature.

iv. *Effect of Segment Size Distribution.*   In the styrene–diene–styrene systems, the effect of segment size distribution has been investigated by Morton and co-workers (H222). The only comparable study in the field of linear polyurethanes is the elegant investigation by Harrell (H50). In his terminology, the model polyether–polyurethane shown in Table 7-18 had the designations listed. The stress–strain behavior and extension set properties are illustrated in Fig. 7-23 and 7-24. It is interesting to note that the ultimate properties are signifi-

TABLE 7-17

Effect of Diisocyanate Structure on the Properties of Some Thermoplastic Polyurethanes[a]

Diisocyanate	Structure	Tensile strength (psi)	Ultimate elongation (%)	300% modulus (psi)	T_2 (°C)	DMF soluble
A. p-Phenylene diiso-cyanate (p-PDI)	OCN—⬡—NCO	3800	330	3400	152	No
B. Diphenylmethane-p,p'-diisocyanate (MDI)	OCN—⬡—CH₂—⬡—NCO	5500	610	1900	134	Yes
C. m-Phenylene diiso-cyanate (m-PDI)	OCN—⬡(NCO meta)	9000	580	1400	152	Yes
D. 2,4-Tolylene diiso-cyanate (TDI)	OCN—⬡(CH₃, NCO)	4300	680	360	67	Yes

[a] From Schollenberger (H205). *Components:* diisocyanate (2.50 moles), poly(tetramethylene adipate)glycol (1.00 mole), 1,4-bis(2-hydroxyethoxy)benzene (1.50 moles).

TABLE 7-18

$$-O + CH_2CH_2CH_2CH_2O + \overset{\overset{O}{\|}}{C} - N \bigcirc N + \overset{\overset{O}{\|}}{CO(CH_2)_4}O\overset{\overset{O}{\|}}{CN} \bigcirc N + \overset{\overset{O}{\|}}{C} -$$

13.7

Molecular Weight Distribution[a]

Soft segment	Hard segment	Designation
Narrow	Narrow	NN
Narrow	Broad	NB
Broad	Narrow	BN
Broad	Broad	BB

[a] From Harrell (H50).

cantly better in the case of a narrow size distribution. This seems to be a particularly important feature for the hard segment. This behavior might be rationalized on the basis of a more perfect physical network

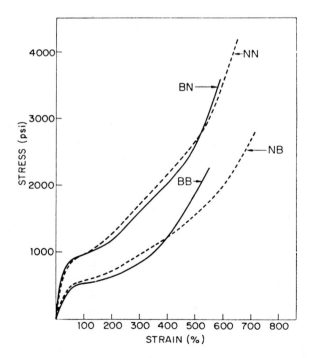

Fig. 7-23.   Stress–strain properties as a function of segment size distribution (H50).

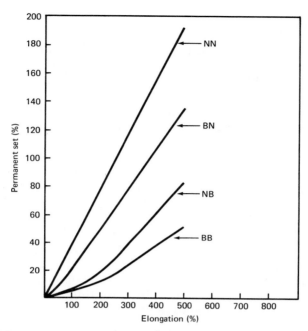

Fig. 7-24. Extension set properties as a function of segment size distribution (H50).

and the enhanced development of crystallization during deformation. On the other hand, recovery or permanent set is better when the size distribution (particularly the hard segment) is broad. This is a more complex property to interpret, but it is entirely possible that the higher degree of order that occurs during deformation of the NN or BN restricts the viscoelastic recovery of the test specimen once the load is removed. It should also be noted that these results are entirely consistent with those cited earlier (H219) on systems where the crystallizability of the soft block was varied.

v. *Miscellaneous Studies of Network Behavior.* Various studies of network behavior have been reported, including chemical stress relaxation (H220), stress relaxation and interaction (H223, H231), swelling (H224), dielectric relaxation (H225, H226), photoelasticity (H227–H230), hydrostatic tensile fracture (H232), chemical cross-linking and selective plasticization (H233), permeability (H234, H235), and crystallization (H146, H147, H236–H240, H243). While many of these investigations are of interest, it is deemed beyond the scope of the present book to discuss them in depth, particularly since several have been covered in a review (H22).

6. Polyurethane Fibers

The development and growth of polyurethane (and/or urea) spandex fibers has been phenomenal. A discussion of the associated fiber technology will not be attempted here. We will only point out many of the excellent reviews in this important area and also discuss some of the basic structural features in these systems. A list of review articles and useful patents are given in Table 7-19.

Elastomeric polyurethane fibers are characterized by a very high elongation at break, very low modulus, and high recovery from large deformations. These properties distinguish them from stretchable hard fibers that lack one or more of these features. For instance (H248), crimped fibers do not have high elongations, and cold-drawn hard fibers may have high elongation but do not show recovery properties. The generic name "spandex" applies to elastomeric fibers with a composition of at least 85% segmented polyurethane.

The chemistry of the "spandex" fibers is thus basically the same as described earlier in Section H,2. The "chain extenders" are normally diamines rather than glycols, since the resulting urea segments show enhanced intermolecular forces (and better recovery) or stress decay

TABLE 7-19

Reviews of Polyurethane Elastic Fibers

Title	Reference	Title	Reference
Elastomeric (spandex) fibers	H249, H251	Use of polyurethanes in the textile industry	H257
Polymers as fibers	H250	Polyurethanes as the bases for elastomeric fibers	H258
Structure of polyurethane elastomeric fibers	H280		
Elastomeric fibers	H248	Use of elastofibers	H259
Elastomeric fibers based on polyurethanes	H252, H253	Research and development of elastic fibers	H260
		Elastomeric polyester urethane fibers	H261–H265
Spandex fibers—A new concept for the textile industry	H254	Stabilized fiber compositions	H100, H104, H109,
Spandex elastomeric fibers	H255		H110, H113,
Elastomeric fiber yarns	H256, H267		H114
		Anti-block agents	H279

(H266) than the carbamate hard blocks. The high melting points of the urea segments are also attractive, although they do necessitate solution spinning rather than melt spinning.

A model for the elastomeric behavior of spandex fibers has been presented (H249) and is shown in Fig. 7-25. The picture is, of course, analogous to the morphological picture put forth for the styrene–diene block copolymer systems described earlier (see Chapter 6). The hard segments of the spandex fibers are stiff and have a tendency to cling to each other through strong hydrogen bonding. In this relaxed state, very little orientation or crystallinity is indicated by X-ray diffraction patterns (H249, H274).

When the fiber is stretched, it is believed that the folded or coiled portions of the flexible segments are extended. During the extention, the urea segments are brought close together so that they attract each other as a crystalline lattice and thereby inhibit further stretching. X-Ray diffraction patterns for the stretched spandex fibers show evidence for this increased orientation and crystallinity, although it is much less marked than for stretched natural rubber.

The recovery after a deformation is a rate-dependent phenomenon strongly affected by time, temperature, extension ratio, etc., as well as by the polymer structure (H275, H276). For example, the number of carbon atoms in the diamine extender has been shown to influence the modulus, tensile strength, and elongation of linear soluble polyesterurethane fibers (H277). The soft segment concentration has a marked effect on the elastic recovery and stress relaxation. The recovery results of Ferguson and Patsavoudis (H278) are shown in Figure 7-26.

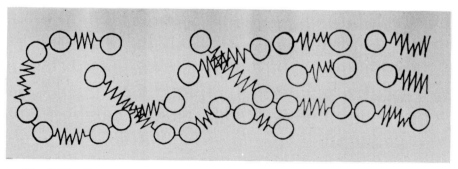

Fig. 7-25. Diagrammatic arrangement for spandex molecules in a relaxed thread. Zigzags represent coiled or folded chains of polyols. The molecules have a large degree of disorder (H249).

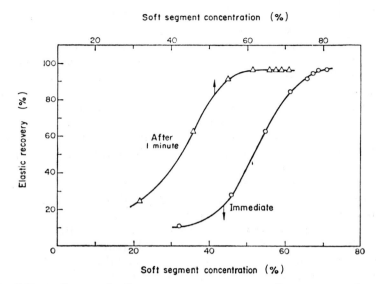

Fig. 7-26. Influence of soft segment concentration on elastic recovery from 100% extension (H278).

I. SILOXANES

The copolymers of this category contain one or more segments of polydialkylsiloxane, the alkyl groups being either methyl or phenyl. Some of these copolymers contain only siloxane blocks. Others contain polysilarylenesiloxane, poly(alkylene ether), poly(arylene ether) (e.g., polysulfone), polyvinyl, polyester, polycarbonate, polyamide, polyurethane, polyurea, or polyimide segments in addition to the polysiloxane segments. The various siloxane-containing {A-B}$_n$ block copolymers reported are listed in Table 7-20.

The polysiloxane segment is a very attractive one due to the low T_g, excellent stability, and high gas permeability properties of silicone polymers. The high degree of incompatibility of silicones with most organic polymers gives rise to some very interesting morphological and rheological properties and surface chemistry in siloxane–organic block copolymers.

1. Siloxane–Siloxane

The synthesis and properties of dimethylsiloxane–diphenylsilox-ane–dimethylsiloxane A-B-A block copolymers were discussed in

TABLE 7-20

{A-B}$_n$ Block Copolymers Containing Polysiloxane Segments

Block A	Block B	Synthetic method	Temperature (°C)	Reference
Diphenylsiloxane	Methylphenyl-siloxane (C block = di-methyl-siloxane)	C_6H_5 | Li⊕⊖OSiO⊖Li⊕ | C_6H_5	100–200	I1, I2
	Dimethylsiloxane	$N(CH_2CH_2O⊖Li⊕)_3$	25–200	I3
	Dimethylsiloxane	C_6H_5 C_6H_5 | | Li⊕⊖OSi—O—SiO⊖Li⊕ | | C_6H_5 C_6H_5 HMPA complex + coupling	25–150	I4, I128
	Dimethylsiloxane	Silanol end-group condensation in presence of stannous octoate	150	I6
Phenylmethyl-siloxane	Dimethylsiloxane	≡SiOH + ≡SiH oligomer condensation	70	I15
Phenylsilsesquioxane	Dimethylsiloxane	End group condensation	150	I16
Tetramethyl-p-sil-phenylenesiloxane	Dimethylsiloxane	SiOH condensation [tetra-methylguanidine bis(2-ethylhexoate) catalyst]	50–200	I17–I22
	Diphenylsiloxane	KOH; silanol condensation	—	I17, I24
Tetramethyl-1,4-naphthalene-siloxane	Dimethylsiloxane	—	—	I17, I25
Tetramethyl-1,3-tetra-fluorophenylene-siloxane	Dimethylsiloxane	—	—	I17, I26
Alkylene ethers	Siloxanes	—	—	(see text)
Polysulfone	Dimethylsiloxane	Oligomer condensation via silylamine–hydroxyl reaction	25–120	I64–I85 J7
Poly(phenylene oxide)	Dimethylsiloxane	Oligomer condensation via silylamine–hydroxyl reaction	200	I86, I87
Isoprene	Dimethylsiloxane	Coupling of silanolate-terminated A-B-A with $(CH_3)_2SiCl_2$	0–25	I88
Styrene	Dimethylsiloxane	Coupling of silanolate-terminated A-B-A with $(C_6H_5)_2Si(OOCH_3)_2$ or R_2SiCl_2	0–55	I89–I96
Styrene	Dimethylsiloxane	Dehydrocondensation of SiOH-terminated A-B-A	—	I97
α-Methylstyrene	Dimethylsiloxane	Oligomer condensation via silylamine–hydroxyl reaction	120–180	I75, I178, 180, I198, I99

TABLE 7-20 (*Continued*)

Block A	Block B	Synthetic method	Temperature (°C)	Reference
α-Methylstyrene–styrene	Dimethylsiloxane	Dehydrocondensation of SiOH-terminated A-B-A	—	I102, I103
Bisphenol A carbonate	Dimethylsiloxane	Phosgenation and oligomer condensation	25–55	I104–I129
9,9-Bis(4-hydroxyphenyl)fluorene carbonate	Dimethylsiloxane	Phosgenation and oligomer condensation	25–55	I130–I132
Tetrabromobisphenol A carbonate	Dimethylsiloxane	Phosgenation and oligomer condensation	25–55	I119
2,2,4,4-Tetramethyl-1,3-cyclobutylene carbonate	Dimethylsiloxane	Oligomer condensation via silylamine–hydroxyl reaction	130	I75, I78–I81, I133, I134
Bisphenol A isophthalate	Dimethylsiloxane	Oligomer condensation via silylamine–hydroxyl reaction	130–150	I75, I78–I80, I135–I137
Bisphenol A isophthalate	Phenylmethylsiloxane	Oligomer condensation	100–220	I138–I140
Bisphenol A terephthalate	Dimethylsiloxane	Oligomer condensation via silylamine–hydroxyl reaction	180	I75, I78–I80, I135–I137
Bisphenol A terephthalate	Phenylmethylsiloxane	Oligomer condensation	100–220	I138, I139
Hexamethylene terephthalate	Dimethylsiloxane	Oligomer condensation via silylamine–hydroxyl reaction	120	I75, I78–I80, I135–I137
γ-Benzyl L-glutamate	Dimethylsiloxane	Oligomer condensation via silylamine–hydroxyl reaction	120	I141, I142
Nylon 6	Dimethylsiloxane	NaH	120	I143, I144
Urethane	Dimethylsiloxane	Oligomer condensation	50	I145, I146, I148–I151
Urea	Dimethylsiloxane	Oligomer condensation	75	I152
Urea	Aluminosiloxane	Oligomer condensation	65	I152, I153
Imide	Dimethylsiloxane	—	—	I154, I155

Chapter 6. Briefly, these were synthesized by the sequential polymerization of hexamethylcyclotrisiloxane and hexaphenylcyclotrisiloxane in the presence of the initiator, dilithiodiphenylsilanolate, and a promoter such as tetrahydrofuran. The same technique was used to prepare an A-B-C-C-B-A copolymer with five rather than three segments by employing a third monomer, 2,4,6-trimethyl-2,4,6-triphenylcyclotrisiloxane (Scheme 7-15) (I1, I2).

Branched block copolymers were prepared by the sequential addition of hexamethylcyclotrisiloxane and hexaphenylcyclotrisiloxane to

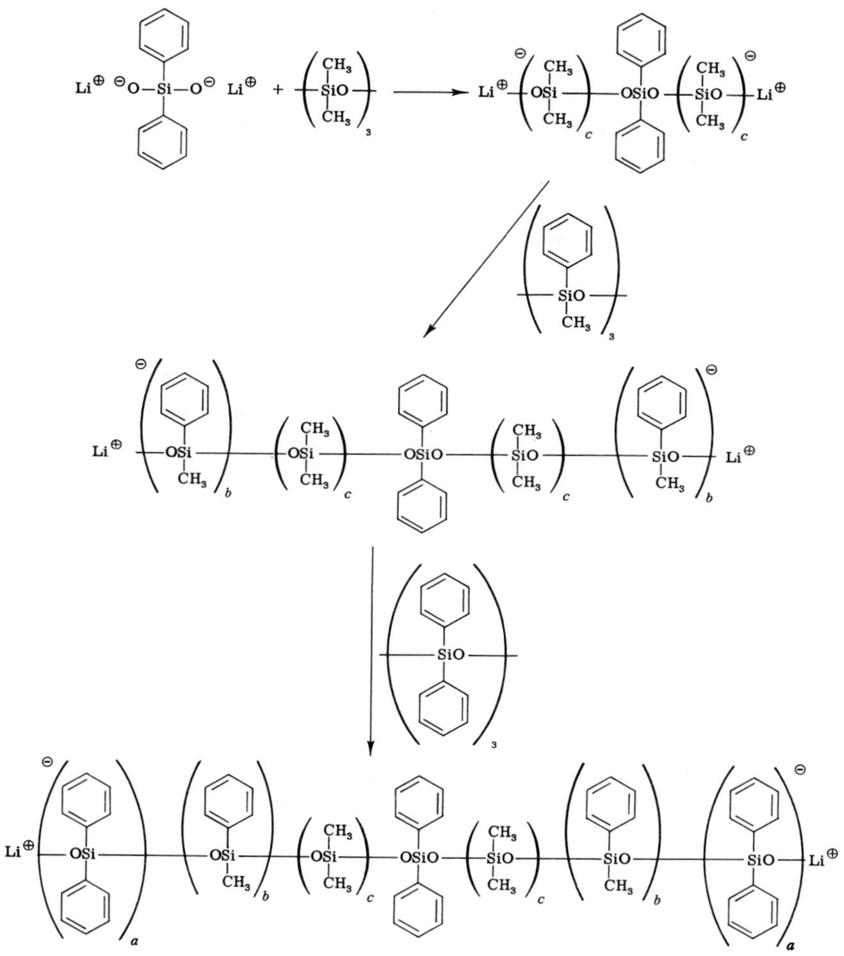

Scheme 7-15

a trifunctional initiator (Scheme 7-16) (I3). Both of these block copolymers, when cured with peroxides, were claimed to be useful in electrical insulation applications.

Multiblock copolymers of dimethylsiloxane–diphenylsiloxane were synthesized by employing a combination of the above sequential addition technique and a coupling condensation step (I4). The initiator used in this approach was a complex of dilithio-1,1,3,3-tetraphenyldisiloxanediolate with hexamethylphosphoramide (HMPA). Hexamethylcyclotrisiloxane was polymerized at room tem-

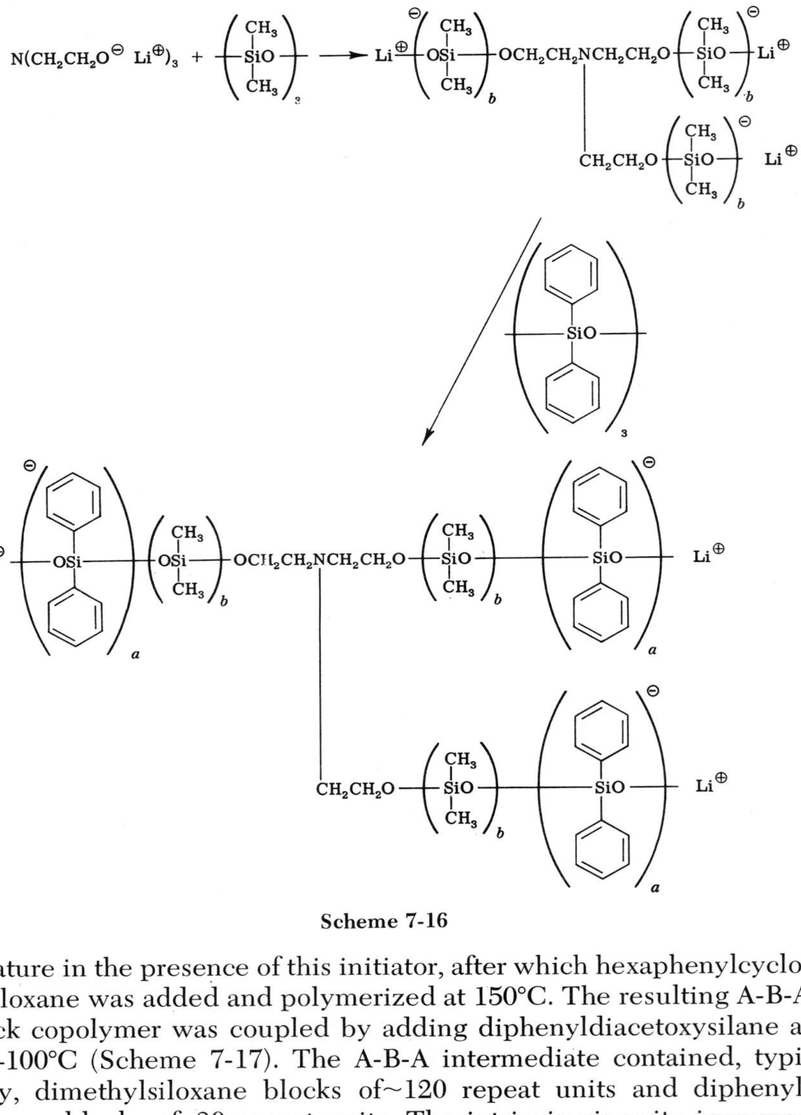

$$N(CH_2CH_2O^{\ominus} Li^{\oplus})_3 + \left(\begin{array}{c} CH_3 \\ | \\ SiO \\ | \\ CH_3 \end{array}\right)_3 \longrightarrow Li^{\oplus}\left(\begin{array}{c} \ominus \\ OSi \\ | \\ CH_3 \end{array}\begin{array}{c} CH_3 \\ | \\ \\ | \\ CH_3 \end{array}\right)_b OCH_2CH_2NCH_2CH_2O\left(\begin{array}{c} CH_3 \\ | \\ SiO \\ | \\ CH_3 \end{array}\right)_b^{\ominus} Li^{\oplus}$$

$$CH_2CH_2O\left(\begin{array}{c} CH_3 \\ | \\ SiO \\ | \\ CH_3 \end{array}\right)_b^{\ominus} Li^{\oplus}$$

$$\left(\begin{array}{c} C_6H_5 \\ | \\ SiO \\ | \\ C_6H_5 \end{array}\right)_3$$

$$Li^{\oplus}\left(\begin{array}{c} \ominus \\ OSi \\ | \\ \end{array}\right)_a \left(\begin{array}{c} CH_3 \\ | \\ OSi \\ | \\ CH_3 \end{array}\right)_b OCH_2CH_2NCH_2CH_2O\left(\begin{array}{c} CH_3 \\ | \\ SiO \\ | \\ CH_3 \end{array}\right)_b \left(\begin{array}{c} \ominus \\ SiO \\ | \\ \end{array}\right)_a Li^{\oplus}$$

$$CH_2CH_2O\left(\begin{array}{c} CH_3 \\ | \\ SiO \\ | \\ CH_3 \end{array}\right)_b \left(\begin{array}{c} \ominus \\ SiO \\ | \\ \end{array}\right)_a Li^{\oplus}$$

Scheme 7-16

perature in the presence of this initiator, after which hexaphenylcyclotrisiloxane was added and polymerized at 150°C. The resulting A-B-A
block copolymer was coupled by adding diphenyldiacetoxysilane at
25°–100°C (Scheme 7-17). The A-B-A intermediate contained, typi-
cally, dimethylsiloxane blocks of~120 repeat units and diphenylsiloxane blocks of~30 repeat units. The intrinsic viscosity increased
from 0.8 to 2.5 dl/gm upon addition of the acetoxysilane, indicating
that an appreciable amount of coupling had occurred. After peroxide
curing, the multiblock product displayed better elastomer properties
than did the peroxide-cured A-B-A copolymer. The A-B-A and multi-

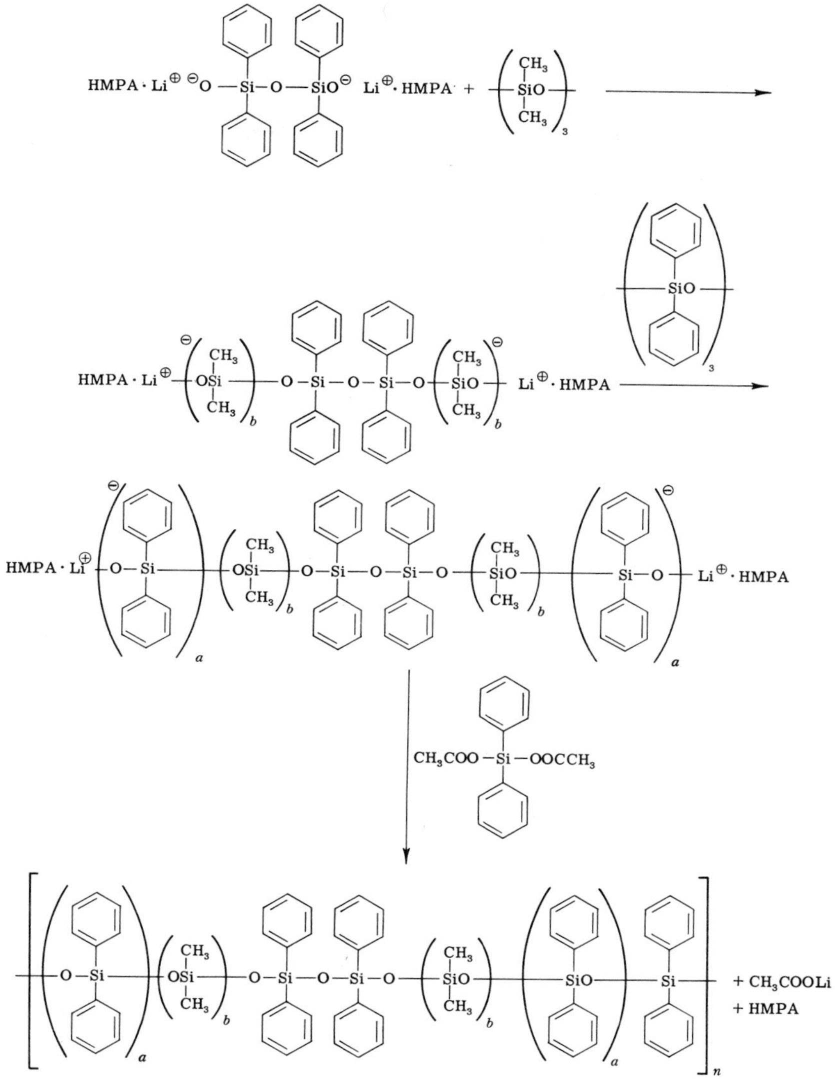

$$\text{HMPA} \cdot \text{Li}^{\oplus} {}^{\ominus}\text{O}-\underset{}{\text{Si}}-\text{O}-\underset{}{\text{SiO}}^{\ominus} \ \text{Li}^{\oplus} \cdot \text{HMPA} \ + \ \left(\underset{\text{CH}_3}{\overset{\text{CH}_3}{\text{SiO}}}\right)_3 \longrightarrow$$

$$\text{HMPA} \cdot \text{Li}^{\oplus}\left(\underset{\text{CH}_3}{\overset{\text{CH}_3}{\text{OSi}}}\right)_b -\text{O}-\underset{}{\text{Si}}-\text{O}-\underset{}{\text{Si}}-\text{O}\left(\underset{\text{CH}_3}{\overset{\text{CH}_3}{\text{SiO}}}\right)_b^{\ominus} \text{Li}^{\oplus} \cdot \text{HMPA} \longrightarrow \left(\underset{}{\text{SiO}}\right)_3$$

$$\text{HMPA} \cdot \text{Li}^{\oplus}\left(\text{O}-\underset{}{\text{Si}}\right)_a^{\ominus}\left(\underset{\text{CH}_3}{\overset{\text{CH}_3}{\text{OSi}}}\right)_b-\text{O}-\underset{}{\text{Si}}-\text{O}-\underset{}{\text{Si}}-\text{O}-\left(\underset{\text{CH}_3}{\overset{\text{CH}_3}{\text{SiO}}}\right)_b\left(\underset{}{\text{Si}}-\text{O}\right)_a^{\ominus} \text{Li}^{\oplus} \cdot \text{HMPA}$$

$$\text{CH}_3\text{COO}-\underset{}{\text{Si}}-\text{OOCCH}_3$$

$$\left[\left(\text{O}-\underset{}{\text{Si}}\right)_a\left(\underset{\text{CH}_3}{\overset{\text{CH}_3}{\text{OSi}}}\right)_b-\text{O}-\underset{}{\text{Si}}-\text{O}-\underset{}{\text{Si}}-\text{O}\left(\underset{\text{CH}_3}{\overset{\text{CH}_3}{\text{SiO}}}\right)_b\left(\underset{}{\text{SiO}}\right)_a\underset{}{\text{Si}}\right]_n \ + \text{CH}_3\text{COOLi}$$
$$+ \ \text{HMPA}$$

Scheme 7-17

block copolymers display similar morphological features (15), but the higher molecular weight of the latter diminishes the rate of domain formation due to higher viscosity. The systems contain lamellae that are a few hundred angstroms thick and threads that are several microns long and~200 Å in diameter. Crystalline regions of the diphenyl-siloxane block are dispersed in the continuous dimethylsiloxane phase.

Another approach has been reported for synthesizing diphenyl-siloxane–dimethylsiloxane block copolymers (I6). This technique involved the interaction of poly(diphenylsiloxane) and poly(dimethyl-siloxane) via condensation of their disilanol end groups in the presence of stannous octoate. Obviously, this approach will give products of ill-defined structure.

Several branched, cross-linkable copolymers containing segments of polymethylsiloxane and segments of polyphenylsiloxane have been reported in the literature (I7–I15). These were prepared by condensing the silanol groups of dihydroxyl-terminated dimethylsiloxane oligomers with phenylsiloxane, diphenylsiloxane, or phenylmethyl-siloxane intermediates containing functional groups such as

$$\equiv\text{Si—ON}{=}\text{C}\underset{\text{CH}_3}{\overset{\text{C}_2\text{H}_5}{<}}\ , \qquad \equiv\text{Si—OCH}_3, \qquad \equiv\text{Si—O}\overset{\overset{\text{O}}{\|}}{\text{C}}\text{CH}_3, \qquad \equiv\text{Si—Cl}, \qquad \equiv\text{Si—H}$$

Some of these multiblock copolymers can be cross-linked at room temperature via reaction of residual functional groups (either as end groups or as pendant groups) with atmospheric moisture. Others are vulcanizable at elevated temperature in the presence of acid or base catalysts via condensation of two silanol groups. The cross-linked products are claimed to be useful in elastomer, adhesive, coating, and electrical applications.

Block copolymers have been reported in which one segment is a phenylsilsesquioxane ladder polymer and the other is poly(dimethyl-siloxane) (I16). These were prepared by condensing silanol-terminated silsesquioxane oligomers with chlorine-terminated dimethyl-siloxane oligomers. The products were claimed to have good electrical insulation and surfactant properties.

2. Siloxane–Silarylenesiloxane

Polysilarylenesiloxanes, which are polysiloxanes in which part of the oxygen atoms are replaced by arylene groups, are more crystalline and more thermally stable than polysiloxanes and are synthesized by the self-condensation of p-bis(dimethyldroxysilyl)benzene (Reaction 7-16) (I17).

$$HO-\underset{\underset{R}{|}}{\overset{\overset{R}{|}}{Si}}-Ar-\underset{\underset{R}{|}}{\overset{\overset{R}{|}}{Si}}-OH \xrightarrow[\text{catalyst}]{-H_2O} \left[\underset{\underset{R}{|}}{\overset{\overset{R}{|}}{Si}}-Ar-\underset{\underset{R}{|}}{\overset{\overset{R}{|}}{Si}}-O\right]_x \qquad (7\text{-}16)$$

Merker *et al.* (I18–I21) synthesized block copolymers containing segments of crystalline poly(tetramethyl-*p*-silphenylenesiloxane) and segments of poly(dimethylsiloxane) by carrying out this condensation reaction in the presence of hydroxyl-terminated poly(dimethylsiloxane) oligomers. The reaction was catalyzed by compounds that produce little or no polysiloxane equilibration, e.g., tetramethylguanidine-bis(2-ethylhexoate) or *n*-hexylamine-2-ethylhexoate (Reaction 7-17).

$$HO-\underset{\underset{CH_3}{|}}{\overset{\overset{CH_3}{|}}{Si}}-\underset{}{\bigcirc}-\underset{\underset{CH_3}{|}}{\overset{\overset{CH_3}{|}}{Si}}-OH \;+\; HO-\left(\underset{\underset{CH_3}{|}}{\overset{\overset{CH_3}{|}}{Si}}-O\right)_b H$$

$$\downarrow \qquad\qquad\qquad\qquad (7\text{-}17)$$

$$\left[\left(\underset{\underset{CH_3}{|}}{\overset{\overset{CH_3}{|}}{Si}}-\underset{}{\bigcirc}-\underset{\underset{CH_3}{|}}{\overset{\overset{CH_3}{|}}{Si}}-O\right)_a \left(\underset{\underset{CH_3}{|}}{\overset{\overset{CH_3}{|}}{Si}}-O\right)_b\right]_n$$

High molecular weight block copolymer products ($[\eta]$ up to 2.1) were obtained that were soluble in aromatic hydrocarbon solvents, indicating that they were linear structures.

It is obvious from this reaction scheme that, since both starting materials bear the same type of silanol functional group, condensation will occur indiscriminantly between both species. As a result, the length of the dimethylsiloxane segment in the block copolymer product will be greater than length of the oligomeric starting material. The *average* length of both blocks will depend on the ratio of the two starting materials and the molecular weight of the dimethylsiloxane oligomer. The statistical nature of the condensation polymerization reaction dictates that the molecular weight of the silphenylenesiloxane block will increase with increasing dimethylsiloxane oligomer molecular weight. In a variation of the above synthesis process, the silphenylene monomer can be partially self-condensed to build up its molecular weight before adding the dimethylsiloxane oligomer. This approach was used to prepare the compositions containing long silphenylenesiloxane segments. Table 7-21 shows that the melting points of the block copolymers increased with increasing length of the sil-

TABLE 7-21

Effect of Relative Molecular Weight of Polysilphenylenesiloxane
Segment on Block Copolymer Melting Point[a]

Degree of polymerization of poly(dimethylsiloxane) oligomer	Weight percent silphenylene-siloxane	Melting point (°C)
18	90	139
43	70	133
18	70	123
400	50	139
43	50	127
18	50	100
43	30	98
18	30	<25

[a] Taken from Merker et al. I18.

phenylenesiloxane segments, approaching the 148°C melting point exhibited by the homopolymer.

The crystallinity of the silphenylenesiloxane blocks enhances the mechanical properties by virtue of forming reinforcing domains, dispersed in the dimethylsiloxane continuous phase, which function as "physical cross-link" sites. These materials display tensile strengths of 100–2700 psi and elongations of 500–1000%. By comparison, unfilled, uncured dimethylsiloxane homopolymers display much poorer properties. Thermomechanical studies indicated the presence of four transitions for two-phase block copolymer compositions (I22). X-Ray and microscopy studies indicated that the amorphous segments do not enter into the crystalline lattice of the crystallizable component (I21).

Block copolymers of this type are claimed (I23) to be good thermoplastic adhesives for polyorganosiloxane membranes employed in artificial lung applications.

Similar block copolymers have also been reported that contain diphenylsiloxane rather than dimethylsiloxane blocks, and others that contain substituted silphenylenesiloxane segments (see Table 7-20) (I17, I24–I26).

3. Siloxane–Alkylene Ether

Copolymers of polysiloxanes and poly(alkylene ethers) have been cited in the patent literature and have been referred to as "block copolymers." However, although a few of them are truly linear block copolymers as defined in this book, most of them are not. The latter materials are nonlinear and should perhaps more properly be called graft copolymers, "comb" copolymers, or branched block copolymers. As such, they do not formally fall within the scope of this book. They will therefore be discussed here only briefly. Some references to these materials are cited for the interested reader.

The linear block copolymers reported in the literature are synthesized by the interaction of difunctionally terminated siloxane oligomers with dihydroxyl-terminated alkylene ether oligomers (Reaction 7-18). The reactive functional end groups employed were acetoxy

$$X \overbrace{}^{\text{Siloxane}} X \ + \ HO \underset{\text{ether}}{\overset{\text{alkylene}}{—}} OH \longrightarrow \left(\overbrace{}^{\text{Siloxane}} O — \underset{\text{ether}}{\overset{\text{alkylene}}{—}} O \right)_{\!n} + \ HX$$

$$(7\text{-}18)$$

(I27), alkoxy (I28, I29), and dialkylamino (I30). The amine-terminated siloxane technique is discussed in detail in Section I,4.

The great majority of siloxane–alkylene ether copolymers reported in the literature are nonlinear materials. The first compositions of this type were reported in 1955 by Bailey and O'Conner (I31, I32). Their lack of linearity is due to the presence of more than two reactive groups in the siloxane starting materials. Some of these copolymers were prepared by the interaction of silyl chloride groups present on the siloxane polymers with the hydroxyl groups of the alkylene ether oligomers in the presence of catalysts such as pyridine (Reaction 7-19)

$$\overbrace{}^{\text{Polysiloxane}} \overset{|}{\underset{|}{Si}} — Cl \ + \ HO — \overset{|}{\underset{|}{C}} \overbrace{}^{\text{Polyether}} \longrightarrow \overbrace{}^{\text{Polysiloxane}} \overset{|}{\underset{|}{Si}} O C \overbrace{}^{\text{Polyether}} + \ HCl$$

$$(7\text{-}19)$$

(I33–I37). The HCl by-product of this reaction, if not completely removed from the product, can cause cleavage of the Si—O—C linkage, thus presenting a distinct disadvantage to this approach. Other copolymers were synthesized by a similar route via alkoxysilane reactive groups in the presence of trifluoroacetic acid catalyst (I38, I39). Another route has been claimed (I40, I41) using silyl hydride functional groups and piperidine or stannous octoate catalysts (Reaction 7-20).

$$\overbrace{}^{\text{Polysiloxane}} \overset{|}{\underset{|}{Si}} — H \ + \ HO — \overset{|}{\underset{|}{C}} \overbrace{}^{\text{Polyether}} \longrightarrow$$

$$(7\text{-}20)$$

$$\overbrace{}^{\text{Polysiloxane}} \overset{|\ |}{\underset{|\ |}{Si}} O C \overbrace{}^{\text{Polyether}} + \ H_2$$

All of these reactions give rise to a Si—O—C linkage between the siloxane and polyalkylene ether segments. Products with superior hydrolytic stability are claimed to result when coupling is achieved via silyl hydride addition to a vinyl group in the presence of catalysts such as chloroplatinic acid to give a Si—C linkage (Reaction 7-21)

$$\underset{\text{Polysiloxane}}{\wwwww}\ \underset{|}{\overset{|}{Si}}-H\ +\ CH_2{=}CHCH_2O\ \underline{\text{Polyether}}\ \longrightarrow$$

$$(7\text{-}21)$$

$$\underset{\text{Polysiloxane}}{\wwwww}\ \underset{|}{\overset{|}{Si}}-CH_2CH_2CH_2O\ \underline{\text{Polyether}}$$

(142–153). Si—C-linked block copolymers have also been prepared by reacting hydroxyl-terminated alkylene oxide oligomers with siloxane oligomers bearing organofunctional end groups, e.g., epoxy groups (154).

Internal pressure measurements on these polysiloxane–poly(alkylene ether) copolymers led to the conclusion that they exhibit an associated structure, i.e., domain aggregation, due to the dissimilarity and incompatibility of the segments (155).

These copolymers are widely used as surfactant additives for urethane foams (156, 157). When added at the ~1% level to foam formulations, they enable cell size and uniformity to be controlled during the foam-forming process, resulting in even-textured foams with reproducible physical properties. The surfactant is believed to concentrate at the polymer–gas interface with the poly(alkylene ether) segments oriented inward toward the film, in which it is soluble, and with the siloxane segments extending into the gas phase. The low surface tension of the surfactant tends to stabilize the cell walls and to induce flow of the urethane to maintain uniform surface energy. Further discussions of the surface tension characteristics of poly(alkylene ether)–polysiloxane block copolymers can be found in the literature (158–161). Block copolymer surfactants containing Si—O—C linkages are most commonly used due to their lower cost. However, in certain applications requiring greater hydrolytic stability, e.g., where urethane catalysts are employed that can cause cleavage of the Si—O—C bond, the more expensive Si—C-linked block copolymers are used. The copolymers are also claimed to be useful in fiber antistatic, antifoam, mold release, lubricant, and wetting agent applications (158, 159, 162, 163).

4. Siloxane–Arylene Ether

a. Polysulfone

Block copolymers containing high molecular weight segments of polysulfone, which is a poly(arylene ether), and poly(dimethylsiloxane) have been reported by Noshay et al. (I64–I80, J7). These block copolymers were synthesized by the interaction of preformed dihydroxyl-terminated polysulfone oligomers and bis-(dimethylamine)-terminated poly(dimethylsiloxane) oligomers in chlorobenzene solution (Scheme 7-18). The evolution of gaseous dimethylamine by-product leaves an essentially pure solution of block copolymer product. The copolymer condensation reaction is very facile due to the high reactivity of the Si—N bonds of the siloxane oligomer toward the phenolic end groups of the polysulfone oligomer. The reaction is complete in 20 minutes at 70°C and in 45 minutes when carried out at 25°C. The synthesis procedure produces high molecular weight block copolymers, up to $\overline{M}_w$ of 238,000 as measured by ultracentrifugation. This novel synthesis technique is also applicable to other block copolymers, as will be discussed below. The chemistry of the silylamine–hydroxyl condensation reaction and its use in the synthesis of a variety of organosiloxane block copolymers is discussed in a review by Noshay et al. (I75).

The structure of the block copolymer is well defined and strictly controlled, since the oligomers are mutually reactive, i.e., they can react with each other but not with themselves. As a result, the structures of the segments in the copolymers are essentially identical to those of the oligomer starting materials. The structure shown in the Scheme 7-18 is idealized. In practice, the end blocks are both polysulfone or both poly(dimethylsiloxane), depending on the actual final stoichiometry of the reaction.

The thermal and thermal oxidative stability of the block copolymers was found to be good, as expected from the known properties of the parent homopolymers. Thermogravimetric analysis in nitrogen indicated an onset of weight loss at 400°C. After aging in an air oven at 170°C for 2 weeks, the block copolymer had retained 90% of its initial reduced viscosity.

These copolymers are also characterized by good hydrolytic stability even though the segments are linked by Si—O—C bonds, which are hydrolytically unstable in small molecules, (i.e., 80% retention of reduced viscosity after immersion in boiling water for 2 weeks). This

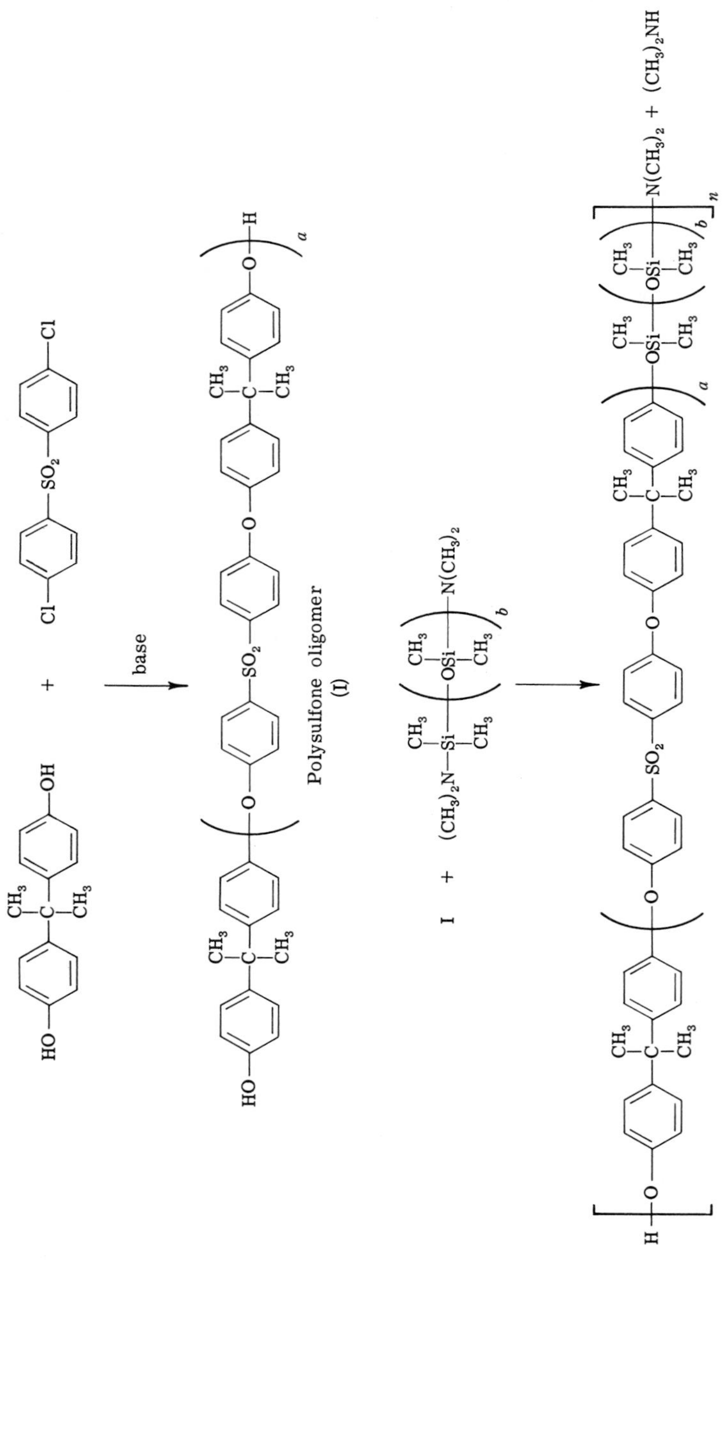

Scheme 7-18

hydrolytic stability is believed to be due to a combination of four factors: (a) the hydrophobic nature of the siloxane blocks, (b) segmental steric hindrance of the Si—O—C linkage, (c) a low concentration of Si—O—C bonds along the backbone, and (d) the two-phase morphology of these compositions (I74).

The morphology of polysulfone–poly(dimethylsiloxane) block copolymers can be controlled by varying the molecular weight ($\overline{M}_n$) of the oligomers. At a constant polysulfone $\overline{M}_n$ of 4700, poly(dimethylsiloxane) oligomers of $\overline{M}_n$ <1700 gave single-phase block copolymers, while ≥5000 $\overline{M}_n$ siloxane oligomers produced two-microphase systems due to domain formation. The latter copolymers displayed two glass transition temperatures—one at −120°C due to the dimethylsiloxane phase and another at +160°C due to the polysulfone domains. This is illustrated in Fig. 7-27. Copolymers containing high $\overline{M}_n$ (e.g.,

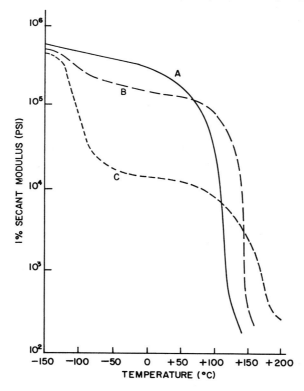

Fig. 7-27. Temperature–modulus curves for block copolymers of polysulfone with poly(dimethylsiloxane). Block $\overline{M}_n$; polysulfone, 4700; poly(dimethylsiloxane), (A) 350, (B) 1700, (C) 5100 (I72).

20,000) siloxane blocks also displayed a siloxane crystalline melting point at −55°C.

The extensive phase separation displayed by these systems at such low block molecular weight levels (≥5000) is unusual. This phenomenon is interpreted as being due to the high degree of incompatibility of polysulfone with poly(dimethylsiloxane) (solubility parameters equal to 10.6 and 7.3, respectively). In contrast to this behavior, much higher block $\overline{M}_n$ is required to attain domain formation in styrene–butadiene block copolymers [5000–10,000 and ∼50,000, respectively (I156)] in which the solubility parameters of the blocks are much closer [8.8 and 8.4, respectively (I157)].

Mechanical properties vary widely depending upon composition, which depends primarily on the oligomer molecular weight ratio. The copolymers described in Table 7-22 range from rigid, low-elongation materials to very flexible, high-elongation compositions.

Copolymers of high siloxane content displayed substantial recovery properties and behaved like cross-linked silicone rubbers, even though they are not chemically cross-linked. This is the result of their two-phase morphology, the polysulfone domains acting as "anchor" sites to produce a "pseudo" physically cross-linked system. These compositions were stronger than unfilled, cross-linked silicone rubbers due to the reinforcement effect of the polysulfone domains, which presumably act as "filler" particles. Films of these block copolymers are transparent due to the small size of these domains, about 150 Å as determined by electron microscopy.

Poly(dimethylsiloxane) is highly permeable to common gases such as oxygen and nitrogen, while polysulfone is relatively inpermeable. Robeson *et al.* (I77) obtained permeabilities over a wide range by using block copolymers of varying composition. For example, copolymers ranging from 24% to 79% in siloxane content displayed oxygen permeabilities of 1×10^{-10} to 4×10^{-8} cm³ (STP)/cm² sec cm Hg.

These workers also found that films of a copolymer containing 5000 $\overline{M}_n$ polysulfone and poly(dimethylsiloxane) blocks did not display significant differences in permeability when cast from different solvents. However, this same copolymer did exhibit large differences in modulus, i.e., 35,000 psi versus 9000 psi for films cast from polysulfone-preferred and siloxane-preferred solvents, respectively. This behavior was explained based on the contribution of each of the blocks to the continuous phase of the films. The contributions were calculated from a model that is an extension of Maxwell's analysis of heterogeneous systems applied to permeability and Kerner's analysis for modulus. The conclusions drawn from this treatment were that, although this

TABLE 7-22

Effect of Block $\overline{M}_n$ on Block Copolymer Properties[a]

Block $\overline{M}_n$		Siloxane wt%	Reduced viscosity[b]	Tensile modulus (psi)	Tensile strength (psi)	Elongation (%)	T_g (°C)[c]
Polysulfone	Poly(dimethyl-siloxane)						
4700	350	10	0.4	240,000	6000	5	+125
4700	1,700	28	1.1	170,000	4700	12	+140
9300	4,900	41	0.4	29,000	2700	150	-110; +170
4700	5,100	55	0.8	20,000	2400	350	-120; +160
4700	9,200	67	1.5	2,000	1300	500	-120; +160
6500	25,000	79	1.3	300	900	550	-120; +160

[a] From Noshay et al. (172).
[b] 0.2 gm/dl in CH_2Cl_2 at 25°C.
[c] Determined from temperature–modulus and resilience measurements.

copolymer contained approximately equal amounts by weight of polysulfone and polysiloxane, the polysulfone blocks make only a minor contribution (7–29% in the two solution cast films) to the continuous phase. A change from 7% to 29% in polysulfone contribution would be expected to result in a considerable change in modulus. However, a change from 71% to 93% in siloxane contribution would not be expected to give a large change in permeability.

Polysulfone–poly(dimethylsiloxane) block copolymers containing segments of $\geq$5000 $\overline{M}_n$ display astonishingly poor melt processability. Compression molding leads to poor quality films. Furthermore, extrusion at 340°C (180°C above the T_g of the polysulfone phase) produces a *powdery* "extrudate." This behavior is due to extremely high melt viscosity and extensive melt fracture. This, in turn, is believed to be caused by a significant degree of retention of the physical network structure even in the "melt." Improved melt processability was observed in copolymers containing shorter polysulfone segments (2000 $\overline{M}_n$ rather than 5000 $\overline{M}_n$). However, this phenomenon, which is probably due to reduced chain entanglement and greater phase blending, results in a 20°C sacrifice in the T_g of the polysulfone phase.

The high degree of physical network retention in the melt has been attributed by Matzner *et al* (I78, I81) to a large difference in the solubility parameters of the polysulfone and siloxane segments. These workers evaluated the melt processability of a series of organosiloxane block copolymers containing organic segments of varying structure (and therefore solubility parameter), as is shown in Table 7-23. Their

TABLE 7-23

Melt Processability of Organosiloxane Block Copolymers Correlated with the Differential Solubility Parameter (I78, I81)

Hard block	Solubility parameter	Δ^a	Melt processability
Polysulfone	10.6	3.3	Poor
Poly(bisphenol A terephthalate)	9.9	2.6	Poor
Poly(bisphenol A isophthalate)	9.9	2.6	Poor
Poly(α-methylstyrene)	8.9	1.6	Good
Poly(2,2,4,4-tetramethyl-1,3-cyclobutylene carbonate)	6.8	0.5	Excellent

^{a} Δ = Difference between solubility parameters of hard block and poly(dimethylsiloxane) (7.3).

conclusion was that the differential solubility parameter (Δ) between the segments should be ≤ 1 in order to obtain good melt processability. The ideal composition would have a Δ value of zero and would contain at least one crystallizable segment to provide a driving force for phase separation on cooling after melt processing. This material would be expected to display excellent melt processability and good two-phase mechanical properties after fabrication. It was predicted that these principles should also be applicable to block copolymers other than those containing siloxane segments, and to graft copolymers as well.

Polysulfone–poly(dimethylsiloxane) block copolymers were said to be useful per se in film and elastomer applications. Their utility in high temperature elastomer applications has recently been reviewed by Matzner *et al.* (179). They are also claimed to improve stress crack resistance when blended with polysulfone and to act as smoke depressants in these blends (182).

A particularly interesting application for the polysulfone–poly(dimethylsiloxane) block copolymer is its use as an impact modifier additive for polysulfone homopolymer (165, 167, 168, 176, 183). The incorporation of only 5% of the block copolymer into polysulfone increased the notched Izod impact strength from ~1 foot-pound per inch to ~20 foot-pounds per inch. This unique behavior is believed to result from two important characteristics of the block copolymer— compatibility and rheology. The borderline compatibility of the block copolymer with the polysulfone matrix resin, which is due to the presence of the polysulfone segment in the copolymer, results in efficient dispersion and good interphase adhesion. The unusual rheological behavior of the copolymer, which prevents it from being melt processable on its own, serves an important beneficial function in the blend: it discourages excessive particle breakdown during high shear blending and melt fabrication operations. In this respect, the block copolymer resembles the behavior of the styrene–diene graft systems formed *in situ* in impact-modified polystyrene systems, with the important exception that it is *not* cross-linked.

The foregoing discussion pertains to copolymers prepared by Noshay *et al.* by the interaction of hydroxyl-terminated polysulfone and amine-terminated siloxane oligomers. Copolymers similar to these but containing low molecular weight segments have also been synthesized by Strachan *et al.* (184, 185) by the interaction of NaO-terminated polysulfone and $BrCH_2Si$-terminated siloxane oligomers or by the interaction of $CH_2=CHCH_2O$-terminated polysulfone and HSi-terminated siloxane oligomers.

b. Poly(Phenylene Oxide)

A modification of the silylamine–hydroxyl reaction has recently been employed to prepare poly(phenylene oxide)–poly(dimethylsiloxane) block copolymers (I86, I87). Hydroxyl-terminated poly(phenylene oxide) oligomers were reacted with isopropylamine-terminated low molecular weight poly(dimethylsiloxane) oligomers at 200°C to produce products that were claimed to be useful as heat-shrinkable films, electrical insulators, and surface-tension depressant additives.

5. Siloxane–Vinyl

The block copolymers of this category contain segments of poly(dimethylsiloxane) together with segments of polyisoprene, polystyrene, or poly(α-methylstyrene).

a. Isoprene

Morton et al. (I88) prepared A-B-A block copolymers in which A is poly(dimethylsiloxane) and B is polyisoprene by the initiation of isoprene polymerization with potassium naphthalene followed by addition of octamethylcyclotetrasiloxane. Coupling of this copolymer with dimethyldichlorosilane to give a higher molecular weight product was cited as evidence for the formation of a multiblock copolymer, presumably one with an {A-B-A}$_n$ structure (Reaction 7-22).

(7-22)

b. Styrene

Polystyrene–poly(dimethylsiloxane) block copolymers were synthesized later by Dean (I89, I90) and by Saam et al. (I91–I96) by a similar route. Styrene was polymerized via dilithium naphthalene initiation in the presence of a promoting solvent such as tetrahydrofuran, after which the cyclic trimer of dimethylsiloxane (hexamethylcyclo-trisiloxane) was added. As was discussed earlier in Chapters 5 and 6, the cyclic trimer is superior to the tetramer as a monomer in that it polymerizes more rapidly and with much less siloxane redistribution, thus resulting in greater structure control. The resulting disilanolate-terminated A-B-A block copolymer was then coupled with diphenyl-diacetoxysilane, dialkyldichlorosilane, or trifluoropropylmethyl-dichlorosilane to produce the {A-B-A}$_n$ product shown in Scheme 7-19.

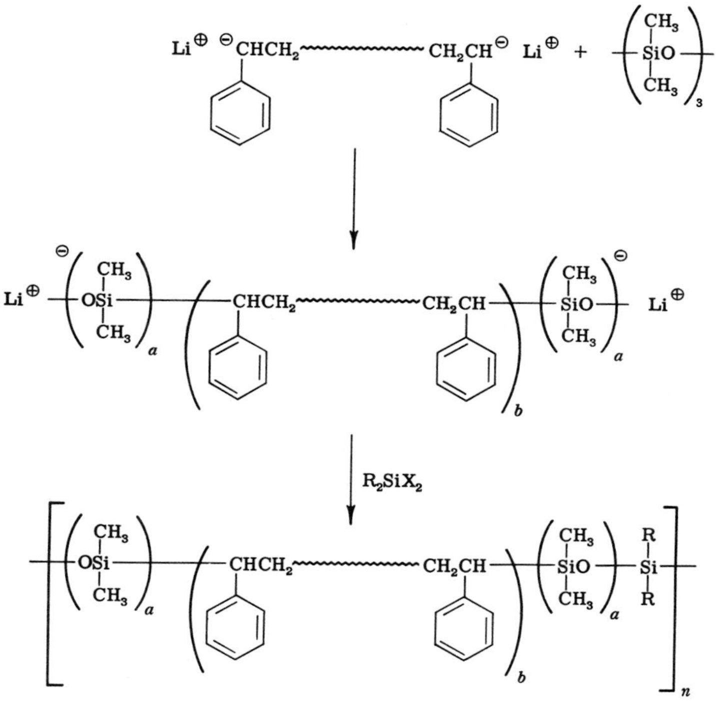

Scheme 7–19

The n values of these block copolymers ranged from 2 to 10, and overall $\overline{M}_n$ (as measured by osmometry) ranged from 10,000 to 300,000. Another approach used by Saam *et al.* (197) was to prepare disilanol-terminated siloxane–styrene–siloxane block copolymers and to couple these by dehydrocondensation with stannous octoate catalysts to form the {A-B}$_n$ products.

Electron microscopy studies (192, 193) on these {A-B-A}$_n$ copolymers indicated that morphology did not change over a broad range of composition. The use of solvents that selectively solvate either of the blocks gave cast films of 50% siloxane content, which showed better definition of the microphases than was obtained with nonselective solvents, but gross changes in morphology were not observed. This behavior is in contrast to that of A-B block copolymers (of ~60% siloxane content), which did show morphological differences with varying casting solvents.

The two-phase nature of the {A-B-A}$_n$ block copolymers is further demonstrated by the modulus–temperature relationship shown in Fig. 7-28, which was obtained by torsional braid analysis (I92). The curve indicates that the block copolymer has two glass transition temperatures—one at about −120°C due to the siloxane block and another at 50°C due to the styrene block.

The composition of styrene–siloxane block copolymers controls their mechanical properties. Copolymers containing 50% siloxane resemble polyethylene, but compositions containing 80% siloxane are rubbery (I91, I92). The tensile strength–elongation properties of the former were 2300 psi/130%, while those of the latter were 400 psi/400% for copolymers of ~300,000 and ~150,000 $\overline{M}_n$, respectively. The degree of condensation (n) of the A-B-A units and the $\overline{M}_n$ of the styrene blocks are also critical factors in determining properties. The n value must be in excess of 2 to obtain any significant tensile strength, and little further improvement was observed when n was >8. The minimum polystyrene block $\overline{M}_n$ required to obtain useful properties was found to be 8000, and values of >12,000 gave little further improvement.

It was reported that the presence of poly(dimethylsiloxane)

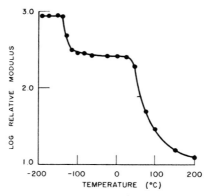

Fig. 7-28.   Styrene–siloxane block copolymer temperature–modulus curve (I92).

homopolymer as an impurity in the block copolymer is very detrimental to properties (I91). The incorporation of 20% of the homopolymer into a block copolymer of 50% siloxane content completely obliterated the mechanical strength of the block copolymer whose original tensile strength was 2300 psi. This behavior, however, does not agree with that observed with styrene–butadiene block copolymers systems (I158) or with polysulfone–siloxane block copolymers.

High siloxane-content block copolymers were claimed to be thermoplastic elastomers (I92). Molding was carried out at 150°C, and repeated molding did not alter properties.

The oxygen and nitrogen permeability of a block copolymer containing 80% siloxane was said to be equivalent to that of a silica-filled silicone elastomer (I91). Oxygen permeability decreased linearly (from 49×10^{-9} to 4×10^{-9} cm^3 cm/cm^2 sec cm Hg at 25°C) with decreasing siloxane content with an inflection at the 50% siloxane level. The inflection was concluded to be due to a predominance of the less permeable polystyrene in the continuous phase of the microdisperse two-phase system.

The styrene–siloxane block copolymers were also claimed to have useful electrical properties (600–700 volts/mil dielectric strength in $\frac{1}{16}$ inch plaques) and an unexpected degree of weatherability (I91).

c. α-Methylstyrene

Poly(α-methylstyrene)–poly(dimethylsiloxane) ${A-B}_n$ multiblock copolymers have been prepared by Noshay et al. (I75, I98, I99). α-Methylstyrene–siloxane multisequence copolymers have also been reported by other workers (I100, I101), but these compositions contain very short segments (tetramers, pentamers) and therefore more closely resemble the properties of random rather than block copolymers. The synthesis route employed by Noshay et al. is similar to the one used to prepare polysulfone–poly(dimethylsiloxane) block copolymers (I72, I75), i.e., condensation of preformed dihydroxyl-terminated poly(α-methylstyrene) oligomers with bis(dimethylamine)-terminated poly (dimethylsiloxane) oligomers (Reaction 7-23). The end blocks and end groups of the copolymer will depend on the actual final stoichiometry of the condensation reaction rather than as depicted above for perfect 1 : 1 stoichiometry. The copolymerization reaction is carried out in chlorinated aromatic solvents at 120°–180°C.

The α-methylstyrene oligomers were synthesized by polymerizing α-methylstyrene anionically via sodium initiation followed by capping with ethylene oxide and terminating with acetic acid (Reaction 7-24). This oligomer synthesis was carried out in nonaqueous disper-

$$
HOCH_2CH_2\left(\underset{a}{\overset{CH_3}{\underset{|}{\overset{|}{C}}}CH_2 \sim\sim\sim CH_2\overset{CH_3}{\underset{|}{\overset{|}{C}}}}\right)CH_2CH_2OH + (CH_3)_2N\left(\underset{b}{\overset{CH_3}{\underset{CH_3}{\overset{|}{Si}O}}}\right)\overset{CH_3}{\underset{CH_3}{\overset{|}{Si}}}-N(CH_3)_2
$$

$$
H\left[\left(OCH_2CH_2\left(\underset{a}{\overset{CH_3}{\underset{|}{\overset{|}{C}}}CH_2 \sim\sim\sim CH_2\overset{CH_3}{\underset{|}{\overset{|}{C}}}}\right)CH_2CH_2O\left(\underset{b}{\overset{CH_3}{\underset{CH_3}{\overset{|}{Si}O}}}\right)\overset{CH_3}{\underset{CH_3}{\overset{|}{Si}}}\right)N(CH_3)_2\right]_n + (CH_3)_2NH
$$

$$(7\text{-}23)$$

sion medium in order to avoid the large viscosity build up (due to association) in the ethylene oxide capping step, which is characteristic of this reaction when carried out in solution.

$$
CH_2{=}\overset{CH_3}{\underset{|}{C}} \xrightarrow{Na} Na^{\oplus}{}^{\ominus}\overset{CH_3}{\underset{|}{C}}{-}CH_2 \sim\sim\sim CH_2{-}\underset{a}{\overset{CH_3}{\underset{|}{C}}}{}^{\ominus}{-}Na^{\oplus}
$$

$$
\begin{array}{l}(1)\ H_2C\overset{O}{\diagup\!\!\diagdown}CH_2\\(2)\ H^{\oplus}\end{array}
$$

$$(7\text{-}24)$$

$$
HOCH_2CH_2\left(\underset{a}{\overset{CH_3}{\underset{|}{\overset{|}{C}}}{-}CH_2 \sim\sim\sim CH_2{-}\overset{CH_3}{\underset{|}{\overset{|}{C}}}}\right)CH_2CH_2OH
$$

The block copolymer condensation reaction takes place at a much slower rate than is observed in the preparation of polysulfone-poly(dimethylsiloxane) block copolymers. This is due to the lower acidity (and therefore lower reactivity with the siloxane oligomer) of the aliphatic hydroxyl end groups of the α-methylstyrene oligomer, compared to the phenolic end groups of polysulfone oligomers. However, in spite of the slow rate of reaction, high molecular weight products were obtained by slow addition of the siloxane oligomer to the α-methylstyrene oligomer to approach the stoichiometric end point gradually. The low reactivity of aliphatic hydroxyl-terminated oligomers can be increased by carrying out the copolymerization in the presence of carbon dioxide, which catalyzes the reaction.

Block copolymers were prepared from oligomers ranging in $\overline{M}_n$ from ~3,000 to ~20,000 to obtain compositions varying from 20% to 73% siloxane content. The former composition was relatively rigid with a tensile strength of 4500 psi and an elongation of 2%. The latter was an elastomeric material with a tensile strength of 1300 psi and an elongation of 935%.

The two-phase morphology of these block copolymers is illustrated by the modulus–temperature curves shown in Fig. 7-29. The curves indicate the presence of two glass transition temperatures (T_g), corre-

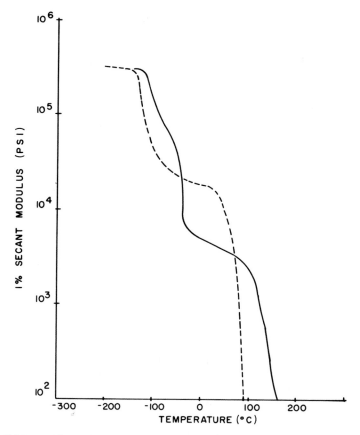

Fig. 7-29. Modulus–temperature relationship of α-methylstyrene–siloxane block copolymer; broken line = 2900–5200 PαMS–PSX, solid line = 7300–20,500 PαMS–PSX (199).

sponding to the α-methylstyrene and siloxane blocks. The curves also illustrate the effect of α-methylstyrene block $\overline{M}_n$ on T_g. The copolymer with a 2900 $\overline{M}_n$ α-methylstyrene block had a T_g of 80°C, while the one with a 7300 $\overline{M}_n$ block had a T_g of 130°C. Both compositions gave siloxane T_g values at about −100°C. The copolymer with a 20,000 $\overline{M}_n$ siloxane block, due to its length, also displayed a crystalline siloxane melting point at −45°C. It is to be noted that the α-methylstyrene–siloxane block copolymer has a considerably wider useful temperature range than the styrene–siloxane block copolymers discussed earlier by virtue of its higher upper T_g, i.e., +130°C versus +50°C.

The α-methylstyrene–siloxane block copolymer displayed good compression moldability, as was demonstrated by the observation that the properties of molded films were essentially equivalent to those of solution-cast films, which should reflect ultimate properties. The melt processability of these compositions was compared to that of other organosiloxane block copolymers by Matzner *et al.* (I78) (see Section I,4).

An {α-methylstyrene–styrene–dimethylsiloxane}$_n$ block copolymer was prepared by Saam *et al.* (I102) by (a) initiating α-methylstyrene polymerization with dilithiostilbene, (b) addition of a small amount of styrene to prevent depolymerization and to form a short styrene segment, (c) addition of hexamethylcyclotrisiloxane, (d) neutralization to form a disilanol-terminated intermediate, which was then (e) dehydrocondensed in the presence of stannous octanoate to form the final multiblock product. The tensile properties of this material improved with increasing α-methylstyrene block length up to a point, beyond which physical properties deteriorated. This was ascribed to enlargement of the α-methylstyrene domains to a size at which brittle failure could occur, thereby initiating crack formation in the bulk material (I103).

6. Siloxane–Ester

The block copolymers of this category contain polyester segments and poly(dimethylsiloxane) segments. The polyester blocks are of three types—bisphenol A carbonate, tetramethyl cyclobutylene carbonate, and arylene (or alkylene) phthalates.

a. Bisphenol A Carbonate

Bisphenol A carbonate–dimethylsiloxane block copolymers were first synthesized by Vaughn (I104–I108) by the phosgenation of a mixture of bisphenol A and an α,ω-dichloro-terminated dimethylsiloxane oligomer in the presence of pyridine (Scheme 7-20). The copolymerization is carried out at 25°–55°C in methylene chloride solution. The

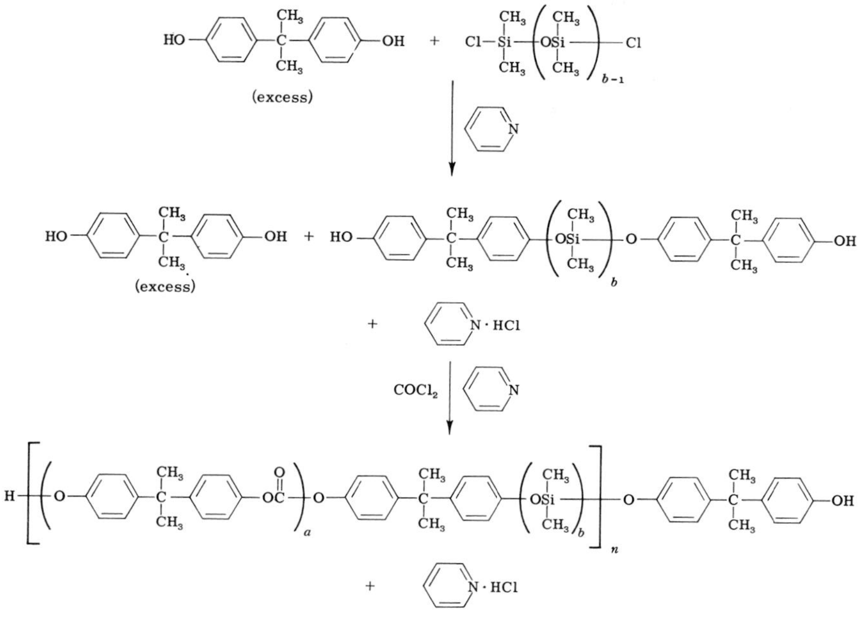

Scheme 7-20

pyridine hydrochloride by-product is removed by filtration before isolating the product by coagulation. Three reactions occur simultaneously during the copolymerization: (a) bisphenol A reacts with phosgene to form polycarbonate blocks; (b) the bisphenol A-capped siloxane oligomer reacts with phosgene and bisphenol A to form carbonate–siloxane linkages; and (c) two bisphenol A-capped siloxane oligomers react with phosgene to effectively form higher molecular weight siloxane blocks containing within them the

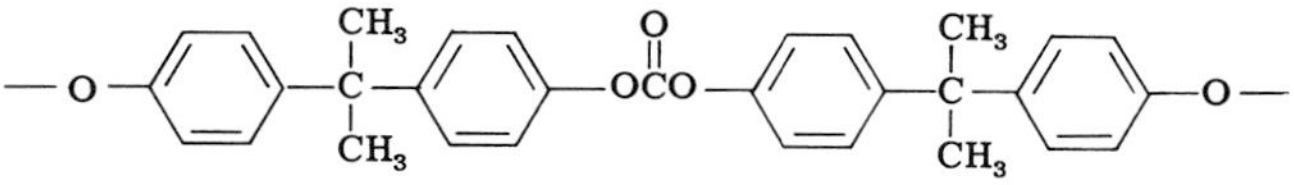

linkage. The resulting copolymer is referred to by Vaughn (I108) as an alternating random block copolymer in which the blocks are polydisperse and of fairly low degree of polymerization.

The degree of polymerization of the siloxane blocks (*b*) was determined by chlorine end group analysis of the siloxane oligomer; the degree of polymerization of the carbonate blocks (*a*) was calculated from the overall composition of the copolymer. The (*a*) and (*b*) values ranged from 3 to 100, and the polycarbonate contents ranged from 35% to 85%. Copolymers with intrinsic viscosities up to ~1.0 (in chloroform) were obtained corresponding to a $\overline{M}_w$ of 50,000–100,000 (I109). Tensile strength increased from 1000 to 6000 psi and elongation decreased from 800% to ~100% with increasing polycarbonate content.

Other workers (I110, I111) have synthesized similar polycarbonate–siloxane block copolymers by the interaction of preformed chlorine-terminated siloxane oligomers and hydroxyl-terminated bis A carbonate oligomers in the presence of a tertiary amine. This technique should produce a more controlled block sequence structure.

The copolymers can be solution cast or molded into optically clear films, and they can be extruded at 550°–625°F. Thermogravimetric analysis in nitrogen indicated that decomposition begins at 390°C, somewhat lower than that of bisphenol A carbonate homopolymer. It was reported that at constant composition, tensile strength increases and elongation decreases with increasing average block length. It was also reported (I107, I112) that longer blocks gave increased modulus and decreased dielectric loss.

These copolymers were claimed to have useful electrical properties, corona resistance, and permeability properties (I107, I113). They are less permeable to gases than are silicone homopolymers, but relative rates of permeation of various gases are about the same. Pinhole-free solution-cast films are claimed to be good membranes for preparing oxygen-rich air (I114). The polymers are also claimed to be useful in coating and adhesive applications (I104).

The two-phase nature of these copolymers was demonstrated by torsion pendulum studies (I109, I115, I116). The data obtained on a block copolymer containing 51 wt% of polycarbonate and composed of carbonate and siloxane blocks of 6 and 20 repeat units, respectively, are illustrated in Fig. 7-30. Two major relaxations are evident—one at −110°C due to the T_g of the siloxane block and another at 72°C, due to the T_g of the relatively short carbonate block. The low-temperature transition displayed little dependence on siloxane block length, and crystallization of the siloxane segment (at −40°C) was not observed even in copolymers containing siloxane blocks of up to 40 repeat units. On the other hand, the T_g of the polycarbonate block was quite dependent upon molecular weight, ranging from 60°C at 5 repeat units to

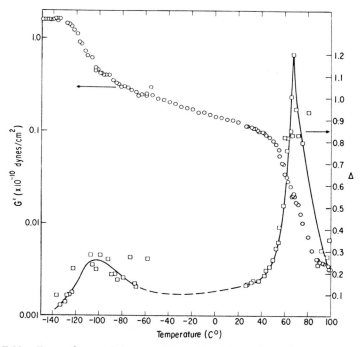

Fig. 7-30. Dependence of shear modulus G' and logarithmic decrement Δ on temperature at about 1 Hz for a dimethylsiloxane–bisphenol A carbonate alternating block polymer (51 wt% BPAC, $\bar{n} = 20$, $\bar{m} = 6$) (I115).

140°C at 46 repeat units (see Fig. 7-31). This effect was said to be analogous to the dependence of T_g on molecular weight in homopolymers, and the siloxane segments, due to their high degree of mobility, exert little restraint on the carbonate segments. It was therefore concluded that the carbonate blocks behave to a large extent as though they were free-ended. It was estimated that the polycarbonate domains contain no more than 4% poly(dimethylsiloxane) segments.

Shear modulus–temperature relationship studies (I117, I118) indicated that longer blocks (40 versus 20 repeat units) gave higher T_g values and higher rubbery moduli. This work also showed that higher molecular weight copolymers displayed a rubbery region beyond the T_g of the carbonate block, while lower molecular weight compositions did not display this behavior. This behavior might be interpreted as being due to the partial retention of the physical network in the melt. An increase in flow observed in the 220°–230°C region was attributed to melting of crystallized portions of the copolymer.

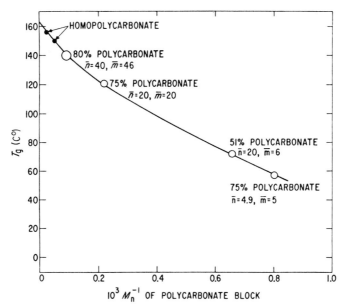

Fig. 7-31. Dependence of upper relaxation temperature on reciprocal $\overline{M}_n$ of polycarbonate block (○) and comparison with relaxation temperatures of two homopolycarbonates (●) (I115).

Low-angle X-ray and electron microscopy studies (I115, I116, I109) indicated that these copolymers contain polycarbonate aggregates with an average center-to-center distance of 100–200 Å. Two-phase systems are obtained in these block copolymers at much lower block molecular weight levels than is the case in styrene–diene block copolymers. This is due to the much greater disparity in the solubility parameters of the carbonate and siloxane blocks, i.e., 7.5 and 10.0, compared to 8.4 and 8.8 for polybutadiene and polystyrene.

The component present in the highest concentration forms the continuous phase with the other component forming discrete aggregates throughout the former. At intermediate concentrations, both carbonate and siloxane continua exist. There is also some degree of phase blending, depending upon the composition and conditions.

The morphology of these block copolymers has also been studied (I109, I112, I115, I116, I119–I122) by low-angle and wide-angle x-ray scattering, birefringence, infrared dichroism, and proton magnetic resonance techniques. In these investigations, interference was observed at distances ranging from 50 Å to 500 Å due to a difference in electron density between the different blocks. Swelling with hexane (a siloxane-preferred solvent) caused the scattering centers to move apart while scattering intensity increased. This was interpreted as evidence for scattering due to the distance between centers rather than due to the size of the centers themselves. The size of the scattering entities was estimated to be 10–50 Å. Scattering behavior was found to be independent of copolymer molecular weight. The morphological model was described as one in which some but not all of the bisphenol A carbonate blocks are associated to form domains that act as "cross-links." However, the continuous matrix is composed of both siloxane and carbonate blocks, since the mechanical behavior of the matrix is distinctly different from that of either homopolymer. This feature distinguishes these materials from the styrene–diene block copolymers and is probably due to the method of synthesis, which, as pointed out above, takes place in such a way that very short carbonate segments can be incorporated into the siloxane block.

A comparison was made of the mechanical and dielectric relaxation behavior of these copolymers (I123). Master curves constructed using the time–temperature superposition principle showed that considerable discrepancies exist, thereby demonstrating the thermorheological complexity of these systems.

The mechanical properties of the copolymers range from elastomeric to rigid as the siloxane content is decreased (I109). This is illustrated by the stress-elongation curves shown in Fig. 7-32. At high siloxane contents, the ultimate elongation is high and largely elastic. At lower siloxane contents, a yield stress is observed and recovery decreases.

The properties of solution-cast films are dependent upon casting solvent due to the effect of the solvent on phase continuity in the resulting film (I109). Films cast from a mixture of methylene chloride (good solvent for both blocks) and hexane (siloxane-preferred solvent) displayed greater rubbery character than did films cast from methylene chloride alone. This behavior is due to the tendency of the carbonate blocks to coacervate early in the drying process, while the siloxane blocks remain extended in the solvent medium. Heating of the resulting films above the upper T_g temperature results in an increase in yield stress to a level closer to that obtained by casting from methylene chloride alone. This behavior is presumably due to the fusing of some of the polycarbonate domains to increase their continuity. Another indication of coacervation of the carbonate blocks was the observation that intrinsic viscosity values for a given polymer were lower when measured in the mixed solvent system than in methylene chloride alone.

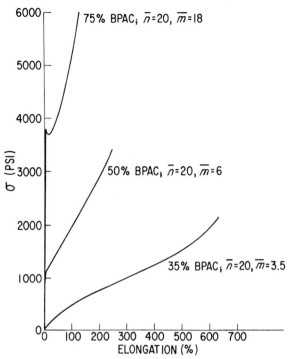

Fig. 7-32. Engineering stress (σ) versus elongation for three dimethylsiloxane–bisphenol A carbonate block polymers of the compositions indicated. Instron crosshead speed = 2 inches/minute (I109).

Single crystals of bisphenol A carbonate–dimethylsiloxane block copolymers have been reportedly (I124) grown in solvent–nonsolvent mixtures. X-Ray diffraction studies indicated that they had a hexagonal unit cell, compared to orthorhombic and monoclinic unit cell structures, respectively, for the carbonate and siloxane homopolymers. These materials displayed higher melting points (257°C versus 233°C) and lower heats of fusion ($\leq$14 cal/gm versus 32 cal/gm) than crystallized carbonate homopolymer. It was concluded that the high melting point was the result of a high degree of molecular conformation and crystal packing, which lowers both the heat and entropy of fusion.

Crystallization was claimed to be induced by contacting preformed block copolymer articles with methyl ethyl ketone liquid or vapor, resulting in improved mechanical properties (I125).

The properties of bisphenol A carbonate homopolymer can be modified by blending it with a low ($\leq$4%) concentration of the carbonate–

siloxane block copolymer (I121, I126, I127). Such blends have re-duced wettability and frictional properties. This is due to concentra-tion of the copolymer at the surface of the molded or solution-cast blend as a result of the low surface energy and incompatibility of the siloxane component of the block copolymer.

Cross-linkable block copolymers were prepared by end capping with allyl groups (I105, I106, I129). These compositions can be cross-linked with mercaptan-bearing compounds, or they can be further modified to convert the allyl end groups to diacetoxysilyl end groups, which can be cross-linked by reaction with atmospheric moisture.

Block copolymers of a similar nature, but with higher upper-use temperatures, have been prepared by substituting the bisphenol A with 9,9-bis(4-hydroxyphenyl) fluorene (I130—I132):

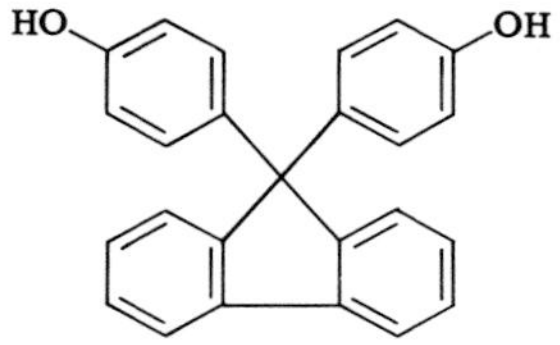

The resulting carbonate–siloxane block copolymers at 30% siloxane content in the form of a 17 unit siloxane block displayed a T_g at $-100°C$ due to the siloxane phase and another T_g at 215°C due to the carbonate phase. The T_g of the carbonate homopolymer at high molecular weight is 280°C, compared to 150°C for bisphenol A carbonate homopolymer. This block copolymer is not readily melt processable without degrada-tion. It is claimed to be potentially useful in high-speed aircraft canopies and in autoclavable membrane applications.

b. Tetramethyl-1,3-cyclobutylene Carbonate

Block copolymers containing segments of 2,2,4,4-tetramethyl-1,3-cyclobutylene carbonate and dimethylsiloxane were synthesized by Matzner et al. (I75, I78–I81, I133, I134) by the oligomer condensation technique discussed earlier for the preparation of polysulfone–siloxane block copolymers. The reaction entails the interaction of pre-formed, characterized carbonate oligomers with bis(dimethylamine)-terminated dimethylsiloxane oligomers. The carbonate oligomer was synthesized by preparing the dichloroformate of the diol with excess phosgene and then reacting this with a slight excess of the diol (Reac-tion 7-25). The oligomer contained approximately equal amounts of cis and trans isomers. The block copolymers were synthesized by adding the siloxane oligomer in several increments (over a 2 day period) to a

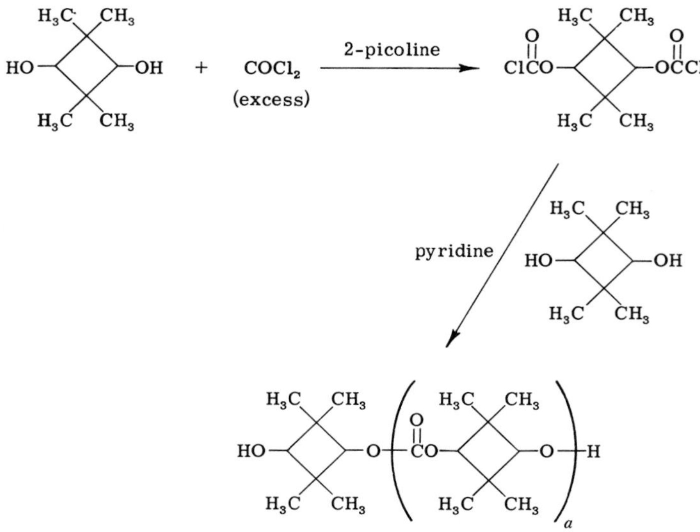

$$
\text{HO}-\underset{\substack{\text{H}_3\text{C}\quad\text{CH}_3}}{\overset{\substack{\text{H}_3\text{C}\quad\text{CH}_3}}{\diamond}}-\text{OH} \;+\; \underset{\text{(excess)}}{\text{COCl}_2} \;\xrightarrow{\text{2-picoline}}\; \text{ClCO}-\underset{\substack{\text{H}_3\text{C}\quad\text{CH}_3}}{\overset{\substack{\text{H}_3\text{C}\quad\text{CH}_3}}{\diamond}}-\text{OCCl}
$$

$$
\xrightarrow[\text{pyridine}]{\quad \text{HO}-\overset{\substack{\text{H}_3\text{C}\quad\text{CH}_3}}{\underset{\substack{\text{H}_3\text{C}\quad\text{CH}_3}}{\diamond}}-\text{OH}\quad} \tag{7-25}
$$

$$
\text{HO}-\underset{\substack{\text{H}_3\text{C}\;\;\text{CH}_3}}{\diamond}-\text{O}\left(\!\!\overset{\text{O}}{\underset{}{\text{CO}}}-\underset{\substack{\text{H}_3\text{C}\;\;\text{CH}_3}}{\diamond}-\text{O}\!\!\right)_{a}\!\!\text{H}
$$

refluxing chlorobenzene solution of the carbonate oligomer (Reaction 7-26). The end blocks and end groups of the copolymer will depend on

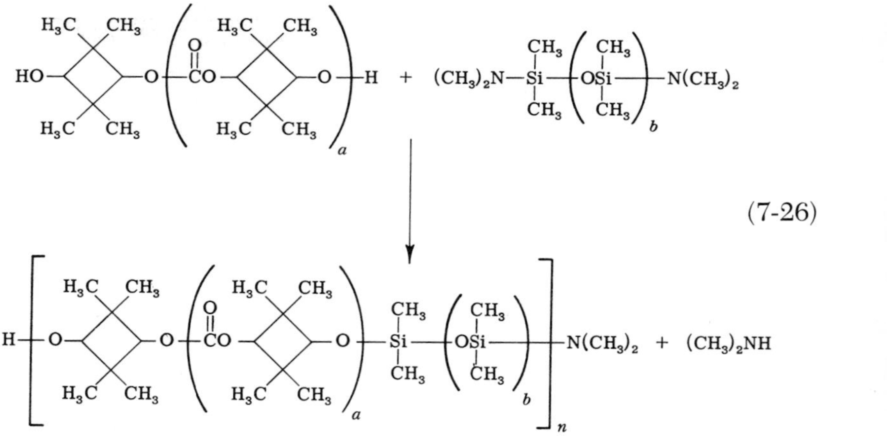

$$
\tag{7-26}
$$

the actual final stoichiometry of the condensation polymerization reaction rather than as depicted above for perfect 1 : 1 stoichiometry. The reaction rate is slower than that of the polysulfone–siloxane block copolymer reaction due to the lower reactivity of the aliphatic, sterically hindered hydroxyl end groups of the carbonate oligomer.

Copolymers were prepared from 4400 and 6400 $\overline{M}_n$ carbonate oligomers and siloxane oligomers of 400–10,000 $\overline{M}_n$. The $\overline{M}_n$ of both oligomers was determined by end group analysis. Mechanical properties varied from high modulus (190,000 psi), high tensile strength (5200 psi), low elongation (5%) materials at 7% siloxane content to low modulus (2000 psi), moderate tensile strength (1700 psi), high elongation (600%) materials at 63% siloxane content. The latter were elastomeric with good recovery properties. The copolymers were readily compression molded at 270°C and they displayed good thermal, hydrolytic, and ultraviolet stability. The melt processability of these compositions was compared to that of other organosiloxane block copolymers (I78, I81) (see Section I,4). The excellent processability of the cyclobutylene carbonate–siloxane system was attributed to the similarity of the solubility parameters of the carbonate and siloxane segments. Crystallization of the carbonate block provides the driving force necessary to produce phase separation after fabrication, resulting in good mechanical properties.

Temperature–modulus studies on these block copolymers indicated that compositions containing $>5000\ \overline{M}_n$ siloxane blocks display a T_g at $-120°C$ due to the siloxane block, a T_g at 40°–80°C due to the carbonate block, and a crystalline melting point (T_m) at 200°–230°C due to the carbonate block.

c. Aryl–Alkyl Phthalates

Block copolymers containing dimethylsiloxane segments and segments of alkyl or aryl phthalates were also reported by Matzner *et al.* (I75, I78–I80, I135–I137). These were synthesized by the oligomer condensation technique discussed above for the preparation of polysulfone–siloxane copolymers. Hydroxyl-terminated polyester oligomers were reacted with bis(dimethylamine)-terminated dimethylsiloxane oligomers. Two types of polyester oligomer were used: bisphenol A isophthatate (terephthalate) and hexamethylene terephthalate. The oligomers were prepared by condensation of a

Bisphenol A isophthalate or terephthalate

Hexamethylene terephthalate

slight deficiency of the phthaloyl chlorides with bisphenol A or 1,6-hexanediol in refluxing trichlorobenzene or monochlorobenzene solution, respectively. The block copolymers were synthesized by adding 5000 $\overline{M}_n$ poly(dimethylsiloxane) oligomers to chlorinated aromatic solutions of the polyester oligomers at 120°–180°C (Reaction 7-27).

$$
\text{HO—RO}\left(\!\!\begin{array}{c}\text{O}\\ \|\\ \text{C}\end{array}\!\!-\text{C}_6\text{H}_4\!\!\begin{array}{c}\text{O}\\ \|\\ \text{C}\end{array}\!\!\text{ORO}\right)_a\!\!\text{H} \;+\; (\text{CH}_3)_2\text{N}-\underset{\underset{\text{CH}_3}{|}}{\overset{\overset{\text{CH}_3}{|}}{\text{Si}}}\!\!\left(\!\!\text{OSi}\!\!\right)_b\!\!-\text{N(CH}_3)_2
$$

(7-27)

$$
\text{H}\!\!\left[\text{ORO}\left(\!\!\begin{array}{c}\text{O}\\ \|\\ \text{C}\end{array}\!\!\text{C}_6\text{H}_4\!\!\begin{array}{c}\text{O}\\ \|\\ \text{C}\end{array}\!\!\text{ORO}\right)_a\!\!\underset{\underset{\text{CH}_3}{|}}{\overset{\overset{\text{CH}_3}{|}}{\text{Si}}}\!\!\left(\!\!\text{OSi}\!\!\right)_b\right]_n\!\!\text{N(CH}_3)_2 \;+\; (\text{CH}_3)_2\text{NH}
$$

The block copolymer products obtained with the bis A phthalate oligomers contained ~65 wt% siloxane and were elastomeric with ~1500 psi tensile strengths, 3000–12,000 psi tensile moduli and 100–500% elongations. They displayed T_g values at −125°C due to the siloxane blocks and at +120 to +135°C for the polyester blocks. They also displayed polyester block melting point transitions at 260°C and 295°C for the isophthalate and terephthalate compositions, respectively. The hexamethylene terephthalate–siloxane block copolymer displayed 500 psi tensile strength, 1000 psi tensile modulus, 150% elongation, and a melting point transition at 120°C.

Other reports discuss the synthesis of polyester copolymers with short siloxane segments by the condensation of a mixture bisphenol A-terminated tetramer of methylphenylsiloxane with additional bisphenol A (or phenolphthalein) and iso- or terephthaloyl chloride (I138–I140). Due to the short length of the siloxane segment, these materials display primarily random copolymer properties. Softening temperature decreased and crystallinity increased with increasing siloxane content.

7. Miscellaneous

Several references report the preparation of copolymers containing segments of poly(dimethylsiloxane) and segments of other polymers, such as poly(α-amino acid amides), polycaprolactam, polyurethanes, polyureas, polyimides, and polyaluminosiloxanes.

Block copolymers containing α-amino acid amide and siloxane blocks have been prepared by the condensation of dihydroxyl-terminated amide oligomers with bis(dimethylamine)-terminated dimethylsiloxane oligomers (I141, I142). The amide oligomers were prepared by polymerizing N-carboxy-γ-benzyl L-glutamate anhydride followed by end capping to provide dihydroxyl termination. This was then reacted with the siloxane oligomer in refluxing chlorinated aromatic solvents to produce the block copolymers. The products are claimed to be stable to heat and light.

Nylon-6 (polycaprolactam)—poly(dimethylsiloxane) block copolymers were reportedly prepared by polymerizing caprolactam with NaH initiator and phenyl isocyanate promotor in the presence of α,ω-bis(aminopropyl)polydimethysiloxane oligomers (I143, I144). This material was claimed to be useful as an additive for nylon to lower its surface tension.

Urethane–siloxane block copolymers were prepared by reacting poly(propylene glycol) urethanes or poly(ethylene terephthalate) urethanes with acetoxy functional polydimethylsiloxanes (I145, I146). These materials, however, were branched rather than linear since the siloxanes were multifunctional. The products were claimed to be blood-compatible and therefore useful for the storage and handling of blood. Another reference (I147) claims the preparation of siloxane-urethanes by the reaction of α,ω-bishydroxymethyl polysiloxanes with aromatic diisocyanates. These products had 50°–70°C lower softening points than comparable polyether urethanes. Another synthetic route involves the interaction of α,ω-dihydroxypoly(dimethylsiloxanes) with preformed macrodiisocyanates prepared from a diisocyanate and hydroxyl-terminated polyisoprene (I148) or with a mixture of a diisocyanate and a polycaprolactonediol (I149). Urethane-containing block copolymers were also prepared by condensing hydroxy-terminated poly(tetramethylene oxide) and α,ω-bis(3-hydroxypropyl)poly(dimethylsiloxane) oligomers with toluene diisocyanate (I150). Thiourethane–siloxane block copolymers have also been reported in the literature (I150, I151).

Polyurea–dimethylsiloxane block copolymers were reportedly prepared by reacting a siloxane oligomer, a diisocyanate, and 4,4'-methylenebisaniline (I152). The product was claimed to give elastomeric fibers by solution spinning. Copolymers containing segments of polyurea and segments of polyaluminosiloxane, i.e.,

$$\left(\begin{array}{c} R \quad\ OR \\ | \qquad | \\ OSi-OAl \\ | \\ R \end{array} \right)_x$$

have also been reported to give elastomeric materials which display good stability (I152, I153).

Polyimides containing short dimethylsiloxane segments were prepared by reacting aminopropyl-terminated dimethylsiloxane tetramers with pyromellic dianhydride (I154).

J. CROSS-LINKED EPOXY RESIN SYSTEMS

The previous chapters and sections of this book have dealt almost exclusively with linear block copolymers. The importance of segment structure and molecular weight in determining the morphology and properties of these materials is well recognized. It is not so well recognized that these same principles can also apply to thermosetting systems. Noshay *et al.* (J1, J2, J7) demonstrated this in thermosetting epoxy–modifier block copolymer systems.

Since cured epoxy resins are inherently brittle materials, modification is necessary to make them useful in many applications. Improved toughness can be achieved by adding flexibilizing low molecular weight polymeric modifiers to the epoxy–hardener formulation. However, if these modifiers are fully compatible with the cured epoxy, they act as plasticizers and cause a considerable reduction in the glass transition temperature (T_g) of the system. This results in toughened systems at an appreciable sacrifice in heat distortion temperature (HDT) and, therefore, elevated temperature properties. The plasticizing effect of these modifiers was minimized by controlling the morphology of the cured composition.

The curing of epoxy resins with anhydrides in the presence of linear, hydroxyl-terminated polyester or polyether modifiers [e.g., polycaprolactone or poly(propylene oxide)] produced block copolymer structures. In these systems, one "block" is the cross-linked epoxy–anhydride network and the other block is linear modifier segment, as shown in Scheme 7-21.

The morphology of the cured system is dependent upon modifier molecular weight. The critical molecular weight of polycaprolactone and poly(propylene oxide) was found to be 3000–5000. Below this level, single morphological phase systems were obtained, but two-microphase systems resulted above this level. This is illustrated by the mechanical loss curves, shown in Fig. 7-33, of a series of systems prepared with polycaprolactones (PCL) of varying molecular weight ($\overline{M}_n$).

In the 1300 $\overline{M}_n$ PCL-modified system, the PCL was completely compatible with (soluble in) the cross-linked epoxy–anhydride matrix,

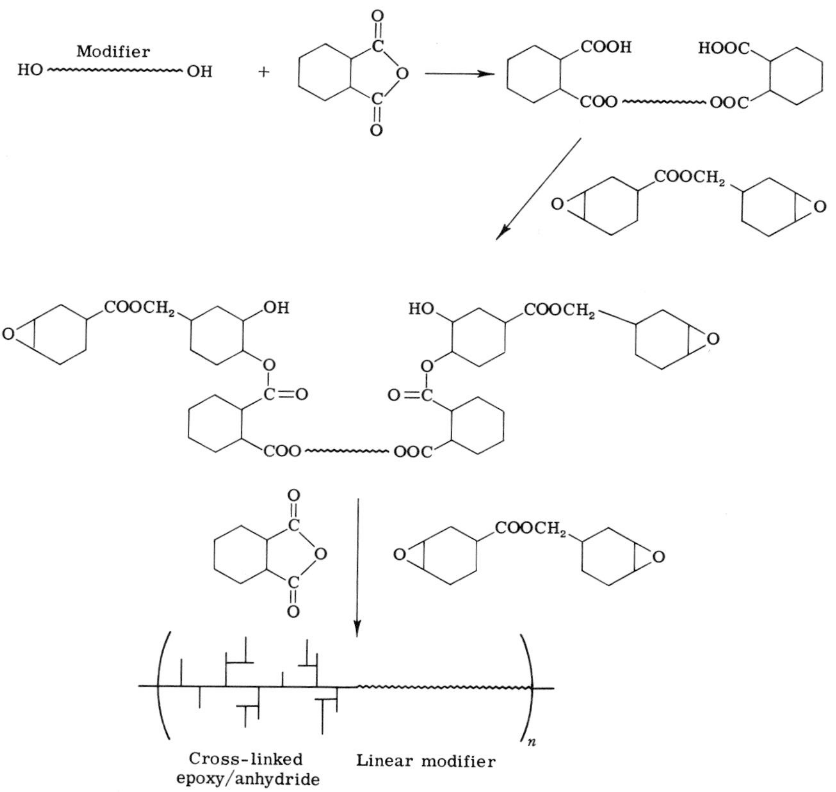

Scheme 7-21

as can be seen from the presence of a single T_g at 70°C. This is intermediate to the T_g of the unmodified cured resin (180°C) and that of polycaprolactone homopolymer (−60°C). The peak at −140°C is a low-temperature polycaprolactone transition ascribed to the $(CH_2)_n$ chain in which $n > 4$. In comparison to the single-phase system obtained with 1300 $\overline{M}_n$ PCL, the 2000 $\overline{M}_n$ PCL-modified system showed some sign of incipient phase segregation, and definite microphase segregation was observed with 5000 $\overline{M}_n$ and 10,000 $\overline{M}_n$ PCL. The latter two systems each displayed two distinct T_g values, one at −20°C to −30°C due to the PCL phase and another at +150 to +170°C due to the epoxy–anhydride phase.

In all cases, the PCL modifier was soluble in the epoxy–anhydride system before curing. It is only upon curing, during which the epoxy–

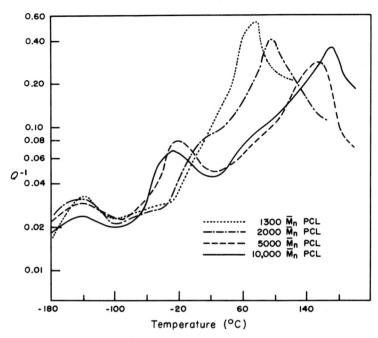

Fig. 7-33. Effect of PCL molecular weight on mechanical loss properties of PCL-modified, cured epoxy–anhydride systems; $\overline{M}_n$ PCL: ($\cdots$) 1300, ($-\cdot-\cdot-$) 2000, ($---$) 5000, (———) 10,000 (J1).

anhydride matrix increases in molecular weight, that incompatibility and the resulting phase segregation develops. The cured two-phase system containing 5000 $\overline{M}_n$ PCL was transparent, attesting to the small (less than the wavelength of visible light) size of the phase domains. However, the 10,000 $\overline{M}_n$ PCL system was opaque, presumably due to at least some degree of macroincompatibility or crystallization of the PCL segment.

The behavior of poly(propylene oxide)-modified resins was similar to that of the above-described PCL-modified systems. Modification with dihydroxy-terminated poly(propylene oxide) of 1000 $\overline{M}_n$ resulted in a single-phase system with one T_g (at 60°C), while a 4000 $\overline{M}_n$ modified system displayed two T_g values one at −50°C and another at +170°C.

This effect of modifier segment molecular weight on morphology appears to be a general one. These workers showed that the effect was observed when the following three different epoxy resin structures were modified with PCL:

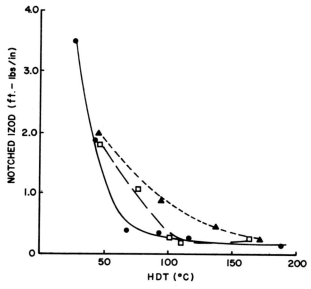

Furthermore, the toughening of epoxy resins observed by other work-ers (J3–J6), who used ≥ 2000 $\overline{M}_n$ carboxyl terminated butadiene–acrylonitrile copolymers as modifiers, may also be due to this phenomenon.

The morphological considerations described above lead to the ex-pectation that two-phase systems should display higher heat distortion temperatures (HDT) than single-phase systems at comparable mod-ifier concentrations. HDT is, in essence, the temperature at which the modulus falls to ~80,000 psi. A single-phase system will fall to that modulus level at a lower temperature than a two-phase system will by virtue of the high upper T_g of the two-phase system. From a practical

Fig. 7-34. Izod impact strength versus HDT in PCL-modified epoxy resin systems: ● = 1300 $\overline{M}_n$ PCL; □ = 5000 $\overline{M}_n$ PCL; ▲ = 10,000 $\overline{M}_n$ PCL (J1).

point of view, this means that a two-phase system should display a higher HDT than a single-phase system at comparable levels of toughness. The overall result would thus be tough systems with superior elevated temperature capabilities. This is indeed the case, as is demonstrated by the curves in Fig. 7-34, which relate HDT to notched Izod impact strength. These curves show that at intermediate HDT levels (50°–150°C), the two-phase systems, especially those based on 10,000 $\overline{M}_n$ PCL, gave a superior balance of HDT and toughness. Thermal shock resistance measurements on these systems also led to the same conclusion.

REFERENCES

A1. Schmolka, I. R., *in* "Nonionic Surfactants" (M. J. Schick, ed.), Chapter 10, p. 300. Dekker, New York, 1967.

A2. French Patent 1,557,086 (Badische Anilin-und Soda-Fabrik) (1969); *C.A.* **71**, 92859d (1969).

A3. Levchenko, D. N., and Nikolaeva, N. M., *Poverkhn. Aktiv. Veshchestva Sint. Zhirozamen.* **1**, 4 (1967); *C.A.* **69**, 36652c (1968).

A4. Halek, G. W., and Berardinelli, F. M., Belgian Patent 666,162 (Celanese Corp.) 1965; *C.A.* **65**, 17169b (1966).

A5. Davtyan, S. P., Rozenberg, B. A., and Enikolopyan, N. S., *Vysokomol. Soedin., Ser. A* **11**(9), 2051 (1969); *C.A.* **72**, 55960e (1970).

A6. Allport, D. C., British Patent 1,318,892 (Imperial Chemical Industries Ltd.) (1973); *C.A.* **80**, 15438z (1974).

A7. McGrath, J. E., Robeson, L. M., Matzner, M., and Barclay, R., Jr., "Rigid Block Copolymers," presented at the 8th Regional ACS Meeting, Akron, Ohio, May, 1976; *J. Polymer Sci.* Part C (in press).

B1. Richards, D. H., and Szwarc, M., *Trans. Faraday Soc.* **55**, 16 (1959).

B2. Finaz, G., Rempp, P., and Parrod, J., *Bull. Soc. Chim. Fr.* p. 262 (1962); *C.A.* **57**, 1055f (1962).

B3. Fabris, H. J., and Milkovich, R., German Offen. 1,931,996 (General Tire and Rubber Co.) (1970); *C.A.* **72**, 91139c (1970).

B4. Shimura, Y., and Lir, W. S., *J. Polym. Sci., Part A-1* **8**(8), 2171 (1970); *C.A.* **73**, 77664j (1970).

B5. Yamashita, Y., Hirota, M., Matsui, H., Hirao, A., and Nobutoki, K., *Polym. J.* **2**(1), 43 (1971); *C.A.* **75**, 21128m (1971).

B6. Yamashita, Y., Hirota, M., Nobutoki, K., Nakamura, Y., Hirao, A., Kozawa, S., Chiba, K., Matsui, H., and Hattori, G., *J. Polym. Sci., Part B* **8**(7), 481 (1970).

B7. Yamashita, Y., Nobutoki, K., Nakamura, Y., and Hirota, M., *Macromolecules* **4**(5), 548 (1971); *C.A.* **76**, 34703r (1972).

B8. McGrath, J. E., Robeson, L. M., Matzner, M., and Barclay, R., Jr., "Rigid Block Copolymers," presented at the 8th Regional ACS Meeting, Akron, Ohio, May, 1976.

C1. Coleman, D., *J. Polym. Sci.* **14**, 15 (1954).

C2. Coleman, D., British Patent 682,866 (Imperial Chemical Industries Ltd.) (1952); *C.A.* **47**, 7227e (1953).

C3. Leibnitz, E., and Reinisch, G., *Faserforsch. Textiltech.* **21**(10) 426 (1970); *C.A.* **74**, 43365t (1971).

C4. Snyder, M. D., U.S. Patent 2,744,087 (du Pont) 1956; *C.A.* **50**, 13464h (1956).

C5. Brooks, T. W., and Daffin, C. L., *Polym. Prepr., Am. Chem. Soc., Div. Polym. Chem.* **10**(2), 1174 (1969).

C6. Lyman, D. J., *Trans. N.Y. Acad. Sci.* [2] **30**, 113 (1967).

C7. Okazaki, K., Toyama, S., Nakata, M., and Kishimoto, K., German Offen. 2,006,810 (Toray Industries) (1970); *C.A.* **73**, 121456u (1970).

C8. Shima, T., Nawata, K., Tsunawaki, K., Yamahide, M., Kuwahara, M., and Sakakibara, S., Japanese Patent 69/13,271 (Teijin Ltd.) (1969); *C.A.* **72**, 13759r (1970).

C9. Sangen, O., Yamamoto, Y., and Imamura, R., *Sen'i Gakkaishi* **29**(11), T472 (1973); *C.A.* **80**, 83814u (1974).

C10. Sumoto, M., Komagata, H., Matsumoto, H., and Furusawa, H., German Offen. 1,961,005 (Toyo Spinning Co.) (1970); *C.A.* **73**, 78454c (1970).

C11. Froehlich, H., and Brinkman, L., S. African Patent 69/04,819 (Farbewerke Hoechst. A.-G.) (1970); *C.A.* **73**, 99711z (1970).

C12. French Patent 2,012,585 (Farbewerke Hoechst A.G.) (1968).

C13. Kobayashi, H., and Sasaguri, K., Japanese Paten 69/20,469 (Asahi Chemical Industry Co., Ltd.) (1969); *C.A.* **72**, 13757p (1970).

C14. Kobayashi, H., and Sasaguri, K., Japanese Patent 69/20,471 (Asahi Chemical Industry Co., Ltd.) (1969); *C.A.* **72**, 4193z (1970).

C15. Kobayashi, H., and Sasaguri, K., Japanese Patent 69/20,472 (Asashi Chemical Industry Co., Ltd.) (1969); *C.A.* **72**, 4194a (1970).

C16. Kobayashi, H., and Sasaguri, K., Japanese Patent 69/20,475 (Asahi Chemical Industry Co., Ltd.) (1969); *C.A.* **72**, 13758q (1970).

C17. Wolfe, J. R., Jr., U.S. Patent 3,775,373 (du Pont) (1973); *C.A.* **80**, 122131r (1974).

C18. Hoeschele, G. K., German Offen. 2,263,046 (duPont) (1973); *C.A.* **80**, 48949v (1974).

C19. Nishimura, A. A., and Komagata, H., *J. Macromol. Sci., Chem.* **1**(4), 617 (1967); *C.A.* **67**, 109450f (1967).

C20. Witsiepe, W. K., *Polym. Prepr., Am. Chem. Soc., Div. Polym. Chem.* **13**(1), 588 (1972).

C21. Witsiepe, W. K., German Offen. 2,240,801 (du Pont) (1973); *C.A.* **79**, 19440w (1973).

C22. Witsiepe, W. K., *Am. Chem. Soc., Rubber Div., Contrib.* No. 270 (1973).

C23. Witsiepe, W. K., *Adv. Chem. Ser.* **129**, 39 (1973); *C.A.* **80**, 97042c (1974).

C24. Hoeschele, G. K., and Witsiepe, W. K., *Angew. Makromol. Chem.* **29/30**, 267 (1973); *C.A.* **79**, 80048r (1973).

C25. Witsiepe, W. K., German Offen. 2,213,128 (du Pont) (1972); *C.A.* **78**, 17337y (1973).

C26. Brown, M., German Offen. 2,301,115 (du Pont) (1973); *C.A.* **80**, 15736v (1974).

C27. Brown, M., Hoeschele, G. K., and Witsiepe, W. K., German Offen. 2,210,119 (du Pont) (1972); *C.A.* **78**, 17364e (1973).

C28. Knox, J. B., *Polym. Age* **4**, p. 357 (1973).

C29. Buck, W. H., and Cella, R. J., *Polym. Prepr., Am. Chem. Soc., Div. Polym. Chem.* **14**(1), 98 (1973).

C30. Cella, R. J., *J. Polym. Sci., Polym. Symp.* **42**, 727 (1973).

C31. West, J. C., Lilaonitkul, A., Cooper, S. L., Mehra, U., and Shen, M., *Polym. Prepr., Am. Chem. Soc., Div. Polym. Chem.* **15**(2), 186 (1974).

C32. Brown, M., and Witsiepe, W. K., *Rubber Age (N.Y.)* **104**, 35 (1972).

C33. Knox, J. B., *SGF Publ.* **41**, II (1972); *C.A.* **78**, 44706z (1973).

C34. Wright, J., *U.S. N.T.I.S., AD Rep.* **AD-768711/4GA** (1973); *C.A.* **80**, 134495a (1974).

C35. Whitlock, K. H., *Plastica* **26**(10), 438 (1973); *C.A.* **80**, 97039g (1974).

C36. Whitlock, K. H., *Ned. Rubberind.* **34**(16), 1 (1973); *C.A.* **80**, 60772q (1974).

C37. Hoeschele, G. K., German Offen. 2,302,654 (du Pont) (1973); *C.A.* **80**, 84144n (1974).

C38. German Patent 2,354,757 (du Pont) (1972).

C39. Hoeschele, G. K., German Offen. 2,313,903 (du Pont) (1973); *C.A.* **80**, 97067q (1974).

C40. Ghaffar, A., Goodman, I., and Hall, I. H., *Br. Polym. J.* **5**(4), 315 (1973); *C.A.* **80**, 60475v (1974).

C41. Smith, J. G., and Sublett, B. J., U.S. Patent 3,523,923 (Eastman Kodak) (1970); *C.A.* **73**, 88875e (1970).

C42. Riches, K. M., and Haward, R. N., *Polymer* **9**(2), 103 (1968); *C.A.* **68**, 78644m (1968).

C43. Abbott, T. A., McPherson, N. S., Clachan, M. L., and Tatchell, K. R., British Patent 918,046 (Bexford Ltd.) (1963); *C.A.* **58**, 10357g (1963).

C44. Korshak, V. V., Vinogradova, S. V., and Papava, G. Sh., U.S.S.R. Patent 166,830 (1964); *C.A.* **62**, 7896c (1965).

C45. Papava, G. Sh., Khitarishvili, I. S., Vinogradova, S. V., and Korshak, V. V., *Soobshch. Akad. Nauk Gruz. SSR* **66**(3), 597 (1972); *C.A.* **77**, 127129u (1972).

C46. Sangen, O., Yamamoto, Y., and Imamura, R., *Sen'i Gakkaishi* **25**(3), 115 (1969); *C.A.* **72**, 13688s (1970).

C47. Horio, M., Imamura, R., and Sangen, H., Japanese Patent 69/20,466 (Research Institute for Synthetic Fibers, Japan) (1969); *C.A.* **72**, 13738h (1970).

C48. Sangen, O., Yamamoto, Y., and Imamura, R., *Sen'i Gakkaishi* **25**(10), 445 (1969); *C.A.* **73**, 4786r (1970).

C49. Horio, M., Imamura, R., and Sangen, H., Japanese Patent 69/20,465 (Research Institute for Synthetic Fibers, Japan) (1969); *C.A.* **72**, 13736f (1970).

C50. Horio, M., Imamura, R., Matsui, C., Seizo, T., Miken, H., and Yamada, Y., Japanese Patent 69/07,747 (Research Institute for Synthetic Fibers, Japan) (1969); *C.A.* **71**, 125900r (1969).

C51. Senoo, T., and Onaya, Y., U.S. Patent 3,381,047 (Nitto Boseki Co., Ltd.) (1968); *C.A.* **69**, 20321h (1968).

C52. Galin, M., and Galin, J. C., *Makromol. Chem.* **160**, 321 (1972); *C.A.* **78**, 4680e (1973).

C53. McGrath, J. E., Robeson, L. M., Matzner, M. and Barclay, R., Jr. "Rigid Block Copolymers," presented at the 8th Regional ACS Meeting, Akron, Ohio, May, 1976; *J. Polymer Sci,* Part C (in press).

C54. Papava, G. S., Khitarishvili, I. S., Vinogradova, S. V., and Korshak, V.V., *Soobshch. Akad. Nauk Gruz. SSR* **68**(3) 593 (1972); *C.A.* **78**, 136713c (1973).

C55. Mueller, E., German Offen. 2,210,839 (Bayer A.-G.) (1973); *C.A.* **80**, 96754z (1974).

C56. Hayashi, I., Taniguchi, S., Hamada, H., and Minami, T., Japanese Patent (Kokai) 73/28,078 (Toyobo Co., Ltd.) (1973); *C.A.* **79**, 54537e (1973).

C57. Okazaki, K., and Shimizu, T., Japanese Patent 72/09,421 (Toray Industries, Inc.) (1972); *C.A.* **77**, 49992d (1972).

C58. Date, M., and Wada, M., Japanese Patent (Kokai) 73/29,845 (Toyobo Co., Ltd.) (1973); *C.A.* **79**, 54441u (1973).

C59. Hayashi, I., Taniguchi, S., Hamada, H., and Hasegawa, Y., Japanese Patent (Kokai) 73/28,077 (Toyobo Co., Ltd.) (1973); *C.A.* **79**, 54536d (1973).

C60. Sumoto, M., Komagata, H., and Kijima, Y., Japanese Patent (Kokai) 73/55,235 (Toyobo Co., Ltd.) (1973); *C.A.* **80**, 4774e (1974).

C61. Wolle, J. R., Jr., U.S. Patent 3,775,375 (1973); *C.A.* **80**, 134037c (1974).

C62. Witsiepe, W. K., German Offen. 2,035,333 (duPont) (1971); *C.A.* **74**, 143109f (1971).

C63. Masai, T., Sumoto, M., Shindo, S., and Itahana, Y., Japanese Patent 71/38,911 (Toyo Spinning Co., Ltd.) (1971); *C.A.* **77**, 115376j (1972).

C64. French Demande 2,121,662 (Dunlop Co., Ltd.) (1972); *C.A.* **78**, 112560z (1973).

C65. Crawford, R. W., and Witsiepe, W. K., German Offen. 2,224,678 (du Pont) (1972); *C.A.* **78**, 125446s (1973).

C66. Hoeschele, G. K., German Offen. 2,311,849 (du Pont) (1973); *C.A.* **80**, 71863w (1974).

C67. Riches, K. M., British Patent 1,058,389 (Shell Internationale) (1967); *C.A.* **66**, 66111f (1967).

C68. Sangen, O., Yamamoto, Y., and Imamura, R., *Sen'i Gakkaishi* **27**(4), 153 (1971); *C.A.* **76**, 114610y (1972).

C69. Khitarishvili, I. S., Papava, G. Sh., Korshak, V. V., and Vinogradova, S. V., *Soobshch. Akad. Nauk Gruz. SSR* **65**(3), 597 (1972); *C.A.* **77**, 35007z (1972).

C70. Bell, A., Kibler, C. J., and Smith, J. G., U.S. Patent 3,234,413 (Eastman Kodak) (1966); *C.A.* **64**, 19943g (1966).

C71. Nishimura, A., Komagata, H., Matsumoto, H., Ohkuchi, M., and Shindo, S. Japanese Patent 72/45,198 (Toyo Spinning Co., Ltd.) (1972); *C.A.* **79**, 147366p (1973).

D1. Coffey, D. H., and Meyrick, T. J., *Rubber Chem. Technol.* **30**, 283 (1957).

D2. Grieveson, B. M., *Polymer* **1**, 499 (1960).

D3. Iwakura, Y., Taneda, Y., and Uchida, S., *J. Appl. Polym. Sci.* **5**(13), 108 (1961).

D4. Flory, P. J., U.S. Patent 2,691,006 (Goodyear Tire and Rubber Co.) (1954); *C.A.* **49**, 2120d (1955).

D5. Quisenberry, R. K., U.S. Patent 3,316,326 (du Pont) (1967); *C.A.* **67**, 3635e (1967).

D6. British Patent 1,034,194 (Borg-Warner Corp.) (1966).

D7. Frazer, A. H., U.S. Patent 3,037,960 (du Pont) (1962); *C.A.* **57**, 7484d (1962).

D8. Papava, G. Sh., Agladze, L. D., Tsiskarishvili, P. D., Korshak, V. V., and Vinogradova, S. V., *Soobshch. Akad. Nauk Gruz. SSR* **43**(3), 593 (1966); *C.A.* **66**, 3102x (1967).

D9. Vinogradova, S. V., Korshak, V. V., Papava, G. Sh., and Tsiskarishvili, P. D., *Izv. Akad. Nauk SSSR, Ser. Khim.* **4**, 895 (1967); *C.A.* **67**, 33227r (1967).

D10. Goldberg, E. F., and Scardiglia, F., French Patent 1,419,852 (Borg-Warner Corp.) (1965).

D11. Morimoto, S., *Man-Made Fibers* **2**, 21 (1968).

D12. Lenz, R. W., and Go, S., *J. Polym. Sci., Polym. Chem. Ed.* **11**(11), 2927 (1973); *C.A.* **80**, 83809w (1974).

D13. Lenz, R. W., and Go, S., *J. Polym. Sci., Polym. Chem. Ed.* **12**(1), 1 (1974); *C.A.* **80**, 121480s (1974).

D14. Kovalenko, V. I., and Maklakov, L. I., *Zh. Prikl. Spektrosk.* **13**(1), 111 (1970); *C.A.* **73**, 88230r (1970).

D15. Cusano, C. M., Dunigan, E. P., and Weiss, P., *J. Polym. Sci., Part C* **4**, 743 (1963); *C.A.* **60**, 3108b (1964).

D16. Charch, W. H., and Shivers, J. C., *Text. Res. J.* **29**, 536 (1959).

D17. Weissermel, K., Brinkman, L., and Guenther, G., S. African Patent 68/06,268 (Farbwerke Hoechst A.-G.) (1969); *C.A.* **71**, 92575h (1969).

D18. British Patent 1,060,401 (Kurashiki Rayon Co., Ltd.) (1967); *C.A.* **67**, 12432p (1967).

D19. Imanaka, H., and Sumoto, M., Japanese Patent (Kokai) 73/00,991, (Toyobo Co., Ltd.) 1973; *C.A.* **78**, 137334s (1973).

D20. Quisenberry, R. K., U.S. Patent 3,265,762 (du Pont) (1966); *C.A.* **65**, 15608e (1966).

D21. French Patent 1,303,888 (Eastman Kodak Co.) (1962); *C.A.* **58**, 8114e (1963).

D22. Caldwell, J. R., and Gilkey, R., British Patent 982,575 (Eastman Kodak Co.) (1965); *C.A.* **62**, 11937d (1965).

D23. Weissermel, K., Guenther, G., and Brinkmann, L., S. African Patent 68/06,269 (Farbwerke Hoechst A.-G.) (1969); *C.A.* **71**, 102585c (1969).

D24. Quisenberry, R. K., and Pacofsky, E.A., U.S. Patent 3,317,632 (du Pont) (1967); *C.A.* **67**, 22829n (1967).

D25. Netherlands Patent Appl. 6,515,870 (Imperial Chemical Industries, Ltd.) (1966); *C.A.* **65**, 17141h (1966).

D26. Vinogradova, S. V., Korshak, V. V., Okulevich, P. O., Perfilov, Yu. I., and Vasnev, V. A., *Vysokomol. Soedin.*, *Ser. B* **15**(6), 470 (1973); *C.A.* **79**, 105610h (1973).

D27. Onder, K., Peters, R. H., and Spark, L. C., *Polymer* **13**(3), 133 (1972); *C.A.* **77**, 49273v (1972).

D28. O'Malley, J. J., *J. Polym. Sci.*, (Polym. Lett. Ed.) **12**,(7), 381 (1974); *C.A.* **81**, 78445r (1974).

E1. Katayama, S., Horikawa, H., and Masuda, N., German Offen. 2,217,470 (Denki Onkyo Co., Ltd.) (1972); *C.A.* **78**, 30516a (1973).

E2. Witsiepe, W. K., *Nuova Chim.* **48**(10), 79 (1972); *C.A.* **78**, 5138w (1973).

E3. Brinkmann, L., and Froehlich, H., S. African Patent 68/06,236 (Farbwerke Hoechst) (1969); *C.A.* **71**, 82062z (1969).

E4. Brinkmann, L., and Frohlich, H., U.S. Patent 3,598,882 (Farbwerke Hoechst) (1971).

E5. Brinkmann, L., and Herwig, W., S. African Patent 68/06,124 (Farbwerke Hoechst A.-G.) (1969); *C.A.* **71**, 82065c (1969).

F1. Sekine, Y., Ikeda, K., and Taketani, H., *Kogyo Kagaku Zasshi* **73**(2), 429 (1970); *C.A.* **72**, 133496z (1970).

F2. Laakso, T. M., and Buckley, D. A., U.S. Patent 3,038,879 (Eastman Kodak Co.) (1962); *C.A.* **58**, 4707a (1963).

F3. Laakso, T. M., and Petropoulos, M. C., U.S. Patent 3,106,544 (Eastman Kodak Co.) (1963).

F4. Merrill, S. H., *J. Polym. Sci.* **55**, 343 (1961).

F5. Robertson, A. B., Cook, J. A., and Gregory, J. T., *Adv. Chem. Ser.* **128**, 258 (1973).

F6. Merrill, S. H., and Petrie, S. E., *J. Polym. Sci.*, *Part A* **3**(6), 2189 (1965); *C.A.* **63**, 8511h (1965).

F7. Goldberg, E.P., *J. Polym. Sci.*, *Part C* **4**, 707 (1964); *C.A.* **60**, 4313h (1964).

F8. Alabran, D. M., *U.S. Dep. Comm.*, *C.F.S.T.I.*, *AD Rep.* **AD-664756** (1967); from *U.S. Gov. Res. Dev. Rep.* **68**(7), 100 (1968); *C.A.* **69**, 52812k (1968).

F9. Faoro, R. A., *U.S. Dep. Comm.*, *C.F.S.T.I.*, *AD Rep.* **AD- 687252** (1969); from *U.S. Gov. Res. & Dev. Rep.* **69**(14), 93 (1969); *C.A.* **71**, 82335r (1969).

F10. Faoro, R. A., *U.S. Dep. Comm.*, *C.F.S.T.I.*, *AD Rep.* **AD-701875** (1969); from *U.S. Gov. Res. & Dev. Rep.* **70**(8), 71 (1970); *C.A.* **73**, 56922b (1970).

F11. Perry, K. P., Jackson, W. J., Jr., and Caldwell, J. R., *J. Appl. Polym. Sci.* **9**(10), 3451 (1965); *C.A.* **64**, 3807a (1966).

F12. Caldwell, J. R., and Jackson, W. J., U.S. Patent 3,367,993 (Eastman Kodak) (1968); *C.A.* **68**, 60371k (1968).

F13. Kobayashi, H., Sasaguri, K., and Makita, M., Japanese Patent 69/15,430 (Asahi Chemical Industry Co., Ltd.) (1969); *C.A.* **71**, 102904f (1969).

F14. Kobayashi, H., Sasaguri, K., and Makita, M., Japanese Patent 69/15,433 (Asahi Chemical Industry Co., Ltd.) (1969); *C.A.* **71**, 102946w (1969).

F15. Berger, M. N., Boulton, J. J. K., and Brooks, B. W., *J. Polym. Sci., Part A-1* **7**(5), 1339 (1969); *C.A.* **71**, 39466x (1969).

F16. Matzner, M., U.S. Patent 3,641,200 (Union Carbide Corp.) (1972); *C.A.* **76**, 142100a (1972).

F17. Matzner, M., U.S. Patent 3,639,503 (Union Carbide Corp.) (1972); *C.A.* **76**, 142098f (1972).

F18. Kobayashi, H., Sasaguri, K., and Makita, M., Japanese Patent 69/02,999 (Asahi Chemical Industry Co., Ltd.) (1969); *C.A.* **70**, 107322d (1969).

F19. Kobayashi, H., Sasaguri, K., and Makita, M., Japanese Patent 69/03,000 (Asahi Chemical Industry Co., Ltd.) (1969); *C.A.* **70**, 107323e (1969).

F20. Kobayashi, H., Sasaguri, K., and Makita, M., Japanese Patent 69/04,120 (Asahi Chemical Industry Co., Ltd.) (1969); *C.A.* **70**, 97793s (1969).

F21. Kobayashi, H., Sasaguri, K., and Makita, M., Japanese Patent 69/15,434 (Asahi Chemical Industry Co., Ltd.) (1969); *C.A.* **71**, 103041j (1969).

F22. Kobayashi, H., Sasaguri, K., and Makita, M., Japanese Patent 69/15,431 (Asahi Chemical Industry Co., Ltd.) (1966); *C.A.* **71**, 102905g (1969).

F23. Netherlands Patent Appl. 6,514,132 (du Pont) (1966); *C.A.* **65**, 15627e (1966).

F24. Goldberg, E. P., Strause, S. F., and Munro, H. E., *Polym. Prepr., Am. Chem. Soc., Div. Polym. Chem.*, **5**(1), 233 (1964); *C.A.* **64**, 2170f (1966).

F25. Netherlands Patent Appl. 6,514,747 (Farbenfabriken Bayer A.G.) (1966); *C.A.* **65**, 17144c (1966).

F26. McGrath, J. E., Robeson, L. M., Matzner, M. and Barclay, R., Jr. "Rigid Block Copolymers," presented at the 8th Regional ACS Meeting, Akron, Ohio, May, 1976; *J. Polymer Sci*, Part C (in press).

F27. Kolesnikov, G. S., Smirnova, O. V., Mikitaev, A. K., and Gladyshev, V. M., *Vysokomol. Soedin, Ser. A* **12**(6), 1424 (1970); *C.A.* **73**, 56658v (1970).

F28. Laakso, T. M., and Buckley, D. A., British Patent 938,788 (Kodak Ltd.) (1959).

F29. Ikeda, K., and Sekine, Y., *Ind. Eng. Chem., Prod. Res. Dev.* **12**(3), 202 (1973); *C.A.* **80**, 3818k (1974).

F30. Serini, V., Vernaleken, H., and Schnell, H., German Offen. 2,211,956 (Bayer A.-G.) (1973); *C.A.* **80**, 83884s (1974).

F31. Laakso, T. M., and Petropoulos, M. C., U.S. Patent 3,038,880 (Eastman Kodak Co.) (1962); *C.A.* **58**, 6948g (1963).

F32. Memon, N.A., and Williams, H. L., *J. Appl. Polym. Sci.* **17**(5), 1361 (1973); *C.A.* **79**, 67559p (1973).

F33. Kobayashi, H., Sasaguri, K., and Makita, M., Japanese Patent 69/15,432 (ASAHI Chemical) (1969); *C.A.* **71**, 114069a (1969).

G1. Brubaker, M. M., Coffman, D. D., and McGrew, F. C., U.S. Patent 2,339,237 (E.I. du Pont de Nemours and Co.) (1944); *C.A.* **38**, 3853q (1944).

G2. Sweeny, W., and Zimmerman, J., *Encycl. Polym. Sci. Technol.* **10**, 483 (1969); *C.A.* **72**, 44148v (1970).

G3. Jones, D. C., and White, T. R., *Kinet. Mech. Polym.* **3**, 41 (1972); *C.A.* **80**, 10885h (1974).

G4. Hongo, T., Suzuki, H., Murano, M., Shirasugi, K., and Yamadera, R., *Kobunshi Kagaku* **28**(313), 407 (1971); *C.A.* **75**, 152206h (1971).

G5. Kenney, J. F., *Polym. Eng. Sci.* **8**(3), 216 (1968); *C.A.* **69**, 59637r (1968).

G6. Allen, S. J., *J. Text. Inst.* **44**, 286 (1953).

G7. Bruck, S. D., and Thadani, A., *Polym. Prepr., Am. Chem. Soc., Div. Polym. Chem.* **10**(2), 897 (1969); *C.A.* **75**, 6571g (1971).

G8. Bruck, S. D., and Levi, A. A., *J. Macromol. Sci., Chem.* **1**(6), 1095 (1967).

G9. Beste, L. F., and Houtz, R. C., *J. Polym. Sci.* **8**(4), 395 (1952).

G10. British Patent 918,637 (E. I. du Pont de Nemours and Co.) (1963); *C.A.* **58**, 11514g (1963).

G11. Zimmerman, J., Pearce, E. M., Miller, I. K., Muzzio, J. A., Epstein, I. G., and Hosegood, E. A., *J. Appl. Polym. Sci.* **17**, 849 (1973).

G12. Soma, I., Fukuma, N., Chiaya, M., Okamoto, T., and Kitago, T., Japanese Patent 70/22,348 (Asahi, Chemical Industry Co., Ltd.) (1970); *C.A.* **73**, 88622v (1970).

G13. Champetier, G., and Lauth, B., *Makromol. Chem.* **92**, 170 (1966).

G14. Saotome, K., and Sato, K., *Makromol. Chem.* **102**(1), 105 (1967); *C.A.* **66**, 116031c (1967).

G15. French Patent 1,512,966 (Inventa A.-G. fuer Forschung und Patentverwertung) (1968); *C.A.* **70**, 68855h (1969).

G16. Saotome, K., and Sato, K., Japanese Patent 69/28,318 (Asahi, Chemical Industry Co., Ltd.) (1969); *C.A.* **72**, 67766j (1970).

G17. Saotome, K., Sato, K., and Daito, T., Japanese Patent 69/28,317 (Asahi Chemical Industry Co., Ltd.) (1969); *C.A.* **72**, 56107u (1970).

G18. French Patent 1,529,089 (Asahi Chemical Industry Co., Ltd.) (1968); *C.A.* **70**, 107008f (1969).

G19. Watson, J., British Patent 1,141,023 (Courtaulds Ltd.) (1969); *C.A.* **70**, 58824h (1969).

G20. Kobayashi, S., and Fujie, H., Japanese Patent 71/27,824 (Teijin Ltd.) (1971); *C.A.* **76**, 25798y (1972).

G21. Fedotova, O. Ya., Trezvov, V. V., and Kolesnikov, G. S., *Vysokomol. Soedin., Ser. A* **11**(7), 1437 (1969); *C.A.* **71**, 91960t (1969).

G22. Kiyotsukuri, T., Banba, T., and Nishikawa, Y., *Kobunshi Kagaku* **30**(10), 587 (1973); *C.A.* **80**, 83699k (1974).

G23. Bruck, S. D., *Polymer* **10**(12), 939 (1969); *C.A.* **72**, 79999a (1970).

G24. Yakubovich, A. Ya., Flerova, A. N., Yakubovich, V. S., Shalygin, G. F., Naumov., V. S., and Sokolov, L. B., *Vysokomol. Soedin., Ser. A* **14**(8), 1838 (1972); *C.A.* **77**, 152626m (1972).

G25. Charch, W. H., and Shivers, J. C., *Text. Res. J.* **29**, 536 (1959).

G26. British Patent 1,243,238 (Matsushita Electric Works, Ltd.) (1971); *C.A.* **76**, 35052q (1972).

G27. Kobayashi, H., Sasaguri, K., and Fujimoto, Y., Japanese Patent 71/26,979 (Asahi Chemical Industry Co., Ltd.) (1971); *C.A.* **76**, 15680q (1972).

G28. Curnuck, P. A., and Jones, M. E. B., *Br. Polym. J.* **5**(1), 21 (1973); *C.A.* **79**, 92629u (1973).

G28a. Okazaki, K., Nakagawa, A., Nakayama, Y., and Ichikawa, Y., British Patent 1,211,118 (Toray Industries, Inc.) (1970); *C.A.* **74**, 23578m (1971).

G29. Okazaki, K., Nakagawa, A., and Sugii, K., British Patent 1,230,195 (Toray Industries, Inc.) (1971); *C.A.* **75**, 37758c (1971).

G30. French Patent 1,459,484 (Asahi Chemical Industry Co., Ltd.) (1966); *C.A.* **67**, 11953x (1967).

G31. Okazaki, K., Nakagawa, T., and Nakata, M., Japanese Patent 70/07,559 (Toyo Rayon Co., Ltd.) (1970); *C.A.* **73**, 4578z (1970).

G32. Okazaki, K., Nakagawa, A., Ichikawa, Y., and Nakayama, Y., U.S. Patent 3,522,329 (1970); *C.A.* **73**, 67589z (1970).

G33. Okazaki, K., and Nakada, M., Japanese Patent 72/08,934 (Toray Industries, Inc.) (1972); *C.A.* **77**, 153713z (1972).

G34. Iwasaki, N., Kojima, T., Takizawa, T., Nakamura, K., and Ikenaga, Y., Japanese Patent (Kokai) 73/80,177 (Toray Industries, Inc.) (1973); *C.A.* **80**, 84658h (1974).

G35. Baguley, M. E., British Patent 1,134,213 (Courtaulds Ltd.) (1968); *C.A.* **70**, 29813y (1969).

G36. Kiyotsukuri, T., *Kobunshi Kagaku* **28**(312), 302 (1971); *C.A.* **75**, 98919c (1971).

G37. Korshak, V. V., Vinogradova, S. V., Vasnev, V. A., Perfilov, Yu. I., and Okulevich, P. O., *J. Polym. Sci., Polym. Chem. Ed.* **11**, 2209 (1973).

G38. Kayiyama, S., Iijima, K., Kitsuda, Y., and Kusushita, T., Japanese Patent 73/24,037 (Matsushita Electric Works, Ltd.) (1973); *C.A.* **80**, 84466u (1974).

G39. Okazaki, K., and Shimizu, T., Japanese Patent 71/35,377 (Toray Industries, Inc.) (1971); *C.A.* **76**, 73610c (1972).

G40. Kiyotsukuri, T., and Shimomura, Y., *Kobunshi Kagaku* **28**(315), 563 (1971); *C.A.* **76**, 4388z (1972).

G41. Japanese Patent (Ohmi Kenshi Spinning Co.) (1974).

G42. Kivetsukuri, T., and Shimomura, Y., *Kobunshi Kagaku* **28**(314), 516 (1971); *C.A.* **75**, 152163s (1971).

G43. Okazaki, K., and Nakagawa, T., Japanese Patent 69/18,931 (Toyo Rayon Co., Ltd.) (1969); *C.A.* **72**, 13765q (1970).

G44. Okazaki, K., and Nakagawa, T., Japanese Patent 69/11,668 (Toyo Rayon Co., Ltd.) (1969); *C.A.* **71**, 81942z (1969).

G45. Japanese Patent 69/11,668 (Toyo Rayon Co., Ltd.) (1969).

G46. Japanese Patent 69/11,669 (Toyo Rayon Co. Ltd.) (1969).

G47. Netherlands Patent Appl. 6,410,530 (Firestone Tire and Rubber Co.) (1965); *C.A.* **63**, 10153a (1965).

G48. British Patent 1,141,118 (Toyo Rayon Co., Ltd.) (1969); *C.A.* **70**, 69220j (1969).

G49. Martins, J. G., and Ashley, K. F., German Offen. 2,129,476 (Monsanto Co.) (1971); *C.A.* **76**, 114155d (1972).

G50. Jasse, B., *C.R. Hebd. Seances Acad. Sci., Ser. C* **268**(4), 319 (1969); *C.A.* **70**, 78667n (1969).

G51. Okazaki, K., and Nakagawa, T., Japanese Patent 69/11,669 (Toyo Rayon Co., Ltd.) (1969); *C.A.* **71**, 103056t (1969).

G52. Ateya, K., *Angew. Makromol. Chem.* **7**, 79 (1969); *C.A.* **71**, 91955v (1969).

G53. Korshak, V. V., Vinogradova, S. V., Teplyakov, M. M., and Maksimov, A. D., *Izv. Vyssh. Uchebn. Zaved., Khim. Khim. Tekhnol.* **10**(6), 688 (1967); *C.A.* **68**, 88054h (1968).

G54. Jasse, B., *Bull. Soc. Chim. Fr.* **10**, 3699 (1969); *C.A.* **72**, 44251y (1970).

G55. Okazaki, K., and Shimokawa, Y., Japanese Patent 70/03,392 (Toyo Rayon Co., Ltd.) (1970); *C.A.* **73**, 15692h (1970).

G56. Inoshita, K., Terakawa, M., and Yasuda, T., Japanese Patent 72/28,916 (Toyo Spinning Co., Ltd.) (1972); *C.A.* **78**, 30750x (1973).

G57. British Patent 1,140,463 (I.C.I. Ltd.) (1969).

G58. Netherlands Patent Appl. 6,609,101 (Imperial Chemical Industries, Ltd.) (1967); *C.A.* **67**, 12438v (1967).

G59. Lyssy, T., Keller, E., Heller, H., and Mueller, H., S. African Patent 67/06,198 (Geigy, J. R., A.-G.) (1968); *C.A.* **70**, 58749n (1969).

G60. Belgian Patent 702824 (Farbwerke Hoechst A.G.) (1968).

G61. Shimura, Y., and Ikeda, N., *J. Polym. Sci., Polym. Chem. Ed.* **11**, 1271 (1973).

G62. Shima, T., Asami, A., and Yamashiro, S., Japanese Patent 70/12,150 (Teijin Ltd.) (1970); *C.A.* **73**, 56625g (1970).

G63. Crivello, J. V., and Juliano, P. C., *Polym. Prepr., Am. Chem. Soc., Div. Polym. Chem.* **14**(2), 1220 (1973); *C.A.* **83**, 10942m (1975).

G64. Vogel, H. A., and Oien, H. T., French Patent 1,559,971 (Minneosta Mining and Manufacturing Co.) (1969); *C.A.* **71**, 81912q (1969).

G65. Yakubovich, A. Ya., Flerova, A. N., and Yakubovich, V. S. *Vysokomol. Soedin., Ser. A* **13**(5), 994 (1971); *C.A.* **75**, 64386m (1971).

G66. Yakubchik, A. I., Tikhomirov, B. I., Polyakov, Yu. N., and Troshkova, O. K., *Vysokomol. Soedin., Ser. A* **11**(11), 2481 (1969); *C.A.* **72**, 55938d (1970).

G67. De Winter, W. F., and Preston, J., U.S. Patent 3,621,076 (1971); *C.A.* **76**, 73490p (1972).

G68. Kobayashi, F., Sakata, H., Mizoguchi, T., and Suyama, N., Japanese Patent 69/27,674 (Japan Rayon Co., Ltd.) (1969); *C.A.* **72**, 56598e (1970).

G69. Bach, H. C., U.S. Patent 3,644,283 (Monsanto Co.) (1972); *C.A.* **77**, 6029c (1972).

G70. Jablonski, R. J., Witzel, J. M., and Kruh, D., *J. Polym. Sci., Part B* **8**(3), 191 (1970); *C.A.* **73**, 15953u (1970).

G71. Goulay, M., and Marechal, E., *Bull. Soc. Chim. Fr.* **3**, 854 (1971); *C.A.* **75**, 36806e (1971).

G72. Mori, S., *Kobunshi Kagaku* **30**(341), 546 (1973).

G73. Miller, I. K., *J. Polym. Sci.* (Polym. Chem. Edit), **14**, 1403 (1976).

H1. Bayer, O., Muller, E., Petersen, S., Piepenbrink, H. F., and Windemuth, E., *Angew. Chem.* **62**, 57 (1950).

H2. Wilkes, G. L., and Samuels, S. L., *J. Biomed. Mater. Res.* **7**, 541 (1973); *C.A.* **80**, 121583c (1974).

H3. Kim, S. W., and Lyman, D. J., *Appl. Polym. Symp.* **22**, 289 (1973).

H4. Marsh, H. E., Jr., Hsu, G. C., Wallace, C. J., and Blankenhorn, D. H., *Am. Chem. Soc., Div. Org. Coat. Plast. Chem., Pap.* **33**, No. 2, 327 (1973); see also *Polymer Sci. and Technol.* **8**, (1975).

H5. Pigott, K. A., *Kirk-Othmer Encycl. Chem. Technol., 2nd Ed.* vol. 21, p. 56 (1970); *C.A.* **74**, 23250y (1971).

H6. Buist, J. M., and Gudgeon, H., eds., "Advances in Polyurethane Technology." Wiley (Interscience), New York, 1968; *C.A.* **70**, 106966e (1969).

H7. Hadobas, F., and Janik, Z., *Plast. Hmoty Kauc.* **7**, 1 (1970); *C.A.* **72**, 101578m (1970).

H8. Lyman, D. J., *Kinet. Mech. Polym.* **3**, 95 (1972); *C.A.* **80**, 108886j (1974).

H9. Nishijima, Y., and Fukushima, M., *Kobunshi Kagaku* **30**, 13 (1973); *C.A.* **78**, 136773x (1973).

H10. Hill, B. G., *Chem. Technol.* Pg. 613, October, (1973).

H11. Frisch, K. C., *Rubber Chem. Technol.* Pg. 1442 (1972).

H12. Matzner, M., Noshay, A., Schober, D. L., and McGrath, J. E., *Ind. Chim. Belge* **38**(11), 1104 (1973); *C.A.* **80**, 108884g (1974).

H13. Buist, J. M., and Lowe, A., *Br. Polym. J.* **3**, 104 (1971).

H14. Kallert, W., *Kautsch. Gummi, Kunstst.* **19**, 363 (1966).

H15. Lyman, D. J., *Rev. Macromol. Chem.* **1**, 191 (1966); *C.A.* **65**, 5533b (1966).

H16. Saunders, J. H., *High Polym.* **23**(pt. 2), 727 (1968); *C.A.* **71**, 82326p (1969).

H17. Trappe, G., *in* "Advances in Polyurethane Technology" (J. M. Buist, ed.), p. 25. Wiley, New York, 1968; *C.A.* **72**, 56442z (1970).

H18. Bruins, P. F., ed., "Polyurethane Technology." Wiley (Interscience), New York, 1969.

H19. Frisch, K. C., *in* "Polyurethane Technology" (P. F. Bruins, ed.), p. 1. Wiley (Interscience), New York 1969; *C.A.* **72**, 13241r (1970).

H20. Saunders, J. H., and Frisch, K. C., "Polyurethanes," Vol. I. Wiley (Interscience), New York, 1964.

H21. Saunders, J. H., and Frisch, K. C., "Polyurethane Technology," Vol. II. Wiley (Interscience), New York, 1964.

H22. Estes, G. M., Cooper, S. L., and Tobolsky, A. V., *J. Macromol. Sci., Rev. Macromol. Chem.* 4(2), 313 (1970).

H23. Strassel, A., *Double-Liaison* 149, 51 (1968); *C.A.* 68, 87961q (1968).

H24. Frisch, K. C., and Vogt, H. C., in "Chemical Reaction of Polymers" (E. M. Fettes, ed.), p. 927. Wiley (Interscience), New York, 1964.

H25. Brushwell, W., *Farbe U. Lack* 78, 141 (1972); *C.A.* 77, 6187c (1972).

H26. Whittaker, R. E., *J. Appl. Polym. Sci.* 15, 1205 (1971).

H27. Womack, H. G., *SGF Publ.* 19, 8 (1962); *C.A.* 60, 4299g (1964).

H28. Schollenberger, C. S., U.S. Patent 3,015,650 (B. F. Goodrich Co.) (1962); *C.A.* 58, 10359e (1962).

H29. Gable, C. L., *Plast. Des. Process.* 8, 16 (1968); *C.A.* 69, 20086k (1968).

H30. Bedoit, W. C., Jr., *J. Cell. Plast.* 10, 78 (1974).

H31. Stewart, F. D., and Schollenberger, C. S., U.S. Patent 3,769,245 (B. F. Goodrich Co.) (1973); *C.A.* 80, 84116e (1974).

H32. Mark, H. F., in "Polyurethane Technology" (P. F. Bruins, ed.), p. 181. Wiley (Interscience), New York, 1969.

H33. Consaga, J. P., and French, D. M., *J. Appl. Polym. Sci.* 15(12), 2941 (1971); *C.A.* 76, 86827d (1972).

H34. Ryan, P. W., *Br. Polym. J.* 3, 145 (1971); *C.A.* 75, 50185h (1971).

H35. Makhmurov, A. G., Kipnis, Yu. B., Sinaiskii, A. G., Shtern, I. A., and Pavlov, N. N., *Kozh.-Obuvn. Prom. SSSR* 13, 47 (1971); *C.A.* 75, 64794t (1971).

H36. Ryan, P. W., *J. Elastoplast.* 3, 57 (1971); *C.A.* 74, 127098w (1971).

H37. Spirin, Y. L., Lipatov, Y. S., Grishchenko, V. K., Sergeeva, M., Kercha, Y. Y., Bin'kevich, N. I., and Bakhmaleva, V. L., *Vysokomol. Soedin., Ser. A* 10, 263 (1968); *C.A.* 68, 96610b (1968).

H38. Hsieh, H. L., U.S. Patent 3,175,997 (to Phillips Petroleum Co.) (1965); *C.A.* 63, 3136c (1965).

H39. Moore, R. A., Kuncl, K. L., and Gower, B. G., *Rubber World* 159, 55 (1969); *C.A.* 70, 97756g (1969).

H40. Dole-Robbe, J. P., *Bull. Soc. Chim. Fr.* 3, 1078 (1967); *C.A.* 67, 11774q (1967).

H41. Gosnell, R., and Hollander, J., *J. Macromol. Sci., Phys.* 1, 831 (1967); *C.A.* 69, 19645k (1968).

H42. Manton, J. E., and Brock, D. J., Canadian Patent 792,805 (Polymer Corp. Ltd.) (1968); *C.A.* 69, 87868k (1968).

H43. Hollander, J., Trischler, F. D., and Harrison, E. S., *Polym. Prepr., Am. Chem. Soc., Div. Polym. Chem.* 8, 1149 (1967); *C.A.* 70, 97293d (1969).

H44. Bryk, M. T., Fil, T. I., Lantukh, G. V., and Namanson, Z. M., *Vysokomol. Soedin., Ser. A* 14, 472 (1972); *C.A.* 76, 155169e (1972).

H45. Dunbar, R. A., Thesis, University of Delaware, Newark (1967); *C.A.* 70, 97294e (1969).

H46. Fielding, H. C., and Ray, N. H., British Patent 982,932 (Imperial Chemical Industries Ltd.) (1965); *C.A.* 62, 10623d (1965).

H47. Fogiel, A. W., *Macromolecules* 2, 581 (1969); *C.A.* 72, 32301b (1970).

H48. Meisert, E., Knipp, U., Stelte, B., Hederich, M., Awater, A., and Erdmenger, R., German Offen. 1,964,834 (Farbenfabriken Bayer A.-G.) (1971); *C.A.* 75, 152745h (1971).

H49. Snyder, C. E., and Lovell, J. A., *Polym. Prepr., Am. Chem. Soc., Div. Polym. Chem.* **9**, 1541 (1968); *C.A.* **72**, 112466z (1970).

H50. Harrell, L. L., Jr., *Macromolecules* **2**, 607 (1969); *C.A.* **72**, 32368d (1970).

H51. Harrell, L. L., Jr., U.S. Patent 3,541,053 (du Pont de Nemours, E. I. and Co.) (1970); *C.A.* **74**, 65438w (1971).

H52. Harrell, L. L., Jr., *Block Polym., Proc. Symp. 1969* p. 213 (1970); *C.A.* **75**, 7083m (1971).

H53. Panov, N. V., *Vysokomol. Soedin., Ser. B* **13**, 129 (1971); *C.A.* **74**, 112947y (1971).

H54. Dieterich, D., Keberle, W., and Witt, H., *Angew. Chem., Int. Ed. Engl.* **9**, 40 (1970); *C.A.* **72**, 79707d (1970).

H55. Trischler, F. D., and Hollander, J., U.S. Patent 3,463,761 (Whittaker Corp.) 1969; *C.A.* **71**, 82193t (1969).

H56. Axelrood, S. L., and Lajiness, W. J. *U.S. C.F.S.T.I., AD Rep.* **AD-619376** (1965); Army Contract No. DA 20-018-ORD-24883.

H57. Matei, I., Cocea, E., and Petrus, A., *Acad. Repub. Pop. Rom., Fil. Cluj, Stud. Cercet. Chim.* **13**, 231 (1962); *C.A.* **60**, 4315d (1962).

H58. Ebihara, S., *Yuki Gosei Kagaku Kyokai Shi* **26**, 278 (1968); *C.A.* **69**, 27957d (1968).

H59. Iwakura, Y., *Makromol. Chem.* **95**, 217 (1966); *C.A.* **65**, 7287e (1966).

H60. Lipatova, T. E., Bakalo, L. A., Sirotinskaya, A. L., and Syutkina, O. P., *Vysokomol. Soedin., Ser. A* **10**, 859 (1968); *C.A.* **69**, 19559k (1968).

H61. Oberth, A. E., and Bruenner, R. S., *J. Phys. Chem.* **72**, 845 (1968).

H62. Datskevich, L. A., Miller, V. K., and Kolesnikov, G. S., *Tr. Mosk. Khim.-Tekhnol. Inst.* **52**, 205 (1967); *C.A.* **68**, 69450t (1968).

H63. Gianatasio, P. A., *Rubber Age (N.Y.)* **101**, 51 (1969); *C.A.* **71**, 82341q (1969).

H64. Aleksandrova, Y. V., and Tarakanov, O. G., *Sint. Fiz.-Khim. Poliuretanov* p. 80 (1967); *C.A.* **69**, 97210d (1968).

H65. Entelis, S. G., *Kinet. Mech. Polyreactions, IUPAC Int. Symp. Macromol. Chem., Plenary Main Lect., 1969* p. 89 (1971); *C.A.* **77**, 34948v (1972).

H66. Wolgemuth, L. G., and Burk, E. H., *J. Polym. Sci., Polym. Lett. Ed.* **11**, 285 (1973); *C.A.* **79**, 53832k (1973).

H67. Kim, S., Stannett, V. T., and Gilbert, R. D., *J. Polym. Sci., Polym. Lett. Ed.* **11**, 731 (1973); *C.A.* **80**, 121585e (1974).

H68. Yeh, Y.-H., Ph.D. Thesis, North Carolina State University, Raleigh (1972).

H69. Ward, R. J., and Critchfield, F. E., U.S. Patent 3,761,439 (Union Carbide Corp.) (1973); *C.A.* **80**, 15685c (1974).

H70. Koleske, J. V., and Magnus, G., U.S. Patent 3,746,665 (Union Carbide Corp.) (1973); *C.A.* **79**, 79886f (1973).

H71. Reed, S. F., Jr., *Polym. Prepr., Am. Chem. Soc., Div. Polym. Chem.* **15**(2), 46 (1974).

H72. Tanabe, T., Tanaka, I., and Kobayashi, M., Japanese Patent 69/11,154 (Teijin Ltd.) (1969); *C.A.* **71**, 71732s (1969).

H73. Ambrose, R. J., and Hergenrother, W. L., *Polym. Prepr., Am. Chem. Soc., Div. Polym. Chem.* **15**(2), 180 (1974).

H74. Nishijima, Y., and Ogawa, Y., Japanese Patent (Kokai) 73/101,495 (Kanebo Co., Ltd.) (1973); *C.A.* **81**, 4444t (1974).

H75. Chawng, W.-H., and Kaman, A. J., U.S. Patent 3,764,457 (PPG Industries, Inc.) (1973); *C.A.* **80**, 71784w (1974).

H76. Kaman, A. J., and Ammons, V. G., German Offen. 2,237,501 (PPG Industries, Inc.) (1973); *C.A.* **80**, 19645s (1974).

H77. Rueter, G., French Patent 1,551,926 (Elastomer A.-G.) (1969); *C.A.* **71**, 40009g (1969).

H78. Kolycheck, E. G., French Patent 1,523,720 (Goodrich, B. F., Co.) (1968); *C.A.* **71**, 13766v (1969).

H79. British Patent 1,149,771 (Elastomer A.-G.) (1969); *C.A.* **71**, 13996v (1969).

H80. Reuter, G., Netherlands Patent Appl. 6,404,809 (1965); *C.A.* **64**, 5276d (1965).

H81. von Bornhaupt, B., Lorenz, O., and Wirth, E. F., German Offen. 2,013,316 (Koelnische Gummifaeden-Fabrik Co.) (1970); *C.A.* **74**, 4285x (1971).

H82. Rappoport, L. Y., Estrin, A. S., Sidorovich, E. A., and Sochilin, E. G., *Kauch. Rezina* **27**, 2 (1968); *C.A.* **69**, 36846u (1968).

H83. Dietrich, D., and Mueller, E., German Offen. 1,801,403 (Farbenfabriken Bayer A.-G.) (1970); *C.A.* **73**, 35945w (1970).

H84. Calamari, J. A., Jr., U.S. Patent 3,164,565 (Dow Chemical Co.) (1965); *C.A.* **62**, 10623e (1965).

H85. Stewart, F. D., British Patent 1,110,818 (B. F. Goodrich Co.) (1968); *C.A.* **69**, 20212y (1968).

H86. French Patent 1,488,017 (Farben, Bayer A.G.) (1967); *C.A.* **68**, 60073q (1968).

H87. Suzuki, I., Ichikawa, K., and Fuji, S., German Offen. 1,955,725 (Asahi Chemical Industry Co., Ltd.) (1970); *C.A.* **73**, 26526h (1970).

H88. British Patent 968,079 (Goodyear Tire and Rubber Co.) (1964); *C.A.* **62**, 10649b (1965).

H89. Suzuki, H., and Ono, H., *Kogyo Kagaku Zasshi* **72**(7), 1593 (1969); *C.A.* **71**, 113882y (1969).

H90. Reegen, S. L., and Frisch, K. C., *J. Polym. Sci., Part C* **16**(5), 2733 (1967); *C.A.* **68**, 22411r (1968).

H91. Kanavel, G. A., Koons, P. A., and Lauer, R. E., *Rubber Chem. Technol.* **39**, 1338 (1966).

H92. Frisch, K. C., Reegen, S. L., and Robertson, E. J., *N.A.S.—N.R.C., Publ.* **1462**, 75 (1967); *C.A.* **67**, 82624b (1967).

H93. Netherlands Patent Appl. 6,512,161 (B. F. Goodrich Co.) (1966); *C.A.* **65**, 9146c (1966).

H94. Hippe, Z., and Jablonski, H., *Polimetry* **12**, 261 (1967); *C.A.* **68**, 69525w (1968).

H95. Ossefort, Z. T., and Testroet, F. B., *Rubber Chem. Technol.* **39**, 1308 (1966).

H96. Magnus, G., Dunleavy, R. A., and Critchfield, F. E., *Rubber Chem. Technol.* **39**, 1328 (1966).

H97. British Patent 1,147,852 (Continental Gummi-Werke A.-G.) (1969); *C.A.* **70**, 116085f (1969).

H98. Stephens, R. W., Vervolet, C., and Monnee, H. C. W., German Offen. 2,312,960 (Shell Internationale Research Maatschappij N.V.) (1973); *C.A.* **80**, 71877d (1974).

H99. Schollenberger, C. S., and Stewart, F. D., *J. Elastoplast.* **3**, 28 (1971); *C.A.* **74**, 127103u (1971).

H100. Britain, J. W., U.S. Patent 3,401,144 (Mobay Chemical Co.) (1968); *C.A.* **69**, 87759a (1968).

H101. French Demande 2,006,788 (Farbenfabriken Bayer A.-G.) (1970); *C.A.* **73**, 16041g (1970).

H102. Koenig, K., Mueller, E., Mühlhausen, C., and Dobereiner, U. J., German Offen. 1,964,998 (Farbenfabriken Bayer A.-G.) (1971); *C.A.* **75**, 99154m (1971).

H103. Koning, K., Mueller, E., Mühlhausen, C., and Dobereiner, U. J., U.S. Patent 3,758,443 (Bayer A.-G.) (1973); *C.A.* **80**, 48952r (1974).

H104. French Patent 1,566,435 (Farbenfabriken Bayer A.-G.) (1969); *C.A.* **71**, 113691k (1969).

H105. Schollenberger, C. A., and Stewart, F. D., *Adv. Urethane Sci. Technol.* **1**, 65 (1971); *C.A.* **76**, 35040j (1972).

H106. Schollenberger, C. S., and Stewart, F. D., *J. Elastoplast.* **4**, 294 (1972); *C.A.* **78**, 44408d (1973).

H107. Berger, S. E., *Conf. Elastoplast. Technol.* [*Pap.*], *1965* (1965); *C.A.* **67**, 117849h (1966).

H108. Chang, E. Y. C., and Kaizerman, S., U.S. Patent 3,503,927 (American Cyanamid Co.) (1970); *C.A.* **72**, 122647y (1970).

H109. French Patent 1,546,262 (Farbenfabriken Bayer A.-G.) (1968); *C.A.* **71**, 51211t (1969).

H110. French Patent 1,564,518 (Farbenfabriken Bayer A.-G.) (1969); *C.A.* **71**, 92296t (1969).

H111. Megna, I. S., and Sullivan, F. A. V., German Offen 1,912,179 (American Cyanamid Co.) (1969); *C.A.* **72**, 4302j (1970).

H112. French Patent 1,534,536 (Imperial Chemical Industries Ltd.) (1968); *C.A.* **71**, 14115u (1969).

H113. Collardeau, G., and Robin, J., French Patent 1,499,459 (Societe des Usines Chimiques Rhone-Poulenc) (1967); *C.A.* **69**, 44478r (1968).

H114. French Patent 1,565,486 (Société des Usines Chimiques Rhône-Poulenc) (1969); *C.A.* **71**, 102921j (1969).

H115. Schultze, H., *Makromol. Chem.* **172**, 57 (1973); *C.A.* **80**, 71316v (1974).

H116. David, D. J., and Staley, H. B. *High Polym.* **16**, Part 3, (1969).

H117. Mulder, J. L., *Anal. Chim. Acta* **38**, 563 (1967).

H118. Jackson, M. B., and Solomon, D. H., *Anal. Chem.* **44**(6), 1074 (1972).

H119. Suzuki, H., and Ono, H., *Makromol. Chem.* **132**, 305 (1970); *C.A.* **72**, 101589r (1970).

H120. Beachell, H. C., and Blumstein, R., *J. Polym. Sci., Part C* **16**(3), 1403 (1967); *C.A.* **68**, 40128h (1968).

H121. Ahad, E., *J. Appl. Polym. Sci.* **17**(2), 365 (1973); *C.A.* **79**, 5736k (1973).

H122. Groom, T., Babiec, J. S., Jr., and Van Leuwen, B. G., *J. Cell. Plast.* **10**(1), 43 (1974); *C.A.* **81**, 4406g (1974).

H123. Ramey, K. C., Hayes, M. W., and Altenau, A. G., *Macromolecules* **6**(5), 795 (1973); *C.A.* **79**, 146935m (1973).

H124. Kirret, O., and Kullik, E., Z. *Gesamte Textilind.* **71**(3), 169 (1969); *C.A.* **70**, 88733v (1969).

H125. Chokki, Y., Nakabayashi, M., and Sumi, M., *Makromol. Chem.* **153**, 189 (1972); *C.A.* **77**, 6859y (1972).

H126. Sumi, M., Chokki, Y., Nakai, Y., Nakabayashi, M., and Kanzawa, T., *Makromol. Chem.* **78**, 146 (1964); *C.A.* **61**, 16278f (1964).

H127. Suzuki, H., Ono, H., and Hongo, T., *Makromol. Chem.* **132**, 309 (1970); *C.A.* **72**, 101306w (1970).

H128. Suzuki, H., and Ono, H., *Bull. Chem. Soc. Jpn.* **43**(3), 687 (1970); *C.A.* **72**, 122625q (1970).

H129. Suzuki, H., *J. Polym. Sci., Part A-1* **9**(2), 387 (1971); *C.A.* **74**, 127092q (1971).

H130. Suzuki, H., and Ono, H., *Bull. Chem. Soc. Jpn.* **43**(3), 682 (1970); *C.A.* **72**, 122624p (1970).

H131. Suzuki, H., and Ono, H., *Bull. Chem. Soc. Jpn.* **43**(3), 687 (1970); *C.A.* **72**, 122625q (1970).

H132. Suzuki, H., *Bull. Chem. Soc. Jpn.* 43(12), 3870 (1970); *C.A.* **74**, 42818f (1971).

H133. Nachinkin, O. I., Ageev, A. I., and Ruban, I. G., *Vysokomol. Soedin., Ser. B* **15**(7), 520 (1973); *C.A.* **80**, 37534e (1974).

H134. Onder, K., Peters, R. H., and Spark, L. C., *Polymer* **13**(3), 133 (1972); *C.A.* **77**, 49273v (1972).

H135. Vasil'ev, B. V., and Tarakanov, O. G., *Vysokomol. Soedin.* **6**(12), 2193 (1964); *C.A.* **62**, 9302h (1965).

H136. Vasil'ev, B. V., and Tarakanov, O. G., *Vysokomol. Soedin.* **6**(12), 2189 (1964); *C.A.* **62**, 9302g (1965).

H137. Kercha, Y. Y., and Lipatov, Y. S., *Ukr. Khim. Zh. (Russ. Ed.)* **34**(2), 158 (1968); *C.A.* **68**, 96308j (1968).

H138. MacKnight, W. J., Yang, M., and Kajiyama, T., *Anal. Calorimetry, Proc. Am. Chem. Soc. Symp., 1968* p. 99 (1968); *C.A.* **70**, 88494t (1969).

H139. Saunders, J. H., and Backus, J. K., *Rubber Chem. Technol.* **39**(2), 461 (1966); *C.A.* **65**, 7372d (1966).

H140. Bebchuk, T. S., Golubkov, G. E., Karina, T. L., and Sokolov, N. N., *Tr. Vses. Elektrotekh. Inst.* **74**, 223 (1966); *C.A.* **68**, 78906y (1968).

H141. Ingham, J. D., and Rapp, N. S., *Polym. Eng. Sci.* **6**(1), 36 (1966); *C.A.* **64**, 11377f (1966).

H142. Becker, R., and Neumann, G., *Plaste Kautsch.* **20**(11), 809 (1973); *C.A.* **80**, 71482w (1974).

H143. Bonart, R., Morbitzer, L., and Hentze, G., *J. Macromol. Sci., Phys.* **3**(2), 339 (1969); *C.A.* **71**, 92423g (1969).

H144. Seymour, R. W., and Cooper, S. L., *J. Polym. Sci., Part B* **9**(9), 689 (1971); *C.A.* **76**, 4714c (1972).

H145. Miller, G. W., and Saunders, J. H., *J. Appl. Polym. Sci.* **13**(6), 1277 (1969); *C.A.* **71**, 92399d (1969).

H146. Yagfarov, M. Sh., and Guvanov, E. F., *Vysokomol. Soedin., Ser. A* **12**(5), 1155 (1970); *C.A.* **73**, 25889s (1970).

H147. Guvanov, E. F., Sinaiskii, A. G., Apukhtina, N. P., and Teitel'baum, B. Ya., *Dokl. Akad. Nauk SSSR* **163**(5), 1151 (1965); *C.A.* **63**, 16485g (1965).

H148. Gurenkov, M. S., and Maklakov, A. I., *Sb. Aspir. Rab., Kazan. Gos. Univ., Mat., Mekh., Fiz.* p. 111 (1970); *C.A.* **76**, 113997t (1972).

H149. Illinger, J. L., Schneider, N. S., and Karasz, F. E., *Polym. Eng. Sci.* **12**(1), 25 (1972); *C.A.* **76**, 114473f (1972).

H150. Samuels, S. L., and Wilkes, G. L., *Polym. Prepr., Am. Chem. Soc., Div. Polym. Chem.* **12**(2), 694 (1971); *C.A.* **78**, 148768n (1973).

H151. Samuels, S. L., and Wilkes, G. L., *J. Polym. Sci., Polym. Phys. Ed.* **11**(4), 807 (1973); *C.A.* **79**, 5749s (1973).

H152. Samuels, S. L., and Wilkes, G. L., *Polym. Prepr., Am. Chem. Soc., Div. Polym. Chem.* **14**(2), 1226 (1973).

H153. Clough, S. B., and Schneider, N. S., *U.S. Dep. Comm., C.F.S.T.I., AD Rep.* **AD-666753** (1967); from *U.S. Gov. Res. Dev. Rep.* **68**(10), 78 (1968); *C.A.* **69**, 97559z (1968).

H154. Clough, S. B., Schneider, N. S., and King, A. O., *J. Macromol. Sci., Phys.* **2**(4), 641 (1968); *C.A.* **70**, 20827y (1969).

H155. Clough, S. B., and Schneider, N. S., *J. Macromol. Sci., Phys.* **2**(4), 553 (1968); *C.A.* **70**, 29875v (1969).

H156. Palyutkin, G. M., Sokolov, A. R., Vasil'ev, B. V., and Tarakanov, O. G., *Vysokomol. Soedin., Ser. A* **13**(10), 2286 (1971); *C.A.* **76**, 46623c (1972).

H157. Apukhtina, N. P., Teitel'baum, B. Y., Cherkasova, L. A., Yagfarova, T. A., and Palikhov, N. A., *Vysokomol. Soedin., Ser. A* **13**(11), 2481 (1971); *C.A.* **76**, 86861k (1972).

H158. Vrouenraets, C. M. F., *Polym. Prepr., Am. Chem. Soc., Div. Polym. Chem.* **13**(1), 529 (1972); *C.A.* **80**, 4625g (1974).

H159. Lipatov, Y. S., Kosenko, L. A., Kercha, Y. Y., and Lipantnikov, N. A., *Vysokomol. Soedin., Ser. A* **15**(5), 1057 (1973); *C.A.* **79**, 79680j (1973).

H160. Vasil'ev, B. V., Tarakanov, O. G., Demina, A. I., and Shirobokova, A. I., *Vysokomol. Soedin.* **8**(5), 938 (1966); *C.A.* **65**, 4029b (1966).

H161. Gurenkov, M. S., and Maklakov, A. I., *Sb. Aspir. Rab., Kazan. Gos. Univ. Tochn. Nauki: Mekh., Fiz.* **2**, 40 (1970); *C.A.* **76**, 46806q (1972).

H162. Morbitzer, L., and Bonart, R., *Kolloid Z. & Z. Polym.* **232**(2), 764 (1969); *C.A.* **71**, 92451q (1969).

H163. Huh, D. S., and Cooper, S. L., *Polym. Eng. Sci.* **11**(5), 369 (1971); *C.A.* **76**, 4688x (1972).

H164. Kercha, Y. Y., Lipatov, Y. S., Krafchik, S. S., and Privalko, V. P., *Vysokomol. Soedin., Ser. A* **15**(6), 1297 (1973); *C.A.* **79**, 92698r (1973).

H165. Bonart, R., *J. Macromol. Sci., Phys.* **2**(1), 115 (1968); *C.A.* **68**, 87984z (1968).

H166. Wilkes, C. E., and Yusek, C. S., *J. Macromol. Sci., Phys.* **7**(1), 157 (1973); *C.A.* **78**, 31126k (1973).

H167. Gogolewski, S., *Kolloid Z. & Z. Polym.* **251**(7), 502 (1973); *C.A.* **80**, 71226r (1974).

H168. Paik Sung, C. S., Schneider, N. S., Matton, R. W., and Illinger, J. L., *Polym. Prepr., Am. Chem. Soc., Div. Polym. Chem.* **15**(1), 620 (1974).

H169. Seymour, R. W., Estes, G. M., and Cooper, S. L., *Polym. Prepr., Am. Chem. Soc., Div. Polym. Chem.* **11**(2), 867 (1970); *C.A.* **77**, 6838r (1972).

H170. Estes, G. M., Seymour, R. W., and Cooper, S. L., *Polym. Prepr., Am. Chem. Soc., Div. Polym. Chem.* **11**(2), 516 (1970); *C.A.* **77**, 6885d (1972).

H171. Cooper, S. L., Estes, G. M., and Seymour, R. W., *Macromolecules* **4**(4), 452 (1971); *C.A.* **75**, 152720w (1971).

H172. Cooper, S. L., Seymour, R. W., and Estes, G. M., *Macromolecules* **3**(5), 579 (1970); *C.A.* **73**, 131787s (1970).

H173. Nakayama, K., Ino, T., and Matsubara, I., *J. Macromol. Sci., Chem.* **3**(5), 1005 (1969); *C.A.* **71**, 31178j (1969).

H174. Zharkov, V. V., and Rudnevskii, N. K., *Vysokomol. Soedin., Ser. B* **10**, 29 (1968); *C.A.* **68**, 69516u (1968).

H175. Teitel'baun, B. Y., Murtazina, I. O., Anoshina, N. P., Makalakov, L. I., Apukhtina, N. P., and Sinaiskii, A. G., *Dokl. Akad. Nauk SSSR* **166**(4), 887 (1966); *C.A.* **66**, 14277d (1966).

H176. Koutsky, J. A., Hien, N. V., and Cooper, S. L., *J. Polym. Sci., Part B* **8**(5), 353 (1970); *C.A.* **73**, 110643b (1970).

H177. Samuels, S. L., and Wilkes, G. L., *Polym. Lett.* **9**, 761 (1971).

H178. Macknight, W. J., and Yang, M., *J. Polym. Sci., Polym. Symp.* **42**, 817 (1973).

H179. Wilkes, G. L., *J. Macromol. Sci., Rev. Macromol. Chem.* **10**(2), 149 (1974); *C.A.* **80**, 146559n (1974).

H180. Samuels, S. L., and Wilkes, G. L., *J. Polym. Sci., Polym. Symp.* **43**, 149 (1973).

H181. Paik Sung, C. S., and Schneider, N. S., *Polym. Prepr., Am. Chem. Soc., Div. Polym. Chem.* **15**(1), 625 (1974).

H182. Seymour, R. W., Allegrezza, A. E., and Cooper, S. L., *Polym. Prepr., Am. Chem. Soc., Div. Polym. Chem.* **15**(1), 631 (1974).

H183. Seymour, R. W., Allegrezza, A. E., and Cooper, S. L., *Macromolecules* **6**(6), 896 (1973); *C.A.* **80**, 122077c (1974).

H184. Chang, Y. J., Chen, C. T., and Wilkes, G. L., *Polym. Prepr., Am. Chem. Soc., Div. Polym. Chem.* **14**(2), 1277 (1974).

H185. Seymour, R. W., Estes, G. M., Huh, D. S., and Cooper, S. L., *J. Polym. Sci., Part A-2* **10**(8), 1521 (1972); *C.A.* **77**, 102563t (1972).

H186. Ng, H. N., Allegrezza, A. E., Seymour, R. W., and Cooper, S. L., *Polymer* **14**(6), 255 (1973); *C.A.* **79**, 32637k (1973).

H187. Maklakov, L. I., Kovalenko, V. I., Apukhtina, N. P., and Sinaiskii, A. G., *Zh. Prikl. Spektrosk.* **7**(1), 99 (1967); *C.A.* **68**, 87708n (1968).

H188. Rustad, N. E., and Krawiec, R. G., *J. Appl. Polym. Sci.* **18**(2), 401 (1974); *C.A.* **80**, 146810n (1974).

H189. Work, J. L., *Polym. Prepr., Am. Chem. Soc., Div. Polym. Chem.* **14**(2), 1040 (1973).

H190. Ferguson, J., Hourston, D. J., Meredith, R., and Patsavoudis, D., *Eur. Polym. J.* **8**, 369 (1972).

H191. Seymour, R. W., and Cooper, S. L., *Rubber Chem. Technol.* **47**, 19 (1974).

H192. Klosner, J. M., Segal, A., and Franklin, H. N., *J. Appl. Phys.* **39**(1), 15 (1968).

H193. Critchfield, F. E., Koleske, J. V., Magnus, G., and Dodd, J. L., *J. Elastoplast.* **4**, 22 (1972).

H194. Shibayama, K., and Kodama, M., *Kobunshi Kagaku* **21**(236), 737 (1964); *C.A.* **62**, 13320h (1965).

H195. Sotnikova, E. N., Cherkasova, L. A., Sodorovich, E. A., and Moshtitskava, N. L., *Kauch. Rezina* **31**(2), 7 (1972); *C.A.* **76**, 155210m (1972).

H196. Cluff, E. F., Gladding, E. K., and Rogan, J. B., *J. Appl. Polym. Sci.* **5**, 80 (1961).

H197. Guvanov, E. F., Teitel'baum, B. Y., Apukhtina, N. P., and Sinaiskii, A. G., *Sint. Fiz.-Khim. Polim.* **5**, 168 (1968); *C.A.* **70**, 4898s (1969).

H198. Apukhtina, A. G., Sinaiskii, A. G., Teitel'baum, B. T., Sidorovich, E. A., and Lubanov, E. F., *Plaste Kautsch.* **13**(4), 212 (1966).

H199. Gubanov, E. F., Teitel'baum, B. Y., Sinaiskii, A. G., and Apukhtina, N. P., *Dokl. Akad. Nauk SSSR* **179**(3), 621 (1968); *C.A.* **69**, 10785y (1968).

H200. Lyman, D. J., Heller, J., and Barlow, M., *Makromol. Chem.* **84**, 64 (1965); *C.A.* **63**, 7120b (1965).

H201. Magnusson, A. B., *J. Appl. Polym. Sci.* **11**(11), 2175 (1967); *C.A.* **68**, 40235r (1968).

H202. Critchfield, F. E., Koleske, J. V., and Dunleavy, R. A., *Rubber World* **164**(5), 61 (1971); *C.A.* **75**, 152696t (1971).

H203. Tanaka, T., Yokoyama, T., Yamaguchi, Y., Yoshitake, N., and Yano, T., *Kyushu Daigaku Kogaku Shuho* **41**(3), 519 (1968); *C.A.* **70**, 115668m (1969).

H204. Kimura, I., Suzuki, H., Saito, K., Watanabe, K., Ishihara, H., and Ono, H., *Kogyo Kagaku Zasshi* **73**(7), 1541 (1970); *C.A.* **74**, 13965w (1971).

H205. Schollenberger, C. S., *in* "Polyurethane Technology" (P. F. Bruins, ed.), p. 197. Wiley (Interscience), New York, 1969; *C.A.* **72**, 13234r (1970).

H206. Bonart, R., and Morbitzer, L., *Kolloid Z. & Z. Polym.* **240**(1-2), 807 (1970); *C.A.* **74**, 54844z (1971).

H207. Apukhtina, N. P., Korotkina, D. S., Livshits, K. S., and Sidorovich, E. A., *Kauch. Rezina* **26**(12), 11 (1967); *C.A.* **68**, 50761y (1968).

H208. Flocke, H. A., *Kunststoffe* **56**(5), 328 (1966); *C.A.* **65**, 18807c (1966).

H209. Magnus, G., *Rubber Age (N.Y.)* **97**(4), 86 (1965); *C.A.* **63**, 11830b (1965).

H210. Smith, C. H., *Ind. Eng. Chem., Prod. Res. Dev.* **4**(1), 9 (1965); *C.A.* **62**, 9308b (1965).

H211. Schollenberger, C. S., and Dinberg, K., *J. Elastoplast.* **5**, 222 (1973); *C.A.* **80**, 48934m (1974).

H212. Gianatasio, P. A., *Rubber Age (N.Y.)* **8**(101), 57 (1969); *C.A.* **71**, 92408f (1969).

H213. Mitchell, D. C., *J. IRI* **2**(1), 37 (1968); *C.A.* **70**, 107279v (1969).

H214. Cumming, A. P. C., and Wright, P., *J. IRI* **2**(1), 29 (1968); *C.A.* **70**, 107262j (1969).

H215. Carvey, R. M., and Stetz, T. T., *Adhes. Age* **11**(9), 35 (1968); *C.A.* **69**, 87774b (1968).

H216. Ellegast, K., *Kunststoffe* **55**(5), 306 (1965); *C.A.* **65**, 8586g (1965).

H217. Gable, C. L., *Appl. Polym. Symp.* **11**, 47 (1969); *C.A.* **72**, 32798u (1970).

H218. Bonk, H. W., Sardanopoli, A. A., Ulrich, H., and Sayigh, A. A. R., *J. Elastoplast.* **3**, 157 (1971); *C.A.* **76**, 35000w (1972).

H219. Morbitzer, L., and Hespe, H., *J. Appl. Polym. Sci.* **16**, 2697 (1972).

H220. Apukhtina, N. P., Zimina, M. G., Novoselok, F. B., and Myuller, B. E., *Tr. Mezhdunar. Konf. Kauch. Rezine, 1969* p. 129 (1971); *C.A.* **77**, 76372h (1972).

H221. Pedley, K. A., and Ross, J. A., *Rubber Plast. Age* **45**, 417 (1964); *C.A.* **64**, 802e (1964).

H222. Morton, M., McGrath, J. E., and Juliano, P. C., *J. Polym. Sci., Part C* **26**, 99 (1969).

H223. Wang, T. T., and Klosner, J. M., *Trans. Soc. Rheol.* **13**(2), 193 (1969); *C.A.* **71**, 92409g (1969).

H224. Cooper, S. L., *J. Polym. Sci., Part A-1* **7**(7), 1765 (1969); *C.A.* **71**, 71678d (1969).

H225. North, A. M., Reid, J. C., and Shortall, J. B., *Eur. Polym. J.* **5**(4), 565 (1969); *C.A.* **71**, 71683b (1969).

H226. North, A. M., and Reid, J. C., *Eur. Polym. J.* **8**(10), 1129 (1972); *C.A.* **78**, 17265y (1973).

H227. Miyano, Y., Tamura, T., and Kunio, T., *Bull. JSME* **12**(49), 26 (1969); *C.A.* **71**, 13483a (1969).

H228. Estes, G. M., Seymour, R. W., and Cooper, S. L., *Polym. Eng. Sci.* **9**(6), 383 (1969); *C.A.* **72**, 32964v (1970).

H229. Estes, G. M., Seymour, R. W., and Cooper, S. L., *Soc. Plast. Eng., Tech. Pap.* **15**, 20 (1969); *C.A.* **71**, 125304z (1969).

H230. Estes, G. M., Huh, D. S., and Cooper, S. L., *Block Polym., Proc. Symp.*, 1969 p. 225 (1970); *C.A.* **75**, 7071f (1971).

H231. Bonart, R., and Morbitzer, L., *Kolloid Z. & Z. Polym.* **241**(1–2), 909 (1970); *C.A.* **75**, 77868v (1971).

H232. Lindsey, G. H., *Diss. Abstr.* **26**(12), 7189 (1966); *C.A.* **65**, 12385d (1966).

H233. Cooper, S. L., and Tobolsky, A. V., *Polym. Prepr., Am. Chem. Soc., Div. Polym. Chem.* **8**(1), 52 (1967); *C.A.* **66**, 116421e (1967).

H234. Ziegel, K. D., *J. Macromol. Sci., Phys.* **5**(1), 11 (1971); *C.A.* **74**, 23490b (1971).

H235. Lipatov, Y. S., Sergeeva, L. M., and Kovalenko, G. F., *Vysokomol. Soedin., Ser. B* **10**(3), 205 (1968); *C.A.* **69**, 3386r (1968).

H236. Anoshina, N. P., Apukhtina, N. P., Guvanov, E. F., Murtazina, I. O., Sinaiskii, A. G., and Tietelbaum, B. Y., *Vysokomol. Soedin., Ser. A* **9**(4), 815 (1967); *C.A.* **67**, 54520w (1967).

H237. Apukhtina, N. P., Marei, A. I., Novikoya, G. E., and Myuller, B. E., *Vysokomol. Soedin.* **7**(6), 1117 (1965); *C.A.* **63**, 8586h (1965).

H238. Gubanov, E. F., Teitel'baum, B. Y., and Sinaiskii, A. G., *Vysokomol. Soedin., Ser. B* **12**(8), 623 (1970); *C.A.* **73**, 110260z (1970).

H239. Rausch, K. W., Jr., and Farrissey, W. J., Jr., *J. Elastoplast. 114 (1970); C.A.* **73**, 15617n (1970).

H240. Teitel'baum, B. Y., Yagfarov, M. S., Anoshina, N. P., Palikhov, N. A., Men'shova, R. K., Anukhtina, N. P., and Myuller, B. E., *Vysokomol. Soedin., Ser. A* **13**(10), 2291 (1971); *C.A.* **76**, 47106y (1972).

H241. Schneider, N. S., Dusablon, L. V., Snell, E. W., and Prosser, R. A., *J. Macromol. Sci., Phys.* **3**(4), 623 (1969); *C.A.* **72**, 32345u (1970).

H242. Kovalenko, G. F., Lipatov, Y. S., Kosenko, L. A., and Sergeeva, L. M., *Vysokomol. Soedin., Ser. B* **15**(9), 651 (1973); *C.A.* **80**, 84057m (1974).

H243. Seibert, G., *Melliand Textilber. Int.* **53**(2), 124 (1972); *C.A.* **76**, 142013z (1972).

H244. Cherkasova, L. A., Sidorovich, E. A., Apukhtina, N. P., and Marei, A. I., *Vysokomol. Soedin., Ser. A* **14**(4), 735 (1972); *C.A.* **77**, 21148j (1972).

H244a. Makhmurov, A. G., Chuikova, L. F., Shaposhnikova, T. K., Sinaiskii, A. G., and Pavlov, N. N., *Vysokomol. Soedin., Ser. A* **15**(6), 1262 (1973); *C.A.* **79**, 67204u (1973).

H245. Kousaka, S., *Sen'i Gakkaishi* **11**, T482 (1973); *C.A.* **80**, 71844r (1974).

H246. Shchepetkina, N. I., and Treskunov, E. I., *Sint. Fiz.-Khim. Poliuretanov* p. 154 (1967); *C.A.* **69**, 87660m (1968).

H247. Seymour, R. W., and Cooper, S. L., *Macromolecules* **6**(1), 48 (1973); *C.A.* **78**, 98744d (1973).

H248. Ibrahim, S. M., and Ultee, A. J., *Encycl. Polym. Sci. Technol.* **6**, 573 (1967); *C.A.* **69**, 11270p (1968).

H249. Cuthbertson, G. R., Kelly, R. J., Logan, L. R., Porter, D. S., and Brass, P. D., *Man-Made Fibers* **3**, 401 (1968); *C.A.* **70**, 38734a (1969).

H250. Tippetts, E. A., and Zimmerman, J., *J. Appl. Polym. Sci.* **8**, 2465 (1964).

H251. Gritsenko, T. M., and Popov, I. A., *Usp. Khim. Poliuretanov* p. 81 (1972); *C.A.* **78**, 5238d (1973).

H252. Rinke, H., *Angew. Chem., Int. Ed. Eng.* **1**, 419 (1962).

H253. Rinke, H., *Rubber Chem. Technol.* **36**(3), 719 (1963); *C.A.* **60**, 707f (1964).

H254. Schmutzler, M., *Text. Prax.* **19**(8), 836 (1964); *C.A.* **61**, 14831g (1964).

H255. Rose, L., *Rep. Prog. Appl. Chem.* **51**, 609 (1966); *C.A.* **69**, 36913p (1968).

H256. Meredith, R., *Rev. Text. Prog.* **17**, 220 (1971); *C.A.* **69**, 20261p (1968).

H257. Weber, K. A., *Text.-Prax.* **23**(5), 323 (1968); *C.A.* **69**, 20262q (1968).

H258. Rinke, H., *Chimia* **22**(4), 164 (1968).

H259. Lemmens, J., *Text. Chim.* **24**(5), 275 (1968); *C.A.* **69**, 52854a (1968).

H260. Tanabe, T., and Yamashiro, S., *Sen'i To Kogyo* **2**(1), 27 (1969); *C.A.* **70**, 116122r (1969).

H261. Netherlands Patent 6,501,927 (I.C.I. Fibres Ltd.) (1965); *C.A.* **64**, 5276e (1964).

H262. Carvey, R. M., and Kolycheck, E. G., British Patent 1,025,970 (B. F. Goodrich Co.) (1966); *C.A.* **64**, 19948d (1966).

H263. Netherlands Patent 291,290 (Polythane Corp.) (1965); *C.A.* **64**, 8456g (1966).

H264. French Patent 1,522,611 (Kurashiki Rayon Co., Ltd.) (1968); *C.A.* **70**, 116175k (1969).

H265. Jones, A. P., Jr., and Bryant, G. M., French Patent 1,555,129 (Union Carbide Corp.) (1969); *C.A.* **71**, 71840a (1969).

H266. Britain, J. W., and Nardo, N. R., *Polym. Prepr., Am. Chem. Soc., Div. Polym. Chem.* **9**(2), 1536 (1968); *C.A.* **72**, 112609y (1970).

H267. Worthington, J. V., British Patent 1,038,167 (1966); *C.A.* **65**, 13862e (1966).

H268. British Patent 1,104,127 (E. I. du Pont de Nemours and Co.) (1968); *C.A.* **68**, 88168y (1968).

H269. French Patent 1,559,394 (Farbenfabriken Bayer A.-G.) (1969); *C.A.* **71**, 82403m (1969).

H270. Polestak, W. J., U.S. Patent 3,417,043 (Celanese Corp.) (1968); *C.A.* **70**, 38813a (1969).

H271. Frazer, A. H., and Shivers, J. C., Jr., U.S. Patent 3,071,557 (E. I. du Pont de Nemours and Co.) (1963); *C.A.* **58**, 11555e (1963).

H272. Kawaguchi, T., Matsubayashi, K., and Tanabe, K., Japanese Patent 67/27,079 (Kurashiki Rayon Co., Ltd.) (1967); *C.A.* **69**, 20331m (1968).

H273. Tate, C. W., and Talbert, T. L., U.S. Patent 3,383,365 (Monsanto Co.) (1968).

H274. Heidemann, G., Jellinek, G., and Ringens, W., *Kolloid Z. & Z. Polym.* **221**(2), 119 (1967); *C.A.* **68**, 50872k (1968).

H275. Wegener, W., and Wulfhorst, B., *Z. Gesamte Textilind.* **71**(10), 683 (1969); *C.A.* **71**, 125806q (1969).

H276. Wilson, N., *J. Text. Inst.* **58**(12), 611 (1967); *C.A.* **68**, 40855t (1968).

H277. Heikens, D., Meijers, A., and von Reth, P. H., *Polymer* **9**(1), 15 (1968).

H278. Ferguson, J., and Patsavoudis, D., *Eur. Polym. J.* **8**, 385 (1972).

H279. Matsubayashi, K., and Yasui, T., British Patent 1,112,497 (Kurashiki Rayon Co., Ltd.) (1968).

H280. Kimura, I., Ishihara, H., and Ono, H., *IUPAC, Prepr., 23rd, 1971* Vol. 1, p. 525 (1971).

H281. Seymour, R. W., and Cooper, S. L., *Polym. Prepr., Am. Chem. Soc., Div. Polym. Chem.* **14**(2), 1046 (1973).

I1. Bostick, E. E., *Polym. Prepr., Am. Chem. Soc., Div. Polym. Chem.* **10**(2), 877 (1969).

I2. Bostick, E. E., U.S. Patent 3,337,497 (General Electric Co.) (1967); *C.A.* **67**, 82684w (1967).

I3. Bostick, E. E., U.S. Patent 3,378,521 (General Electric Co.) (1968); *C.A.* **69**, 3464q (1968).

I4. Bostick, E. E., and Fessler, W. A., U.S. Patent 3,578,726 (General Electric Co.) (1971).

I5. Fritsche, A. K., and Price, F. P., *Polym. Prepr., Am. Chem. Soc., Div. Polym. Chem.* **10**, (2); 893 (1969).

I6. British Patent 1,039,445 (General Electric Co.) (1966); *C.A.* **65**, 20336b (1966).

I7. Antonen, R. C., German Offen. 2,059,110 (Dow Corning Corp.) (1971); *C.A.* **75**, 141747j (1971).

I8. Hartlein, R. C., Vincent, H. L., and Kiles, J. D., German Offen. 2,059,111 (Dow Corning Corp.) (1971); *C.A.* **75**, 141748k (1971).

I9. Hartlein, R. C., and Olson, C. R., German Offen. 2,059,112 (Dow Corning Corp.) (1971); *C.A.* **75**, 141749m (1971).

I10. McKellar, R. L., and Howden, R. C., U.S. Patent 3,576,905 (Dow Corning Corp.) (1971); *C.A.* **75**, 37620b (1971).

I11. Netherlands Patent Appl. 6,611,358 (Dow Corning Corp.) (1967); *C.A.* **67**, 54966w (1967).

I12. McVannel, D. E., French Patent 1,575,067 (Dow Corning Corp.) (1969); *C.A.* **72**, 67770f (1970).

I13. Antonen, R. C., U.S. Patent 3,294,718 (Dow Corning Corp.) (1964); *C.A.* **66**, 38459g (1967).

I14. Vincent, H. L., French Patent 1,388,207 (Dow Corning Corp.) (1965); *C.A.* **62**, 16471e (1965).

I15. Andrianov, K. A., Nogaideli, A. I., Tkeshelashvili, R. Sh., Nakaidze, L. I., and Dzhashiashvili, T. K., *Soobshch. Akad. Nauk, Gruz SSR* **60**(3), 581 (1970); *C.A.* **74**, 76721j (1971).

I16. Krantz, K. W., U.S. Patent 3,294,738 (General Electric Co.) (1966); *C.A.* **66**, 46907s (1967).

I17. Dunnavant, W. R., *Inorg. Macromol. Rev.* **1**(3), 165 (1971); *C.A.* **74**, 76689e (1971).

I18. Merker, R. L., Scott, M. J., and Haberland, G. G., *J. Polym. Sci., Part A* **2**, 31 (1964).

I19. British Patent 925,433 (Dow Corning Corp.) (1963); *C.A.* **58**, 8950e (1963).

120. David, M. P., and Merker, R. L., French Patent 1,523,068 (Dow Corning Corp.) (1969); *C.A.* **71**, 13514m (1969).

121. Kojima, M., Magill, J. H., and Merker, R. L., *J. Polym. Sci., Polym. Phys. Ed.* **12**, 317 (1974).

122. Karpova, T. N., Sidorovich, A. V., Dolgoplosk, S. B., Sviridova, N. G., and Kuvshinskii, E. V., *Vysokomol. Soedin., Ser. A* **15**(4), 859 (1973); *C.A.* **79**, 19283x (1973).

123. Stebleton, L. F., German Offen. 1,955,657 (Dow Corning Corp.) (1970); *C.A.* **73**, 26240k (1970).

124. Andrianov, K. A., Pakhomov, V. I., Gel'perina, V. M., and Mukhina, D. N., *Vysokomol. Soedin.* **8**(9), 1618 (1966).

125. Andrianov, K. A., Nogaidelli, A. I., and Tkeshelashvili, R. Sh., *Vysokomol. Soedin.* **8**(11), 1917 (1966).

126. Yuzhelevskiy, Yu. A., Kagan, Y. E., Klebanskiy, A. K., Zevakin, I. A., and Kharalamova, A. V., *Vysokomol. Soedin.* **11**(11), 854 (1969).

127. Netherlands Patent Appl. 6,409,104 (Union Carbide Corp.) (1965); *C.A.* **63**, 1963e (1965).

128. British Patent 802,688 (Union Carbide Corp.) (1955).

129. British Patent 880,022 (Union Carbide Corp.) (1958).

130. Prokai, B., and Kanner, B., German Offen. 2,323,398 (Union Carbide Corp.) (1973); *C.A.* **80**, 134043b (1974).

131. Bailey, D. L., and O'Conner, F. M., British Patent 804,369 (Union Carbide Corp.) (1955).

132. Bailey, D. L., and O'Conner, F. M., British Patent 892,136 (Union Carbide Corp.) (1958).

133. Haluska, L. A., French Patent 1,374,032 (Dow Corning Corp.) (1964); *C.A.* **62**, 13330e (1965).

134. Nitzche, S., and Pirson, E., German Patent 1,301,576 (Wacker-Chemie G.m.b.H.) (1969); *C.A.* **72**, 22291j (1970).

135. Caldwell, D. S., and Saunders, F. C., British Patent 1,098,646 (Midland Silicones Ltd.) (1968); *C.A.* **68**, 60071n (1968).

136. Netherlands Patent Appl. 6,601,245 (Th. Goldschmidt A.-G.) (1966).

137. Rossmy, G., Koerner, G., and Wassermeyer, J., British Patent 1,115,897 (Goldschmidt, Th., A.-G.) (1968); *C.A.* **69**, 20057b (1968).

138. Wheeler, M. L., Jr., French Patent 1,323,121 (Union Carbide Corp.) (1963); *C.A.* **59**, 10257d (1963).

139. Koebner, A., and Pitt, C. G., British Patent 981,812 (Midland Silicones Ltd. and Marchon Products Ltd.) (1965); *C.A.* **62**, 10632b (1965).

140. French Patent 1,458,459 (Imperial Chemical Industries, Ltd.) (1966).

141. Bailey, D. L., and O'Connor F. M., German Patent 1,140,712 (Union Carbide Corp.) (1957); *C.A.* **58**, 14215f (1963).

142. Pikula, J. E., U.S. Patent 3,445,276 (Union Carbide Corp.) (1969); *C.A.* **71**, 51183k (1969).

143. Netherlands Patent Appl. 6,601,584 (Union Carbide Corp.) (1966); *C.A.* **66**, 19220f (1967).

144. Delaval, J. C. A., Guinet, P. A. E., Morel, J. M. E., and Puthet, R. R., French Patent 1,520,444 (Société des Usines Chimiques Rhône-Poulenc) (1968); *C.A.* **70**, 107091c (1969).

145. British Patent 1,151,960 (Rhone-Poulenc S.A.) (1969); *C.A.* **71**, 71445a (1969).

146. Morehouse, E. L., German Offen. 1,922,595 (Union Carbide Corp.) (1969); *C.A.* **72**, 32756d (1970).

147. Boudreau, R. J., U.S. Patent 3,483,240 (General Electric) (1965).

148. Johnson, G. C., and Pikula, J. E., French Patent 1,356,976 (Union Carbide Corp.) (1964); *C.A.* **62**, 9291d (1965).

149. Marlin, L., German Offen. 2,001,355 (Union Carbide Corp.) (1970); *C.A.* **73**, 99648j (1970).

150. Bennett, E. W., German Offen. 2,215,393 (Union Carbide Corp.) (1972); *C.A.* **78**, 17121y (1973).

151. French Patent 1,590,602 (Rhône-Poulenc) (1970); *C.A.* **74**, 23352h (1971).

152. British Patent Amended 1,151,960 (Rhône-Poulenc) (1970); *C.A.* **76**, 25846n (1972).

153. Rick, E. A., and Omietanski, G. M., German Offen. 2,301,789 (Union Carbide Corp.) (1973); *C.A.* **80**, 109290x (1974).

154. British Patent 983,851 (Union Carbide Corp.) (1961).

155. Rossi, C., Bianchi, U., and Bianchi, E., *J. Polym. Sci., Part C* **4**, 699 (1964).

156. Lichtenwalner, H. K., and Sprung, M. N., *Encycl. Polym. Sci. Technol.* **12**, 464 (1970); *C.A.* **74**, 42634t (1971).

157. Kendrick, T. C., Kingston, B. M., Lloyd, N.C., and Owen, M. J., *J. Colloid Interface Sci.* **24**(2), 135 (1967); *C.A.* **67**, 91304z (1967).

158. Bailey, D. L., Peterson, I. H., and Reid, W. G., *Proc. Int. Congr., Surf. Act., 4th, 1964* (1967).

159. Kanner, B., Reid, W. G., and Peterson, I. H., *Ind. Eng. Chem., Prod. Res. Dev.* **6**(2), 88 (1967).

160. Kanner, B., and Decker, T. G., *J. Cell. Plast.* Jan., 1969, pg. 32.

161. Boudreau, R. J., *Mod. Plast.* **44**(5), 133 (1967).

162. British Patent 802,467 (Union Carbide Corp.) (1957).

163. French Patent 1,581,196 (ICI) (1967).

164. Noshay, A., Matzner, M., and Merriam, C. N., German Offen. 1,913,749 (Union Carbide Corp.) (1969); *C.A.* **71**, 125475f (1969).

165. Noshay, A., Matzner, M., Barth, B. P., and Walton, R. K., German Offen. 1,913,907 (Union Carbide Corp.) (1969); *C.A.* **72**, 44528u (1970).

166. Noshay, A., Matzner, M., and Merriam, C. N., German Offen. 1,913,908 (Union Carbide Corp.) (1969); *C.A.* **72**, 22289q (1970).

167. Noshay, A. B., Matzner, M. E., Barth, B. P., and Walton, R. K., German Offen. 1,927,787 (Union Carbide Corp.) (1970); *C.A.* **72**, 79930w (1970).

168. Noshay, A., Matzner, M., Barth, B. P., and Walton, R. K., U.S. Patent 3,536,657 (Union Carbide Corp.) (1970).

169. Noshay, A., Matzner, M., and Merriam, C. N., U.S. Patent 3,539,656 (Union Carbide Corp.) (1970).

170. Noshay, A., Matzner, M., and Merriam, C. N., U.S. Patent 3,539,657 (Union Carbide Corp.) (1970).

171. Noshay, A., Matzner, M., and Merriam, C. N., *Polym. Prepr., Am. Chem. Soc., Div. Polym. Chem.* **12**(1), 247 (1971).

172. Noshay, A., Matzner, M., and Merriam, C. N., *J. Polym. Sci., Part A-1* **9**(11), 3147 (1971); *C.A.* **76**, 100111z (1972).

173. Noshay, A., Matzner, M., and Merriam, C. N., *in* "Macromolecular Syntheses." Wiley, New York (in press).

174. Noshay, A., and Matzner, M., *Angew. Makromol. Chem.* **37**, 215 (1974).

175. Noshay, A., Matzner, M., and Williams, T. C., *Ind. Eng. Chem., Prod. Res. Dev.* **12**(4), 268 (1973); *C.A.* **80**, 109042t (1974).

176. Noshay, A., Matzner, M., Barth, B. P., and Walton, R. K., *Am. Chem. Soc., Div.*

Org. Coat. Plast. Chem., Pap. **34**(2), 217 (1974); also see *Adv. Chem. Ser.* (in press).

I77. Robeson, L. M., Noshay, A., Matzner, M., and Merriam, C. N., *Angew. Makromol. Chem.* **29/30**, 47 (1973); *C.A.* **79**, 19957v (1973).

I78. Matzner, M., Noshay, A., and McGrath, J. E., *Polym. Prepr., Am. Chem. Soc., Div. Polym. Chem.* **14**(1), 68 (1973); *Trans. Soc. Rheol.*, in press (1976).

I79. Matzner, M., Noshay, A., Robeson, L. M., Merriam, C. N., Barclay, R., Jr., and McGrath, J. E., *Appl. Polym. Symp.* **22**, 143 (1973); *C.A.* **80**, 71813e (1974).

I80. Matzner, M., Noshay, A., Schober, D. L., and McGrath, J. E., *Ind. Chim. Belge* **38**, 1104 (1973).

I81. Matzner, M., and Noshay, A., Presented at *Gordon Conf. Polym., 1974* (1974).

I82. Johnston, N. W., and Joesten, B. L., *J. Fire Flammability* **3**, 274 (1972); *C.A.* **78**, 30693f (1973).

I83. French Patent 2,050,554 (Union Carbide Corp.) (1971); *C.A.* **76**, 114135x (1972).

I84. Strachan, J. D., and Williams, T. C., German Offen. 1,913,751 (Union Carbide Corp.) (1969); *C.A.* **72**, 22292k (1970).

I85. Strachan, J. D., and Williams, T. C., U.S. Patent 3,539,655 (Union Carbide Corp.) (1970).

I86. Krantz, K. W., U.S. Patent 3,668,273 (General Electric Co.) (1972); *C.A.* **77**, 76029b (1972).

I87. Clark, R. F., and Krantz, K. W., U.S. Patent 3,696,137 (General Electric Co.) (1972); *C.A.* **78**, 44290j (1973).

I88. Morton, M., Rembaum, A. A., and Bostick, E. E., *J. Appl. Polym. Sci.* **8**(6), 2707 (1964); *C.A.* **62**, 2839h (1965).

I89. Dean, J. W., *J. Polym. Sci., Part B* **8**(10), 677 (1970); *C.A.* **74**, 23474z (1971).

I90. Dean, J. W., U.S. Patent 3,760,030 (General Electric Co.) (1973).

I91. Saam, J. C., Ward, A. H., and Fearon, F. W. G., *Polym. Prepr., Am. Chem. Soc., Div. Polym. Chem.* **13**(1), 524 (1972).

I92. Saam, J. C., and Fearon, F. W. G., *Ind. Eng. Chem., Prod. Res. Dev.* **10**, 10 (1971).

I93. Saam, J. C., and Fearon, F. W. G., *Polym. Prepr., Am. Chem. Soc., Div. Polym. Chem.* **11**(2), 455 (1970).

I94. Saam, J. C., and Fearon, F. W. G., German Offen. 2,142,595 (Dow Corning Corp.) (1972).

I95. Saam, J. C., Ward, A. H., and Fearon, F. W. G., *Adv. Chem. Ser.* **129**, 239 (1973); *C.A.* **80**, 121373j (1974).

I96. Saam, J. C., Ward, A. H., and Fearon, F. W. G., *J. IRI* **7**(2), 69 (1973); *C.A.* **79**, 54573p (1973).

I97. Saam, J. C., and Fearon, F. W. G., German Offen. 2,142,594 (Dow Corning Corp.) (1972); *C.A.* **77**, 20367t (1972).

I98. Noshay, A., Matzner, M., Karoly, G., and Stampa, G. B., *Polym. Prepr., Am. Chem. Soc., Div. Polym. Chem.* **13**(1), 292 (1972).

I99. Noshay, A., Matzner, M., Karoly, G., and Stampa, G. B., *J. Appl. Polym. Sci.* **17**, 619 (1973).

I100. Gallot, Y., and Marsiat, A., *C.R. Hebd. Seances Acad. Sci., Ser. C* **272**(17), 1474 (1971); *C.A.* **75**, 36741e (1971).

I101. Lee, C. L., and Johannson, O. K., *J. Polym. Sci., Part A-1* **4**, 3013 (1966).

I102. Saam, J. C., and Fearon, F. W. G., German Offen. 2,142,664 (Dow Corning Corp.) (1972); *C.A.* **77**, 21164m (1972).

I103. Ward, A., Kendrick, T., and Saam, J., *Polym. Prepr., Am. Chem. Soc., Div. Polym. Chem.* **15**(1), 183 (1974).

I104. Vaughn, H. A., British Patent 989,379 (General Electric Co.) (1965); *C.A.* **63**, 1904e (1965).

I105. Vaughn, H. A., Jr., U.S. Patent 3,419,635 (General Electric Co.) (1968); *C.A.* **70**, 48439x (1969).

I106. Vaughn, H. A., Jr., U.S. Patent 3,419,634 (General Electric Co.) (1968); *C.A.* **70**, 48235c (1969).

I107. Vaughn, H. A., Jr., *Am. Chem. Soc., Div. Org. Coat. Plast. Chem., Pap.* **29**(1), 133 (1969); *C.A.* **74**, 4096m (1971).

I108. Vaughn, H. A., *J. Polym. Sci., Part B* **7**(8), 569 (1969); *C.A.* **71**, 91965y (1969).

I109. Kambour, R. P., *in* "Block Polymers" (S. L. Aggarwal, ed.), p. 263. 1970; *C.A.* **75**, 22098p (1971).

I110. French Demande 2,163,700 (Bayer A.-G.) (1973); *C.A.* **80**, 60397w (1974).

I111. Buechner, W., Noll, W., and Bressel, B., German Offen. 2,162,418 (Bayer A.-G.) (1973) *C.A.* **79**, 92984f (1973).

I112. Magila, T. L., and LeGrand, D. G., *Polym. Eng. Sci.* **10**(6), 349 (1970); *C.A.* **73**, 120987f (1970).

I113. Ward, W. J., Kimura, S. G., and Neulander, C. K., German Offen. 1,948,645 (General Electric Co.) (1970); *C.A.* **72**, 133569a (1970).

I114. Lundstrom, J. E., U.S. Patent 3,767,737 (General Electric Co.) (1973); *C.A.* **80**, 84363h (1974).

I115. Kambour, R. P., *J. Polym. Sci., Part B* **7**(8), 573 (1969); *C.A.* **71**, 92001t (1969).

I116. Kambour, R. P., *Polym. Prepr., Am. Chem. Soc., Div. Polym. Chem.* **10**(2), 885 (1969).

I117. Narkis, M., and Tobolsky, A. V., *J. Macromol. Sci., Phys.* **4**(4), 877 (1970); *C.A.* **73**, 110645d (1970).

I118. Narkis, M., and Tobolsky, A., *U.S. Dep. Comm., C.F.S.T.I., AD Rep.* **AD-683 671** (1969); *C.A.* **71**, 51030z (1969).

I119. LeGrand, D. G., *J. Polym. Sci., Part B* **7**(8), 579 (1969); *C.A.* **71**, 102380g (1969).

I120. LeGrand, D. G., *Polym. Prepr., Am. Chem. Soc., Div. Polym. Chem.* **11**(2), 434 (1970).

I121. LeGrand, D. G., *Trans. Soc. Rheol.* **15**(3), 541 (1971); *C.A.* **75**, 141290e (1971).

I122. LeGrand, D. G., *Polym. Sci. Technol.* **1**, 81 (1973).

I123. Kaniskin, V. A., Kaya, A., Ling, A., and Shen, M., *J. Appl. Polym. Sci.* **17**(9), 2695 (1973); *C.A.* **80**, 83752x (1974).

I124. LeGrand, D. G., *J. Polym. Sci., Part B* **9**(2), 145 (1971); *C.A.* **74**, 112563v (1971).

I125. LeGrand, D. G., U.S. Patent 3,679,774 (General Electric Co.) (1972); *C.A.* **77**, 115366f (1972).

I126. Gaines, G. L., Jr., and LeGrand, D. G., German Offen. 2,033,608 (General Electric Co.) (1971); *C.A.* **75**, 21772s (1971).

I127. LeGrand, D. G., and Gaines, G. L., Jr., *Polym. Prepr., Am. Chem. Soc., Div. Polym. Chem.* **11**(2), 442 (1970); *C.A.* **77**, 6273c (1972).

I128. Bostick, E. E., Gaines, G. L., Jr., and LeGrand, D. G., U.S. Patent 3,640,943 (General Electric Co.) (1972); *C.A.* **76**, 141879z (1972).

I129. Viventi, R. V., U.S. Patent 3,600,288 (General Electric Co.) (1971); *C.A.* **75**, 142028n (1971).

I130. Kantor, S. W., and Juliano, P. C., German Offen. 2,257,206 (General Electric Co.) (1973); *C.A.* **79**, 67278w (1973).

I131. Kambour, R. P., and Niznik, G. E., General Electric Company, "Synthesis and Properties of Bisphenol Fluorenone Polycarbonate and BPF Carbonate-Dimethylsiloxane Block Polymers," Final Rep., Contract N00019-73-C-0152. Naval Air Systems Command, Department of the Navy, 1972–1973.

1132. Juliano, P. C., "New Silicone Elastoplastics," Rep. No. 74CRD172. General Electric Company, Corporate Research and Development, Schenectady, N.Y. 1974, Presented at U. of Detroit conf. series.

1133. Matzner, M., and Noshay, A., German Offen. 2,001,958 (Union Carbide Corp.) 1970; *C.A.* **73**, 88663j (1970).

1134. Matzner, M., and Noshay, A., U.S. Patent 3,579,607 (Union Carbide Corp.) (1971).

1135. Matzner, M., Noshay, A., and Barclay, R., Jr., German Offen. 2,001,945 (Union Carbide Corp.) (1970); *C.A.* **73**, 121207p (1970).

1136. Matzner, M., Noshay, A., and Barclay, R., Jr., Canadian Patent 864020 (Union Carbide Corp.) (1971).

1137. Matzner, M., Noshay, A., and Barclay, R., Jr., U.S. Patent 3,701,815 (Union Carbide Corp.) (1972); *C.A.* **78**, 4952v (1973).

1138. Vinogradova, S. V., Korshak, V. V., Papava, G. Sh., and Tsiskarishvili, P. D., *Polym. Sci. USSR (Engl. Transl.)* **8**, 141 (1966).

1139. Papava, G. Sh., Maisuradze, N. A., Tsiskarishvili, P. D., Korshak, V. V., and Vinogradova, S. V., *Soobshch. Akad. Nauk Gruz. SSR* **43**(2), 349 (1966); *C.A.* **65**, 18776f (1966).

1140. Vinogradova, S. V., Korshak, V. V., Andrianov, K. A., Papava, G. Sh., and Khitarishvili, I. S., *Vysokomol. Soedin., Ser. A* **15**(6), 1215 (1973); *C.A.* **79**, 105607n (1973).

1141. Chow, S. W., and Byck, J. S., U.S. Patent 3,562,353 (United States Dept. of Health, Education, and Welfare) (1971); *C.A.* **74**, 142637q (1971).

1142. Chow, S. W., and Byck, J. S., U.S. Patent 3,562,353 (United States Dept. of Health, Education and Welfare) (1971); *C.A.* **76**, 127734s (1972).

1143. Thompson, J., and Owen, M. J., German Offen. 2,120,961 (Midland Silicones Ltd.) (1971); *C.A.* **76**, 100644a (1972).

1144. Thompson, J., and Owen, M. J., German Offen. 2,121,787 (Midland Silicones Ltd.) (1971); *C.A.* **76**, 100378s (1972).

1145. Nyilas, E., U.S. Patent 3,562,352 (Avco Corp.) (1971).

1146. Nyilas, E., German Offen. 1,944,969 (Avco Corp.) (1970); *C.A* **72**, 133546r (1970).

1147. Tanaka, T., Tanaka, R., and Ryoke, H., *Kogyo Kagaku Zasshi* **71**(12), 2072 (1968); *C.A.* **70**, 48366w (1969).

1148. Fil, T. I., Bryk, M. T., and Natanson, E. M., *Fiz.-Khim. Mekh. Liofil'nost Dispersnykh Sist.* **4**, 102 (1973); *C.A.* **80**, 27738k (1974).

1149. Perperot, A., French Demande 2,168,221 (Pechiney Ugine Kuhlmann) (1973); *C.A.* **80**, 122129w (1974).

1150. Mileshkevich, V. P., Novikova, N. F., and Karlin, A. V., *Vysokomol. Soedin., Ser. B* **14**(9), 682 (1972); *C.A.* **78**, 44717d (1973).

1151. Novikova, N. F., Mileshkevich, V. P., Karlin, A. V., and Novoselok, F. B., U.S.S.R. Patent 239,560 (1969); *C.A.* **71**, 50833k (1969).

1152. Atlas, S. M., German Patent 1,238,157 (1967); *C.A.* **68**, 30990u (1968).

1153. Atlas, S. M., German Patent 1,494,766 (1969).

1154. Greber, G., *Angew. Makromol. Chem.* **4/5**, 212 (1968).

1155. Holub, F. F., U.S. Patent 3,392,144 (General Electric Co.) (1968).

1156. Meier, D. J., *J. Polym. Sci., Part C* **26**, 81 (1969).

1157. Schwab, F. C., Ph.D. Thesis, University of Akron, Akron, Ohio (1970).

1158. Morton, M., *Encycl. Polym. Sci. Technol.* **15**, 508 (1971); *C.A.* **76**, 100848V (1972).

J1. Noshay, A., and Robeson, L. M., *J. Polym. Sci., Polym. Chem. Ed.* **12**(3), 689 (1974); *C.A.* **81**, 121632p (1974).

J2. Noshay, A., and Robeson, L. M., *Polym. Prepr., Am. Chem. Soc., Div. Polym. Chem.* **15**(1), 613 (1974).

J3. McGarry, F. J., Willner, A. M., and Sultan, J. N., Research Report R69-35. Dept. of Civil Engineering, Material Research Laboratory, Massachusetts Institute of Technology, Cambridge, 1969.

J4. Sultan, J. N., Laible, R. C., and McGarry, F. G., *Appl. Polym. Symp.* **16**, 127 (1971).

J5. Kalfoglou, N. K., and Williams, H. L., *J. Appl. Polym. Sci.* **17**, 1377 (1973).

J6. Soldatos, A. C., and Burhans, A. S., *Polym. Prepr., Am. Chem. Soc., Div. Polym. Chem.* **11**(1), 360 (1970); *Ind. Eng. Chem., Prod. Res. Dev.* **9**, 296 (1970); *Adv. Chem. Ser.* **97**, 531 (1970).

J7. Noshay, A., Matzner, M. and Robeson, L. M., "Morphology and Impact Resistance in Linear and Crosslinked Block Copolymers," Presented at the *8th* Regional ACS Meeting, Akron, Ohio, May, 1976; *J. Polymer Sci.*, Part C (in press).

Supplemental References

A list of recent references is given here to provide the most up-to-date coverage of the literature. Also included are some earlier references which are not cited in the text. The references are arranged into five structural categories: (A) Styrene-Dienes, (B) Polyolefins, (C) Polyesters, (D) Polyurethanes, and (E) Other Systems. The compositions are not organized according to block copolymer architecture, and therefore may pertain to A-B, A-B-A, and/or $(A-B)_n$ sequence structures. Titles of the references are given to help identify the subject matter.

A. STYRENE-DIENES

Aggarwal, S. L., Block polymers as a model for reinforced rubbers. *Colloq. Int. C.N.R.S.* p. 231, (1975); *C.A.* **84**, 32317j (1976).

Akutin, M. S., Andrianov, B. V., Kulyamin, V. S., Zisman, D. O., and Babanova, L. A., Rheological properties of polystyrene-thermoplastic elastomer composition. *Plast. Massy* (4), 44 (1975); *C.A.* **83**, 28985n (1975).

Bares, J., Glass transition of the polymer microphase. *Macromolecules* **8**(2), 244 (1975); *C.A.* **83**, 11080r (1975).

Beudouin, L., Organized and oriented polystyrene-polyisoprene-polystyrene block copolymers. II. Mechanical properties. *Rev. Gen. Caoutch. Plast.* **52**(3), 173 (1975); *C.A.* **83**, 180077a (1975).

Bi, L. K., Synthesis, mechanical properties and morphology of star block-copolymers. Ph.D. Thesis, University of Akron, Akron, Ohio, 1975; *C.A.* **84**, 45109t (1976).

Burchard, W., and Eschwey, H., Polyfunctional alkali metal-organic compounds. Ger. Offen. 2,427,955 (BASF) (1976); *C.A.* **84**, 106507v (1976).

Childers, C. W., and Clark, E., Suspension polymerization process. U.S. Patent 3,903,201 (Phillips Petroleum Co.) (1975); *C.A.* **83**, 194392m (1975).

Childers, C. W., Kraus, G., Gruver, J. T., and Clark, E., Formation of high impact polysterene from blends of polystyrene with butadiene/styrene block copolymers. *Colloid. Morphol. Behav. Block Graft Copolym., Proc. Am. Chem. Soc. Symp., 1970* p. 193 (1971); *C.A.* **83**, 60204t (1975).

Childers, C. W., Clark, E., and Johnson, W. D., Preparation of impact plastic compositions. U.S. Patent 3,935,136 (Phillips Petroleum Co.) (1976); *C.A.* **84**, 136549e (1976).

Cole, W. M., and Futamura, S., Process for producing multiblock SBR copolymer and products thereby. U.S. Patent 3,937,760 (Firestone Tire and Rubber Co.) (1976); *C.A.* **84**, 136960e (1976).

Conio, O., Orlandini, D., and Pedemonte, E., Silicon containing butadiene-styrene block copolymers; molecular characterization and morphology. *Rass. Chim.* **27**(3), 135 (1975); *C.A.* **84**, 122682d (1976).

Crossland, R. K., and Harlan, J. T., Jr., Block copolymer adhesive compositions. U.S. Patent 3,917,607 (Shell Oil Co.) (1975); *C.A.* **84**, 32063y (1976).

Damimon, H., Okitsu, H., and Kumanotani, J., Glass transition behavior of random and block copolymers of styrene and cyclododecyl acrylate. *Polym. J.* **7**(4), 460 (1975); *C.A.* **83**, 164712y (1975).

Donatelli, A. A., Thomas, D. A., and Sperling, L. H., Electron microscopy of polystyrene/polybutadiene polyblends; staining and fine structures. *Polym. Prepr., Am. Chem. Soc., Div. Polym. Chem.* **14**(2), 1080 (1973); *C.A.* **83**, 11304s (1975).

Douy, A., Jouan, G., and Gallot, B., Preparation and structural study of polybutadiene-poly-α-methylstyrene block copolymers. *C.R. Hebd. Seances Acad. Sci., Ser. C* **281**(11), 355 (1975); *C.A.* **84**, 17908j (1976).

Douy, A., Jouan, G., and Gallot, B., Preparation of polybutadiene-poly(vinyl-2-Naphthalene) block copolymers. Evidence and study of lamellar and inverted hexagonal structure. *C.R. Hebd. Seances Acad. Sci., Ser. C* **282**(4), 221 (1976); *C.A.* **84**, 151121a (1976).

Du Mee, P. E. J., Applications of thermoplastic rubbers. *Plastica* **28**(3), 98 (1975); *C.A.* **83**, 61185z (1975).

Durst, R. R., Compatible polymer mixture. Ger. Offen 2,342,219 (General Tire and Rubber Co.) (1973); *C.A.* **83**, 29180h (1975).

Durst, R. R., Compatible polymer blends. U.S. Patent 3,906,057 (General Tire and Rubber Company) (1975a); *C.A.* **84**, 18476d (1976).

Durst, R. R., High impact two-component resin blends. U.S. Patent 3,906,058 (General Rubber and Tire Co.) (1975b); *C.A.* **84**, 18475c (1976).

Durst, R. R., Compatible three-component polymer alloys. U.S. Patent 3,907,929 (General Tire and Rubber Co.) (1975c); *C.A.* **84**, 75091y (1976).

Durst, R. R., High impact two-component polystyrene blends. U.S. Patent 3,907,931 (General Tire & Rubber Co.) (1975d); *C.A.* **84**, 5864d (1976).

Durst, R. R., Compatible polymer mixture. Ger. Offen. 2,432,372 (General Tire and Rubber Co.) (1976); *C.A.* **84**, 1229148 (1976).

Eschwey, H., and Burchard, W., Star polymers from styrene and divinylbenzene. *Polymer* **16**(3), 180 (1975a); *C.A.* **83**, 115272W (1975).

Eschwey, H., and Burchard, W., Conditions of gelation in the anionic divinyl benzene-styrene copolymerization. *J. Polym. Sci., Polym. Symp.* **53**, 1 (1975b); *C.A.* **84**, 136114g (1976).

Evans, D. C., Barrie, J. A., and George, M. H., Synthesis and characterization of a cis-1,4-polyisoprene-b-poly(methyl methacrylate) copolymer. *Polymer* **16**(2), 151 (1975); *C.A.* **83**, 10928m (1975).

Fahrbach, G., Seiler, E., and Gerbernding, K., Star rubber modified polystyrene. Ger. Offen. 2,401,629 (BASF A.G.) (1975); *C.A.* **83**, 194116z (1975).

Fetters, L. J., and Bi, L. K., Branched Polymer. Ger. Offen. 2,529,065 (Univ. of Akron) (1975).

Fetters, L. J., and Morton, M., Polystyrene with predictable molecular weight distribution. *Macromol. Synth.* **3**, 77 (1972); *C.A.* **83**, 11114e (1975).

Fodor, L. M., Radial block copolymers stabilized by a urea compound. U.S. Publ. Pat. Appl. B 510,548 (Phillips Petroleum Co.) (1976); *C.A.* **84**, 151576c (1976).

Furukawa, K., Films with good gas barrier property. Japan Kokai 75/143,896 (Toyobo Co. Ltd) (1975); *C.A.* **84**, 165814r (1976).

Futamura, S., Hydrogenated block copolymers of butadiene and isoprene. U.S. Publ. Pat. Appl. B 575,851 (Firestone Tire and Rubber Co.) (1976); *C.A.* **84**, 151891b (1976).

Futamura, S., and Bouton, T. C., Branched block SBR. Ger. Offen. 2,460,009 (Firestone Tire and Rubber) (1975); *C.A.* **83**, 148781q (1975).

Grezlak, J. H., Block and graft copolymers by vinyl polymerization. Thesis, Princeton University, Princeton, New Jersey, 1974; *C.A.* **83**, 164642a (1975).

Grishin, B. S., Tutorskii, I. A., Boikacheva, E. G., and Kochetkova, G. V., Diffusion and solubility of sulphur in block copolymers of butadiene and styrene. *Vysokomol. Soedin., Ser. A* (11), 2481 (1975); *C.A.* **84**, 44911m (1976).

Guyon-Gellin, J. M., Gole, J., and Pascault, J. P., Preparation and study of tri-block copolymers of isoprene and methyl methacrylate. *J. Appl. Polym. Sci.* **19**(12), 3173 (1975); *C.A.* **84**, 31652c (1976).

Haaf, W. R., Composition containing a Poly(phenylene ether) and a hydrogenated block copolymer. Ger. Offen 2,434,848 (General Electric Co.) (1975); *C.A.* **83**, 11523n, (1975).

Hansen, D. R., and Shen, M., Viscoelastic retardation time computations for homogeneous block copolymers. *Macromolecules* **8**(3), 343 (1975a); *C.A.* **83**, 59480y (1975).

Hansen, D. R., and Shen, M., Viscoelastic properties of homogeneous triblock copolymers of styrene-α-methylstyrene and their polyblends with homopolymers. *Macromolecules* **8**(6), 903 (1975b); *C.A.* **84**, 5775a (1976).

Hansen, D. R., and Shen, M., Viscoelastic properties of homogeneous triblock copolymers of styrene/α-methyl styrene and their polyblends with homopolymers. *U.S. N.T.I.S. AD,-A Rep.* 1-38 (1975c); *C.A.* **84**, 18199r (1976).

Harlan, J. T., Jr., Hydroxylated and chlorinated block copolymer blends. U.S. Patent 3,917,742 (Shell Oil Co.) (1975a); *C.A.* **84**, 32064z (1976).

Harlan, J. T., Jr Derivatives of nonhydrogenated or hydrogenated block copolymers. *C.A.* **83**, 148346b (1975).

Hild, G., Haeringer, A., and Rempp, P., Formation of hydrogels by sulfonation of polystyrene block copolymers. *C.R. Hebd. Seances Acad. Sci., Ser. C* **280**, 140S (1975); *C.A.* **83**, 147983v (1975).

Hokonoki, H., Higasa, K., Ugajin, I., Takeuchi, A., and Kusuncse, T., Diene-styrene derivative block copolymer fibers. Japan Kokai 75/094,232 (Asahi Chemical Industry Co. Ltd.) (1975); *C.A.* **84**, 6368g (1976).

Horiie, S., Block copolymers for a sheet or film. U.S. Patent 3,939,224 (Denki Kagaku Kogyo K.K.) (1976); *C.A.* **84**, 165572b (1976).

Horiie, S., Asai, S., and Moriyam, Y., Transparent impact—resistant block copolymers. Japan Kokai 75/136,386 (Denki Kagoku Kogyo K.K.) (1975a); *C.A.* **84**, 106537e (1976).

Horiie, S., Asai, S., and Moriya, Y., Block copolymers. Japan Kokai 75/139,889 (Denki Kagaku Kogyo K.K.) (1975b); *C.A.* **84**, 136552y (1976).

Horiie, S., Asai, S., and Moriya, Y., Block copolymers of α-methyl styrene and conjugated diene. Japan Kokai 75/141,692 (1975c); *C.A.* **84**, 136578m (1976).

Horiie, S., Handa, M., Yamamoto, K., and Nishizawa, T., Block copolymers. Japan Kokai 75/141,693 (Denki Kagaku Kogyo K.K.) (1975d); *C.A.* **84**, 151244t (1976).

Ibaragi, T., Transparent impact—resistant polystyrenes. Japan Kodai 75/95,386 (Asahi Chem Ind. Co. Ltd.) (1975); *C.A.* **83**, 194373f (1975).

Imanaka, H., and Sumoto, M., Blends of block butadiene styrene rubber and polyesters. Japan Kodai 75/82,162 (Toyobo Co. Ltd) (1975); *C.A.* **83**, 180798m (1975).

Ionescu, M. L., and Skoulios, A., Lamellar structure of block copolymers; structural parameters and partition coefficients of solvents in swelling of two block-copolymers polystyrene/polyisoprene. *Makromol. Chem.* **177**(1), 257 (1976); *C.A.* **84**, 74772j (1976).

Ishizu, K., Fukutomi, T., and Kakurai, T., Coupling reaction of poly (α-methyl-styryl) anion with (MMA [methyl-methacrylate]-α-methylstyrene block copolymer. *J. Polym. Sci., Polym. Chem. Ed.* **13**(11), 2453 (1975); *C.A.* **84**, 18006a (1976).

Kaempf, G., Ordered superstructure in single and multiphase amorphous high polymers. *Prog. Colloid Polym. Sci.* **57**, 249 1975; *C.A.* **83**, 179798e (1975).

Kahle, G. R., Kitchen, A. G., and Uraneck, C. A., Methoxy silane coupling of block copolymers. U.S. Patent 3,880,954 (Phillips Petroleum Co.) (1975); *C.A.* **83**, 115709n (1975).

Kamatsu, K., "A-B-C type block copolymer moldings. Japan Patent 75/15,028 (Japan Synthetic Rubber Co.) (1975); *C.A.* **83**, 180257j (1975).

Krause, S., and Reismiller, P. A., Micelle formation in butanane solutions of styrene-butadiene-styrene triblock copolymer. *J. Polym. Sci., Polym. Phys. Ed.* **13**(3) 663 (1975); *C.A.* **83**, 11027d (1975).

Larkin, J. B., Kraton G and elexar. Thermoplastic rubber for use in appliances. *Plast. Appl., Natl. Tech. Conf., SPE* p. 204 (1975); *C.A.* **84**, 123071n (1976).

Leblanc, J. L., Characterization of sequential butadiene styrene copolymers. *Trib. CEBEDEAU* **28**(378) 23 (1975); *C.A.* **83**, 180710b (1975).

Lee, G. F., Jr., Thermoplastic molding compositions from a poly(phenylene ether) resin and interpolymers of an alkenyl aromatic compound and block copolymers. Ger. Offen. 2,523,101 (1976); *C.A.* **84**, 91072y (1976).

Lundberg, R. D., and Makowski, H. S., Fabrication of multiphase plastics from liquid suspension. U.S. Patent 3,925,280 (Exxon Research and Engineering Co.) (1975); *C.A.* **84**, 106702e (1976).

Makowski, H., and Buckley, D., Styrene-tertiary butylstyrene block polymers. Ger. Offen. 2,443,875 (Exxon Research & Engineering Co.) (1975); *C.A.* **83**, 44034g (1975).

Matsuo, M., and Sagaye, S., Micromorphology-property relations in graft and block copolymers. *Colloid. Morphol. Behav. Block Graft Copolym., Proc. Am. Chem. Soc. Symp., 1970* p. 1 (1971); *C.A.* **83**, 60085 (1975).

Mehra, U., Choi, G., Biliyar, K., and Shen, M., Relaxation behavior of blends of homopolymers with block copolymers. *Polym. Prepr., Am. Chem. Soc., Div. Polym. Chem.* **15**(1), 426 (1974); *C.A.* **84**, 17900a (1976).

Mehra, U., Toy, L., Biliyar, K., and Shen, M., Dynamic mechanical and dielectric relaxations in homopolymer-block copolymer blends. *Adv. Chem. Ser.* **142**, 399 (1975); *C.A.* **83**, 206677y (1975).

Meier, D. J., Theory of the interface in block copolymers. *Polym. Prepr., Am. Chem Soc., Div. Polym. Chem.* **15**(1), 171 (1974); *C.A.* **83**, 206706g (1975).

Miki, T., Imai, A., and Kobayashi, K., Transparent block colpolymers. Japan Kokai 75/114,491 (Sumitomo Chemical Co., Ltd.) (1975); *C.A.* **84**, 5885m (1976).

Moore, E. R., Lehrer, R. G., Lyons, C. E., and McKeever, L. D., Impact resistant polymers of a resinous copolymer of an alkenyl aromatic monomer and unsaturated dicarboxylic anhydride. U.S. Patent 3,919,354 (Dow Chemical Co.) (1975); *C.A.* **84**, 75035h (1976).

Morton, M., "Rubber Technology," 2nd ed. 1973; *C.A.* **83**, 11847w (1975).

Morton, M., and Fetters, L. J., Anionic polymerization of vinyl monomers. *Rubber Chem. Technol.* **48**(3), 359 (1975); *C.A.* **83**, 179632w (1975).

Nguyen Vinh Chih, Isaev, À. I., Malkin, A. Ya., Vinogradov, G., and Kircheveskaya, I. Yu., Viscosity and viscoelasticity of mixtures and block copolymers of polybutadiene with polyisoprene. *Vysokomol. Soedin., Ser. A.* **17**(4), 855 (1975); *C.A.* **83**, 59615w (1975).

Odani, H., Taira, K., Yamacuchi, T., Nemoto, N., and Kurata, M., Solubilities of inver gases in S-B-S block copolymers. *Bull. Inst. Chem. Res., Kyoto Univ.* **53**(4), 409 (1975); *C.A.* **84**, 151111x (1976).

Odani, H., Taira, K., Nemoto, N., and Kurata, M., Permeation and diffusion of gases in styrene-butadiene-styrene block copolymers. *Bull. Inst. Chem. Res., Kyoto Univ.* **53**(2), 216, (1975); *C.A.* **84**, 5778d (1976).

Pinazzi, C., Esnault, J., and Pleurdeau, A., Synthesis of block copolymers by reactions of polymers with w-α.w-hydroxyl groups and polymers with w- and α.w-oxychloroformyl groups. *Makromol. Chem.* **177**(3), 663 (1976); *C.A.* **84**, 151037c (1976).

Plestil, J., and Baldrian, J., Determination of the structure parameters of styrene/butadiene block copolymer in heptane by small angle x-ray scattering. *Makromol. Chem.* **176**(4), 1009 (1975); *C.A.* **83**, 43904d (1975).

Prudence, R. J., Three block polymers. Ger. Offen. 2,449,299 (Goodyear Tire and Rubber Company) (1975); *C.A.* **83**, 116543x (1975).

Ptaszynski, B., Terrisse, J., and Skoulious, A., Solubilization of polystyrene in the lamellar phase of a two-block copolymer polystyrene/polyisoprene. *Makromol. Chem.* **176**(11), 3483 (1975); *C.A.* **84**, 17965a (1976).

Riess, G. Structures of systems of heterogeneous polymers. Properties of block and graft copolymers. *Prog. Colloid. Polym. Sci.* **57**, 262 (1975); *C.A.* **83**, 164839h (1975).

Riess, G., and Periard, J., Emulsifying effect of polyisoprene block and graft copolymers. Formation of oil in oil emulsions. *Colloid Polym. Sci.* **253**(5), 362 (1975); *C.A.* **83**, 115159g (1975).

Riess, G., and Jolivet, Y., Rubber-modified polymers. Location of block copolymers in two-phase materials, *Adv. Chem. Ser.* **142**, 243 (1975); *C.A.* **83**, 194202z (1975).

Riess, G., Periard, J., and Banderet, A., Emulsifying effects of block and graft copolymers. Oil in oil emulsions. *Colloid. Morphol. Behav. Block Graft Copolym., Proc. Am. Chem. Soc. Symp., 1970* p. 173 (1971); *C.A.* **83**, 60086f (1975).

Riess, G., Bordeaux, M., Brie, M., and Jouquet, G., Surface treatment of carbon fibers with alternating block and block copolymers. *Plast. & Polym., Conf. Suppl.* **6**, 52 (1974); *C.A.* **84**, 45084f (1976).

Robertson, W. J., Crossland, R. K., and Harlan, J. T., Jr., Process for preparation of pressure-sensitive adhesives. U.S. Patent 3,935,338 (Shell Oil Co.) (1976); *C.A.* **84**, 123059t (1976).

Robeson, L. M., Matzner, M., Fetters, L. J., and McGrath, J. E., Styrene-α-methyl styrene AB block copolymers and alloys. *Polym. Prepr., Am. Chem. Soc., Div. Polym. Chem.* **14**(2), 1063 (1973); *C.A.* **83**, 43880t (1975).

Robeson, L. M., Pilato, L. A., and Godlewski, R. E., Impact-modified blends of vinyl chloride polymers, lactone graft copolymers, with styrene-diene-styrene block copolymers. U.S. Patent 3,825,622 (Union Carbide Corp.) (1974); *C.A.* **83**, 60534a (1975).

Schepers, J. J. A., Block copolymers containing α-methyl styrene. Ger. Offen. 2,442,849 (Stamicarbon B.V.) (1975); *C.A.* **83**, 11865a (1975).

Schulz, D. N., Halasa, A. F., and Oberster, A. E., Anionic polymerization initiators containing protected functional groups and functionally terminated diene polymers. *Polym. Prepr., Am. Chem. Soc., Div. Polym. Chem.* **14**(2), 1215 (1973); *C.A.* **83**, 10975z (1975).

Seymour, R. B., Stahl, G. A., and Wood, H., New block copolymers from styrene macroradicals. *Appl. Polym. Symp.* **26**, 249 (1975); *C.A.* **84**, 151035a (1976).

Shen, M., Cirlin, E. H., and Kaeble, D. H., Influence of morphology on the viscoelastic behvior of a solventcast S—B—S— block copolymer. *Colloid. Morphol. Behav. Block Graft Copolym., Proc. Am. Chem. Soc. Symp., 1970* p. 307 (1971); *C.A.* **83**, 61172t (1975).

Shima, M., Ogawa, E., and Konishi, K., Synthesis of (styrene-*p*-cholorostyrene) triblock copolymers and their properties in dilute solution. *Makromol. Chem.* **177**(1), 241 (1976); *C.A.* **84**, 74771h (1976).

Slukin, A. D., and Shutilin, Yu. F., Effect of certain factors on transitions in block copolymers of butadiene and styrene. *Izv. Vyssh. Uchebn. Zaved., Khim. Khim. Teknol.* **18**(5), 800 (1974); *C.A.* **83**, 80136a (1975).

Smith, T. L., Strength of segmented and triblock elastomers. *Polym. Prepr., Am. Chem. Soc., Div. Polym. Chem.* **15**(1), 58 (1974); *C.A.* **84**, 32307f (1976).

Soen, T., Shimonura, M., Uchida, T., and Kawai, H., Grain boundary relaxation phenomena in block and graft copolymers. *Colloid Polym. Sci.* **252**(11), 933 (1974); *C.A.* **83**, 11002s (1975).

Sothman, R. D., and Urwin, J. R., Soluting properties of S—I—S via DSC. *Aust. J. Chem.* **28**(10), 2161 (1975); *C.A.* **83**, 193935d (1975).

Spatorico, A. L., Evaluation of the light-scattering technique for determining compositional heterogeneity in copolymers. *Polym. Prepr., Am. Chem. Soc., Div. Polym. Chem.* **15**(1), 515 (1974); *C.A.* **84**, 5520p (1976).

Szwarc, M., Anionic polymerization. Its past achievements and its present status. *Proc. Int. Symp. Macromol.* p. 153 (1975); *C.A.* **83**, 59282k (1975).

Tabana, M., Mitsumo, T., and Maki, H., Antiblocking agents for butadiene-styrene block copolymers. Japan Kokai 75/155,555 (Sumitomo Chem. Co. Ltd.) (1975); *C.A.* **84**, 106620b (1976).

Takeshita, Y., and Shuto, Y., Block copolymers of butadiene and propylene oxide. Japan Kokai 75/141,694 (Idemitsu Kosan Co. Ltd) (1975); *C.A.* **84**, 151263y (1976).

Tarnat, J., and Meyer, G., Reaction between N-methyl maleimide and S—I—S. *Bull. Soc. Chim. Fr.* (7-8, Part 2), 1969 (1975); *C.A.* **83**, 206866j (1975).

Teraoka, T., Satake, K., and Sone, T., Block butadiene styrene rubber films for food packaging. Japan Kokai 76/12,845 (Asaki Chem Ind. Co. Ltd.) (1976); *C.A.* **84**, 36961f (1976).

Toy, L., Niinomi, M., and Shen, M., Dynamic mechanical and morphological studies of homopolymer/block copolymer blends. *J. Macromol. Sci., Phys.* **11**(3), 281 (1975a); *C.A.* **84**, 6209f (1976).

Toy, L., Niinomi, M., and Shen, M., Dynamic mechanical and morphological studies of homopolymer/block copolymer blends. *Gov. Rep. Announce. (U.S.)* **75**(7), 182 (1975b); *C.A.* **83**, 115220c (1975).

Tritscher, G., and Meyer, G., Grafting methacrylonitrile onto S/I/S. *C.R. Hebd. Seances Acad. Sci., Ser. C* **282**(9), 421 (1976); *C.A.* **84**, 165247h (1976).

Tung, W. H., Influence of triblock copolymer upon the mutual compatibility of polystyrene/polybutadiene blends. Thesis, University of Maryland, College Park, 1974; C.A. 83, 206958r (1975).

Tutorskik, I. A., Boikacheva, E. G., Bukanova, E. F., Gueseva, T. V., and Bukanov, I. G., Distribution of monomer units in butadiene-styrene copolymers studied by a pyrolytic gas chromatographic method. Izv. Vyssh. Uchebn. Zaved., Khim. Khim. Tekhol. 18(3), 460 (1975); C.A. 83, 28957e (1975).

Vestner, S., Jenne, H., and Benker, K., Self extinguishing styrene polymer composition. Ger. Offen. 2,312,804 (BASF A.-G.) (1974); C.A. 83, 60554g (1975).

Walker, J. H., Printable antiblocking resinous block-copolymer. U.S. Patent 3,896,086 (Phillips Petroleum Co.) (1975); C.A. 84, 75081V (1976).

Warrach, W., Bonding styrene-butadiene block copolymers to other surfaces. U.S. Patent 3,919,035 (Stein, Hall, Ltd.) (1975); C.A. 84, 45811r (1976).

Worsfold, D. J., Block copolymers from p-divinyl-benzene capped isoprene, α-methyl styrene, or styrene polymers. Can. Patent 973,295 (National Research Council, Canada) (1975); C.A. 84, 18056s (1976).

B. POLYOLEFINS

Bina, J., and Lagerova, A., Thermoplastic rubber. Elektroizolacna Kablova Tech. 28(3), 199 (1975); C.A. 84, 151807d (1976).

Campbell, D. G., Effects of ethylene-propylene block copolymers on melt blended linear polyethylene-isotactic polypropylene polyblends. Thesis, University of Maryland, College Park, 1974; C.A. 83, 148208h (1975).

Castagna, E. G., Ethylene-propylene block copolymer-polyethylene blends. Ger. Offen. 2,414,508 (Dart Industries, Inc.) (1975); C.A. 83, 116275m (1975).

Castagna, E. G., Process for the production of high impact compositions of polyethylene and polypropylene block copolymers. U.S. Patent 3,937,758 (Dart Industries, Inc.) (1976); C.A. 84, 165558k (1976).

Cogswell, F. N., and Hanson, D. E., Anomalous melt elasticity of isotoctic ethylene-propylene block copolymers. Polymer 16(12), 936 (1975); C.A. 84, 151108b (1976).

Dankovics, A., and Kissin, Y., Copolymerization of propylene and some vinyl monomers on the titanium trichloride-triethylaluminum catalyst. Copolymerization kinetics and structure of copolymers. Acta Chim. Acad. Sci. Hung. 86(3), 307 (1975); C.A. 84, 17856r (1976).

Gehrke, K., and Schimmel, C., Block copolymers of olefins and styrene. Ger. (East) Patent 108,547 (1974); C.A. 83, 61238u (1975).

Hartman, P. F., Grafted block copolymers of synthetic rubbers and polyolefins. U.S. Patent 3,909,463 (Allied Chemical Corp.) (1975); C.A. 84, 45197v (1976).

Imai, K., Moldability of propylene block copolymer. Japan Patent 75/17,217 (Chisso Corp.) (1975); C.A. 84, 165698f (1976).

Jezl, J. L., Crystalline olefin block polymers. U.S. Patent 3,873,642 (Avisun Corp.) (1975); C.A. 83, 28831j (1975).

Johnson, J. R., Extruding uniroyal TPR thermoplastic rubbers. Soc. Plast. Eng., Tech. Pap. 21, 502 (1975); C.A. 83, 148707v (1975).

Mitra, B. C., and Katti, M. R., Ethylene-propylene block copolymers. Pop. Plast. 20(12), Suppl. 5 (1975); C.A. 84, 60280k (1976).

St. Clair, D. J., and Harlan, J. T., Styrene-olefin block copolymers. Versatile new base for sealants. Adhes. Age 18(11), 39 (1975); C.A. 84, 46166w (1976).

Sogolova, T. I., Akutin, M. S., Tsvankin, D. Ya., Kerber, M. L., Mudziri, B. G., and

Cherdabaev, A. Sh., Modification of the supramolecular structure and properties of polyethylene by thermoplastic elastomers. *Vysokomol. Soedin., Ser. A* **17**(11), 2505 (1975); *C.A.* **84**, 75197n (1976).

C. POLYESTERS

Allegrezza, A. E., Lenz, R. W., Cornibert, J., and Marchessault, R. H., Crystalline, thermal and mechanical properties of block and random copolymers of pivalolactone and (DL) a-methyl- -propyl- -propiolactone. *Polym. Prepr., Am. Chem. Soc., Div. Polym. Chem.* **14**(2), 1232 (1973); *C.A.* **83**, 4388w (1975).

Brooks, T. W., Bledsoe, C., and Rodriguez, J., Urethane block copolymer from polyoxymethylene and poly(propylene adipate). *Macromol. Synth.* **4**, 1 (1972); *C.A.* **84**, 4535r (1976).

Cella, R. J., and Buck, W. H., Segmented polyester thermoplastic elastomers. *Polym. Prepr., Am. Chem. Soc., Div. Polym. Chem.* **15**(1), 159 (1974); *C.A.* **84**, 18879n (1976).

Dinov, K., and Georgiev, J., Interaction exchange and compatibility in melt dispersions of polycaproamide and poly(ethyleneterephthalate). *Faserforch. Textiltech.* **26**(10), 479 (1975); *C.A.* **84**, 75464x (1976).

Eastman, E. F., Melt-stabilized segmented copolyester adhesive. Ger. Offen. 2,428,251 (duPont de Nemours, E. I. and Co.) (1975a); *C.A.* **83**, 60620a (1975).

Eastman, E. F., Thermally stabilized block copolyester adhesive. Ger Offen. 2,507,321 (duPont de Nemours, E. I. and Co.) (1975b); *C.A.* **84**, 18416j (1976).

Foss, R. P., Copolymers of pivalolactone and isoprene or butadiene. U.S. Patent 3,907,933 (duPont de Nemours, E. I. and Co.) (1975); *C.A.* **84**, 18921v (1976).

Foss, R. P., Jacobson, H. W., Cripps, H. N., and Sharkey, W. H., Block and graft copolymers of pivololactone. II. ABA and ABA-g-A copolymers with dienes. *Macromolecules* **9**(2), 373 (1976); *C.A.* **84**, 165897 v (1976).

Froix, M. F., and Goedde, A. O., Nuclear spin relaxation and molecular motions in sebacate polyesters. *J. Macromol. Sci., Phys.* **11**(3), 345 (1975); *C.A.* **84**, 44807g (1976).

Georgiev, Y., Dimov, K., and Garvanska, R., Length of polycaproamide blocks in nylon 6—PET block copolymers. *Vysokomol. Soedin., Ser. A* **18**(2), 401 (1976); *C.A.* **84**, 122744a (1976).

Hespe, H., Rempel, D., and Morbitzer, L., Modified elastomeric, segmented mixed polyester. Ger. Offen. 2,363,512 (Bayer A.G.) (1975); *C.A.* **83**, 148804z (1975).

Hsieh, H. L., Lactone containing polymers. U.S. Patent 3,880,955 (Phillips Petroleum Co.) (1975); *C.A.* **83**, 115934p (1975).

Huet, J. M., Polycarbonate-polycaprolactone sequenced copolycondensates from oligomers with reactive extremities. Fr. Demande 2,235,965 (Ugine Kuhlmann) (1975); *C.A.* **83**, 148293g (1975).

Imanaka, H., Heat stabilizers for polyether-polyesters. Japan Kokai 75/10,849 (Toyobo Co., Ltd.) (1975); *C.A.* **83**, 165106r (1975).

Kato, Y., Imanaka, H., and Sumoto, M., Polyester elastic films. Japan Kokai 74/119,949 (Toyobo Co. Ltd.) (1974); *C.A.* **83**, 29537e (1975).

Koseki, T., Block poly(ether esters) with good whiteness. Japan Kokai 75/139,194 (Toray Ind. Inc.) (1975); *C.A.* **84**, 122589d (1976).

Light, R. R., Gray, T. F., and Joyner, F. B., Elastomeric poly(ether-ester) molding composition. Ger. Offen. 2,502,905 (Eastman Kodak Co.) (1975); *C.A.* **84**, 6277b (1976).

Minami, T., Polyester polyether block copolymers. Japan Kokai 75/91,694 (Toyobo Co. Ltd.) (1975); *C.A.* **83**, 148970a (1975).

Minami, T., Tsuji, S., and Sumoto, M., Polyester polyether block copolymers. Japan Kokai 75/127,994 (Toyobo Co., Ltd.) (1975); *C.A.* **84**, 75583K (1976).

Neslon, J. P., Copolyester (ethylene terephtholate/tetramethylene dibromoterephthalate), block copolymers. U.S. Patent 3,883,611 (Standard Oil Co. of Indiana) (1975); *C.A.* **83**, 115735z (1975).

Nishi, T., Kwei, T. K., and Wang, T. T., Physical properties of poly(vinyl chloride)-copolyester thermoplastic elastomer mixtures. *J. Appl. Phys.* **46**(10), 4157 (1975); *C.A.* **84**, 45717g (1976).

O'Malley, J. J., Crystalline isomeric polyester block copolymers. Scanning calorimetry and density measurements. *J. Polym. Sci., Polym. Phys. Ed.* **13**(7), 1353 (1975); *C.A.* **83**, 164723c (1975).

Seeger, N. V., and Kaman, A. J., Impact resistant thermoplastic polyester urethanes. U.S. Patent 3,931,113 (PPG Industires) (1976); *C.A.* **84**, 165618e (1976).

Smith, S., and Hubin, A. J., Difunctional cationic polyethers and polyesters. U.S. Patent 3,824,197 (Minnesota Mining and Manufg. Co.) (1974); *C.A.* **83**, 59891h (1975).

Sumoto, M., Tsuji, K., and Furusawa, H., Polyester-polyether block copolymer. Japan Patent 74/48,195 (Toyobo Co. Ltd.) (1974); *C.A.* **83**, 44011x (1975).

Sumoto, M., Watanabe, K., Hattori, A., and Sawaki, R., Compositions of polyester polyether block copolymers. Japan Kokai 75/140,554 (Toyobo Co., Ltd.) (1975); *C.A.* **84**, 75298w (1976).

Tsuji, S., Polyester-polyether block copolymers. Japan Kokai 75/45,895 (Toyobo Co. Ltd.) (1975); *C.A.* **83**, 116767y (1975).

Tsuji, S., Imanaka, H., Minami, T., and Sumoto, M., Compounding of block polyester elastomers with additives. Japan Kokai 75/092,345 (Toyobo Co., Ltd.) (1975a); *C.A.* **84**, 6274y (1976).

Tsuji, S., Minami, T., and Sumoto, M., Stabilized block poly(ester ether) compositions. Japan Kokai 75/098,955 (Toyobo Co., Ltd.) (1975b); *C.A.* **84**, 5962j (1976).

VanderVoort, H. G. P., Esters of alkyl-or alkenylsuccinic acid with polyhydroxy alcohols. Ger. Offen. 2,443,538 (Shell Internationale Research Maatschappij B.V.) (1975); *C.A.* **83**, 59954f (1975).

Wolfe, J. R., Jr., Segmented thermoplastic copolyester elastomers. Ger. Offen. 2,456,536 (DuPont) (1975); *C.A.* **83**, 180756w (1975).

D. POLYURETHANES

Allegrezza, A. E., Jr., Seymour, R. W., and Cooper, S. L., Segmental orientation of polyurethane block polymers. *Polym. Prepr., Am. Chem. Soc., Div. Polym. Chem.* **15**(1), 631 (1974); *C.A.* **84**, 17903d (1976).

Alliger, G., McGillvary, D. R., and Hayes, R. A., Liquid systems for rubber products including tires. *Pure Appl. Chem.* **39**(1-2), 45 (1974); *C.A.* **83**, 207329y (1975).

Borkent, G., Kinetics and mechanism of urethane and urea formation. *Adv. Urethane Sci. Technol.* **3**, 1 (1974); *C.A.* **83**, 10976a (1975).

Brooks, T. W., Bledsoe, C., and Rodriguez, J., Urethane block copolymers from polyoxymethylene and poly(propylene adipate). *Macromol. Synth.* **3**, 1 (1972); *V.A.* **83**, 59375t (1975).

Chang, Y. J., Chen, C. T., and Wilkes, G. L., Optical anisotropy in segmented polyurethanes. *Polym. Prepr., Am. Chem. Soc., Div. Polym. Chem.* **14**(2), 1277 (1973); *C.A.* **83**, 60108g (1975).

Chang, Y. J. P., and Wilkes, G. L., Superstructure in segmented polyether-urethanes. *J. Polym. Sci., Polym. Phys. Ed.* **13**(3), 455 (1975); *C.A.* **83**, 11026c (1975).

Dieter, J. A., Frisch, K. C., Shanafelt, G. K., and Devanney, M. T., Structure-properties relations in thermoplastic urethane elastomers based on poly(azelate) glycols. *Adv. Urethane Sci. Technol.* **3**, 197 (1974); *C.A.* **83**, 61180u (1975).

Dieterich, D., and Reiff, H., Polyurethane dispersions by the melt-dispersion method. *Adv. Urethane Sci. Technol.* **4**, 112 (1975); *C.A.* **84**, 165217y (1976).

Estes, G. M., Seymour, R. W., Borchert, S. J., and Cooper, S. L., Infrared dichrosism of segmented polyurethane elastomers. *Collod. Morphol. Behav. Block Graft Copolymers, Proc. Am. Chem. Soc. Symp., 1970* p. 159 (1971); *C.A.* **83**, 61171s (1975).

Fulcher, K. U., and Corbett, G. E., Effect of formulation on properties and morphology of polyurethane elastomers. *Br. Polym. J.* **7**(4), 225 (1975); *C.A.* **84**, 136932x (1976).

Halliwell, A., Factors affecting polyurethane stability. *J. Elastomers Plast.* **7**(3), 258 (1975); *C.A.* **83**, 148510a (1975).

Hayashi, I., Tsjui, K., Yamane, Y., Miyake, K., and Hitomi, C., Polyester block copolymer films with improved slip property. Japan Kokai 75/023,445 (Toyobo Co., Ltd.) (1975); *C.A.* **83**, 80270q (1975).

Hepburn, C., and Reynolds, R. J. W., Chemistry and technology of polyurethanes. *Mol. Behav. Dev. Polym. Mater.* p. 238 (1975); *C.A.* **83**, 59287r (1975).

Kolycheck, E. G., Poly(caprolactone)-based polyurethanes having improved blocking characteristics. U.S. Patent 3,923,747 (Goodrich, B. F. Co.) (1975); *C.A.* **84**, 60479g (1976).

Mayashi, I., Yamane, Y., Miyake, K., and Hitomi, C., Block copolyester films with improved frictional property. Japan Kokai 75/23,446 (Toyobo Co., Ltd.) (1975); *C.A.* **83**, 80271r (1975).

O'Shea, F. X., Poly(oxypropylene)glycol based polyurethane elastomers suitable for automative body parts. U.S. Patent 3,915,937 (Uniroyal, Inc.) (1975); *C.A.* **84**, 45764c (1976).

Pavlov, V. I., and Kipatnikov, N. A., Effect of the oligomer block on the mechanical properties of some linear polyurethanes. *Sint. Fiz.-Khim. Polim.* **13**, 117 (1974); *C.A.* **83**, 59512k (1975).

Peebles, L. H., Jr., Hard block length distribution in segmented block copolymers. *Macromolecules* **9**(1), 58 (1976); *C.A.* **84**, 106251 (1976).

Rieser, R. G., and Chabal, J., Laminated safety windshields. U.S. Patent 3,881,043 (PPG Industries, Inc.) (1975); *C.A.* **83**, 116379y (1975).

Samuels, S. L., and Wilkes, G. L., Controlled morphologies in segmented polyurethane-polyether copolymer films. *Polym. Prepr., Am. Chem. Soc., Div. Polym. Chem.* **14**(2), 1226 (1973); *C.A.* **83**, 44080u (1975).

Saunders, J. H., Bioconstituent fibers from segmented polyurethanes and nylon 6. *J. Appl. Polym. Sci.* **19**(5), 1387 (1975); *C.A.* **83**, 44540u (1975).

Schneider, N. S., Desper, C. R., Illinger, J. L., King, A. O., and Barr, D., Structural studies of crystalline MDI-based polyurethanes. *J. Macromol. Sci., Phys.* **11**(4), 527 (1975); *C.A.* **84**, 106202s (1976).

Schollenberger, C. S., and Dinberg, K., Thermoplastic polyurethane molecular weight property relations. *Adv. Urethane Sci. Technol.* **3**, 36 (1974); *C.A.* **83**, 28717b (1975).

Schollenberger, C. S., and Dinberg, K., Thermoplastic urethane chemical crosslinking effects. *J. Elastomers Plast.* **7**(1), 65 (1975); *C.A.* **83**, 11810d (1975).

Seefried, C. G., Jr., Koleske, J. V., and Critchfield, F. E., Thermoplastic urethane elastomers. I. Effects of soft segment variations. *J. Appl. Polym. Sci.* **19**(9), 2493 (1975a); *C.A.* **83**, 116458y (1975).

Seefried, C. G., Jr., Koleske, J. V., and Critchfield, F. E., Thermoplastic urethane elastomers. II. Effects of variations in hard segment concentrations. *J. Appl. Polym. Sci.* **19**(9), 2503 (1975b); *C.A.* **83**, 116459z (1975).

Seefried, C. G., Jr., Koleske, J. V., and Critchfield, F. E., Thermoplastic urethane elastomers. III. Effects of variations in isocyanate structure. *J. Appl. Polym. Sci.* **19**(12), 3185 (1975c); *C.A.* **84**, 75365r (1976).

Seefried, C. G., Jr., Koleske, J. V., Critchfield, F. E., and Dodd, J. L., Thermoplastic urethane elastomers. IV. Effects of cycloaliphatic chain extender on dynamic mechanical properties *Polym. Eng. Sci.* **15**(9), 646 (1975d); *C.A.* **83**, 180744n (1975).

Seymour, R. W., and Cooper, S. L., Orientation studies of polyurethane block polymers. *Polym. Prepr., Am. Chem. Soc., Div. Polym. Chem.* **14**(2), 1046 (1973); *C.A.* **83**, 44463w (1975).

Seymour, R. W., and Cooper, S. L., Viscoelastic properties of polyurethane block polymers. *Adv. Urethane Sci. Technol.* **3**, 66 (1974); *C.A.* **83**, 28718c (1975).

Slowikowska, I., and Kosinska, M., Preparation of crosslinked polyurethanes from polystyrene segments. *Polimery (Warsaw)* **20**(7), 325 (1975); *C.A.* **84**, 6192v (1976).

Sung, C. P., Schneider, N. S., Matton, R. W., and Illinger, J., Thermal transition behavior of polyurethanes based on toluene diisocyanate. *Polym. Prepr., Am. Chem. Soc., Div. Polym. Chem.* **15**(1), 620 (1974); *C.A.* **84**, 17902C (1976).

Tobolsky, A. V., Block copolymers with carbamate linkages between a polyvinyl chain and a polymer having active hydrogens. U.S. Patent 3,865,898 (1975); *C.A.* **83**, 79965p (1975).

Whittaker, R. E., Mechanical properties of segmented polyurethane elastomers. *Rheol. Acta* **13**(4-5), 675 (1974); *C.A.* **83**, 61182w (1975).

Wilde, A. F., Matton, R. W., Rogers, J. M., and Wentworth, S. E., Preparation and ballistic evaluation of transparent polyurethane block copolymers based on 2, 4-toluene diisocyanate. *U.S.N.T.I.S., AD Rep.* pp. 1–16 (1973); *C.A.* **83**, 60106n (1975).

Wilde, A. F., Matton, R. W., Rogers, J. M., and Wentworth, S. E., Synthesis and ballistic evaluation of selected transparent polyurethane block copolymers. II. Further changes in formulation. *U.S. N.T.I.S., AD-A Rep.* **AD-A012207**, 1-18 (1975); from *Gov. Rep. Announce. (U.S.)* **75**(18), 142 (1975); *C.A.* **84**, 45031m (1976).

Wilkes, G. L., and Wildnauer, R., Kinetic behavior of the thermal and mechanical properties of segmented urethanes. *J. Appl. Phys.* **46**(10), 4148 (1975); *C.A.* **84**, 32308g (1976).

Wilkes, G. L., Bagrodia, S., Humphries, W., and Wildnauer, R., Time dependence of the thermal and mechanical properties of segmented urethanes following thermal treatment. *J. Polym. Sci., Polm. Lett. Ed.* **13**(6), 321 (1975); *C.A.* **83**, 148164u (1975).

E. OTHER SYSTEMS

Andrianov, K. A., Synthesis and study of siloxane-carbonate copolymers. *Vysokomol. Soedin., Ser. A* **17**(1), 84 (1975); *C.A.* **83**, 164932 (1975).

Ashman, P. C., and Booth, C., Crystallinity and fusion of ethylene oxide-propylene oxide block copolymers. 1. Type PE copolymers. *Polymer* **16**(12), 889 (1975); *C.A.* **84**, 151102v (1976).

Ashman, P. C., Booth, C., Cooper, D. R., and Price, C., Crystallinity and fusion of ethylene oxide-propylene oxide block copolymers. 2. Type PEP copolymers. *Polymer* **16**(12), 897 (1975); *C.A.* **84**, 151103w (1976).

Barnabeo, A. E., Creasy, W., and Robeson. L. M., Permeability of methacrylonitrile-styrene block and random copolymers. *J. Polym. Sci., Polym. Chem. Ed.* **13**(9), 1979 (1975); *C.A.* **83**, 179815h (1975).

Birshtein, T. M., Skvortsov, A. M., and Sariban, A. A., Conformation characteristics of a block copolymer molecule in solution studied by the Monte-Carlo method. *Vysokomol. Soedin., Ser. A* **17**(11), 2558 (1975); *C.A.* **84**, 44834p (1976).

Bobovich, B. B., and Krasnov, B. Ya., Rheological properties of ternary block copolymers, thermoplastic elastomers. *Kozh. Obuvna. Promst.* **17**(9), 48 (1975); *C.A.* **84**, 18883j (1976).

Borghi, I., Foschi, S., and Galli, P., Thermoplastic Rubbers. Ger. Offen. 2,505,825 (Montedison s.p.A.) (1975); *C.A.* **84**, 18918z (1976).

Botham, R. A., Shank, C. P., and C. Thies, Adsorption behavior of polystyrene-poly (methyl methacrylate) mixtures containing block copolymers. *Colloid. Morphol. Behav. Block Graft. Copolym., Proc. Am. Chem. Soc. Symp., 1970* p. 247 (1971); *C.A.* **83**, 59520m (1975).

Bunk, A. J. N., Block copolymers. *Chem. Tech. (Amsterdam)* **30**(7), A17 (1975); *C.A.* **83**, 114964m (1975).

Buser, H., Friedrich, K., and Grolimund, K., Polyamide-polyimide block copolymers as ferrales in gas chromatography. *J. Chromatogr.* **108**(1), 181 (1975); *C.A.* **83**, 44323a (1975).

Casale, A., and Porter, R. S., Mechanical synthesis of block and graft copolymers. *Adv. Polym. Sci.* **17** 1 (1975); *C.A.* **83**, 80034v (1975).

Ceresa, R. J., The synthesis of block and graft copolymers of poly(vinyl chloride). *In* "Block and Graft Copolymerization" (R. J. Ceresa, ed.), Vol. 2, p. 273. Wiley, New York, 1976; *C.A.* **84**, 90837q (1976).

Chalykh, A. E., and Avgonov, A., Diffusion and sorption of low molecular weight compounds by block copolymers. *Vysokomol. Soedin., Ser. A* **17**(6), 1291 (1975); *C.A.* **83**, 115209f (1975).

Clark, D. T., Peeling, J., and O'Malley, J. J., Application of ESCA to polymer chemistry. VIII. Surface structures of AB block copolymers of poly(dimethylsiloxane) and polystyrene. *J. Polym. Sci., Polym. Chem. Ed.* **14**(3), 543 (1976); *C.A.* **84**, 15149r (1976).

Crivello, J. V., and Juliano, P. C., Polyimidothioether-polysulfide block copolymers. *Polym. Prepr., Am. Chem. Soc., Div. Polym. Chem.* **14**(2), 1220 (1973); *C.A.* **83**, 10942m (1975).

Crivello, J. V., and Juliano, P. C., Polyimidothioether-polysulfide block polymers. *J. Polym. Sci., Polym. Chem. Ed.* **13**(8), 1819 (1975); *C.A.* **83**, 194802y (1975).

Crystal, R. C., Surface morphologies of styrene-ethylene oxide block copolymers. *Colloid. Morphol. Behav. Block Graft Copolym., Proc. Am. Chem. Soc. Symp., 1970* p. 279 (1971); *C.A.* **83**, 59521n (1975).

Davies, W. G., and Jones, D. P., Synthesis and dilute solution properties of styrene-siloxane ABA block copolymers. *Colloid. Morphol. Behav. Block Graft Copolym., Proc. Am. Chem. Soc. Symp., 1970* p. 63 (1971); *C.A.* **83**, 39519t (1975).

Dean, J. W., Block copolymers of silicones with vinyl pyridine. U.S. Patent 3,875,254 (Gen. Elect. Co.) (1975); *C.A.* **83**, 148280a (1975).

Dekking, H. G. G., Block copolymerization with azoamidino compounds as initiator. U.S. Publ. Pat. Appl. B 292,140 (General Tire and Rubber Co.) (1975); *C.A.* **84**, 44969m (1976).

Dondos, A., Rempp, P., and Benoit, H., Segregation and conformational transitions in triblock copolymers in dilute solution. 3. Viscometric investigations in solvent mixtures. *Polymer* **16**(10), 698 (1975); *C.A.* **84**, 31627y (1976).

Eaves, D. E., and Stokes, A., Structures of ethylene sulfide crystallites in block copolymers. *Eur. Polym. J.* **11**(3), 215 (1975); *C.A.* **83**, 11009z (1975).

Endo, R., Surface properties of block and graft copolymers. *Kobunshi* **24**(9), 612 (1975); *C.A.* **83**, 179,639d (1975).

Falender, J. K., Kendrick, T. C., Lindsey, S. E., and Ward, A. H., Modification of the permeability of polymeric materials. Belgian Patent 818,374 (Dow Corning Corp.) (1975); *C.A.* **83**, 116013z (1975).

Fugawa, I., Satake, K., Yamada, T., and Sakamoto, K., Transparent acrylic thermoplastic elastomers. Japan Kokai 75/142,692 (Asahi, Chem. Ind. Co. Ltd.) (1975); *C.A.* **84**, 136959m (1976).

Fukawa, S., Satake, K., Yamada, T., and Sakamoto, K., High impact-resistant polymethacrylate. Japan Kokai 75/141,691 (Asahi Chem. Ind. Co. Ltd.) (1975); *C.A.* **84**, 136297n (1976).

Funt, J. M., Degradation of siloxane copolymers. *J. Polym. Sci., Polym. Chem. Ed.* **13**(9), 2181 (1975); *C.A.* **83**, 179870x (1975).

Furukawa, K., Tsukamoto, C., and Nagai, H., Block copolymers containing xylylene groups. Japan Kokai 75/144,799 (Toyobo Co., Ltd.) (1975); *C.A.* **84**, 136301r (1976).

Gaines, G. L., Jr., Monolayers of dimethylsiloxane-containing block copolymers. *Adv. Chem. Ser.* **144**, 338 (1975); *C.A.* **84**, 5559h (1976).

Gallot, B., and Russo, S., Structure and properties of random, alternating and block copolymers. The uv spectra of styrene-methylmethacrylate copolymers. *Adv. Chem. Ser.* **142**, 85 (1975); *C.A.* **84**, 17889d (1976).

Guilbault, L. J., Water-soluble, vinyl-pyrrolidone-acrylamide block copolymers. U.S. Patent 3,907,927 (Calgon Corp.) (1975)4 *C.A.* **84**, 5863c (1976).

Guyot, A., Ceyesson, M., Michel, A., and Revillon, A., New technique for preparation of block copolymers in two radical stages. *Inf. Chim.* **116**, 127 (1973); *C.A.* **83**, 59413d (1975).

Hake, P. T., and Pope, G. A., Preparation and properties of BAB block copolymers based on polyethylene sulfide and polyisoprene. *Eur. Polym. J.* **11**(10), 677 (1975); *C.A.* **84**, 106786h (1976).

Hall, W. F., and Dewames, R. E., Analytical relations for block copolymer relaxation times in the Rouse model. *Macromolecules* **8**(3), 349 (1975); *C.A.* **83**, 43962w (1975).

Hedrick, R. M., and Gabbert, J. D., Catalytic process for imide-alcohol condensation. U.S. Patent 3,922,254 (Monsanto Co.) (1975); *C.A.* **84**, 74880t (1976).

Helfand, E., Block copolymers, polymer-polymer interfaces and the theory of inhomogeneous polymers. *Acc. Chem. Res.* **8**(9), 295 (1975); *C.A.* **83**, 193747u (1975).

Helfand, E., Theory of inhomogeneous polymers. Lattice model for poly-polymer interfaces. *J. Chem. Phys.* **63**(5), 2192 (1975); *C.A.* **83**, 179810c (1975).

Hergenrother, W. L., Block polymers of polysiloxanes and polybutadiene. U.S. Patent 3,928,490 (Firestone Tire and Rubber Co.) (1975); *C.A.* **84**, 75408g (1976).

Hergenrother, W. L., Lactam-butadiene Block Copolymers of Controlled Molecular Weight. U.S. Patent 3,940,372 (Firestone Tire and Rubber Co.) (1976); *C.A.* **84**, 165392b (1976).

Hergenrother, W. L., and Ambrose, R. J., Block polymers from isocyanate terminated intermediates. IV. Properties of cured butadiene-ε-caprolactam block copolymers. *J. Appl. Polym. Sci.* **19**(12), 3225 (1975); *C.A.* **84**, 31839 (1976).

Hill, M. P. L., Millard, P. L., and Owen, M. J., Migration phenomena in silicone modified polystyrene. *Am. Chem. Soc., Div. Org. Coat. Plast. Chem., Pap.* **34**(1), 334 (1974); *C.A.* **84**, 31814g (1976).

Hirata, E., Domain structure and domain formation mechanism in A—B—A and A—B block copolymers of ethylene oxide and isoprene. *Adv. Chem. Ser.* **142**, 288 (1975); *C.A.* **83**, 194203a (1975).

Hoelle, H. J., and Lehnen, B. R., Preparation and characterization of poly(dimethyl-siloxanes) with narrow molecular weight distribution. *Eur. Polym. J.* **11**(9); 663 (1975); *C.A.* **84**, 12237q (1976).

Hoffman, D. K., ABA block copolymers with dimethylsiloxane center blocks. Thesis, University of Akron, Akron, Ohio, 1975; *C.A.* **84**, 44740 (1976).

Holtschmidt, N., Bressel, B., Buechner, W., and DeMontigney, A., Silicone-polycarbonate combination. Ger. Offen 2,343,275 (Bayer A.G.) (1975); *C.A.* **83**, 11389y (1975).

Hortte, S., Nishizawa, T., and Aido, E., Antifogging agents for styrene copolymer films. Japan Kokai 74/130,447 (Denki Kagaku Kogyo K.K.) (1974); *C.A.* **83**, 29117t (1975).

Illinger, J. L., Lewis, R. W., and Barr, D. B., Effect of interlayer on impact resistance of acrylic/polycarbonate laminates. *U.S.N.T.I.S., AD Rep.* **AD-`**, 1–16 (1972); *C.A.* **83**, 28920n (1975).

Ivanov, P., Levin, E., Dolgoplosk Valestskii, S., and Vinogradova, S. V., Molecular mobility in arylatesiloxane block copolymers. *Dokl. Akda. Nauk. SSSR* **221**(4), 872 (1975); *C.A.* **83**, 43928q (1975).

Iwatani, K., Nishiyama, T., Sakurada, S., Katayama, S., and Serita, H., Block poly(vinyl-sulfone amides). Japan Kokai 75/132,095 (Tohoku Fertilizer Co., Ltd.) (1975); *C.A.* **84**, 60218w (1976).

Kaelble, D. H., Block copolymers as adhesives. *Polym. Sci. Technol.* **9A**, 199 (1975); *C.A.* **84**, 136766w (1976).

Kennedy, J. P., Synthesis, characterization and properties of block and bigraft copolymers. *J. Polym. Sci., Polym. Chem. Ed.* **13**(10), 2213 (1975); *C.A.* **84**, 18136t (1976).

Kennedy, J. P., and Vidal, A., Block and graft copolymers by selective cationic initiation. III. Synthesis and characterization of bigraft copolymers. *J. Polym. Sci., Polym. Chem. Ed.* **13**(8), 1765 (1975); *C.A.* **84**, 31539w (1976).

Kennedy, J. P., Melby, E. G., and Vidal, A., Block and bigraft copolymers by carbocation polymerization. *J. Macromol. Sci., Chem.* **9**(5), 833 (1975); *C.A.* **83**, 179664h (1975).

Kollmeier, H. J., and Rossmy, G., Poly(organosiloxane) block copolymers. Ger. Offen. 2,431,394 (Goldschmidt, Th., A.G.) (1976); *C.A.* **84**, 106508w (1976).

Komarova, T. P., Markelov, M. A., Nenakhov, S., Semenenko, E. I., and Chalykh, A. E., Diffusion and sorption of water in block copolymers. *Vysokomol. Soedin., Ser. A* **18**(2), 264 (1976); *C.A.* **84**, 151865w (1976).

Kotlair, A. M., Block sequence distributions and homopolymer content for condensation polymers. *J. Polym. Sci., Polym. Chem. Ed.* **13**(4), 973 (1975); *C.A.* **83**, 79640d (1975).

Kuo, C., and McIntyre, D., Morphology and physical properties of triblock copolymers with crystalline end blocks. *J. Polym. Sci., Polym. Phys. Ed.* **13**(8), 1543 (1975); *C.A.* **83**, 164748q (1975).

Kuzaev, A. I., Chromatographic separation of tetrahydrofuran-epichlorohydrin block copolymers. *Vysokomol. Soedin., Ser. A* **17**(9), 2120 (1975); *C.A.* **83**, 206700a (1975).

Lazar, M., Borsig, E., and Manasek, Z., Block copolymers Czech Patent 152,011 (1974); *C.A.* **83**, 80022K (1975).

Litt, M. H., and Matsuda, T., Surface properties of imino ether block copolymers. *Adv. Chem. Ser.* **142**, 320 (1975); *C.A.* **83**, 193912u (1975).

Lundsted, L. G., and Schmolka, I. R., Synthesis and properties of block copolymer polyol surfactants. *In* "Block and Graft Copolymers" (R. J. Ceresa, ed.), Vol. 2, p. 1. Wiley, New York, 1976; *C.A.* **84**, 91944 (1976).

Marsiat, A., and Gallot, Y., Synthesis and characterization of the block copolymers: polystyrene/poly(dimethvl siloxane) and polyisoprene/poly(dimethyl siloxane) *Makromol. Chem.* **176**(6), 164 (1975); *C.A.* **83**, 164622u (1975).

Masar, B., and Cefelin, P., Soluble copolymers of lactams as bifunctional prepolymers. *Polyamidy '75 Sb. Prednasek.* p. 136 (1975); *C.A.* **84**, 44802b (1976).

Mehra, U., Toy, L., Shen, M., and Biliyar, K., Dynamic mechanical and dielectric relaxations in blends of homopolymers and block copolymers. *U.S.N.T.I.S., AD Rep.* **AD-**, 1–25 (1974); *C.A.* **83**, 11025b (1975).

Miki, T., and Narisawa, S., Block copolymers. Ger. Offen. 2,446,255 (Sumitomo Chem. Co, Ltd.) (1975); *C.A.* **83**, 29063x (1975).

Miki, T., Narisawa, S., Horiike, H., and Maki, H., Transplant block copolymers. Japan Kokai 75/98,992 (Sumitomo Chemical Co., Ltd.) (1975); *C.A.* **84**, 5858e (1976).

Minoura, Y., and Nakano, A., block copolymerization of methyl methacrylate with poly(ethylene oxide). *Macromol. Synth.* **4**, 25 (1972); *C.A.* **84**, 44907q (1976).

O'Driscoll, K. F., and Sridharan, A. U., Mechanochemical synthesis of homopolymers and block copolymers. *Appl. Polym. Symp.* **26**, 135 (1975); *C.A.* **84**, 165235c (1976).

O'Grady, V. J., Tough polyblends with high flowability and malleability. Ger. Offen. 2,453,110 (Monsanto Co.) (1975); *C.A.* **83**, 115972z (1975).

O'Malley, J. J., and Marchessault, R. H., Block copolymers of styrene and ethylene oxide. *Macromol. Synth.* **4**, 35 (1972); *C.A.* **84**, 17809c (1976).

O'Malley, J. J., and Marchessault, R. H., Block copolymers of styrene and ethylene oxide. *Macromol. Synth.* **3**, 35 (1972); *C.A.* **83**, 28618v (1975).

Prokal, B., and Kanner, B., Organosiloxane polymers useful in preparing foams. Ger. Offen. 2,514,384 (Union Carbide Corp.) (1975); *C.A.* **84**, 32018n (1976).

Ramsh, A. S., and Sidorovich, E. A., Physical phase state of arylate-siloxane block copolymers. *Fiz. Svoistva Elastomerov* p. 120 (1975); *C.A.* **84**, 106812n (1976).

Ramsh, A. S., Sidorovich, E. A., Korshak, V. V., Dolgoplosk, S. B., Valetskii, P. M., Vinogradova, S. V., and Marei, A. I., Phase aggregation state of arylate siloxane block copolymers. *Dokl. Akad. Nauk. SSSR* **221**(2), 361 (1975); *C.A.* **83**, 11064p (1975).

Riess, G., Thermal degradation of copolymers. Comparison of mixtures, random, alternating and block methyl methacrylate styrene copolymers. *Eur. Polym. J.* **11**(5–6) 429 (1975); *C.A.* **83**, 179854v (1975).

Roggero, A., Mazzei, A., Bruzzone, M., and Cernia, E., Block and random copolymerization of episulfides. *Adv. Chem. Ser.* **142**, 330 (1975); *C.A.* **84**, 6190t (1976).

Sefton, M. V., and Merrill, E. W., Infrared spectroscopic analysis of complex polymer systems. *J. Appl. Polym. Sci.* **20**(1) 157 (1976); *C.A.* **84**, 60138v (1976).

Selb, J., and Gallot, Y., Micelle formation in polystyrene-poly(vinyl-n-alkypyridinium bromide) block copolymer solutions in methanol-water mixtures. *J. Polym. Sci., Polym. Lett. Ed.* **13**(10), 615 (1975); *C.A.* **84**, 17917m (1976).

Seow, P. K., Gallot, Y., and Skoulios, A., Synthesis and characterization of poly(alkyl methacrylate)/poly(oxyethylene) block copolymers. *Makromol. Chem.* **176**(11), 3153 (1975); *C.A.* **84**, 31650a (1976).

Seow, P. K., Gallot, Y., and Skoulios, A., Crystallization of poly(oxyethylene) in two block copolymers. 1. Dilatometric study. *Makromol. Chem.* **177**(1), 177 (1976); *C.A.* **84**, 74767m (1976).

Seow, P. K., Gallot, Y., and Skoulios, A., Crystallization of poly(oxylethylene) in two-block copolymers. 2. Study of structure and texture. *Makromol. Chem.* **177**(1), 199 (1976); *C.A.* **84**, 74768n (1976).

Seymour, R. B., Stahl, G. A., Owen, D. R., and Wood, H., Block copolymers of methyl methacrylate. *Adv. Chem. Ser.* **142**, 309 (1975); *C.A.* **83**, 193785e (1975).

Shatalov, V. P., Grigorieva, L. A., Kisterva, A. E., and Alekhin, V. D., Heat-stable block copolymers. Ger. Offen. 2,325,314 (All-Union Scientific Research Institute of Synthetic Rubber, Voronezh.) (1974); *C.A.* **83**, 59823n (1975).

Shatalov, V. P., Grigorieva, L., Kistereva, A. E., Alekhin, V. D., Samotsvetov, A. R., and Kirchevskaya, I. J., Thermostable block-copolymer. British Patent 1,409,956 (All-Union Scientific Research Institute of Synthetic Rubber) (1975); *C.A.* **84**, 75025e (1976).

Shimura, Y., and Hatakeyama, T., Mechanical relaxation in polystyrene/polyethylene oxide block copolymers. *Makromol. Chem.* **176**(7), 2127 (1975); *C.A.* **84**, 31622t (1976).

Shuttleworth, R., and Watson, W. F., Mechanochemical polymerization: natural rubber-methacrylic acid block-graft copolymer. *Macromol. Synth.* **5**, 65 (1974); *C.A.* **83**, 44430h (1975).

Skoulios, A., Mesomorphic properties of block copolymers. *Adv. Liq. Cryst.* **1**, 169 (1975); *C.A.* **84**, 17749h (1976).

Stehlicek, J., and Sebenda, J., Preparation of block copolymers by anionic polymerization of lactams. *Polyamidy '75 Sb. Prednasek.* p. 91 (1975); *C.A.* **84**, 17798y (1976).

Stockmayer, W. H., and Kennedy, J., Viscoelastic spectrum of free-draining block copolymers. *Macromolecules* **8**(3), 351 (1975); *C.A.* **84**, 43963x (1975).

Szwarc, M., Ions and ion pairs in ionic polymerization. *Ions Ion Pairs Org. React.* **2**, 375 (1974); *C.A.* **83**, 43771h (1975).

Szwarc, M., Ionic intermediates in polymerization processes. *Mol. Behav. Dev. Polym. Mater.* p. 1 (1975); *C.A.* **83**, 43227q (1975).

Thompson, R. M., and Stearm, R. S., Polyamide block copolymers. Ger. Offen 2,454,120 (Sun Ventures, Inc.) (1975); *C.A.* **83**, 116760n (1975).

Tsvetanov, K., and Panaiotov, I., On the nature of the active centers in the anionic polymerization of acrylonitrile and methacrylonitrile. *Izv. Khim.* **8**(1), 146 (1975); *C.A.* **84**, 122411q (1976).

Vocel, J., and Stepankova, L., Rheological properties of propylene oxide-ethylene oxide block copolymers. *Vodohospod. Cas.* **23**(3), 268 (1975); *C.A.* **83**, 193889s (1975).

Wang, F. W., Dynamics of block-copolymer molecules in dilute solution. *Macromolecules* **8**(3), 364 (1975); *C.A.* **83**, 43966a (1975).

Wang, F. W., Frictional properties of dilute block-copolymer solutions and homopolymer solutions. Application to molecular weight determination. *Macromolecules* **9**(1), 97 (1976); *C.A.* **84**, 122466m (1976).

Wang, F. W., and Dimarzio, E. A., Dynamics of block-copolymer molecules in solution. Free draining limit. *Macromolecules* **8**(3), 356 (1975); *C.A.* **83**, 43964y (1975).

Ward, A. H., and Kendrick, T. C., Methylstyrene-dimethylsiloxane block-copolymers for films. Ger. Offen. 2,430,957 (Dow Corning Company) (1975); *C.A.* **83**, 11753n (1975).

Ward, A. H., Kendrick, T. C., and Saam, J. C., Poly(α-methylstyrene-dimethylsiloxane) block copolymers. Effects of microstructure on properties. *Adv. Chem. Ser.* **142**, 300 (1975); *C.A.* **83**, 189119r (1975).

Wesslen, B., and Mansson, P., Synthesis and chromatographic separation of poly(styrene-b-ethylene oxide. *J. Polym. Sci., Polym. Chem. Ed.* **13**(11), 2545 (1975); *C.A.* **84**, 17794u (1976).

Wetton, R. E., and Tuminello, W. H., Temperature-independent relaxation in a lamelar block copolymer. *Nature (London)* **257**, 123 (1975); *C.A.* **84**, 31624v (1976).

White, D. M., Bis(polyphenylene oxide)-carbonate block copolymers. U.S. Patent 3,875,256 (Gen. Elect. Co.) (1975); *C.A.* **83**, 164839v (1975).

White, D. M., and Klopfer, H. J., Brominated poly(phenylene oxides). *ACS Symp. Ser.* **6**, 169 (1975); *C.A.* **83**, 164637c (1975).

Yamashita, Y., Iwaya, Y., and Ito, K., Block copolymerization. 9. Polymerization of the NCA of methyl D-glutamate by telechelic polystyrene having glycyl groups as active chain ends. *Makromol. Chem.* **176**(5), 1207 (1975); *C.A.* **83**, 115037y (1965).

Yonemoto, K., and Saito, C., Heat resistant adhesives. Japan Kokai, 75/122,533 (Denki Kagaku Kogyo K. K.) (1975); *C.A.* **84**, 32128y (1976).

Zdrahala, R. J., Block copolymers based on wholly aromatic polyamide and poly (ethylene oxide) or poly(dimethyl siloxane). Ph.D. Thesis, University of Tennessee, Knoxville, 1975; *C.A.* **84**, 122625n (1976).

Zilliox, G., Roovers, J. E. L., and Bywater, S., Preparation and properties of poly(dimethylsiloxane) and its block copolymers with styrene. *Macromolecules* **8**(5), 573 (1975); *C.A.* **84**, 5434p (1976).

AUTHOR INDEX

Numbers in parentheses are reference numbers and indicate that an author's work is referred to although his name is not cited in the text. Numbers in italics show the page on which the complete reference is listed.

SUBJECT INDEX

A

Acetaldehyde, 139, 144, 248
Acrylic polymer, 35, 103–106, 238–240,
 see also Styrene–methyl methacrylate
 polymers
 acrylonitrile, 104
 anionic polymerization, 36, 104, 238–
 240
 methyl methacrylate, 104, 238–240
 methyl methacrylate–α-methyl styrene,
 240
 moderately defined structures, 35,
 238–240
Acrylonitrile, 137,149, 157, 236, 247, 260,
 263, 276
N-Acylalkylene imine polymers, 155
Aldehyde polymers, 141, 256, 257
Aldehyde–vinyl polymers, 147, 256, 257
Alkenyl aromatic–diene polymers, 235–
 237
 1,4-dilithio-1,1,4,4-tetraphenylbutane
 initiator, 235
 α-methyl styrene–diene polymers, 235,
 236
Alkylene adipate, 260
Alkylene oxide polymers, 141, 242, 259,
 306
Allyl chloride, 137
Aluminosiloxane, 395
Amide–amide polymers, 152, 268–274,
 354–360
Amide–ester, 361, 362
Amide–ether, 269–272, 360–361
Amide–siloxane, 284
Amide–vinyl polymers, 154, 268, 361
 styrene–isocyanate, 154
Amylene oxide, 272
Anionic polymerization, 32, 41, 83, 94,
 102, 104, 111, 143, 152, 158, 187–194,
 238–240, 242–259, 269–272, 275–277
 monodisperse blocks, 95, 104, 187–198,
 275
 predictable molecular weight, 83, 95,
 102, 107, 187–198, 238–240, 275–
 276

side reactions, 105, 158, 238–240
Applications, 69–75, *See also* individual
 systems
 adhesives, caulk, and sealants, 73,
 93
 automotive, 72
 biomedical, 78
 elastomers, 70, 93, 228–230
 footwear, 73, 93
 general, 69–75, 78, 163, 228–230
 mechanical goods, 73, 93
 membranes, 68, 78
 property advantages of various block
 copolymer thermoplastic elastomers, 71
 surfactants, 74, 163
 toughened thermoplastic resins, 74,
 121–126, 229
 transparent products, 93, 229, 232
 wire and cable, 93
Architecture, 24–29, 55, 63
 A-B, diblocks, 24, 83–184
 A-B-A, triblocks, 24, 186–304
 [A-B]$_n$, multiblocks, 24, 306–456
 effect on processibility, 28, 63
 on properties, 27–29, 64
 radial or star-shaped, 26, 43, 229–232

B

p-Benzenediethyl terephthalate, 316
γ-Benzyl L-glutamate, 395
Bis A hexahydroterephthalate, 330
Bis A isophthalate, 330, 395
Bis A iso-terephthalate, 330
Bis A iso-terephthalate, 331
Bis A iso-erephthalate–biphenyl sulfonate, 331
3,3-Bis(chloromethyl)oxacyclobutane, 139
1,1-Bis(4-hydroxyphenyl)-1-(3,4-dichloro-
 phenyl)ethane carbonate, 336
9,9-Bis(4-hydroxyphenyl)fluorene carbonate, 395
Bis(4-hydroxyphenyl)naphthalene, 336
1,1-Bis(4-hydroxyphenyl)-1-phenylethane
 carbonate, 336